Green Energy and Technology

Climate change, environmental impact and the limited natural resources urge scientific research and novel technical solutions. The monograph series Green Energy and Technology serves as a publishing platform for scientific and technological approaches to "green"—i.e. environmentally friendly and sustainable—technologies. While a focus lies on energy and power supply, it also covers "green" solutions in industrial engineering and engineering design. Green Energy and Technology addresses researchers, advanced students, technical consultants as well as decision makers in industries and politics. Hence, the level of presentation spans from instructional to highly technical.

Indexed in Scopus.

Indexed in Ei Compendex.

Donglu Shi
Editor

Energy Materials Science and Engineering: From New Devices to AI Power Systems

Springer

Editor
Donglu Shi
The Materials Science and Engineering
Program
Department of Mechanical and Materials
Engineering
University of Cincinnati
Cincinnati, OH, USA

ISSN 1865-3529 ISSN 1865-3537 (electronic)
Green Energy and Technology
ISBN 978-3-031-88549-5 ISBN 978-3-031-88550-1 (eBook)
https://doi.org/10.1007/978-3-031-88550-1

This work was supported by National Science Foundation (NSF 1953009 and NSF 1635089).

This Springer imprint is published by the registered company Springer Nature Switzerland AG
The registered company address is: Gewerbestrasse 11, 6330 Cham, Switzerland

Preface

The urgent global drive for sustainable energy demands a careful balance between technological advancement and environmental stewardship. The book: *Energy Materials Science and Engineering: From New Devices to AI Power Systems* offers a comprehensive exploration of materials and systems shaping the future of energy technologies. By integrating fundamental scientific principles with emerging applications, this book provides a broad perspective on energy materials, energy storage, and AI-driven advancements, highlighting their implications for sustainable development.

This book is structured into four parts, each addressing key aspects of energy materials science and engineering. *The first part, Advanced Energy Materials*, establishes fundamental principles governing energy materials and their applications. It explores electronic, thermal, optical, and electrochemical properties critical to energy harvesting, storage, and conversion. This part lays the groundwork for understanding cutting-edge materials innovations in photovoltaics, energy storage, and capacitive energy systems.

The part begins with Chapter "Recent Progress in Mn-Based Cathode Materials in Energy Storage": Recent Progress in Mn-Based Cathode Materials for Energy Storage, which provides a comprehensive analysis of the structure-property-performance relationships in manganese-based cathodes. The chapter examines their electrochemical mechanisms, including Mn^{3+}/Mn^{4+} redox activity, and addresses key challenges such as the Jahn-Teller effect, structural degradation, and Mn dissolution. Recent advancements in material design, stability enhancement strategies, and the potential of Mn-based cathodes for alternative ion chemistries, such as sodium-ion and potassium-ion batteries, are also discussed.

Chapter "Squaraine-Based Small Molecule BHJ Absorbing Layers for Organic Thin-Film Solar Cells": Squaraine-Based Small Molecule Bulk Heterojunction (BHJ) Absorbing Layers for Organic Thin-Film Solar Cells explores the molecular design of squaraine compounds to optimize solubility, nanomorphology, and charge transport in organic photovoltaics. The chapter highlights how molecular modifications, solvent processing techniques, and structural optimizations influence photovoltaic performance. Advanced characterization methods, including atomic force microscopy and grazing incidence small- and wide-angle X-ray scattering, provide insights into the crystallinity and domain structure of squaraine-based BHJ layers. The findings underscore the impact of charge carrier mobility and recombination effects on device efficiency, paving the way for enhanced organic solar cell technologies.

Chapter "Application of Silicon-Based Thin Films in High-Efficiency Silicon Solar Cells for Carbon Neutrality": Application of Silicon Thin Films in High-Efficiency HJT and TOPCon Solar Cells for Carbon Neutrality investigates the role of silicon thin films in passivated contact technologies that improve the efficiency of crystalline silicon photovoltaics. This chapter introduces fundamental semiconductor principles, p-n junction behavior, and the significance of hydrogenated amorphous silicon and silicon suboxide/polycrystalline silicon passivation layers in heterojunction and tunnel oxide passivated contact solar cells. The discussion extends to hybrid back-contact designs and perovskite/c-Si tandem structures, emphasizing their contributions to achieving carbon neutrality through next-generation solar technologies.

Concluding the part, Chapter "Perovskite Ferroelectric/Relaxor Materials/Systems for Capacitive Energy Storage": Perovskite Ferroelectric and Relaxor Materials for Capacitive Energy Storage investigates the dielectric properties and phase transitions of perovskite-based materials, which are critical for high-energy-density capacitors. The chapter explores lead-based and lead-free relaxor ferroelectrics, emphasizing their unique combination of high dielectric constant and low remnant polarization. Structural characteristics, spontaneous polarization mechanisms, and temperature-dependent phase transitions are analyzed in the context of energy storage applications. The discussion also highlights key advancements and future directions in the field.

By combining foundational scientific principles with state-of-the-art research, this part provides a comprehensive understanding of advanced energy materials, setting the stage for the technological innovations explored in subsequent chapters.

The second part, *Advanced Energy Conversion and Storage Devices*, explores state-of-the-art technologies that are redefining energy systems. This part introduces the next-generation solar cells, including perovskite photovoltaics, tandem solar cells, and organic photovoltaics, which offer improved efficiency and cost-effectiveness. It also examines solid-state batteries as a transformative advancement that enhances battery safety, energy density, and cycle life, alongside fuel cells such as proton exchange membrane and solid oxide fuel cells, which play a crucial role in clean energy generation. Additionally, the part discusses supercapacitors and hybrid energy storage systems, which provide rapid charge-discharge capabilities, making them ideal complements to battery technologies. Material innovations, including novel electrode materials, solid electrolytes, and nanostructured interfaces, lie at the heart of these advancements, significantly enhancing device performance, energy density, and long-term durability.

A key focus of this part is on lithium/sulfur batteries, explored in Chapter "Advancing Lithium/Sulfur (Li/S) Batteries", which addresses the increasing demand for high-energy-density storage driven by the rapid growth of electric vehicles. While lithium-ion batteries (LIBs) have dominated the market, they are approaching their theoretical energy density limits and rely on expensive, geopolitically constrained materials such as cobalt and nickel. Li/S batteries, utilizing Earth-abundant sulfur, present a promising alternative with a potentially higher energy density and lower cost, but challenges such as sulfur dissolution and capacity fading must be overcome for commercialization.

Chapter "Sodium-Ion Batteries" discusses sodium-ion batteries, which operate on a similar "rocking chair" principle as LIBs but use sodium-based compounds for both electrodes. With an emphasis on cost-effective and sustainable energy storage, Na-ion batteries leverage abundant sodium resources while maintaining reasonable electrochemical performance. The chapter covers key material compositions, electrode reactions, and performance metrics essential for advancing Na-ion technology as a viable alternative to lithium-based systems.

Hydrogen production by water electrolysis, a key process in the green energy transition, is the focus of Chapter "Advances in Electrocatalyst Design for Hydrogen Production by Water Electrolysis". Hydrogen plays a vital role in industrial applications and clean energy systems, yet the dominant production method, steam reforming, is heavily dependent on fossil fuels. This chapter explores electrocatalyst design advancements aimed at improving the efficiency of the hydrogen evolution reaction in electrochemical water splitting, a process powered by renewable energy sources. The discussion highlights innovative catalyst materials that enhance reaction kinetics and durability, facilitating large-scale hydrogen production with minimal environmental impact.

In Chapter "Thermoelectric Materials and Devices for Applications Below 250 °C", thermoelectric materials and devices for applications below 250°C are examined, with a focus on waste heat recovery and solid-state cooling. TE materials, capable of directly converting heat into electricity, operate based on the Seebeck, Peltier, and Thomson effects. This chapter provides a comprehensive analysis of TE materials with optimized properties for low-to-medium temperature applications, discussing fundamental principles, material design strategies, and device engineering to maximize efficiency and practical deployment.

By incorporating these cutting-edge research developments, this section offers a comprehensive perspective on the latest breakthroughs in materials engineering that are driving more efficient, reliable, and sustainable solutions for energy conversion and storage.

The third part, *Artificial Intelligence in Energy Systems*, explores how AI and machine learning are transforming energy research, optimization, and management. AI-driven approaches are accelerating materials discovery, enabling predictive modeling, and enhancing energy system performance. Computational algorithms are now capable of identifying and optimizing new materials for energy applications, significantly reducing the time and cost associated with traditional experimental research. Machine learning models are widely employed to forecast battery lifespan, optimize solar panel efficiency, and predict fuel cell performance, providing valuable insights that enhance system reliability.

A critical application of AI is in the management of smart energy grids, where it improves energy distribution, demand forecasting, and system stability while minimizing waste. The autonomous control of renewable energy systems is becoming increasingly prevalent, with AI facilitating the integration of wind, solar, and storage technologies into existing grids. Beyond system optimization, AI accelerates the development of new energy solutions by analyzing large datasets, identifying promising materials, and suggesting optimal configurations for energy devices.

Chapter "Modeling and Stability Analysis of Microgrids Integrated with Power-Electronics-Interfaced Distributed Energy Resources" provides the modeling and stability analysis of microgrids integrated with power-electronics-interfaced distributed energy resources. The growing adoption of inverter-based DERs offers operational flexibility but also introduces challenges in control and stability. This chapter provides a comprehensive examination of these issues, beginning with an introduction to microgrid fundamentals, their operational dynamics, and the challenges posed by power-electronics-based generation. It covers the mathematical modeling of inverters, the widely used double-loop control strategy for voltage and current regulation, and the application of differential-algebraic equations in microgrid modeling. Additionally, the chapter discusses droop control as a decentralized approach to power-sharing and voltage-frequency regulation, as well as small-signal stability analysis to identify and mitigate instabilities. By integrating power electronics, control theory, and stability assessment, this chapter presents a practical framework for understanding and improving microgrid performance, offering valuable insights for researchers and engineers working on next-generation power systems.

The integration of AI into energy systems is driving efficiency, adaptability, and resilience, shaping the future of intelligent and autonomous energy infrastructure.

The final part, *Energy Integration and Sustainable Building Infrastructure*, explores the critical challenge of incorporating renewable energy sources into modern infrastructure. This part examines the development of smart building skins and advanced energy distribution networks that enhance grid reliability and enable real-time energy management to balance supply and demand efficiently. Innovative materials play a crucial role in sustainable energy infrastructure. Emerging technologies such as transparent photothermal and photovoltaic thin films are enabling solar harvesting and energy generation in energy-neutral buildings. Given that buildings account for nearly 40% of global energy consumption, these advancements provide practical solutions for reducing greenhouse gas emissions in urban environments. Photothermal technologies, including multilayer thin films and the Photothermal Solar Tunnel Radiator, allow efficient solar-to-thermal energy conversion, offering heating and cooling solutions that integrate seamlessly with high-rise structures. Additionally, transparent PVs installed on building surfaces and interiors maximize solar energy capture throughout the day and under indoor lighting conditions, addressing the spatial constraints of urban environments.

Beyond material innovations, this part highlights the role of computational techniques in optimizing building energy performance. Numerical methods, including molecular dynamics, finite element analysis, computational fluid dynamics, and machine learning-based predictive modeling, are transforming the integration of nanoscale material features into large-scale architectural applications. Multi-scale simulations facilitate the translation of nanoscale behaviors—such as localized surface-plasmon-resonance-induced photothermal effects—into real-world energy performance assessments. For instance, spectrally selective glazing systems enhanced by plasmonic nanomaterials improve solar heat gain while maintaining optical transparency, contributing to significant heating energy savings.

Urban energy systems and microgrids are gaining prominence as decentralized energy solutions that provide resilience and energy independence to cities, communities, and industrial hubs. In addition to technical advancements, this part explores the policy, economic considerations, and societal impacts that shape the transition to sustainable energy infrastructure. Challenges such as regulatory frameworks, investment barriers, and the broader implications of renewable energy adoption are examined.

By integrating emerging technologies, computational modeling, and policy perspectives, this part provides a comprehensive view of how innovations in materials science and energy systems can collectively drive the development of a cleaner, more resilient global energy network. Beyond these core parts, the book highlights breakthrough advancements that are shaping the future of energy materials science and engineering. Topics include multifunctional nanomaterials designed for energy harvesting, storage, and conversion, as well as hybrid energy systems that combine multiple renewable sources to improve efficiency and reliability. The evolution of solid-state battery technologies is discussed in detail, emphasizing their potential to replace conventional lithium-ion batteries with safer, higher performance alternatives. The book also examines the increasing role of AI in optimizing high-performance energy materials and enabling the rapid discovery and design of next-generation energy solutions. By addressing these topics, the book provides insight into the future of energy technologies and their potential to drive global sustainability.

This book serves as a comprehensive resource for researchers, engineers, and students dedicated to advancing energy technologies. By integrating fundamental materials science, cutting-edge energy devices, AI-driven optimization, and sustainable infrastructure solutions, it provides both breadth and depth in understanding the rapidly evolving energy landscape. Each chapter is structured in a tutorial format, making it accessible to both undergraduate and graduate students. The interdisciplinary nature of the topics—spanning physics, chemistry, materials science, engineering, and emerging technologies—ensures a holistic perspective on energy innovation. We hope this book inspires transformative solutions that foster a cleaner, more sustainable, and energy-efficient future, advancing technology for generations to come.

Cincinnati, USA Donglu Shi

Contents

Advanced Energy Materials

Recent Progress in Mn-Based Cathode Materials in Energy Storage

Johan Nguyen and Yang Yang

Abstract Manganese-based cathode materials have gained significant attention in energy storage systems due to their cost-effectiveness, abundance, and environmental friendliness. This chapter provides a comprehensive overview of the structure-property-performance relationships and electrochemical mechanisms that govern the performance of Mn-based cathodes. Key structural features, including the layered arrangement of transition metal oxides and intercalating ions, are explored in detail to explain their influence on capacity, stability, and ionic conductivity. The electrochemical processes, particularly the Mn^{3+}/Mn^{4+} redox activity and its associated phase transitions are analyzed to highlight fundamental challenges such as the Jahn–Teller effect, structural degradation, and capacity fading caused by Mn dissolution. Building upon this foundation, the chapter delves into recent progress in addressing these challenges. Advances in material design are discussed, aiming at enhancing structural stability and electrochemical performance. Developments in tailoring Mn-based cathodes for alternative ion chemistries, including sodium-ion and potassium-ion batteries, are also presented, showcasing their versatility and potential for large-scale applications. This chapter provides both foundational knowledge and insights into cutting-edge research that drives the development of Mn-based cathode materials for next-generation energy storage technologies.

Keywords Rechargeable batteries · Mn-based cathode · Layered oxide cathodes · Microstructure design · Local chemistry · Structural unit

---

J. Nguyen · Y. Yang (✉)
NanoScience Technology Center, University of Central Florida, Orlando, FL, USA
e-mail: yang.yang@ucf.edu

Y. Yang
Department of Materials Science and Engineering, University of Central Florida, Orlando, FL, USA

Department of Chemistry, University of Central Florida, Orlando, FL, USA

Renewable Energy and Chemical Transformation Cluster, University of Central Florida, Orlando, FL, USA

The Stephen W. Hawking Center for Microgravity Research and Education, University of Central Florida, Orlando, FL, USA

D. Shi (ed.), *Energy Materials Science and Engineering: From New Devices to AI Power Systems*, Green Energy and Technology, https://doi.org/10.1007/978-3-031-88550-1_1

1 Introduction

Recent decades have witnessed a surge in rechargeable battery cathode material development to fulfill the demand for higher energy density, faster charging, and improved sustainability. The development of rechargeable battery technology relies on the basic solid-state chemistry of materials in which structure-composition-property-performance relationships have played a major role in advancing the field [23, 61, 85]. Large-scale high-energy batteries with electrode materials made from Earth-abundant elements are needed to achieve more sustainable energy development [87, 95]. Among available and potential elements for use as electrode materials, electrodes based on Manganese (Mn) have garnered significant attention due to the abundant availability, low cost, suitable operating voltage range, high energy density, and environmental friendliness of Mn. Mn can possess either layered, spinel, or olivine crystal structures with four- and/or sixfold coordination with oxygen anions, anionic clusters, or ligands, enabling multi-electron transfer, resulting in various types of rechargeable batteries with high specific capacity and energy density [39, 58–60].

The history of Mn-based electrodes in batteries traces back to the 1970s with Lithium-Manganese oxide (Li||MnO_2) batteries commercialized by Sanyo [67] and the successful development of spinel $LiMn_2O_4$ [78]. The development of layered $LiMnO_2$ represents a major step forward since $LiMnO_2$ possesses a higher theoretical capacity (285 mAh g^{-1}) compared with spinel $LiMn_2O_4$ (148 mAh g^{-1}). The $LiMnO_2$ materials demonstrate higher specific capacity compared to spinel structures, driven by the reversible Mn^{3+}/Mn^{4+} redox couple. Interest expanded further with the rise of Sodium (Na)-ion and Potassium (K)-ion batteries with layered Mn-based materials such as Na_xMnO_2 [5, 53, 75] and K_xMnO_2 [35, 50, 80] for grid storage and other large-scale applications, driven by the abundance and low cost of Na and K. In addition, the multiple oxidation states of Mn enable multi-electron transfer to achieve high redox potential, resulting in various types of rechargeable batteries with high specific capacity and energy density [43, 46, 99, 101, 106]. Unfortunately, the electrochemical reactions of Mn involve the existence of Mn^{3+} whose electron configuration causes structural distortion and degradation of the active materials as well as dissolution, migration, and deposition of Mn ions, resulting in capacity decay, poor cycling stability, and thus limitation in the practical application of the Mn-based materials [4, 49, 97].

The historical progression underscores manganese's pivotal role in shaping modern cathode materials while revealing the ongoing challenges in optimizing their performance and stability. In order to serve as a resource for researchers and engineers working to unlock the full potential of Mn-based electrodes in modern energy storage technologies, this chapter provides a comprehensive examination of Mn-based electrodes, especially with layered structures, beginning with an exploration of their fundamental properties, including crystal structure, redox chemistry, and electrochemical behavior. The discussion then shifts to the intrinsic and extrinsic challenges faced by these materials, such as the Jahn–Teller effect, structural degradation, and capacity fading. Finally, the chapter reviews recent advancements in material design, synthesis techniques, and interface engineering that aim to overcome these hurdles. By presenting a holistic view of Mn-based electrodes, this chapter.

2 Structure-Properties-Electrochemical Mechanisms of Mn-Based Cathodes

Understanding the intricate relationship between structure, properties, and electrochemical mechanisms, researchers can design Mn-based cathodes that maximize their advantages while mitigating their limitations for advanced energy storage applications. The redox properties of an active material are best considered in terms of their band structure. As a 3d transition metal with an electron configuration of $3d^5 4s^2$, Mn can exist in various oxidation states in oxides and other compounds but involves Mn^{3+} when acts as an electrochemically active center [22]. Figure 1 shows the electron configurations of Mn^{2+}, Mn^{3+}, and Mn^{4+} [60]. Differing from Mn^{2+} and Mn^{4+}, Mn^{3+} has a unique high-spin electron configuration with a single electron occupying the e_g orbital that splits the energy levels of the d-orbitals and creates an asymmetric occupancy state. This asymmetry distorts the six equivalent Mn–O bonds, stretching two of them into longer axial bonds and shortening the other four into equatorial bonds, leading to crystal distortion or the so-called Jahn–Teller effect, accompanied by a reduction in system energy, and the elimination of degeneracy in the electronic system. As a consequence, the Mn-based cathodes suffer several major challenges, including (1) phase diversity and stability, (2) phase transformation, and (3) dissolution, migration, and anode deposition of Mn ions.

2.1 *Phase Diversity and Stability*

Considering the crystal structure of A–Mn–O systems (A = Li, Na, and K), their structures exhibit a wide range of variations, depending on the specific A–Mn ratios and synthesis conditions. The planar layered framework of A_xMnO_2 (A = Li, Na, K) can be divided into two primary polymorphs: O-type and P-type as shown in Fig. 2, in which the alkali ions respectively occupy octahedral and prismatic sites, depending on the local environment of the alkali ions, the number and arrangement of the Mn–O layers, and the resulting distorted phases [19, 40, 41, 93]. Following the success of $LiCoO_2$ in which the monovalent Li^+ and trivalent Co^{3+} ions are ordered on the alternate (111) planes of the rock salt structure with a cubic close-packed array of oxide ions [57], O-type monoclinic $LiMnO_2$ (m-$LiMnO_2$) is considerable commercial and scientific interest due to its structural similarity to $LiCoO_2$. In addition, unlike Co^{3+}/Co^{4+} and Ni^{3+}/Ni^{4+}, the Mn^{3+}/Mn^{4+} band lies well above the top of the O^{2-}:2p band, indicating that Mn does not involve oxygen release from the lattice [55]. Unfortunately, m-$LiMnO_2$ does not crystallize in the same structure as

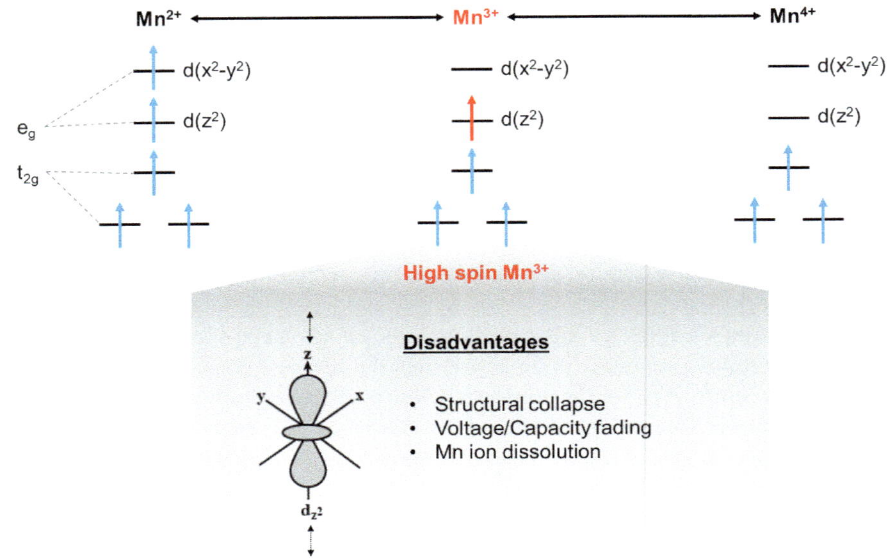

Fig. 1 Electron configurations of Mn^{2+}/Mn^{3+}/Mn^{4+}. Mn^{3+} has a single electron that preferentially occupies the $d(z^2)$ of the e_g orbital, leading to the asymmetry distortion of the Mn^{3+} structure which consequently causes the structural collapse, voltage/capacity fading, and Mn ion dissolution of the active materials [60]

$LiCoO_2$ when synthesized by high-temperature reactions and only can be obtained by ion exchange from α-$NaFeO_2$ [7, 8, 30]. Synthesis of m-$LiMnO_2$ is, therefore, very challenging. Furthermore, m-$LiMnO_2$ can hardly display satisfactory electrochemical performance because of the severe structural change induced by the Jahn–Teller effect. The asymmetry distortion of Mn^{3+} leads to the formation of orthorhombic $LiMnO_2$ (o-$LiMnO_2$) which is more thermodynamically stable but LiO_6 and MnO_6 units are arranged in zigzag layers and thus limits the Li-ion diffusion during cycling [2]. Although o-$LiMnO_2$ can initially provide a relatively high capacity of 210 mAh g^{-1}, it rapidly transforms to the spinel phase, driven by the migration of Mn ions into the vacancies left by the extracted Li ions, resulting in an irreversible capacity loss and capacity fading in the ~3 V region and thus poor capacity retention over time [3, 11, 24, 79].

As shown in Fig. 2, the ion diffusion in P-type structures is generally less complex than in the O-type due to the presence of more direct diffusion pathways with lower hopping energies between adjacent alkali sites. For instance, in the P-type phases, Na ions occupy trigonal prisms that share rectangular faces, creating a direct pathway while, in the O-type phases, Na ions move from one octahedral environment to another equivalent one by crossing an octahedral-tetrahedral face resulting in relatively high activation barrier [93]. The P-type structures are, therefore, more preferable. Unfortunately, the basic planar layered framework of Na_xMnO_2 exhibits a P-type structure when x is in the range of 0.6–0.7 and an O-type structure as x

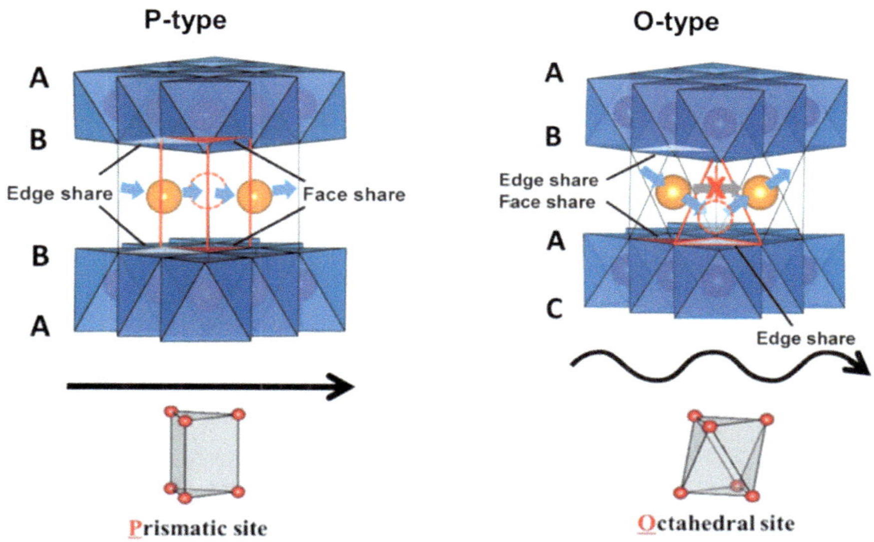

Fig. 2 Typical P- and O-type of A_xMnO_2 (A = Li, Na, K) with corresponding ion diffusion path (Reprinted with permission from [93]. Copyright 2014, American Chemical Society)

approaches 1, strongly depending on the synthesis conditions as shown in Fig. 3 [17, 73, 112]. Within the Na content of ~0.5–0.7, three-layer polymorphs exist, including α-phase with P2-type structure, β-phase with lattice-distorted P′2-type structure, and P3-type layered oxide, due to the different processing temperatures, giving a challenge for a pure phase synthesis. In addition, the low initial Na content in the P-type layered oxides, therefore, raises concerns about Na deficiency, which could pose a challenge for high-capacity delivery for practical use. The incorporation of various elements into this framework can modify the x range while maintaining the P-type structure [102]. However, the high-content Na in the host promotes the oxidation of low-valent cations to their high oxidation states and thus it is required to lower the upper cutoff operating voltage to prevent structural changes. Consequently, it just moderately improves the reversible capacities of the active materials.

2.2 *Phase Transformation During Cycling*

It's worth noting that, due to the presence of Mn^{3+}, the Jahn–Teller distortion can be induced during the electrochemical process, leading to the phase transformation of the Mn-based cathodes during cycling. Gummow et al. [24] observed the phase transformation to spinel during the Li-ion extraction from o-$LiMnO_2$. Later works confirmed that no o-$LiMnO_2$ peaks in the XRD patterns can be detected after a few cycles, providing evidence that the Jahn–Teller distortion triggers the rapid structure

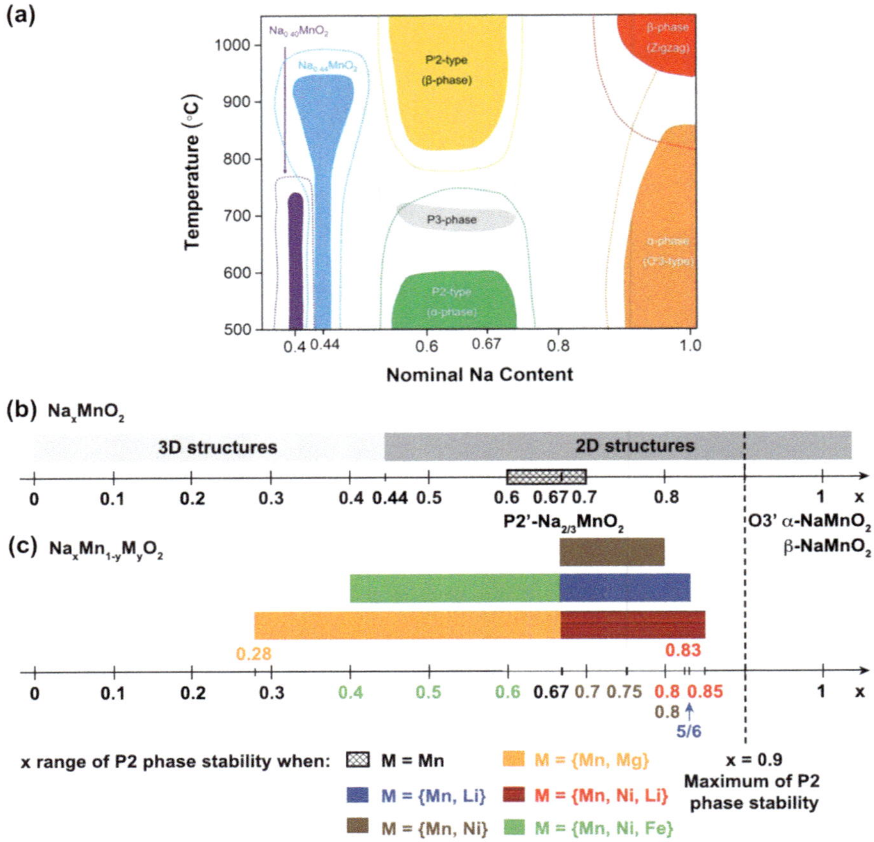

Fig. 3 **a** Stable Na_xMnO_2 phases as a function of the Na content x and the sintering temperature [112]. Reprinted with permission from Zuo and Yang [112]. Copyright 2022, American Chemical Society. Ranges of x in **b** Na_xMnO_2 and **c** $Na_xMn_{1-y}M_yO_2$ for different layered structures [17]

transformation from layered into the spinel or the rock salt upon cycling [79, 94]. Once formed, the structural mismatch between the layer and spinel phases disrupts the Li-ion diffusion pathways, resulting in sluggish kinetics and capacity degradation over repeated cycles [23, 66]. However, the capacities of spinel $LiMn_2O_4$ are limited because of the intrinsic deficiency of Li. When Li ions replace Mn for the formation of ($LiMn_2$)-O sheets, these sheets help to maintain a layered structure but distort the original lattice symmetries, leading to the formation of Li-rich Li_2MnO_3 with non-bonding O orbitals [29, 69, 77] which, in turn, triggers interplanar and intraplanar displacements of the Mn ions, leading to voltage hysteresis and capacity reduction [6]. Compositional Li–Mn–O phase diagrams (Fig. 4) show both reversible and irreversible phase transformations, indicating the complex structural changes that possibly reduce the electrochemical reversibility of the active materials and hinder

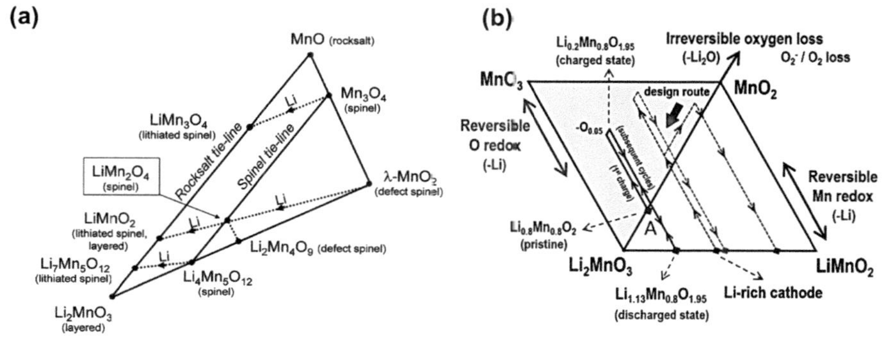

Fig. 4 Compositional Li–Mn–O phase diagram concerning **a** spinel, rock salt, and layered compounds [77] and **b** the electrochemical reaction pathways for Li-rich Mn-based cathodes [10]

the long-term performance of Mn-based cathodes [10, 52, 77]. The compositional Li–Mn–O phase diagram highlights a wide compositional space that exists to exploit structurally integrated layered and spinel Mn-based electrode materials, reflecting the composition of composite Mn electrodes with both spinel and layered structures to exploit the advantages of each component.

Unlike the Li counterpart, where even small deviations from $x = 1$ can induce structural transformation to a spinel phase, the planar layered framework for Na_xMnO_2 and K_xMnO_2 remains relatively stable [31, 36, 93]. However, the Mn–O sheets in P-type Na_xMnO_2 and K_xMnO_2 slide within the a–b plane to prevent close oxygen–oxygen contact during the alkali ion extraction, resulting in different imperfect layered structures with stacking defects [111, 112]. This induces P-type distortions and P–O transformations in both Mn-redox active and Mn-redox inert oxides [38, 86, 92, 109, 110]. Consequently, the charge–discharge curves of the layered Na_xMnO_2 and K_xMnO_2 often show multiple voltage steps, indicating complex phase transitions. In-depth examinations such as in-situ and operado XRD, are required to understand the complex kinetics in the process of the phase transition as shown in Fig. 5. It's worth noting that while spinel phase formation is favorable in $LiMnO_2$ and energetically less favorable in Na_xMnO_2 [53], a dual interphase layer consisting of a spinel interlayer of inactive Mn_3O_4 and a solid-electrolyte interphase film can still form on the surface of P-type $K_{0.67}MnO_2$ active particles [42], leading to a reduction of x in thermodynamic stable P-type K_xMnO_2 and inducing more severe structural changes and distortions [43, 46, 100] and resulting in a significant decrease in power density, underscoring the complexity of developing K-ion batteries.

2.3 Dissolution, Migration, and Anode Deposition of Mn Ions

As discussed in Sect. 2.2, the phase transformation from o-$LiMnO_2$ to spinel during the Li-ion extraction is a spontaneous process. With Li ions being removed from

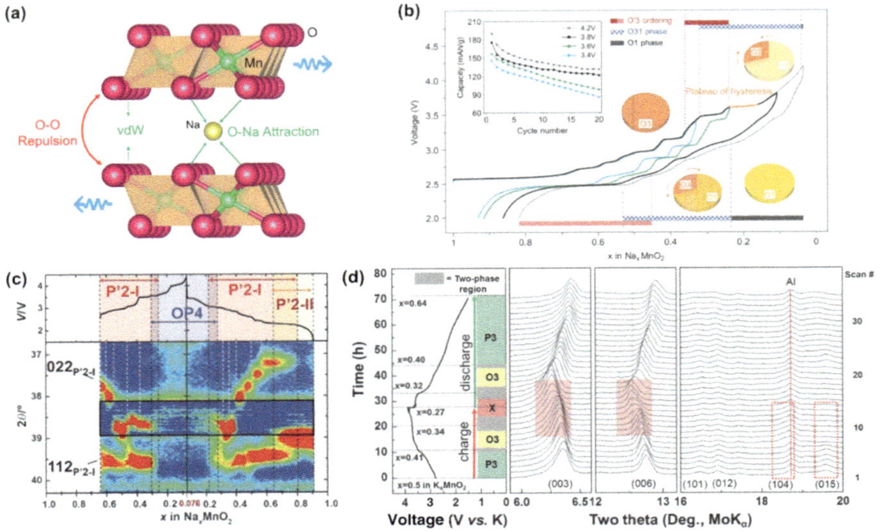

Fig. 5 **a** Schematic illustration of interlayer gliding [112]. Reprinted with permission from Zuo and Yang [112]. Copyright 2022, American Chemical Society. **b** Electrochemical performance reflecting a complicated structural evolution of O-type $NaMnO_2$ [13]. Advanced characterizations such as **c** contour maps of operando XRD patterns showing structural distortion and P–O transformation of P-type $Na_{2/3}MnO_2$ [38] and **d** typical charge–discharge profiles of P3-type $K_{0.5}MnO_2$ with corresponding in-situ XRD pattern, showing the structural changes of the P3-type $K_{0.5}MnO_2$ during charge and discharge with the formation of unknown X-phase [35]

the surface, more Li ions from the bulk diffuse to the surface, leaving tetrahedral vacancies for the migration of Mn^{3+} ions from octahedral sites to the tetrahedral sites, induced by the spontaneous charge disproportionation of Mn^{3+} through the following reaction: $2Mn^{3+}{}_{(octahedra)} \leftrightarrow Mn^{2+}{}_{(tetrahedra)} + Mn^{4+}{}_{(octahedra)}$ [24, 68]. The charge disproportionation not only causes complex chemical and structural transformations at the electrode–electrolyte interface but also induces the formation of dissolvable Mn^{2+} [34, 99, 101, 107]. The amount of Mn^{2+} and the availability of tetrahedral sites are closely related to the Li content in Li_xMnO_2, which further influences the phase transformation of the active materials. Experimental data provide strong evidence that the disproportionation of Mn^{3+} in the cathode is the main reason for Mn dissolution [16, 33]. A small amount of Mn^{2+} (typically at the ppm level) dissolving from the cathode into the electrolyte is noticeably accelerated with an elevated temperature, a larger specific surface area of the active materials, a higher acidity of the electrolyte, and a higher operating potential of the cathode. Once the Mn ions dissolve in the electrolyte, they will migrate to the anode driven by the concentration gradient and/or electric field force and thereby deposit on the anode [72, 96]. The loss of active material from the cathode through Mn dissolution is a straightforward reason for the capacity loss. The displacement of Mn ions, induced by the Li-ion extraction process at moderately high potentials and accompanied by the uncontrolled layered-spinel structure formation, leads to a pronounced hysteresis in the charge/discharge profile. Polarization during cycling can cause voltage drops and hinder efficient charge transfer, diminishing the materials to effectively intercalate Li ions [65]. In addition, the capacity fade of the cell can also be attributed to the accelerated irreversible decomposition of the electrolyte due to the presence of Mn [18]. The migration and decomposition of Mn onto the anode surface interfere with the Li intercalation into the graphite host, the increase in the anode impedance, and thus the capacity fading with cycling.

3 Recent Advancements in Mn-Based Cathodes

To overcome the challenges mentioned in Sect. 2, tremendous efforts have been adopted to improve the electrochemical performance of A_xMnO_2 (A = Li, Na, K) cathode materials. The synthetic route and related process variables play a crucial role in determining crystallinity, particle size, surface morphology, composition, and phase purity. It's worth noting that the electrochemical performance is heavily affected by the morphology and structure of the active materials. By carefully controlling the reaction parameters and utilizing multiple synthetic routes, it is expected to tailor the Mn-based electrode materials with desired compositional stoichiometry and particle morphology with a designed crystal phase to improve their electrochemical properties for practical applications. For example, $LiMnO_2$ nanorods with more favorable 1D electronic transport pathways can deliver a higher discharge capacity (260 mAh g^{-1}) as well as better cyclability compared with $LiMnO_2$ nanoplates (235 mAh g^{-1}), nanospheres (200 mAh g^{-1}), and nanoparticles (200 mAh g^{-1}) [26, 47, 88, 103, 104]. Considering the synthesis of P-type Na_xMnO_2, recent investigations have focused on the optimization of the crystalline structures and morphology engineering by adjusting the precursor ratio, and process variables such as temperature profile [44, 62, 63, 84]. Controlling both the sintering temperature and cooling rate is effective in regulating the concentration of vacancies (denoted as □) in the Na_xMnO_2 materials. The proportion of P′2-type $Na_{0.67}MnO_2$ decreases with the reduction in cooling rates and only P2-type $Na_{0.67}Mn_x\square_{1-x}O_2$ is obtained under natural cooling conditions. A similar trend is observed in other materials, including $Na_{0.67}Li_{0.1}Mn_{0.9}O_2$, $Na_{0.67}Al_{0.1}Mn_{0.9}O_2$, $Na_{0.67}Mg_{0.1}Mn_{0.9}O_2$, and $Na_{0.67}Al_{0.1}Fe_{0.05}Mn_{0.85}O_2$ [51]. The optimized $Na_{0.67}Al_{0.1}Fe_{0.05}Mn_{0.85}O_2$ material exhibits a remarkably high initial capacity of 202 mAh g^{-1} among ever-reported P2-type layered oxides within 2–4 V, a stable capacity retention of 81% after 600 cycles and outstanding rate capability of the specific capacity up to 122 mAh g^{-1} at 1200 mA g^{-1}. Hollow sub-microsphere P-type $K_{0.5}MnO_2$ synthesized by Peng et al. [62, 63] through a facile two-step self-templating strategy can provide an initial discharge capacity of 104 mAh g^{-1} with an acceptable capacity retention of 89.1% after 400 cycles. In addition, the hollow sub-microsphere morphology remains at 200 mA g^{-1} after 200 cycles. Obviously, the selection of appropriate synthesis methods and optimization of reaction conditions are crucial for the development of electrode materials with enhanced electrochemical properties.

The manipulation of interfaces is pivotal for the performance of the active materials since the mass transport and electrochemical reactions at the particle surface are the most complicated. Providing a protective layer on the particle surface can prevent the direct interaction between the active material, reduce phase transformation, and facilitate Li-ion transfer increasing capacity delivery and cycling stability. Cho et al. [14, 15] reported that coating o-$LiMnO_2$ with Al_2O_3 can significantly reduce the capacity loss of the active materials as only 2% capacity reduction after 50 cycles for the coated material compared with 35% loss for the bare one. Most significantly, there are no spinel $Li_2Mn_2O_4$ phases in the Al_2O_3-coated electrode during the process of cycling, implying that the protective coating layer can improve the lattice stability by suppressing the Jahn–Teller distortion. Recently, Zuo et al. [109, 110] took advantage of Al_2O_3 atomic layered deposition coating on $Na_{0.67}Zn_{0.1}Mn_{0.9}O_2$ to mitigate poor electrolyte–electrode interfacial stability (Fig. 6a) and thus benefit the electrochemical performance of the active material.

Improvement of Mn-based cathodes can be achieved through the development of sophisticated gradient, disordered, integrated phase, core–shell structures, or a combination of these approaches [28, 82, 83, 111]. The microstructure design with interwoven spinel and layered domains has provided new insights into the elimination of the Jahn–Teller effect for a $LiMnO_2$ cathode. At the interface between these two domains, the Mn d(z^2) orbitals are oriented perpendicular to each other, resulting in an interfacial orbital order that suppresses the Jahn–Teller distortion and the consequent Mn dissolution. As a result, the heterostructured cathode offers improved structural and electrochemical cycling stability by providing a high Li storage capacity of 254.3 mAh g^{-1} and unprecedented cyclability with ~90.4% capacity retention after 2,000 cycles [108]. Lei et al. [42] improve the performance of P2-$K_{0.67}MnO_2$ layered materials through the creation of a dual interface structure composed of a solid-electrolyte interphase (SEI) and K-poor spinel interlayer to mitigate the structural distortion, facilitate K^+ diffusion, and reduce Mn^{2+} loss (Fig. 6b). The material exhibits a reversible process involving K^+ intercalation/deintercalation and a phase transition, provides an initial capacity of 78.0 mAh g^{-1}, and notably, maintains a high-capacity retention of 90.5% and Coulombic efficiency of 100% after 300 cycles, indicating exceptional stability. In another work, a double-layer phosphate of K_3PO_4/$MnPO_4$-coated $K_{0.5}Mn_{0.8}Co_{0.2}O_2$ cathodes shows a higher electron and K-ion conductivity, less surficial oxygen loss and layered-to-spinel-to-rock salt tri-phase transition, and less internal lattice expansion and contraction during cycling [82, 83]. In addition, first-principles calculations imply that surface modification can effectively prevent Mn ions dissolution and oxygen release. As a result, the coated materials exhibit a highly reversible specific capacity (101.3 mAh g^{-1} at 0.1 A g^{-1}) and excellent rate performance and capacity retention of 80% after 500 cycles at 1 A g^{-1} compared to uncoated one (78.6 mAh g^{-1} at 0.1 A g^{-1} with 67.3% capacity retention).

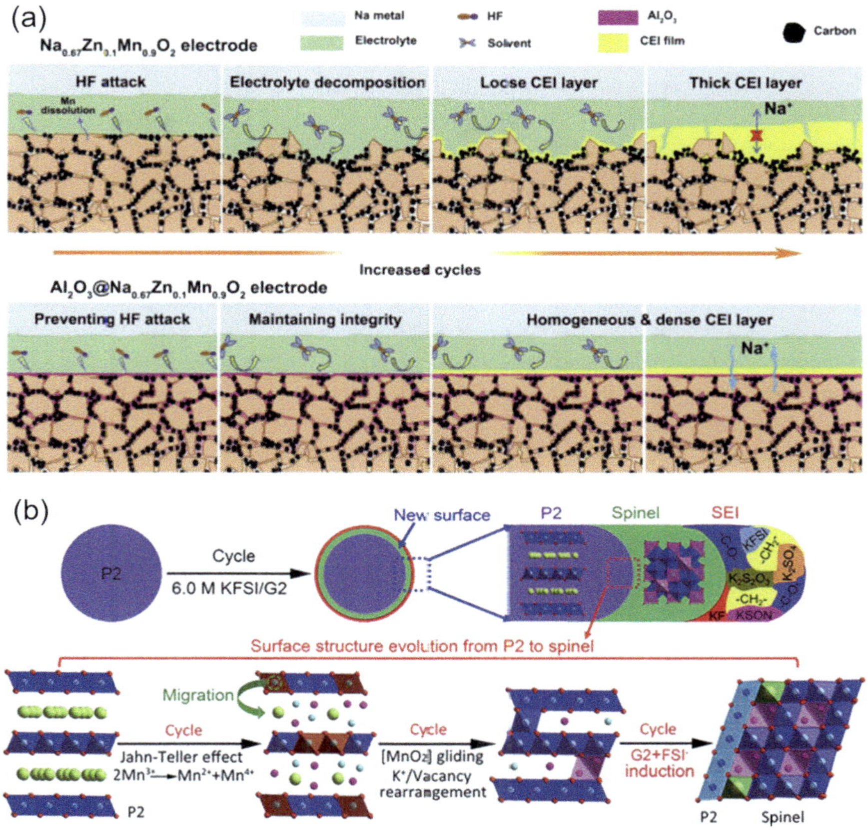

Fig. 6 **a** Schematic illustration of the surficial evolutions of $Na_{0.67}Zn_{0.1}Mn_{0.9}O_2$ and $Al_2O_3@Na_{0.67}Zn_{0.1}Mn_{0.9}O_2$ materials [109]. **b** The formation mechanism of the dual interphase layer composed of the SEI layer and spinel interlayer on P2-$K_{0.67}MnO_2$ [42]

While surface modifications are very useful to resist Mn dissolution and protect the electrode materials, engineering bulk particles are particularly effective in stabilizing the structure through mutual effects between adjacent atoms or layers. Phase transformation in Mn-based cathodes is unlikely to be eliminated when the Mn ions remain in the +3 state throughout the cycling process. Modulation of the electronic structure of Mn is, therefore, the most critical approach to solving the intrinsic drawback of Mn, leading to extensive theoretical and experimental research on various Mn combinations. A series of first-principles calculations have demonstrated that chemical substitutions can offer a powerful approach to influencing the valence state of Mn [27, 37, 64, 71]. Partial density of state calculation shows that the amount of Mn^{4+} content increases after doping Zn into $Na_{0.833}[Li_{0.25}Mn_{0.75}]O_2$, indicating that the Jahn–Teller distortion caused by $2Mn^{3+}$ is effectively suppressed. Further calculation of thermodynamic phase stability shows that doping can reduce the formation energy and rotational decomposition zone of the sample and form a stable intermediate phase, which can suppress the phase transition of cathode materials in the electrochemical process and enhance the cycling stability [98].

The effectiveness of elemental substitution and vacancy formation lies primarily in their ability to adjust the Mn–O–*Sub* electron configuration and lattice parameters. Up to now, various elements have been doped at the alkali metal sites, Mn sites, and O sites to improve the electrochemical properties of layered A_xMnO_2, including Na, Cu, Mg, Zn, Fe, Ni, Al, Co, Cr, Y, In, S, Ti, V, Nb, Ru, B, F, BO_3^{3-}, PO_4^{3-}, and SiO_3^{3-} as comprehensively reviewed [12, 43, 46, 48, 74, 100]. The Mn–O–*Sub* bond can modulate the valance state of Mn to suppress undesirable phase transformations and alter the spacing between alkali metal layers to create favorable ion diffusion pathways. In general, replacing Mn with smaller radius ions can result in a contracted crystal lattice and microstrains due to the shorter and stronger *Sub*–O bond compared to Mn–O. In contrast, larger radius ions are expected to expand the lattice parameters, facilitating the free movement of charged ions within the oxide and potentially increasing the rate capability. Replacing –O with mono atoms such as –S, –F, or polyanion groups like BO_3^{3-}, PO_4^{3-}, and SiO_3^{3-} directly affects the electron configuration and coordination of Mn centers through inductive effects, typically resulting in changes in bond spacing and ionicity, which in turn affect electron conductivity and electrochemical potential [1, 25]. Larger polyanions BO_3^{3-}, PO_4^{3-}, and SiO_3^{3-} dopants can generate an open 3D framework, which enhances the mobility of the alkali ions but increases charge-transfer resistance due to the poor electron conduction [76].

Under the presence of doping elements, a strong preference by Mn for octahedral over tetrahedral coordination should result in reduced mobility for Mn. In this approach, it would be better to maintain Mn in a relatively stable +4 valence state throughout the electrochemical cycle. Reed and Ceder [68] revealed that the substitution of Mn by low fixed valence cations (Li^+, Mg^{2+}, Al^{3+}) and electronegative multivalent cations (Co^{3+}, Ni^{2+}, Cu^{2+}) that preferably oxidize Mn^{3+} ($Mn^{3+} \rightarrow Mn^{4+}$) can stabilize Mn in the octahedral sites and thus increase the resistance against structural transformation while other elemental substitutions with Zr^{4+}, Sn^{4+}, V^{5+} can destabilize the layered structure (Fig. 7a). The calculated qualitative ionization energies for different 3d ions from Reed and Ceder [68] provide insight into predicting substitutions that can stabilize octahedral Mn in a relatively immobile valence state (near +4) throughout the electrochemical cycle (Fig. 7b) and thus can be used as fundamentals to select dopants. While the inactive Mn^{4+} offers better cycling stability, it may partially reduce the initial discharge capacity of the active materials [56, 89, 91, 109, 110]. A combination of an inert cation such as Mg^{2+} and Al^{3+} which stabilizes the Mn–O layers and other redox-active elements that fundamentally broaden the electrochemical properties of the active materials, can lead to improved battery performance in both capacity delivery and cycling stability. Overall electrochemical properties of the active materials are now dominantly generated by the substituted elements, making it crucial to thoroughly investigate the stability of the substitutions [9].

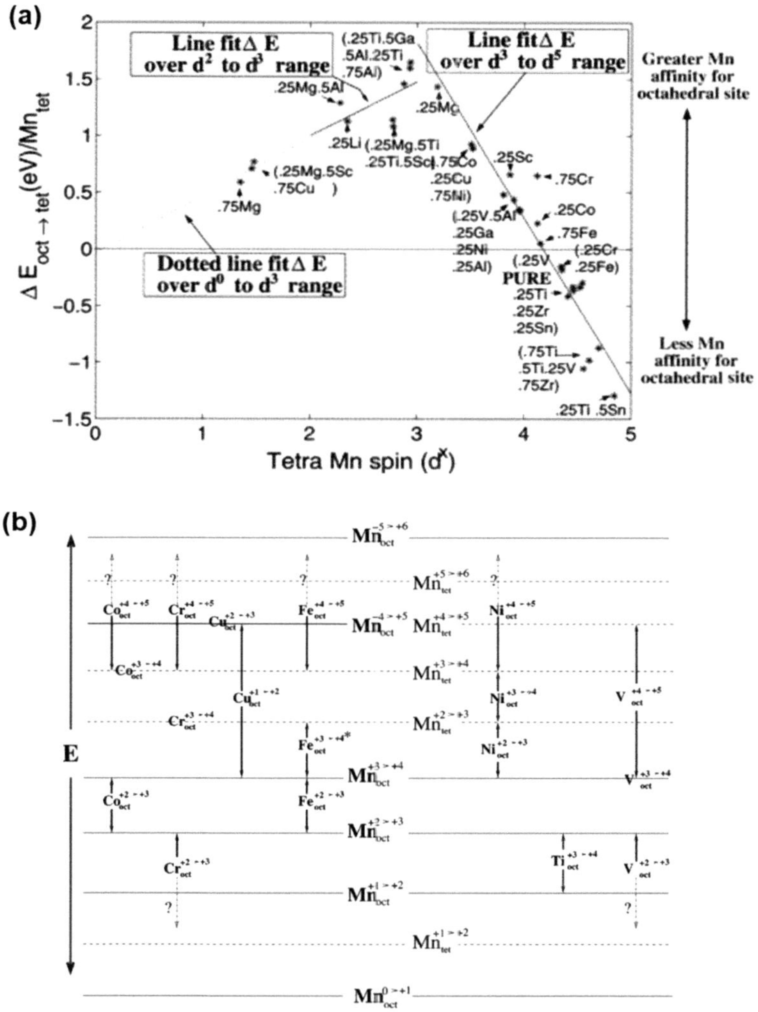

Fig. 7 **a** Energy difference of different Mn substitutions with and without tetrahedral Mn, showing the tendency to migrate between octahedral and tetrahedral sites of Mn under the effects of different substitutions. **b** Qualitative ionization energies for 3d ions at various valences in an oxide framework, in comparison with Mn ionization energies in octahedra and tetrahedra coordination labeled prominently in the center of the figure [68]. Reprinted with permission from Reed and Ceder [68]. Copyright (2004), American Chemical Society.

Advances in theoretical calculations and chemical characterizations enlighten the benefits of modulating the electron configuration of Mn for the improvement of Mn-based materials in rechargeable batteries. As demonstrated in Fig. 8, the Mn average valence can be regulated by the partial substitution of Mn with Ti^{4+} and Mg^{2+} ions [91]. The introduction of Ti^{4+} ions reduces the valence of Mn while the Mg^{2+} ions increase the Mn valence. According to the measured results combined with the calculation of the charge compensation mechanism, the Mn average valence in $K_{0.5}Mn_{0.7}Co_{0.2}Fe_{0.1}O_2$, $K_{0.5}Mn_{0.6}Co_{0.2}Fe_{0.1}Ti_{0.1}O_2$, and $K_{0.5}Mn_{0.6}Co_{0.2}Fe_{0.1}Mg_{0.1}O_2$ are +3.714, +3.667, and +4, respectively. More importantly, the e_g orbitals of Mn in $K_{0.5}Mn_{0.6}Co_{0.2}Fe_{0.1}Mg_{0.1}O_2$ are not split, implying the presence of inactive Mn^{+4}. Upon cycling, no new peaks of other phases are observed, revealing that the $K_{0.5}Mn_{0.6}Co_{0.2}Fe_{0.1}Mg_{0.1}O_2$ experiences a typical reversible single-phase reaction with phase and structure stability. The $K_{0.5}Mn_{0.6}Co_{0.2}Fe_{0.1}Mg_{0.1}O_2$ cathode exhibits outstanding cycling stability with capacity retention of 91% at 0.1 A g^{-1} after 150 cycles, and retained 74% of the initial capacity even after 500 cycles at 1 A g^{-1}. Aside from cations doping, the introduction of –F, a more electronegative element, can drive phase transformations toward more thermodynamically stable structures, lower the energy barrier for alkali ion intercalation, and create additional active sites for ion storage [21, 54, 70, 81, 90]. Although the presence of F does not completely eliminate the high-spin state of Mn, fine-tuning the degree of fluorination along with structural transformation and electrochemical performance provides valuable insights from both atomic and molecular perspectives [20, 32]. The regulation of intrinsic electronic structure by doping reveals the effect of local chemistry regulation on the material system and opens up a new method of combining local chemistry with orbital hybridization to the theoretical design and experimental synthesis of structural units for the development of advanced Mn-based cathodes [45, 89, 91, 105].

4 Conclusion and Outlook

Mn-based cathode materials have emerged as promising candidates for energy storage applications due to their abundance, cost-effectiveness, and environmental friendliness. Their layered structures, characterized by flexible intercalation pathways, enable high specific capacities and compatibility with various alkali-ion chemistries such as lithium, sodium, and potassium. Despite their advantages, these materials face significant challenges, including structural instability caused by Jahn–Teller distortions, Mn dissolution into the electrolyte, phase transitions during cycling, and capacity fading. It's worth noting that those challenges come from the unique electron configuration of Mn^{3+} which is involved in the electrochemical reactions. Recent advancements, such as elemental doping, surface coatings, and optimized synthesis techniques, have made significant strides in addressing these issues, enhancing the stability, conductivity, and overall performance of Mn-based cathodes.

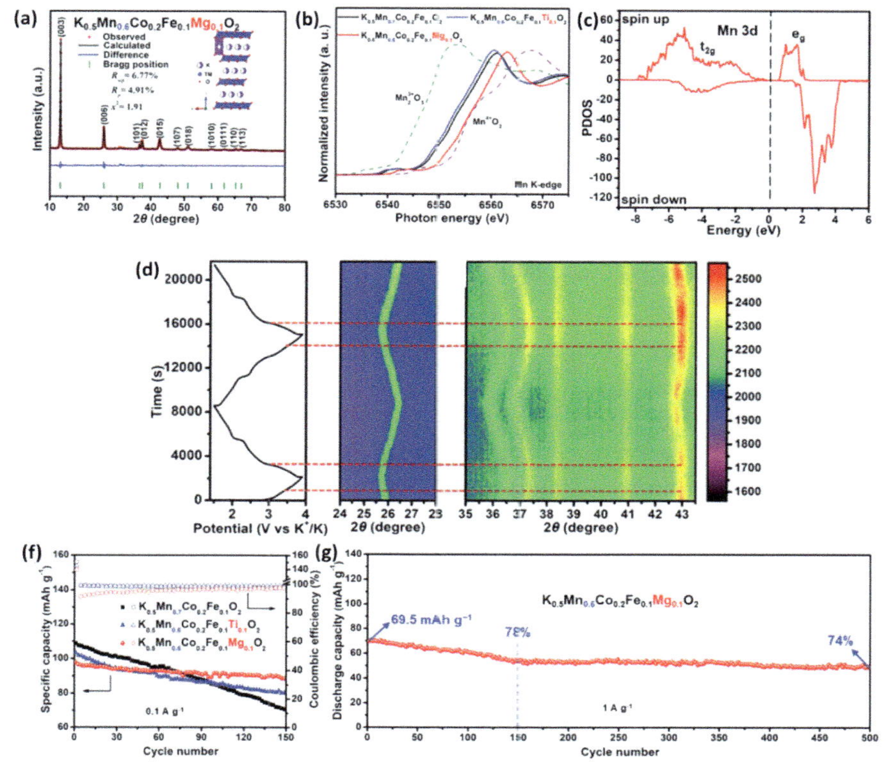

Fig. 8 **a** XRD Rietveld refinement of $K_{0.5}Mn_{0.5}Co_{0.2}Fe_{0.1}Mg_{0.1}O_2$. **b** XANES spectra of Mn K-edge for $K_{0.5}Mn_{0.7}Co_{0.2}Fe_{0.1}O_2$, $K_{0.5}Mn_{0.6}Co_{0.2}Fe_{0.1}Ti_{0.1}O_2$, and $K_{0.5}Mn_{0.6}Co_{0.2}Fe_{0.1}Mg_{0.1}O_2$. **c** Partial density of states of Mn 3d electronic orbital with **d** In situ XRD characterization of $K_{0.5}Mn_{0.6}Co_{0.2}Fe_{0.1}Mg_{0.1}O_2$ during cycling. **e** Cycling property tested at 0.1 A g^{-1} of $K_{0.5}Mn_{0.7}Co_{0.2}Fe_{0.1}O_2$, $K_{0.5}Mn_{0.6}Co_{0.2}Fe_{0.1}Ti_{0.1}O_2$, and $K_{0.5}Mn_{0.6}Co_{0.2}Fe_{0.1}Mg_{0.1}O_2$ and **h** long-term cycling performance of $K_{0.5}Mn_{0.6}Co_{0.2}Fe_{0.1}Mg_{0.1}O_2$ at 1 A g^{-1} [91]

Several unresolved challenges remain, requiring focused research and innovation. Achieving long-term cycling stability is critical, with particular attention needed to suppress Mn dissolution and mitigate oxygen loss during cycling. Improving electronic conductivity through material design and composite engineering is another key area for development. Exploring new synthesis methods, such as solid-state routes and high-entropy approaches, could yield more robust and scalable materials. Furthermore, adapting Mn-based cathodes for multi-valent ion chemistries, such as magnesium and zinc, presents exciting opportunities for expanding their application in next-generation batteries. The integration of advanced in situ characterization techniques can also provide deeper insights into structural dynamics and degradation mechanisms, guiding the development of more durable materials.

The future of Mn-based cathode materials depends on interdisciplinary efforts that combine materials science, electrochemistry, and engineering. Researchers are encouraged to explore innovative solutions, such as designing hybrid structures, optimizing electrolyte formulations, and leveraging machine learning for materials discovery. Collaboration between academia, industry, and policymakers is essential to translate these advancements into scalable and sustainable energy storage technologies. By addressing the remaining challenges and pushing the boundaries of material design, Mn-based cathodes can play a pivotal role in advancing energy storage systems that meet the growing demands of a sustainable energy future.

Discussion Questions

1. What types of phase transformations occur in layered Mn-based electrodes during cycling, and how do they influence electrochemical performance?
2. What are the primary degradation mechanisms (e.g., structural collapse, Mn dissolution, electrolyte decomposition) in Mn-based layered cathodes, and what strategies can mitigate these issues?
3. What approaches could make Mn-based layered cathodes more sustainable and cost-effective for large-scale production, especially compared to Co- and Ni-based alternatives?
4. What novel synthesis methods and materials engineering could be employed to achieve uniform particle morphology, reduce defects, and improve the structural stability, electronic conductivity, and electrochemical performance of Mn-based cathodes?
5. What modifications are needed for Mn-based layered materials to function as efficient cathodes in emerging multi-valent ion batteries (e.g., Mg^{2+}, Zn^{2+}, or Al^{3+})?

References

1. Abakumov AM, Fedotov SS, Antipov EV, Tarascon JM (2020) Solid state chemistry for developing better metal-ion batteries. Nat Commun 11:4976. https://doi.org/10.1038/s41467-020-18736-7
2. Ammundsen B, Paulsen J (2001) Novel lithium-ion cathode materials based on layered manganese oxides. Adv Mater 13:943–956. https://doi.org/10.1002/1521-4095(200107)13:12/13%3c943::AID-ADMA943%3e3.0.CO;2-J
3. Amriou T, Khelifa B, Aourag H, Aouadi SM, Mathieu C (2005) Ab initio investigation of the Jahn-Teller distortion effect on the stabilizing lithium intercalated compounds. Mater Chem Phys 92:499–504. https://doi.org/10.1016/j.matchemphys.2005.01.061
4. Asl HY, Manthiram A (2020) Reining in dissolved transition-metal ions. Science 369:140–141. https://doi.org/10.1126/science.abc5454
5. Billaud J, Clément RJ, Armstrong AR, Canales-Vázquez J, Rozier P, Grey CP, Bruce PG (2014) β-$NaMnO_2$: a high-performance cathode for sodium-ion batteries. J Am Chem Soc 136:17243–17248. https://doi.org/10.1021/ja509704t
6. Boivin E, House RA, Pérez-Osorio MA, Marie JJ, Maitra U, Rees GJ, Bruce PG (2021) Bulk O_2 formation and Mg displacement explain O-redox in $Na_{0.67}Mn_{0.72}Mg_{0.28}O_2$. Joule 5:1267–1280. https://doi.org/10.1016/j.joule.2021.04.006

7. Bruce PG, Armstrong AR (1996) Synthesis of layered $LiMnO_2$ as an electrode for rechargeable lithium batteries. Nature 381 499–500. https://doi.org/10.1038/381499a0
8. Capitaine F, Gravereau P, Delmas C (1996) A new variety of $LiMnO_2$ with a layered structure. Solid State Ion 89:197–202. https://doi.org/10.1016/0167-2738(96)00369-4v
9. Cao MH, Wang Y, Shadike Z, Yue JL, Hu E, Bak SM, Zhou YN, Yang XQ, Fu ZW (2017) Suppressing the chromium disproportionation reaction in O3-type layered cathode materials for high capacity sodium-ion batteries. J Mater Chem A 5:5442–5448. https://doi.org/10.1039/C6TA10818K
10. Cao X, Li H, Qiao Y, Jia M, Li X, Cabana J, Zhou H (2021) Stabilizing anionic redox chemistry in a Mn-based layered oxide cathode constructed by Li-deficient pristine state. Adv Mater 33:2004280. https://doi.org/10.1002/adma.202004280
11. Ceder G, Mishra SK (1999) The stability of orthorhombic and monoclinic-layered $LiMnO_2$. Electrochem Solid State Lett 2:550. https://doi.org/10.1149/1.1390900
12. Chen H, Xia X, Ma J (2024) Comprehensive review of Li-rich Mn-based layered oxide cathode materials for lithium-ion batteries: theories, challenges, strategies and perspectives. ChemSusChem e202401120. https://doi.org/10.1002/cssc.202401120
13. Chen X, Wang Y, Wiaderek K, Sang X, Borkiewicz O, Chapman K, LeBeau J, Lynn J, Li X (2018) Super charge separation and high voltage phase in Na_xMnO_2. Adv Funct Mater 28:1805105. https://doi.org/10.1002/adfm.201805105
14. Cho J, Kim YJ, Kim TJ, Park B (2001) Enhanced structural stability of o-$LiMnO_2$ by sol−gel coating of Al_2O_3. Chem Mater 13:18–20. https://doi.org/10.1021/cm000759+
15. Cho J, Kim YJ, Kim TJ, Park B (2002) Effect of Al_2O_3-coated o-$LiMnO_2$ cathodes prepared at various temperatures on the 55°C cycling behavior. J Electrochem Soc 149:A127. https://doi.org/10.1149/1.1429927
16. Choi W, Manthiram A (2006) Comparison of metal ion dissolutions from lithium ion battery cathodes. J Electrochem Soc 153:A1760–A1764. https://doi.org/10.1149/1.2219710
17. Clément RJ, Bruce PG, Grey CP (2015) Manganese-based P2-type transition metal oxides as sodium-ion battery cathode materials. J Electrochem Soc 162:A2589–A2604. https://doi.org/10.1149/2.0201514jes
18. Delacourt C, Kwong A, Liu X, Qiao R, Yang W, Lu P, Harris S, Srinivasan V (2013) Effect of manganese contamination on the solid-electrolyte-interphase properties in Li-ion batteries. J Electrochem Soc 160:A1099–A1107. https://doi.org/10.1149/2.035308jes
19. Delmas C, Fouassier C, Hagenmuller P (1980) Structural classification and properties of the layered oxides. Physica B+C 99:81–85. https://doi.org/10.1016/0378-4363(80)90214-4
20. Ding F, Wang H, Zhang Q, Zheng L, Guo H, Yu P, Zhang N, Guo Q, Xie F, Dang R, Rong X, Lu Y, Xiao R, Chen L, Hu Y (2023) Tailoring electronic structure to achieve maximum utilization of transition metal redox for high-entropy Na layered oxide cathodes. J Am Chem Soc 145:13592–13602. https://doi.org/10.1021/jacs.3c00879
21. Ganesan BK, Lee YS (2023) Phase transition induced enhanced performance of sodium-rich $Na_{1.2}Mn_{0.8}O_{2-y}F_y$ ($y = 0–0.5$) cathodes. ACS Appl Energy Mater 6:960–968. https://doi.org/10.1021/acsaem.2c03391
22. Ghosh SK (2020) Diversity in the family of manganese oxides at the nanoscale: from fundamentals to applications. ACS Omega 5:25493–25504. https://doi.org/10.1021/acsomega.0c03455
23. Goodenough JB, Park KS (2013) The Li-ion rechargeable battery: a perspective. J Am Chem Soc 135:1167–1176. https://doi.org/10.1021/ja3091438
24. Gummow RJ, Liles DC, Thackeray MM (1993) Lithium extraction from orthorhombic lithium manganese oxide and the phase transformation to spinel. Mater Res Bull 28:1249–1256. https://doi.org/10.1016/0025-5408(93)90172-a
25. Gutierrez A, Benedek NA, Manthiram A (2013) Crystal-chemical guide for understanding redox energy variations of $M^{2+/3+}$ couples in polyanion cathodes for lithium-ion batteries. Chem Mater 25:4010–4016. https://doi.org/10.1021/cm401949n
26. He Y, Li RH, Ding XK, Jiang LL, Wei MD (2010) Hydrothermal synthesis and electrochemical properties of orthorhombic $LiMnO_2$ nanoplates. J Alloy Compd 492:601–604. https://doi.org/10.1016/j.jallcom.2009.11.191

27. Hoang K (2017) First-principles theory of doping in layered oxide electrode materials. Phys Rev Mater 1:075403. https://doi.org/10.1103/PhysRevMaterials.1.075403
28. Hou P, Dong M, Lin Z, Sun Z, Gong M, Li F, Xu X (2023) Alleviating the Jahn–Teller distortion of P3-type manganese-based cathodes by compositionally graded structure for sodium-ion batteries. ACS Sustain Chem Eng 11:10785–10794. https://doi.org/10.1021/acssuschemeng.3c01854
29. Hu E, Yu X, Lin R, Bi X, Lu J, Bak S, Nam KW, Xin HL, Jaye C, Fischer DA, Amine K, Yang XQ (2018) Evolution of redox couples in Li- and Mn-rich cathode materials and mitigation of voltage fade by reducing oxygen release. Nat Energy 3:690–698. https://doi.org/10.1038/s41560-018-0207-z
30. Huang ZF, Du F, Wang CZ, Wang DP, Chen G (2007) Low-spin Mn^{3+} ion in rhombohedral $LiMnO_2$ predicted by first-principles calculations. Phys Rev B 75(5):054411. https://doi.org/10.1103/PhysRevB.75.054411
31. Hwang JY, Myung ST, Sun YK (2017) Sodium-ion batteries: present and future. Chem Soc Rev 46:3529–3614. https://doi.org/10.1039/C6CS00776G
32. Ji H, Wu J, Cai Z, Liu J, Kwon DH, Kim H, Urban A, Papp JK, Foley E, Tian Y, Balasubramanian M, Kim H, Clément RJ, McCloskey BD, Yang W, Ceder G (2020) Ultrahigh power and energy density in partially ordered lithium-ion cathode materials. Nat Energy 5:213–221. https://doi.org/10.1038/s41560-020-0573-1
33. Johnson CS, Li N, Lefief C, Vaughey JT, Thackeray MM (2008) Synthesis, characterization and electrochemistry of lithium battery electrodes: $xLi_2MnO_3{\cdot}(1-x)LiMn_{0.333}Ni_{0.333}Co_{0.333}O_2$ $(0 \leq x \leq 0.7)$. Chem Mater 20:6095–6106. https://doi.org/10.1021/cm801245r
34. Kang S, Cho S, Kang YM (2023) Getting to the bottom of transition metal dissolution. Nat Nanotechnol 18:700–701. https://doi.org/10.1038/s41565-023-01396-1
35. Kim H, Seo DH, Kim JC, Bo SH, Liu L, Shi T, Ceder G (2017) Investigation of potassium storage in layered P3-type $K_{0.5}MnO_2$ cathode. Adv Mater 29:1702480. https://doi.org/10.1002/adma.201702480
36. Kim S, Ma X, Ong SP, Ceder G (2012) A comparison of destabilization mechanisms of the layered Na_xMO_2 and Li_xMO_2 compounds upon alkali de-intercalation. Phys Chem Chem Phys 2012(14):15571–15578. https://doi.org/10.1039/C2CP43377J
37. Kong F, Longo RC, Yeon DH, Yoon J, Park JH, Liang C, Santosh KC, Zheng Y, Doo SG, Cho K (2015) Multivalent Li-site doping of Mn oxides for Li-ion batteries. J Phys Chem C 119:21904–21912. https://doi.org/10.1021/acs.jpcc.5b06844
38. Kumakura S, Tahara Y, Kubota K, Chihara K, Komaba S (2016) Sodium and manganese stoichiometry of P2-type $Na_{2/3}MnO_2$. Angew Chem Int Ed 55:12760–12763. https://doi.org/10.1002/anie.201606415
39. Lee J, Kitchaev DA, Kwon DH, Lee CW, Papp JK, Liu YS, Lun Z, Clément RJ, Shi T, McCloskey BD, Guo J, Balasubramanian M, Ceder G (2018) Reversible Mn^{2+}/Mn^{4+} double redox in lithium-excess cathode materials. Nature 556:185–190. https://doi.org/10.1038/s41586-018-0015-4
40. Lee W, Kim J, Yun S, Choi W, Kim H, Yoon W (2020) Multiscale factors in designing alkali-ion (Li, Na, and K) transition metal inorganic compounds for next-generation rechargeable batteries. Energy Environ Sci 13:4406–4449. https://doi.org/10.1039/D0EE01277G
41. Lee W, Muhammad S, Sergey C, Lee H, Yoon J, Kang Y, Yoon W (2019) Advances in the cathode materials for lithium rechargeable batteries. Angew Chem Int Ed 59:2578–2605. https://doi.org/10.1002/anie.201902359
42. Lei K, Zhu Z, Yin Z, Yan P, Li F, Chen J (2019) Dual interphase layers in situ formed on a manganese-based oxide cathode enable stable potassium storage. Chem 5:3220. https://doi.org/10.1016/j.chempr.2019.10.008
43. Li H, Zhang W, Sun K, Guo J, Yuan K, Fu J, Zhang T, Zhang X, Long H, Zhang Z, Lai Y, Sun H (2021) Manganese-based materials for rechargeable batteries beyond lithium-ion. Adv Energy Mater 11:2100867. https://doi.org/10.1002/aenm.202100867

44. Li J, Wang J, He X, Zhang L, Senyshyn A, Yan B, Muehlbauer M, Cao X, Vortmann-Westhoven B, Kraft V, Liu H, Luerenbaum C, Schumacher G, Paillard E, Winter M, Li J (2019) P2–type $Na_{0.67}Mn_{0.8}Cu_{0.1}Mg_{0.1}O_2$ as a new cathode material for sodium-ion batteries: insights of the synergetic effects of multi-metal substitution and electrolyte optimization. J Power Sources 416:184–192. https://doi.org/10.1016/j.jpowsour.2019.01.086
45. Li X, Chen X, Bai Q, Mo Y, Zhu Y (2024) From atomistic modeling to materials design: computation-driven material development in lithium-ion batteries. Sci China Chem 67:276–290. https://doi.org/10.1007/s11426-022-1506-1
46. Li W, Bi Z, Zhang W, Wang J, Rajagopalan R, Wang Q, Zhang D, Li Z, Wang H, Wang B (2021) Advanced cathodes for potassium-ion batteries with layered transition metal oxides: a review. J Mater Chem A 9:8221–8247. https://doi.org/10.1039/D0TA12129K
47. Liu Q, Mao DL, Chang CK, Huang FQ (2007) Phase conversion and morphology evolution during hydrothermal preparation of orthorhombic $LiMnO_2$ nanorods for lithium ion battery application. J Power Sources 173:538–544. https://doi.org/10.1016/j.jpowsour.2007.03.077
48. Liu S, Wang B, Zhang X, Zhao S, Zhang Z, Yu H (2021) Reviving the lithium-manganese-based layered oxide cathodes for lithium-ion batteries. Matter 4:1511–1527. https://doi.org/10.1016/j.matt.2021.02.023
49. Liu T, Dai A, Lu J, Yuan Y, Xiao Y, Yu L, Li M, Gim J, Ma L, Liu J, Zhan C, Li L, Zheng J, Ren Y, Wu T, Shahbazian-Yassar R, Wen J, Pan F, Amine K (2019) Correlation between manganese dissolution and dynamic phase stability in spinel-based lithium-ion battery. Nat Commun 10:4721. https://doi.org/10.1038/s41467-019-12626-3
50. Liu T, Hou S, Li Y, Xue S, Hu J, Fu H, Yang C, Zhao L (2022) Insight of K-deficient layered K_xMnO_2 cathode for potassium-ions batteries. J Energy Chem 64:335–343. https://doi.org/10.1016/j.jechem.2021.04.062
51. Liu X, Zhong G, Xiao Z, Zheng B, Zuo W, Zhou K, Liu H, Liang Z, Xiang Y, Chen Z, Ortiz GF, Fu R, Yang Y (2020) Al and Fe-containing Mn-based layered cathode with controlled vacancies for high-rate sodium ion batteries. Nano Energy 76:104997. https://doi.org/10.1016/j.nanoen.2020.104997
52. Lu Z, Hao S, Wang Z, Kim H, Wolverton C (2024) Phase stability of Li-rich layered cathodes: insight into the debate over solid solutions vs phase separation. Chem Mater 36:6381–6391. https://doi.org/10.1021/acs.chemmater.2c00927
53. Ma X, Chen H, Ceder G (2011) Electrochemical properties of monoclinic $NaMnO_2$. J Electrochem Soc 158:A1307–A1312. https://doi org/10.1149/2.035112jes
54. Mahara Y, Nagasako N, Oka H, Kondo Y, Kosaka S, Nakano H, Nonaka T, Makimura Y (2022) How fluorine introduction solves the spinel transition, a fundamental problem of Mn-based positive electrodes. ACS Appl Mater Interfaces 14:24321–24331. https://doi.org/10.1021/acsami.2c02868
55. Manthiram A (2020) A reflection on lithium-ion battery cathode chemistry. Nat Commun 11:1550. https://doi.org/10.1038/s41467-020-15355-0
56. Mu L, Xu S, Li Y, Hu YS, Li H, Chen L, Huang X (2015) Prototype sodium-ion batteries using an air-stable and Co/Ni-Free O_3-layered metal oxide cathode. Adv Mater 27:6928–6933. https://doi.org/10.1002/adma.201502449
57. Mizushima K, Jones PC, Wiseman PJ, Goodenough JB (1980) Li_xCoO_2 ($0 < x < 1$): a new cathode material for batteries of high energy density. Mater Res Bull 15:783–798. https://doi.org/10.1016/0025-5408(80)90012-4
58. Moghadam YS, El Kharbachi A, Diemant T, Melinte G, Hu Y, Fichtner M (2021) Toward better stability and reversibility of the Mn^{4+}/Mn^{2+} double redox activity in disordered rocksalt oxyfluoride cathode materials. Chem Mater 33 8235–8247. https://doi.org/10.1021/acs.chemmater.1c02334
59. Nguyen DT, Kim J, Lee Y (2024) Design of interfacial Li-ion transfer channels for practical improvement of a multicomponent high-voltage olivine cathode. Ceram Int 50:12351–12360. https://doi.org/10.1016/j.ceramint.2024.01.140
60. Nguyen J, Lee Y, Yang Y (2024) Suppression of high spin state of Mn for the improvement of Mn-based materials in rechargeable batteries. Small 2410453. https://doi.org/10.1002/smll.202410453

61. Nitta N, Wu F, Lee JT, Yushin G (2015) Li-ion battery materials: present and future. Mater Today 18:252–264. https://doi.org/10.1016/j.mattod.2014.10.040
62. Peng B, Li Y, Gao J, Zhang F, Li J, Zhang G (2019) High energy K-ion batteries based on P3-Type $K_{0.5}MnO_2$ hollow submicrosphere cathode. J Power Sources 437:226913. https://doi.org/10.1016/j.jpowsour.2019.226913
63. Peng B, Sun Z, Jiao S, Wang G, Zhang G (2019) Electrochemical performance optimization of layered P2-type $Na_{0.67}MnO_2$ through simultaneous Mn-site doping and nanostructure engineering. Batter Supercaps 3:147–154. https://doi.org/10.1002/batt.201900126
64. Prasad R, Benedek R, Thackeray MM (2005) Dopant-induced stabilization of rhombohedral $LiMnO_2$ against Jahn-Teller distortion. Phys Rev B 71:134111. https://doi.org/10.1103/PhysRevB.71.134111
65. Qiao Y, Ran Y, Qian Z, Zhao B, Li C, Liu Z (2024) Overcoming polarization and dissolution of manganese-based electrodes to enhance stability in electrochemical lithium extraction. Chem Eng J 497:155009. https://doi.org/10.1016/j.cej.2024.155009
66. Radin MD, Hy S, Sina M, Fang C, Liu H, Vinckeviciute J, Zhang M, Whittingham MS, Meng YS, Van der Ven A (2017) Narrowing the gap between theoretical and practical capacities in Li-ion layered oxide cathode materials. Adv Energy Mater 7:1602888. https://doi.org/10.1002/aenm.201602888
67. Reddy MV, Mauger A, Julien CM, Paolella A, Zaghib K (2020) Brief history of early lithium-battery development. Materials 13:1884. https://doi.org/10.3390/ma13081884
68. Reed J, Ceder G (2004) Role of electronic structure in the susceptibility of metastable transition-metal oxide structures to transformation. Chem Rev 104:4513–4534. https://doi.org/10.1021/cr020733x
69. Seo DH, Lee J, Urban A, Malik R, Kang S, Ceder G (2016) The structural and chemical origin of the oxygen redox activity in layered and cation-disordered Li-excess cathode materials. Nat Chem 8:692–697. https://doi.org/10.1038/nchem.2524
70. Shi WJ, Yan YW, Chi C, Ma XT, Zhang D, Xu SD, Chen L, Wang XM, Liu SB (2019) Fluorine anion doped $Na_{0.44}MnO_2$ with layer-tunnel hybrid structure as advanced cathode for sodium ion batteries. J Power Sources 427:129–137. https://doi.org/10.1016/j.jpowsour.2019.04.038
71. Shishkin M, Kumakura S, Sato S, Kubota K, Komaba S, Sato H (2018) Unraveling the role of doping in selective stabilization of $NaMnO_2$ polymorphs: combined theoretical and experimental study. Chem Mater 30:1257–1264. https://doi.org/10.1021/acs.chemmater.7b04394
72. Shkrob IA, Kropf AJ, Marin TW, Li Y, Poluektov OG, Niklas J, Abraham DP (2014) Manganese in graphite anode and capacity fade in Li ion batteries. J Phys Chem C 118:24335–24348. https://doi.org/10.1021/jp507833u
73. Singh AN, Islam M, Meena A, Faizan M, Han D, Bathula C, Hajibabaei A, Anand R, Nam KW (2023) Unleashing the potential of sodium-ion batteries: current state and future directions for sustainable energy storage. Adv Funct Mater 33:2304617. https://doi.org/10.1002/adfm.202304617
74. Song J, Wang H, Zuo Y, Zhang K, Yang T, Yang Y, Gao C, Chen T, Feng G, Jiang Z, Xiao W, Luo T, Xia D (2023) Building better full manganese-based cathode materials for next-generation lithium-ion batteries. Electrochem Energy Rev 6:20. https://doi.org/10.1007/s41918-023-00184-8
75. Su D, Wang C, Ahn H, Wang G (2013) Single crystalline $Na_{0.7}MnO_2$ nanoplates as cathode materials for sodium-ion batteries with enhanced performance. Chem Eur J 19:10884–10889. https://doi.org/10.1002/chem.201301563
76. Su Z, Ye SH, Yan TY, Gao XP, Shen PW (2008) Synthesis and electrochemical properties of nanosized Li_xMnO_2 as cathode materials for lithium batteries. J Electrochem Soc 155:A839. https://doi.org/10.1149/1.2976208
77. Thackeray MM, Croy JR, Lee E, Gutierrez A, He M, Park JS, Yonemoto BT, Long BR, Blauwkamp JD, Johnson CS, Shin Y, David WIF (2018) The quest for manganese-rich electrodes for lithium batteries: strategic design and electrochemical behavior. Sustain Energy Fuels 2:1375–1397. https://doi.org/10.1039/C8SE00157J

78. Thackeray MM, David WIF, Bruce PG, Goodenough JB (1983) Lithium insertion into manganese spinels. Mater Res Bull 18:461–472. https://doi.org/10.1016/0025-5408(83)90138-1
79. Tu XY, Shu KY (2008) X-ray diffraction study on phase transition of orthorhombic $LiMnO_2$ in electrochemical conversions. J Solid State Electrochem 12:245–249. https://doi.org/10.1007/s10008-007-0384-3
80. Vaalma C, Giffin GA, Buchholz D, Passerini S (2016) Non-aqueous K-ion battery based on layered $K_{0.3}MnO_2$ and hard carbon/carbon black. J Electrochem Soc 163:A1295. https://doi.org/10.1149/2.0921607jes
81. Wang F, Zuo P, Xue Z, Liu Y, Wang C, Chen G (2024) Fluorination effect on lithium- and manganese-rich layered oxide cathodes. ACS Energy Lett 9:1249–1260. https://doi.org/10.1021/acsenergylett.3c02697
82. Wang D, Deng Y-P, Liu Y, Jiang Y, Zhong B, Wu Z, Guo X, Chen Z (2023) Sodium-ion batteries towards practical application through gradient Mn-based layer-tunnel cathode. Nano Energy 110:108340. https://doi.org/10.1016/j.nanoen.2023.108340
83. Wang H, Peng H, Xiao Z, Yu R, Liu F, Zhu Z, Zhou L, Wu J (2023) Double-layer phosphates coated Mn-based oxide cathodes for highly stable potassium-ion batteries. Energy Storage Mater 58:101–109. https://doi.org/10.1016/j.ensm.2023.03.016
84. Wang Y, Kim S, Lu J, Feng G, Li X (2020) A study of Cu doping effects in P2-$Na_{0.75}Mn_{0.6}Fe_{0.2}(Cu_xNi_{0.2-x})O_2$ layered cathodes for sodium-ion batteries. Batter Supercaps 3:376–387. https://doi.org/10.1002/batt.201900172
85. Whittingham MS (2004) Lithium batteries and cathode materials. Chem Rev 104:4271–4301. https://doi.org/10.1021/cr020731c
86. Wu X, Xu GL, Zhong G, Gong Z, McDonald MJ, Zheng S, Fu R, Chen Z, Amine K, Yang Y (2016) Insights into the effects of zinc doping on structural phase transition of P2-type sodium nickel manganese oxide cathodes for high-energy sodium ion batteries. ACS Appl Mater Interfaces 8:22227–22237. https://doi.org/10.1021/acsami.6b06701
87. Xiao J, Shi F, Glossmann T, Burnett C, Liu Z (2023) From laboratory innovations to materials manufacturing for lithium-based batteries. Nat Energy 8:329–339. https://doi.org/10.1038/s41560-023-01221-y
88. Xiao X, Wang L, Wang D, He X, Peng Q, Li Y (2009) Hydrothermal synthesis of orthorhombic $LiMnO_2$ nano-particles and $LiMnO_2$ nanorods and comparison of their electrochemical performances. Nano Res 2:923–930. https://doi.org/10.1007/s12274-009-9094-8
89. Xiao Y, Wang HR, Hu HY, Zhu YF, Li S, Li JY, Wu XW, Chou SL (2022) Formulating high-rate and long-cycle heterostructured layered oxide cathodes by local chemistry and orbital hybridization modulation for sodium-ion batteries. Adv Mater 34:2202695. https://doi.org/10.1002/adma.202202695
90. Xu YS, Qi MY, Zhang QH, Meng FQ, Zhou YN, Guo SJ, Sun YG, Gu L, Chang BB, Liu CT, Cao AM, Wan LJ (2022) Anion doping for layered oxides with a solid-solution reaction for potassium-ion battery cathodes. ACS Appl Mater Interfaces 14:13379–13387. https://doi.org/10.1021/acsami.2c00811
91. Xiao Z, Xia F, Xu L, Wang X, Meng J, Wang H, Zhang X, Geng L, Wu J, Mai L (2022) Suppressing the Jahn-Teller effect in Mn-based layered oxide cathode toward long-life potassium-ion batteries. Adv Funct Mater 32:2108244. https://doi.org/10.1002/adfm.202108244
92. Yabuuchi N, Kajiyama M, Iwatate J, Nishikawa H, Hitomi S, Okuyama R, Usui R, Yamada Y, Komaba S (2012) P2-type $Na_x[Fe_{1/2}Mn_{1/2}]C_2$ made from earth-abundant elements for rechargeable Na batteries. Nat Mater 11:512–517. https://doi.org/10.1038/nmat3309
93. Yabuuchi N, Kubota K, Dahbi M, Komaba S (2014) Research development on sodium-ion batteries. Chem Rev 114:11636–11682. https://doi.org/10.1021/cr500192f
94. Yang JH, Kim H, Ceder G (2021) Insights into layered oxide cathodes for rechargeable batteries. Molecules 26:3173. https://doi.org/10.3390/molecules26113173
95. Yi H, Liang Y, Qian Y, Feng Y, Li Z, Zhang X (2023) Low-cost Mn-based cathode materials for lithium-ion batteries. Batteries 9:246. https://doi.org/10.3390/batteries9050246

96. Zhan C, Lu J, Kropf AJ, Wu T, Jansen AN, Sun YK, Qiu X, Amine K (2013) Mn(II) deposition on anodes and its effects on capacity fade in spinel lithium manganate-carbon systems. Nat Commun 4:2437. https://doi.org/10.1038/ncomms3437
97. Zhan C, Wu T, Lu J, Amine K (2018) Dissolution, migration, and deposition of transition metal ions in Li-ion batteries exemplified by Mn-based cathodes—a critical review. Energy Environ Sci 11:243–257. https://doi.org/10.1039/C7EE03122J
98. Zhang K, Kim D, Hu Z, Park M, Noh G, Yang Y, Zhang J, Lau VW, Chou SL, Cho M, Choi SY, Kang Y (2019) Manganese based layered oxides with modulated electronic and thermodynamic properties for sodium ion batteries. Nat Commun 10:5203. https://doi.org/10.1038/s41467-018-07646-4
99. Zhang N, Wang JC, Guo YF, Wang PF, Zhu YR, Yi TF (2023) Insights on rational design and energy storage mechanism of Mn-based cathode materials towards high performance aqueous zinc-ion batteries. Coord Chem Rev 479:215009. https://doi.org/10.1016/j.ccr.2022.215009
100. Zhang X, Wei Z, Dinh KN, Chen N, Chen G, Du F, Yan Q (2020) Layered oxide cathode for potassium-ion battery: recent progress and prospective. Small 16:2002700. https://doi.org/10.1002/smll.202002700
101. Zhang Y, Hu A, Xia D, Hwang S, Sainio S, Nordlund D, Michel FM, Moore RB, Li L, Lin F (2023) Operando characterization and regulation of metal dissolution and redeposition dynamics near battery electrode surface. Nat Nanotechnol 18:790–797. https://doi.org/10.1038/s41565-023-01367-6
102. Zhao C, Yao Z, Wang Q, Li H, Wang J, Liu M, Ganapathy S, Lu Y, Cabana J, Li B, Bai X, Aspuru-Guzik A, Wagemaker M, Chen L, Hu YS (2020) Revealing high Na-content P2-type layered oxides as advanced sodium-ion cathodes. J Am Chem Soc 142:5742–5750. https://doi.org/10.1021/jacs.9b13572
103. Zhao H, Chen B, Cheng C, Xiong WQ, Wang ZW, Zhang Z, Wang LP, Liu XQ (2015) A simple and facile one-step strategy to synthesize orthorhombic $LiMnO_2$ nano-particles with excellent electrochemical performance. Ceram Int 41:15266–15271. https://doi.org/10.1016/j.ceramint.2015.07.213
104. Zhao H, Liu S, Liu X, Tan M, Wang Z, Cai Y, Komarneni S (2016) Orthorhombic $LiMnO_2$ nanorods as cathode materials for lithium-ion batteries: synthesis and electrochemical properties. Ceram Int 42:9319–9322. https://doi.org/10.1016/j.ceramint.2016.01.207
105. Zhao X, Cao X, Zhou H (2024) Structure design for high-performance Li-rich Mn-based layered oxides, O2- or O3-type cathodes, what's next? Acc Mater Res 5:307–315. https://doi.org/10.1021/accountsmr.3c00217
106. Zhou A, Chi R, Shi Y, Zhao X, Li X, Kou Z, Zhang Z, Zhang X, Nie G (2023) Manganese-based cathode materials for aqueous rechargeable zinc-ion batteries: recent advance and future prospects. Mater Today Chem 27:101294. https://doi.org/10.1016/j.mtchem.2022.101294
107. Zhou G, Sun X, Li Q, Wang X, Zhang J, Yang W, Yu X, Xiao R, Li H (2020) Mn ion dissolution mechanism for lithium-ion battery with $LiMn_2O_4$ cathode. In situ ultraviolet-visible spectroscopy and ab initio molecular dynamics simulations. J Phys Chem Lett 11:3051–3057. https://doi.org/10.1021/acs.jpclett.0c00936
108. Zhu X, Meng F, Zhang Q, Xue L, Zhu H, Lan S, Liu Q, Zhao J, Zhuang Y, Guo Q, Liu B, Gu L, Lu X, Ren Y, Xia H (2021) $LiMnO_2$ cathode stabilized by interfacial orbital ordering for sustainable lithium-ion batteries. Nat Sustain 4:392–401. https://doi.org/10.1038/s41893-020-00660-9
109. Zuo W, Qiu J, Liu X, Zheng B, Zhao Y, Li J, He H, Zhou K, Xiao Z, Li Q, Ortiz GF, Yang Y (2020) Highly-stable P2–$Na_{0.67}MnO_2$ electrode enabled by lattice tailoring and surface engineering. Energy Storage Mater 26:503–512. https://doi.org/10.1016/j.ensm.2019.11.024
110. Zuo W, Ren F, Li Q, Wu X, Fang F, Yu X, Li H, Yang Y (2020) Insights of the anionic redox in P2–$Na_{0.67}Ni_{0.33}Mn_{0.67}O_2$. Nano Energy 78:105285. https://doi.org/10.1016/j.nanoen.2020.105285
111. Zuo W, Liu X, Qiu J, Zhang D, Xiao Z, Xie J, Ren F, Wang J, Li Y, Ortiz GF, Wen W, Wu S, Wang MS, Fu R, Yang Y (2021) Engineering Na^+-layer spacings to stabilize Mn-based layered cathodes for sodium-ion batteries. Nat Commun 12:4903. https://doi.org/10.1038/s41467-021-25074-9

112. Zuo W, Yang Y (2022) Synthesis, structure, electrochemical mechanisms, and atmospheric stability of Mn-based layered oxide cathodes for sodium ion batteries. Acc Mater Res 3:709–720. https://doi.org/10.1021/accountsmr.2c00058

Squaraine-Based Small Molecule BHJ Absorbing Layers for Organic Thin-Film Solar Cells

Vilko Mandić and Floren Radovanović-Perić

Abstract Squaraines function as electron donors and are therefore combined with a model electron acceptor to create photovoltaic devices. The newly synthesized squaraine compounds were designed at the molecular level for use in small-molecule bulk-heterojunction photovoltaic systems. It was found that extending the alkyl chains at the molecular ends enhances the solubility in non-polar solvents, which allows them to be processed from the same solution as commonly used electron acceptors. Following this, thin films with appropriate architecture and processing were developed and assembled into solar cells The most effective solar cells were further optimized to enhance the nanomorphology of the active layer, and thus improve overall device performance, which was accompanied by comprehensive Atomic force microscopy (AFM) and Grazing incidence small and wide-angle X-ray scattering (GISWAXS) characterization. This facilitates the understanding of how domain size, crystallinity, and preferred orientation influence device performance. The highest-performing solar cells were subjected to recombination analysis. This confirmed that the low hole mobility of the squaraine derivatives restricts device performance by creating a significant charge carrier imbalance that impedes transport. Furthermore, it was found that enhancing the crystallinity of the squaraine component within the bulk heterojunction leads to an increase in the short-circuit current of the solar cells, likely due to improved balance between charge carrier mobilities.

Keywords Squaraine · Electron donor · Bulk-heterojunction · Organic solar cell

V. Mandić (✉) · F. Radovanović-Perić
University of Zagreb Faculty of Chemical Engineering and Technology, Trg Marka Marulića 19, 10000 Zagreb, Croatia
e-mail: vmandic@fkit.unizg.hr

D. Shi (ed.), *Energy Materials Science and Engineering: From New Devices to AI Power Systems*, Green Energy and Technology, https://doi.org/10.1007/978-3-031-88550-1_2

1 Introduction

1.1 Context

Over the past few decades, the increasing demand for renewable energy resources has catalyzed the advancement of various innovative technologies aimed at harnessing sustainable, environmentally benign, and inexhaustible energy sources essential for contemporary living. Among these, solar energy has emerged as one of the most significant renewable energy sources. The advent of the third generation of photovoltaic technology has potentially enabled a theoretical breakthrough against the Shockley–Queisser limit, which posits that silicon solar cells cannot achieve efficiencies exceeding approximately 30% due to the limitations of radiative recombination. This generation encompasses a diverse range of technologies and materials characterized by adjustable bandgaps, exceptionally high absorption coefficients and higher open-circuit voltages; all crucial factors for enhancing power conversion efficiency. Within the multitude of systems under investigation, dye-sensitized, perovskite and organic bulk-heterojunction solar cells are particularly noteworthy, each presenting unique benefits and limitations. Organic semiconductor systems show considerable promise for producing low-cost, large-area, flexible, and chemically stable panels suitable for numerous applications, including potential replacements for polymer-based materials, such as solar windows. Nevertheless, despite their potential, these emerging technologies have yet to achieve significant breakthroughs for commercial deployment, primarily due to inherent challenges. Although organic solar cells theoretically possess vast potential for diverse applications, the fundamental physics underlying bulk-heterojunction solar cells remains unclear. Many organic systems experience various energy losses that must be addressed either by developing new systems or by mitigating issues through external influences such as processing and interfacing. Squaraine derivatives represent a relatively underexplored class of materials and have not garnered much attention in recent years regarding solar cell research, yet they exhibit intriguing properties for photovoltaic applications. Their high absorption coefficients, adjustable bandgaps, cost-effective, environmentally friendly synthesis and production methods position them as promising candidates for photovoltaic use, including indoor applications.

1.2 Organic Semiconductors Electronic Structure—Donors, Acceptors, Conductivity, Mobility and Bandgap

Semiconductors can be categorized based on the source of their properties into two main types: intrinsic and extrinsic semiconductors. Intrinsic semiconductors are those that have not been doped and exhibit equal concentrations of electrons in their excited state. Conversely, extrinsic semiconductors are modified by adding impurities, which alter their electronic structure to contain either a greater number of vacancies or an excess of electrons. Consequently, extrinsic semiconductors can be classified as n-doped, exhibiting a higher concentration of electrons relative to holes, or p-doped, showing increased hole concentration compared to electrons, in both their ground and excited states. A significant characteristic of extrinsic, doped semiconductors is their ability to function as electron or hole donors when in contact with other materials. Specifically, n-doped semiconductors serve as electron donors due to their surplus of electrons, while p-doped semiconductors act as electron acceptors owing to their higher electron vacancy count. In intrinsic semiconductors, the electric current that flows when subjected to an electric field arises from both holes and electron movement. In contrast, for extrinsic semiconductors, the surplus electrons or holes result in one type of charge carrier becoming predominant, referred to as the majority charge carrier, while the other is termed the minority charge carrier. Under normal conditions, n-type semiconductors (donors) have electrons as majority carriers, while p-type semiconductors (acceptors) have holes as the majority carriers, leading the current to consist solely of electrons (or holes for p-type semiconductors). At higher temperatures, however, as electrons or holes are excited into the conduction band, the concentration of both carriers' increases, contributing multiple components to the overall current flowing through the material [1].

Recently, organic semiconductors have attracted significant attention due to their unique properties, particularly their suitability for applications requiring economically viable materials over expansive areas. Given their complex and relatively disordered molecular and crystalline structures compared to inorganic crystals, organic semiconductors adhere to the basic principles of semiconductor physics but with notable modifications. It is widely accepted that the ground and excited states of organic semiconductors are best represented by their Highest Occupied Molecular Orbital (HOMO) and Lowest Unoccupied Molecular Orbital (LUMO) energy levels, which correlate to the valence and conduction bands, respectively [2]. This distinction arises because the molecular crystals of organic compounds are bound by weaker interactions (such as van der Waals forces or hydrogen bonding) than those found in inorganic crystals, resulting in band splitting occurring at the molecular rather than supramolecular level, thereby altering the mechanisms of charge carrier transport [3]. The mobility of electrons and holes in organic materials is influenced by the structure of their molecular framework, specifically the extent of sp^2 hybridization present. This is attributable to the relative weakness of π bonds compared to σ bonds, facilitating easier polarization of electrons due to their delocalized characteristics, thus making them more susceptible to excitation in an electric field. Compared to inorganic semiconductors, organic semiconductors exhibit significantly lower charge carrier mobilities for the reasons previously mentioned. While it is relatively straightforward to excite an electron in organic semiconductors, it proves challenging to transport it through an electric field or to prevent it from recombining with a hole [4]. The process, where an energy level is broadened into an energy band that electrons can occupy and traverse freely, is referred to as the formation of an energy band [5]. Between every pair of energy levels lies a gap known as the forbidden region, characterized by the minimum energy required for an electron to transition from a lower to a higher energy state. The occupancy of these energy bands determines the electrical conductivity of materials, a fundamental property derived from their electronic structure, classifying them into three categories: conductors (metals), semiconductors, and insulators [6]. Electrical conductivity refers to the capacity to accelerate electrons within an electric field by promoting them to higher energy states. Conductivity is directly proportional to both the number of free charge carriers in the system and their lifetime. Electricity can be conducted in two ways: through electrons or through holes (defect conductivity), and these two processes may operate simultaneously and independently of one another [7].

2 Concepts Relevant for Squaraine Solar Cells

2.1 Charge Carrier Transport in Organic Semiconductors, Photovoltaic Effect and Solar Cells

When examining the transport phenomena of electrons and holes in organic semiconductors, it is crucial to recognize that these materials consist of ionized molecular states. The generation of an exciton necessitates the removal of an electron from a neutral molecule (M^0), which requires overcoming the attractive force exerted by the resultant radical (M^+). Only after this barrier is surmounted can the electron freely migrate from one molecule to another. During its transit, the same electron will also lead to the formation of negative molecular ions (M^-); all of which are stabilized in the solid state [8]. Additionally, in their condensed phase, organic molecules can form quantum wells that significantly reduce the energy needed to achieve the first excited singlet state, which corresponds to the excitation of the π electron, compared to a free molecule. Furthermore, the polarization (excitation) energy varies with organizational order, being lowest in molecular crystals and highest in amorphous organic compounds. This variance contributes to the characterization of carrier sites, which are often represented by Gaussian density of states distributions instead of precisely defined energy states [9].

Consequently, charge carrier transport mechanisms can be categorized into two extremes: band transport and hopping transport. Band transport typically occurs in molecular crystals, although the charge carrier permeability remains low (up to $10\,\text{cm}\,\text{V}^{-1}\,\text{s}^{-1}$), not reaching the performance levels of inorganic semiconductors. The presence of trap states, which are locations where recombination is statistically more likely, significantly affects this dependency. In contrast, hopping transport exhibits even lower mobilities, not exceeding $10^{-3}\,\text{cm}\,\text{V}^{-1}\,\text{s}^{-1}$. In structures governed by this mechanism, mobility is activated by temperature and is dependent on the electric field.

To comprehensively describe charge carrier transport in organic semiconductors, it is also necessary to consider space charge effects, charge injection, and recombination mechanisms. On a macroscopic scale, the charge carrier current in organic semiconductors is defined similarly to that in inorganic ones; however, this relationship is nonlinear due to both charge carrier mobility and density being functions of the applied electric field. As previously discussed, mobility is highly influenced by the material structure, often determined by the preparation method [4]. In the context of thin films, the deposition techniques and subsequent processing can affect mobility by up to three orders of magnitude. Another critical factor is the intrinsic charge carrier density, which is significantly lower in organic semiconductors than in their inorganic counterparts [10]. Based solely on intrinsic charge carrier density, pure organic semiconductors are unable to conduct electricity unless this density is augmented through external means, such as doping, carrier injection via metal contacts, photogeneration, or field-assisted doping [11].

Photovoltaic materials are defined by their capacity to convert solar energy into electrical energy. This conversion relies on the generation of excitons, or electron–hole pairs when the material is illuminated by light. The third generation of solar cells includes alternative thin-film materials like organic semiconductors (both polymeric and single-molecule) and perovskites [12]. Despite the varying technical implementations across different generations of devices, all these solar cells share a fundamental property: the ability to separate charge and generate current and voltage upon solar irradiation of their surfaces. Thus, an effective photovoltaic material is characterized not only by its potential to produce a significant number of excitons but also by its efficiency in separating charges with minimal loss due to recombination [13].

2.2 *Bulk Heterojunction Organic Solar Cells*

A bulk heterojunction consists of at least two distinct materials, with at least one functioning as a donor and another as an acceptor, arranged in a manner that forms a network throughout the entire material while maintaining a phase separation that is sufficiently large to be observed microscopically. This specific morphology is designed to enable the material to exhibit uniform macroscopic characteristics across its entirety, despite the presence of two clearly defined microscopic phases (donor and acceptor), which are responsible for its bulk properties [14]. Theoretically, this structure replaces the traditional p–n junction as the active photovoltaic layer, and in order to effectively collect the produced current, the solar cell must also incorporate a cathode and an anode. In contemporary organic photovoltaic systems, the extraction of charge is often inadequate due to a significant energy barrier that exists at the metal–semiconductor interface, coupled with the fact that both electrodes collect both types of charge carriers [15]. This challenge has been addressed by the introduction of intermediate energy levels through the deposition of thin films of materials that facilitate unidirectional electron flow while simultaneously blocking hole flow, and vice versa. These layers are referred to as charge transport layers (CTL), commonly designated as electron transport layers (ETL) and hole transport layers (HTL).

When incident light strikes the bulk heterojunction, it results in the formation of excitons by promoting electrons from the HOMO to the LUMO levels in both materials. Following this excitation, the process of charge separation takes place. Electrons migrate from the LUMO of the donor material to the LUMO of the acceptor, subsequently traversing through the acceptor to the charge transport layer, where they are propelled toward the electrodes. Meanwhile, holes move from the HOMO level of the acceptor to that of the donor and progress through the donor material until they reach the hole transport layer, where they are also directed toward the electrodes. Consequently, the mechanism of photocurrent generation resembles that found in silicon solar cells with a notable distinction that organic semiconductors generally demonstrate significantly lower conductivities, attributable to their less-defined atomic structures compared to the orderly arrangement of atoms in inorganic crystal lattices [4, 16]. Despite this limitation, many of these organic materials offset the conductivity issue with their exceptionally high absorption coefficients, leading to the potential production of very high charge carrier densities. This phenomenon can elevate the theoretical voltage generated upon excitation to nearly three times the conventional value of approximately 0.4 V. An increase in voltage results in a diminished net electric field within the depletion region, which occupies a larger volume due to the thin film's morphology. Hence, the current produced is predominantly diffuse. Charge separation and extraction occur primarily through diffusion, and in these systems, the mobility of electrons and holes can vary by up to three orders of magnitude, with electrons typically exhibiting higher mobility. This variance necessitates that both the carrier lifetime and diffusion length be maximized to facilitate effective extraction [17]. Additionally, this is precisely why practical implementations of the bulk heterojunction aim for an optimal interface surface area between donor and acceptor materials, to expedite separation and enhance stability, thereby reducing the likelihood of recombination after crossing the donor–acceptor interface. It is also crucial to highlight that achieving a balance between electron and hole mobilities is a fundamental design criterion for new systems. Given that hole mobility is consistently lower, many strategies focus on enhancing this parameter to align it more closely with that of electron mobility (Fig. 1) [18].

2.3 Typical Constituents, Preparation and Characterization of BHJ Organic Solar Cells

The thin films that constitute a bulk heterojunction organic solar cell can be categorized into three distinct groups: active layers, CTLs and electrodes. The active layer, which serves as the core of the bulk heterojunction for these solar cells, is fundamentally responsible for the photovoltaic effect within the device. This layer must be capable of generating a substantial quantity of excitons, with the donor material typically exhibiting a narrower bandgap than that of the acceptor. Ideal characteristics for the donor material include high hole mobility, stability under

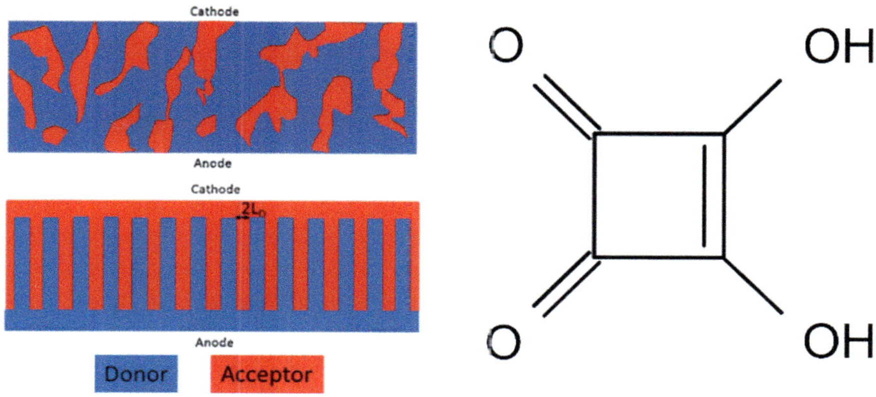

Fig. 1 A schematic representation of a bulk heterojunction layer [19], in an organic solar cell and squaric acid

photochemical and thermal conditions, and complementary absorption spectra relative to the acceptor material, all of which enhance the utilization of the solar spectrum. The charge carrier transport mechanism is reliant on the movement through the LUMO and HOMO orbitals of the materials, which correspond to the conduction and valence bands, respectively. These energy levels provide a representation of the electron density distribution at both molecular and supramolecular scales and are significantly influenced by the molecular structure and degree of crystallinity [20]. Research has indicated that a condensed π–π skeletal structure is most effective for electron transport without recombination, while functional groups containing localized, unstable electrons serve as optimal donor sites [21]. While the theoretical characterization of donor and acceptor properties remains somewhat ambiguous, it is broadly acknowledged that these characteristics are influenced by the oxidation and reduction potentials of the molecules involved. A molecule is considered a strong electron donor if it possesses a high oxidation potential (in absolute terms) since it can readily release an electron; conversely, a good acceptor is characterized by a high reduction potential, as it can easily accept an electron. HOMO and LUMO interdepend upon both molecules [22]. A significant challenge currently faced is the capacity of organic materials to provide high charge carrier mobility; thus far, many of these materials depend on exceptionally high absorption coefficients to generate substantial currents and large differences between HOMO and LUMO levels to generate high open-circuit voltages, even though they are significantly hindered by recombination in comparison to inorganic alternatives. Most assessments rely on experimental data without robust theoretical foundations, leading to the conclusion that factors such as crystallinity, supramolecular self-assembly, and the density of π–π bonds in the system play crucial roles [23].

CTL functions as intermediary layers designed to facilitate the diffusion of one type of charge carrier while impeding or slowing the movement of the other. A wide variety of materials can be employed for these thin films, which are highly system-specific, necessitating the fulfillment of diverse conditions based on the intended application [24].

Electrodes are materials that must meet straightforward criteria to ensure the effective operation of a solar cell. Common examples of such materials include indium-doped tin oxide (ITO) and fluorine-doped tin oxide (FTO) [25, 26].

Fundamentally, all methodologies for synthesizing symmetric and asymmetric aminosquaraines entail the initial preparation of one or more alcohols, which are then condensed with squaric acid in the final step. Various organic chemistry pathways were utilized to synthesize the required constituents, including catalyzed reactions for the formation of ethers, aminization, and dealkylation. Additionally, the condensation of the synthesized alcohols with squaric acid was conducted in a mixture of toluene and isopropanol. Each of these processes involved refluxing at elevated temperatures, followed by cooling to ambient temperature and solvent evaporation to yield the final products. These products were subsequently purified through recrystallization in methanol. Here we consider only the fabrication of the active bulk heterojunction layer. The thin films for the active layer were created by blending filtered solutions of squaraine derivatives and $PC_{71}BM$ in non-polar organic solvents (such as chloroform, chlorobenzene, and 1,2-dichlorobenzene) at different ratios, which were then spin-coated to form thin films. Spin coating is a widely used technique for obtaining thin films from solution or dispersion precursors. It is particularly suited for laboratory applications, although it has limited scalability. Nonetheless, it provides a rapid and cost-effective method for fabricating thin films of varying thicknesses, with both the rotation speed and deposition time adjusted according to the targeted layer thickness. The deposition of other layers employed a variety of techniques, e.g. evaporation, which is a well-established and commercial technique utilized for a broad range of electronic applications. This method offers high precision in thin film deposition, with the potential for scalability, though it does come with the disadvantages of requiring expensive equipment and costly consumables.

Several architectures were tested: ITO was always one electrode / ZnO (30 nm), PFN-Br (10 nm), and PEDOT:PSS (5 nm) were exchanged for ETL / BHJ was constant / MoO_3 (10 nm), PFN-Br (10 nm), PNDIT-F3N-Br (10 nm), and PDINO (10 nm) were exchanged for HTL / Ag (100 nm) was always the other electrode.

There exist very few straightforward characterizations for organic solar cell systems. UV-Vis spectroscopy is a technique that quantifies the amount of monochromatic light that is absorbed, reflected, or transmitted by a sample positioned in front of a light source. This method allows for the determination of optical energy bandgap values. Cyclic voltammetry is an electrochemical technique that can be applied to both solutions and thin films. This potentiodynamic method involves the linear increase and decrease of the voltage at the working electrode, completing a full cycle within a specified voltage range. The positive ramping of voltage against the reference electrode is referred to as a cathodic scan, during which oxidative processes take place; conversely, the negative ramping is known as an anodic scan, associated with reduction processes [27]. External quantum efficiency (EQE) is a methodology that encompasses both UV-Vis and electrical characterization techniques.

More elaborate methods comprise X-ray scattering which involves irradiating particles in a bulk, solution, or thin film with X-rays, thereby generating signals that provide information about the average overall structure, shape, and size of the particles or constituents within the sample. This technique is categorized into small-angle X-ray scattering (SAXS) and wide-angle X-ray scattering (WAXS). Electron microscopy techniques utilize an electron beam to produce a highly magnified image of the sample. Unlike electron microscopy, which may distort the actual size scale of individual features in an image, Atomic force microscopy (AFM) can be used to accurately calculate surface roughness parameters [28]. This technique can generate a two-dimensional image that reflects the resistance distribution across the surface being studied, as well as a distinct topographical image.

3 Developing Squaraine Solar Cell

3.1 Squaraine Derivatives for Photovoltaic Application

Squaraines represent a category of organic compounds derived from squaric acid, which is characterized by its distinct four-membered aromatic ring that exhibits a square-like configuration. Their applications thus far include roles as biosensors, fluorescence probes, xerographic receptors, as well as in non-linear optics and photovoltaic technologies, particularly in dye-sensitized solar cells and bulk heterojunction organic solar cells [29–32]. Typically, these compounds are produced through a nucleophilic attack on the central ring, which is notably electron-deficient. Squaraines are known for their photochemical stability, exceptionally high extinction coefficients, meaning they can absorb light very effectively per mole, with absorption maxima typically occurring at wavelengths greater than 600 nm [33, 34], presenting a promising avenue for photovoltaics. To enhance the photochemical stability of squaraine derivatives and to establish a conjugated backbone that supports semiconductor functionality, it is beneficial to create aryl aminosquaraines by preparing aryl amines and condensing these with squaric acid. The incorporation of aryl groups contributes to a conjugated framework that facilitates efficient electron transfer, while the amino groups significantly bolster the chemical and photochemical stability of the molecule by spreading out the charge distribution from the highly electron-rich oxygen atoms [35]. Their bandgap can be adjusted through molecular design [36], by varying the substituents located near the squaric core. The charge distribution within the structure indicates that the squaric core carries a positive charge, while the oxygen atoms exhibit a negative charge, resembling ionic behavior. Although this representation is not entirely accurate, it reflects a significant localized charge distribution due to the oxygen atoms' high electron affinity, which allows for easy electron release, thereby designating these squaraine derivatives as electron donor materials [37]. Squaraines can manifest as either symmetric or asymmetric, depending on whether the two sides of the squaric core are identical [38]. For asymmetric squaraines, one side of the quadrilateral ring typically requires shielding; the first aryl amine is condensed onto one side, followed by the attachment of a second aryl amine in a subsequent step [39].

The exploration of squaraines within a Schottky barrier solar cell yielding a mere 0.02% efficiency commenced almost fifty years ago. These disappointing results significantly stifled progress in squaraine-based solar cell research until the 1990s, when the advent of bulk heterojunctions rekindled interest in numerous previously overlooked molecules [40]. In 2008, the first squaraine-based solar cell surpassed a 1% power conversion efficiency (PCE), utilizing a structure of ITO/PEDOT:PSS/SQ:$PC_{71}BM$/LiF/Al, wherein Marks et al. illustrated the substantial potential of squaraines for photovoltaic applications [41]. Following this milestone, efficiency records continued to rise due to optimized molecular structures, culminating in 2016 when Yang et al. presented intricate squaraine derivatives featuring two or more cores interconnected by conjugated aromatic rings, achieving approximately 8% PCE within a bulk heterojunction device [42]. Throughout this era, several critical advancements occurred. Kido et al. furthered the field by creating transparent squaraine solar cells with extremely thin electrodes, showcasing the potential for visually appealing solar windows [43]. Namboothiry et al. demonstrated the feasibility of fabricating squaraine solar cells in non-protected atmospheres, underscoring their genuine potential for industrial applications [18]. Most literature reports indicate that squaraine derivatives are typically combined with PC60BM or $PC_{71}BM$, which are widely recognized acceptor materials. Despite their practical limitations, these compounds have become established model systems for investigating newly developed electron donors [44]. Collectively, this body of research has advanced the commercialization prospects of squaraines; however, they remain relatively underexplored in academic literature compared to other organic photovoltaic materials.

Insights gained from more well-studied electron donors and acceptors indicate that the crystallization process is crucial to the nanomorphology of bulk heterojunctions, potentially enhancing solar cell performance by an order of magnitude. Thus, it is critical to comprehend how molecular design influences the crystallization and supramolecular assembly of squaraine-based materials [45]. Furthermore, squaraines have been observed to aggregate in thin films through supramolecular assembly, resulting in shifts in the absorbance spectra. A blue shift in the spectra is indicative of H-aggregation, while a red shift suggests the formation of J-type aggregates. H-aggregation is typically linked to head-to-head and tail-to-tail supramolecular configurations, whereas J-aggregates correspond to a head-to-tail alignment. J-type aggregation is particularly advantageous for photovoltaic applications, as it broadens the range of solar radiation that can be effectively harvested into the near-infrared spectrum [46]. The specific type of aggregation is not easily predictable, as it is significantly influenced by molecular structure and processing conditions. Additionally, the crystallization pathway can vary when blended with PCBM as an acceptor at different ratios, highlighting the importance of the interactions between the two components in determining the final morphology of the film. Here we have synthesized ten compounds by a novel, green synthesis route without the use of expensive catalysts in order to study the influence of molecular design on photovoltaic properties (bandgap, absorption, HOMO and LUMO energy levels), physical properties (solubility in common organic solvents, compatibility with $PC_{71}BM$, thin film crystallization route) and solar cell performance parameters (current, voltage, efficiency). The idea is to establish a relationship between the bulk heterojunction nanomorphology and solar cell parameters. The remaining of the work browses through the development process challenges primarily to raise the bottlenecks and thereby avoiding very comprehensive and specific discussion.

3.2 *Squaraine Development Strategy, Solubility and Absorbance Properties*

The 4, 5, and SSQ series consist of symmetrical squaraines, with the 4 series specifically engineered to stabilize the oxygen atoms through a combination of hydroxide and methoxy groups. In contrast, series 5 utilizes only hydroxide groups for the stabilization of all oxygen atoms. Both series include compounds characterized by condensed π–π skeletal structures and extended alkyl chains that extend from the squaric core. ASQ series includes asymmetric compounds; one side of the squaric core is protected by hydroxide groups, while the opposite side features a piperidine derivative. Solubilities were investigated across a range of solvents due to the solvatochromic nature of photovoltaic squaraine derivatives; their absorption properties vary depending on the solvent in which they are dissolved. Squaraines can be found in forms ranging from monomers (solvated single molecules) to oligomers (solvated aggregates consisting of multiple molecules). Notably, the absorption maxima for the highest wavelength will always correspond to the monomer state, while the aggregated forms (such as dimers and oligomers) will display blue-shifted absorption maxima relative to that of the monomer. The solubility ranges of interest were identified as 0–5 mg mL^{-1}, deemed the minimum necessary for analyzing the compounds in both solution and thin film formats, and up to 30 mg mL^{-1}, recognized as the maximum concentration suitable for thin film processing [47, 48]. Squaraines that feature extended alkyl chains demonstrate solubility in non-polar solvents like chloroform, whereas those lacking such chains are soluble in polar solvents such as acetone. Moreover, chlorobenzene and 1,2-dichlorobenzene are more effective non-polar solvents compared to chloroform and can dissolve all of these compounds at least partially (Fig. 2; Table 1).

Some absorbance spectra observed in solution reveal a limited number of maxima, predominantly within the 600–650 nm range, typically lacking distinct absorption features, with few exceptions. This finding indicates that no optimal solvent has been identified for these compounds, suggesting the presence of multiple-size states of squaraines in the solution. For other series, the solution spectra exhibit clear absorption maxima around ~715 nm without any additional size-dependent behavior,

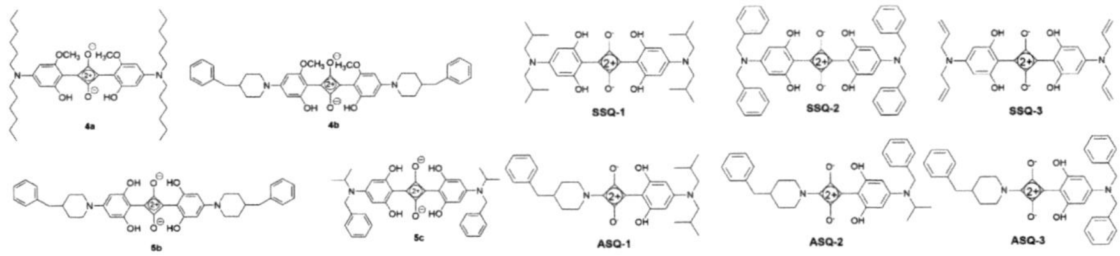

Fig. 2 Scheme of the structures for the synthesized symmetric and asymmetric squaraines

Table 1 Prepared squaraine compound with an approximate visual estimation of its solubility: "No" suggests dispersion with no solution coloration; "Partial" means dispersion with coloration; "Yes" suggests solution

SQ	CF		CLB		o-DCB		EtOH		ACT	
	<5‰	30‰	<5‰	30‰	<5‰	30‰	<5‰	30‰	<5‰	30‰
SQ4a	Yes	Yes	Yes	Yes	Yes	Yes	Partial	Partial	Yes*	Yes*
SQ4b	No	No	No	No	Partial	Partial	Partial	Partial	Yes*	Yes*
SQ5b	Partial	Partial	Partial	Partial	Yes	Yes	Partial	Partial	No	No
SQ5c	Yes	Yes	Yes	Yes	Yes	Yes	Partial	Partial	No	No
SSQ1	Yes	Yes	Yes	Yes	Yes	Yes	Partial	Partial	No	No
SSQ2	Yes	Yes	Yes	Yes	Yes	Yes	Partial	Partial	No	No
SSQ3	No	No	No	No	Partial	Partial	Yes	Yes	Yes	Yes
ASQ1	Yes	Yes	Yes	Yes	Yes	Yes	Yes	Yes	No	No
ASQ2	Yes	Yes	Yes	Yes	Yes	Yes	Yes	Yes	No	No
ASQ3	Partial	Partial	Yes	Yes	Yes	Yes	Yes	Yes	Yes	Yes

*—Suspicious, CF—chloroform, CB—chlorobenzene, o-DCB—1,2-dichlorobenzene, EtOH—ethanol, ACT—acetone

indicating that these compounds are entirely soluble in their respective solvents and are exhibiting absorption characteristics of the monomer state.

In contrast, the situation alters in thin film configurations. The onset of absorption remains comparable to that in solution, but the spectra broaden across a larger wavelength range due to irregular particle sizes during crystallization. This broadening is expected to yield multiple contributions to the optical properties in the solid state. Additionally, the absorption maxima for the majority of these compounds demonstrate a hypsochromic shift, indicative of H-type aggregation. Conversely, the 4 and 5 series exhibit a bathochromic shift, extending into the infrared region, which suggests the formation of J-aggregates that typically assemble in a head-to-tail fashion in the solid state [49]. The ASQ series displays two distinct absorption maxima, with the lower wavelength peak being the more prominent. This behavior has also been noted in other squaraines and is attributed to the formation of "islands" in thin films, comprising extremely large aggregates (ranging from 500 nm to 2 μm), which lead to a separate distinctive absorption band [18]. This phenomenon is unavoidable

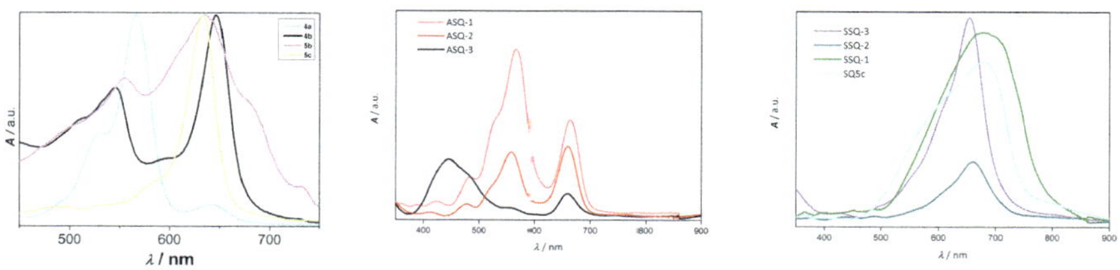

Fig. 3 Representation of differences in UV-Vis absorption spectra for different symmetric and asymmetric squaraines in solution (different solvents) and deposited as thin films

for spin-coating from diluted systems yielding very thin films, in contrast to drop-cast thin films, although this method offers limited process control. Ultimately, all compounds exhibit suitable bandgaps for photovoltaic applications (under different illumination sources). Squaraines with narrower bandgaps are recommended for agrivoltaics, while those with wider bandgaps, such as ASQ-1, are ideally suited for indoor applications, as indoor light spectra generally contain less infrared light compared to sunlight (Fig. 3) [50].

The energy difference between HOMO and LUMO levels corresponds to the material's bandgap, while their respective positions influence the facilitation of charge transport within a bulk heterojunction. It is well established that solvents can significantly affect the positioning of HOMO and LUMO levels, a factor that is not accounted for in methods measuring the optical bandgap. When comparing the optical and electrical bandgaps, the optical bandgap is generally regarded as more reliable for accuracy; however, cyclic voltammetry measurements remain valuable, as UV-Vis spectroscopy provides limited information beyond the bandgap itself. It is evident that nearly all of the synthesized compounds possess suitable energy levels to be classified as electron donors: their LUMO levels sit above that of $PC_{71}BM$, thereby energetically promoting electron transport (whereby electrons transition from the donor's LUMO level to that of the acceptor). Furthermore, their HOMO levels are also positioned above those of $PC_{71}BM$, allowing for hole transport (where holes migrate from the acceptor's HOMO level to that of the donor). The open circuit voltage is directly proportional to the electric field generated by the charge difference between the donor's HOMO level (the site of hole migration following photoexcitation) and the acceptor's LUMO level (the site where electrons reside). Overall, it can be concluded that all these compounds exhibit deep HOMO energy levels (Fig. 4).

3.3 *Squaraine Processing into Thin Films*

Thin film solar cells typically require extremely thin layers, with a root mean square height (Sq), roughness of 1–3 nm. Additionally, it is crucial to examine squaraine and fullerene blend compatibility during the deposition process, through their absorption spectra. Bulk heterojunction (BHJ) solar cells operate effectively only when there

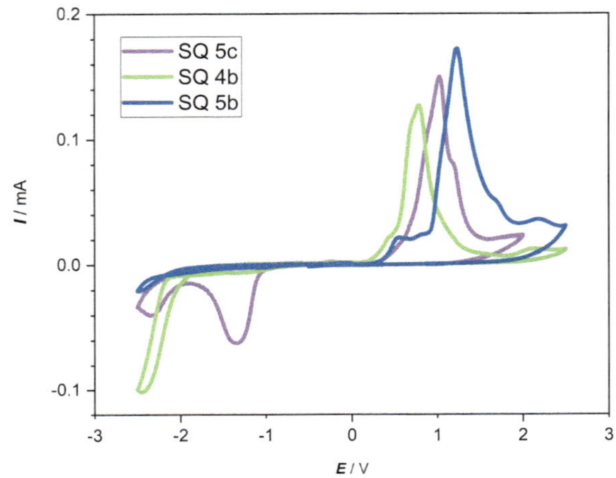

Fig. 4 Cyclic voltammetry of the squaraine compounds in acetonitrile versus an Ag/Ag^+ reference electrode

is an optimal ratio of donor to acceptor, which ensures favorable morphology and charge transport. As long as the squaraine is soluble in the same solvent as $PC_{71}BM$, these three processing parameters alone dictate the film thickness and surface roughness, irrespective of the specific squaraine utilized in the blend. Overall, the surface morphology of the thin films is largely unaffected while the thickness changes with different spin speeds and accelerations. This phenomenon can be attributed to the volatility of chloroform, which results in the formation of a thin film in the initial seconds of the spin-coating process. Given that crystallization cannot be decelerated or controlled, the only remaining factor influencing the shape and size of the particles is the concentration. When films were processed at a concentration of 20 mg mL^{-1}, the initial solutions fell below the saturation curve, leading to the formation of smaller primary nuclei compared to those generated from solutions at 40 mg mL^{-1}. Additionally, the use of 1,2-dichlorobenzene leads to the formation of layers measuring 20–30 nm, attributed to its non-volatile nature. Such thicknesses are typically considered inadequate for photovoltaic (PV) applications. While thinner layers exhibit reduced recombination rates due to the shorter diffusion pathways for charge carriers—thereby increasing their likelihood of reaching the opposite side of the bulk heterojunction—they also restrict the maximum achievable current. Conversely, the open circuit voltage is solely dependent on the concentration of charge carriers, indicating that a specific amount of material, or a particular layer thickness, must be attained to generate an adequate electric field that facilitates the flow and collection of photocurrent at the electrodes. Therefore, the thickness of the active layer must be optimized to ensure that both high open circuit voltage and photocurrent coexist; failing to achieve this balance will result in suboptimal efficiencies for the solar cell. Upon these findings 1,2-dichlorobenzene was not further considered as a solvent rather chloroform was chosen based on estimated thicknesses versus smoothness contribution (Fig. 5).

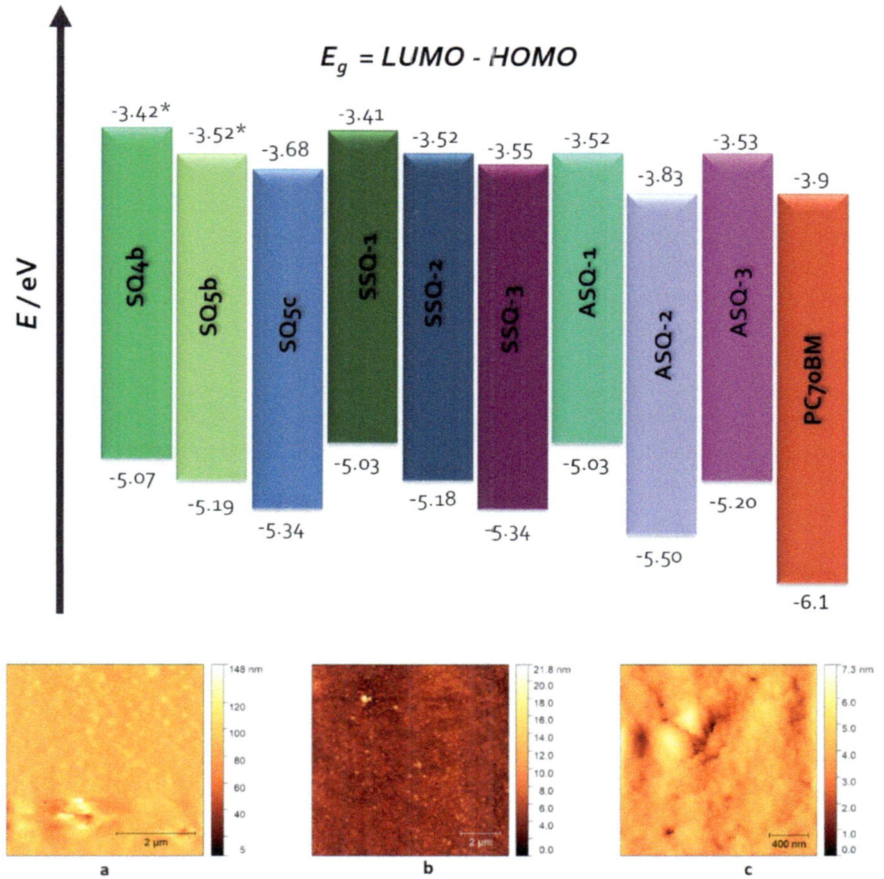

Fig. 5 Schematic energy diagram of the synthesized compounds. Examples of representative AFM z-scan image spin cast SQ films from SQ solutions of different concentrations in different solvents

4 Optimization of Squaraine Solar Cell

4.1 Optimization of Squaraine Solar Cell Architecture

A critical aspect to address is the configuration, which refers to the arrangement of the layers, as it impacts how current traverses the device in relation to incoming light. The configurations are categorized into two types: conventional and inverted. In the conventional design, electrons move in the same direction as the incident light, whereas the inverted design operates in the opposite direction. Two fundamental architectures have been established as starting points: ITO/NiO_x/SQ:$PC_{71}BM$/PFN-Br/Ag for the conventional setup and ITO/ZnO/SQ:$PC_{71}BM$/MoO_3/Ag for the inverted setup. The arrangement of the electron and hole transport layers dictates the flow of current. In the conventional architecture, holes are directed downwards toward the hole transport layer (HTL) while electrons are directed upwards, aligning the current flow with the direction of incoming sunlight; the reverse is true for the inverted configuration. Solar cells were constructed in these configurations, and their current–voltage characteristics were analyzed. The solar cell in the conventional arrangement exhibited a strictly linear current–voltage response, indicating it was short-circuited, functioning merely as a resistor without generating any photocurrent. Although no defects were visually detected on the devices, they were analyzed using scanning electron microscopy, which included cross-sectional cuts made with a focused ion beam. The interface among the layers appeared to be uniform and devoid of morphological anomalies. The thin films of the conventional architecture demonstrated slightly inferior interfacing qualities; however, no definitive causes for the short-circuiting of the solar cells were identified. The bulk heterojunction layer thickness was around 75 nm for the inverted configuration and approximately 60 nm for the conventional configuration. Among the two proposed designs, the inverted architecture appears to perform better with the SQ:$PC_{71}BM$ active layers (Fig. 6).

4.2 Assembly and Characterization of Squaraine Solar Cell

Additionally, solar cells were constructed using a majority of squaraines. Some candidates were excluded due to their insolubility in non-polar solvents and issues with phase separation, which resulted in the formation of resistors instead of functional solar cells. All solar cells were fabricated with the configuration of ITO/ZnO (30 nm)/ bulk heterojunction (BHJ)/MoO_3 (10 nm)/Ag (100 nm). The donor-to-acceptor ratio was the primary parameter adjusted, and previously published ratios typically range from SQ:$PC_{71}BM$ = 1:x, where x ranges from 2 to 6. The solar cells that incorporated SQ4a demonstrated photodiode characteristics in the assembled structure, and the parameters derived from the current–voltage (J–V) curve indicate considerable potential for further optimization. Notably, with a ratio of 1:6, a current was achieved alongside a notably high open-circuit voltage of 0.8 V, which is a significant advantage of organic photoabsorbers due to their adjustable HOMO/LUMO energy levels relative to other conventional and alternative technologies. The dark current curve, which represents the current–voltage characteristics when the solar cell is not illuminated, exhibited some dark current through the device. The current–voltage characteristics of the SQ4c solar cells appeared to have similar traits as those of SQ4a, except for a significantly lower open-circuit voltage, which generally did not exceed 0.5 V. This reduction is likely attributed to the formation of a thinner layer using chlorobenzene as the solvent, necessitated by the insolubility of SQ4b in chloroform. The solar cells made from SQ5b encountered similar problems to those from SQ4b; however, SQ5b's partial solubility in chloroform allowed for a thicker active layer, likely contributing to a slightly higher open-circuit voltage. For SQ5c, the dark current curve approached zero, indicating that all assembled

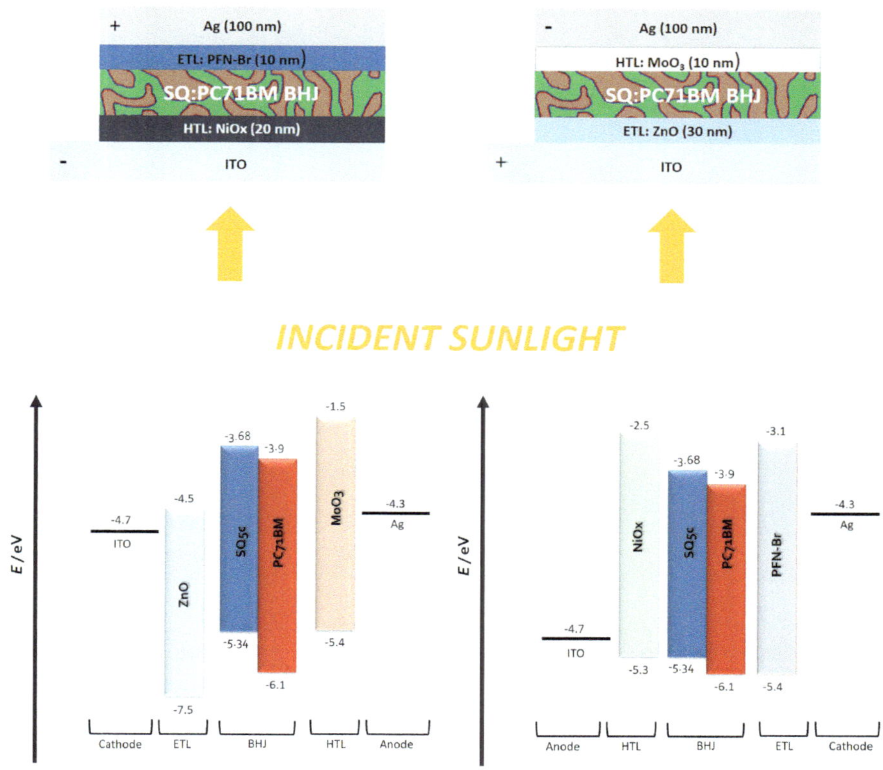

Fig. 6 The proposed starting conventional and inverted architectures. Schematic energy diagrams for the conventional and inverted architectures; HOMO/LUMO levels for SQ were determined experimentally

solar cells exhibited photoresponse, with parameters varying according to the ratios employed. Transitioning from a 1:2 ratio to 1:4 and 1:6 ratios resulted in a substantial increase in open-circuit voltage, along with a moderate rise in short-circuit current, thereby enhancing the power conversion efficiency (PCE) by nearly 400% when comparing the 1:2 and 1:6 ratio solar cells. While the series resistance remained elevated, other parameters appeared to respond significantly to alterations in electron and hole concentrations, as well as changes in nanomorphology, which were likely due to the reduction of electron donor material from 33 to 14.3%. Among all the squaraines synthesized in this study, SQ5c exhibited one of the best solubilities in chloroform, comparable to that of $PC_{71}BM$, which may explain the strong performance of solar cells made from this compound during preliminary tests. The entire ASQ series was developed to enhance solubility in non-polar solvents; hence, these asymmetrical compounds were designed based on the same principles as symmetrical squaraines, incorporating alkyl chains on the outer core to facilitate solubility. Furthermore, the objective was to achieve head-to-tail J-aggregates by modifying the

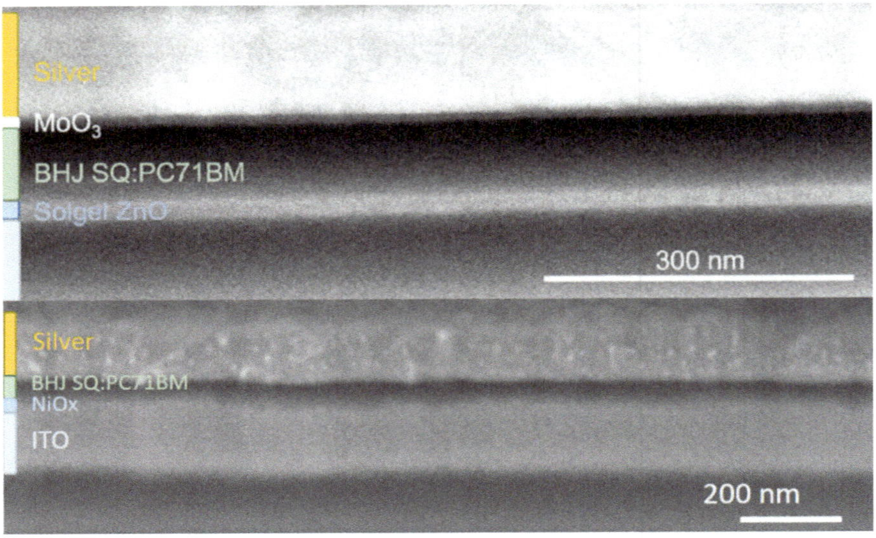

Fig. 7 Cross-section scanning electron microscopy images of the assembled solar cells

charge distribution across the molecule, promoting preferential molecular assembly. However, the assembled solar cells demonstrated negligible photoresponse for any of these compounds, with ASQ-1 performing the best among the three. Notably, both ASQ-1 and ASQ-2 consistently displayed superior fill factors compared to all other solar cells, as evidenced by the shape of the J–V curve (Fig. 7).

4.3 *Squaraine Solar Cell Optimization*

For continued investigation, two compounds were selected: SQ5c and SSQ-1. Adequate thickness is crucial for achieving a high open-circuit voltage, and substantial currents can flow through the solar cell if the active layer is sufficiently thin so that the domain sizes align with the diffusion lengths of the predominant charge carriers, which are electrons and holes. Enhancing current in relation to charge carrier diffusion lengths necessitates the creation of a less resistant medium, which can be achieved by equalizing electron and hole mobilities. Typically, electron mobilities are several orders of magnitude greater than hole mobilities across all known semiconductors; however, this disparity is even more pronounced in organic materials. In the bulk heterojunction, hole mobility is particularly dependent on the squaraine component, as it facilitates the transfer of holes to the hole transport layer, where they stabilize and are subsequently transported to the electrodes. Experimental findings indicate that hole mobility within squaraine thin films is influenced by the degree of crystallinity [17, 46]. Specifically, for bulk heterojunctions, it depends on the orientation of the crystals within the film. Consequently, it is essential for thin films to possess higher crystallinity, which can be achieved either through post-processing or during deposition. Various SQ concentrations and solvent combinations were examined for each SQ solar cell configuration. Chlorobenzene and 1,2-dichlorobenzene were selected due to their slower evaporation rates, attributed to their higher boiling points; the slower evaporation rate extends the film formation period, allowing for better crystallization of the squaraines. When pure 1,2-dichlorobenzene was utilized as the solvent, intriguing results emerged. Lower concentrations yielded solar cells with diminished performance, while higher concentrations led to improved performance. Thin films spun from a 20 mg mL^{-1}, solution displayed poor open-circuit voltages, a consequence of 1,2-dichlorobenzene's elevated boiling point resulting in a layer thickness of approximately 30 nm due to its slower evaporation rate. Nevertheless, the current generated was still superior to that of samples spun from chloroform, indicating enhanced crystallinity. Upon raising the solution concentration to 40 mg mL^{-1}, the spin-coated thin film demonstrated a significant increase in current, along with a modest rise in open-circuit voltage, culminating in a power conversion efficiency (PCE) of 0.75% for the best-performing solar cell in this scenario. Further concentration increase failed, as the solubility approached the saturation limit for this solvent.

The optimization of SSQ-1 commenced using 1,2-dichlorobenzene with SSQ-1:$PC_{71}BM$ ratio at 1:3 across all cells. The findings indicate an enhancement in photovoltaic parameters alongside an increase in short-circuit current, ultimately elevating the total power conversion efficiency (PCE) to 0.85% for the best-performing solar cell. SSQ-1 solar cells also displayed favorable responses to all solvent systems previously evaluated for SQ5c. Consequently, the active layer was also spin-coated using chlorobenzene at two different concentrations, both of which resulted in subpar performance. While 1,2-dichlorobenzene facilitates the formation of the most crystalline films among the three solvents, it also yields the thinnest films, potentially leading to high surface roughness. In contrast, chloroform produces very smooth thin films with lower crystallinity, yet its morphology is optimal for interfacing. It appears that chlorobenzene does not effectively crystallize the films and contributes to increased surface roughness, which introduces surface defects that impair device performance. SSQ-1 demonstrates a tendency to crystallize readily, requiring only a 30% volume substitution of chloroform with chlorobenzene. This configuration resulted in the highest photovoltaic parameters, with a PCE of 1.15% for the leading solar cell. The surface morphology remained intact, defect formation was minimized, and the bulk heterojunction exhibited clear improvements, reflecting reduced energy losses. Although series resistance decreased by more than an order of magnitude as a result of improved morphology, it still remained excessively high, given that the standard for optimized solar cells is generally below 10 Ωcm^2, while non-optimized cells are around 101 Ωcm^2. This issue appears to stem from the interplay between the bulk heterojunction nanomorphology, representing the internal resistance of the semiconductor layer (which can be enhanced), and the solar cell architecture, which contributes to series resistance at the interfaces. The solar cells developed in this study

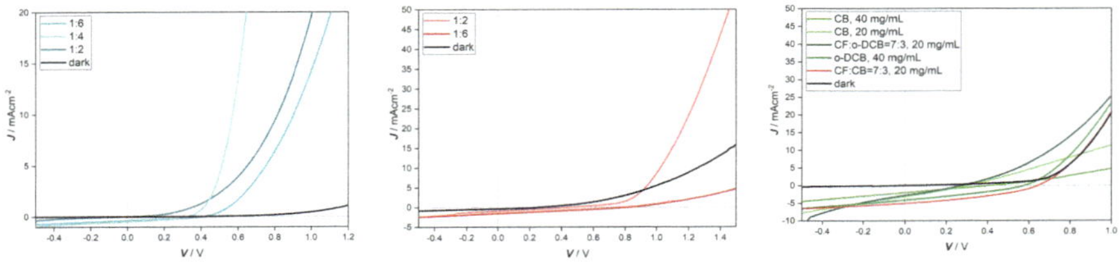

Fig. 8 Current–voltage characteristics of the selected solar cells prepared by varying the BHJ ratio, SQ solution concentration and the type solvent used, among other processing parameters, as well as the type and geometry of the whole solar cell

exhibited significantly reduced photocurrents attributed to high series resistance, particularly within the 10^2–10^3 Ωcm^2 range (Fig. 8) [51].

4.4 In-Depth Analysis

ASQ-2 was selected as a representative from the ASQ series, which exhibited favorable properties for effective devices but registered no response to illumination. Conversely, SQ5c and SSQ-1 were identified as the top-performing solar cells. The dominant phase in the ASQ-2 sample is comprised of spherical agglomerates varying in size from approximately 100–500 nm, attributed to $PC_{71}BM$, which is also the principal phase in the thin film. In contrast, the SQ5c and SSQ-1 solar cells display a considerably more uniform surface where the two components are not visually distinguishable. Furthermore, the root mean square height of these films is an order of magnitude lower than that of the ASQ series.

X-ray scattering (WAXS) was employed to generate diffraction patterns on a two-dimensional detector using synchrotron radiation. This method significantly enhances resolution, enabling the determination of whether a sample is crystalline and allowing for the analysis of crystal orientation relative to the substrate through in-plane and out-of-plane scattering profiles. However, the absence of a resolved crystal structure for all measured compounds precludes any definitive conclusions. The extracted in-plane and out-of-plane scattering profiles reveal shoulder peaks at 7.5 and 13 nm^{-1}, corresponding to $PC_{71}BM$ [20], while other signals can be attributed to the squaraine components. ASQ-2 exhibits some degree of preferred orientation, as evidenced by diffraction peaks appearing in the out-of-plane profile that are absent in the in-plane profile. In contrast, SQ5c does not display any observable crystallinity. Notably, SSQ-1, recognized as the highest-performing squaraine, demonstrates a crystalline structure characterized by multiple peaks, indicative of significant crystal growth. Enhanced crystallinity and preferentially oriented crystals promote superior charge transport, thus increasing the hole mobility of the squaraine material. This results in a more favorable balance between electron and hole mobilities, leading to

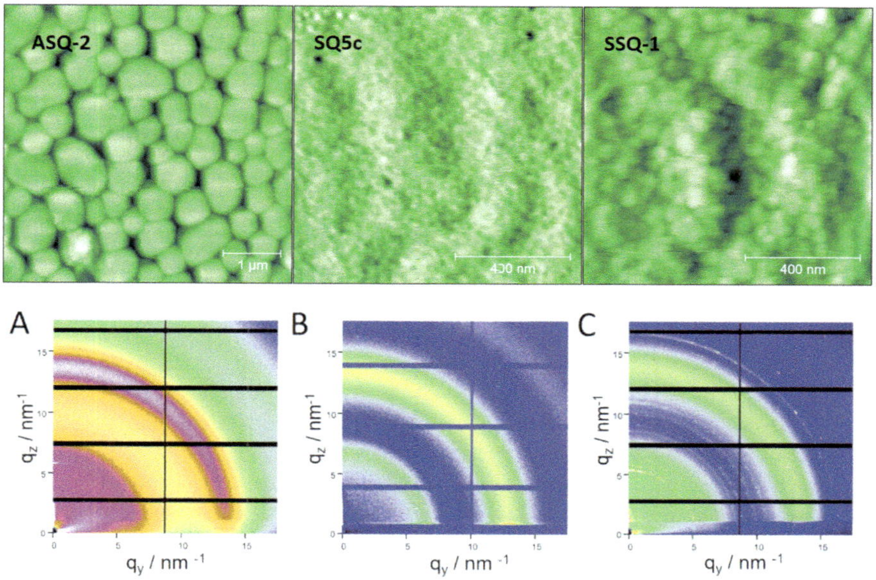

Fig. 9 Selected results of the BHJ layer of squaraine solar cells by atomic force microscopy and X-ray scattering (2D scattering profiles)

a higher photocurrent. This explains why SSQ-1 outperforms SQ5c, despite having a higher bandgap and elevated HOMO level (Fig. 9).

4.5 Recombination Studies

To elucidate the recombination mechanisms influencing device performance, SQ5c and SSQ-1 underwent light intensity assessments, during which their current–voltage characteristics were evaluated at known light intensities diminished by filters. By plotting the resulting open-circuit voltage (V_{oc}) and short-circuit current (J_{sc}) against the light intensity, two critical parameters can be extracted that signify the recombination mechanism in the solar cell: the ideality factor (n) and the α coefficient. The ideality factor indicates the nature of the recombination process, while the α coefficient serves as an indicator of charge collection efficiency [52]. The α coefficient effectively quantifies the extent to which the photogenerated current deviates from the theoretically expected linear relationship with light intensity. Deviations from the ideal coefficient of 1 typically arise due to factors such as bimolecular recombination, space charge effects, and/or an imbalance in charge carrier mobilities. This imbalance is noteworthy because nearly all organic solar cells experience some degree of bimolecular recombination and space charge effects, which can be considered relatively consistent across similar systems [53]. However, variations in charge carrier mobility influence the α coefficient. Notably, the 1:3 ratio solar cell appears to represent an optimal balance, as it effectively curtails squaraine aggregation, thereby maximizing the interface surface area and monomeric absorption while promoting charge carrier separation. Additionally, this ratio balances the amount of squaraine material that enhances exciton generation against the mobilities of donor and acceptor charge carriers. In essence, increasing the concentration of squaraines can enhance performance until excessive aggregation occurs, which in turn compromises both absorption characteristics and interface surface area. The ideality factor is particularly valuable as it assesses how closely the current–voltage characteristics of the solar cell resemble those of an ideal photodiode. An ideality factor of 1 corresponds to Shockley–Read–Hall low-level injection recombination, indicating that recombination is constrained by the minority carrier, whereas a factor of 2 aligns with Shockley–Read–Hall high-level injection recombination, suggesting that both carriers govern recombination. For SQ5c, particularly within the bulk-heterojunction formed from chloroform, the ideality factor is measured at 1.1, suggesting low-level injection recombination where holes (acting as the minority carrier) limit charge carrier separation. In contrast, SSQ-1 exhibits an ideality factor closer to that associated with high-level injection recombination, implying that both charge carriers restrict charge separation. This observation is logical given that SSQ-1 possesses enhanced crystallinity and superior hole mobility compared to SQ5c, leading to more effective charge transport (Fig. 10).

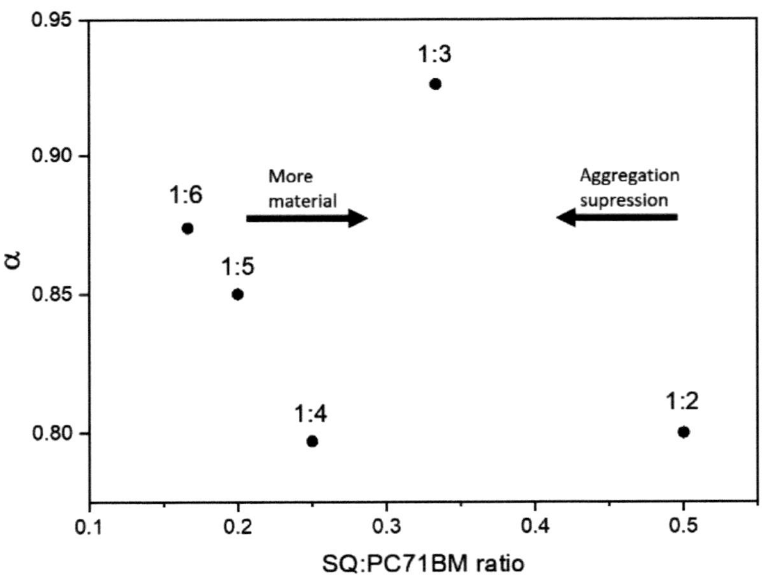

Fig. 10 Calculated alpha coefficients were plotted against the SQ:$PC_{71}BM$ ratio

5 Conclusions

A total of ten innovative squaraine derivatives derived from aminosquaraines were investigated for their potential use in photovoltaic applications. Each compound was characterized using UV-Vis spectroscopy in both solution and thin film forms, alongside cyclic voltammetry, to ascertain their optical bandgaps. The compounds exhibited suitable bandgaps and HOMO/LUMO levels that enable absorption in the visible range of the solar spectrum, thereby promoting energy-efficient transport within a bulk heterojunction configuration with $PC_{71}BM$ as the model acceptor material.

The solubility of each compound, crucial for thin film processing, was empirically determined, and the optimization of thin film fabrication was conducted to create layers with the necessary thickness and surface roughness to conform to the standards of organic solar cells.

Initial optimization of solar cell architecture suggested an inverted arrangement of ITO/ZnO/BHJ/MoO_3/Ag, although this configuration is known to present challenges, such as elevated series resistance stemming from interface issues between inorganic charge transport layers and organic semiconductors.

Solar cells, except for SSQ-3, were constructed using various donor-to-acceptor ratios with $PC_{71}BM$ to identify optimal performance. All assembled solar cells exhibited significantly high series resistance of over 1000 Ωcm^2, which is known to adversely affect short-circuit current. Notably, some compounds demonstrated unsatisfactory performance due to inadequate solubility.

Further optimization of SQ5c and SSQ-1 yielded power conversion efficiencies of 0.75% for SQ5c, spin coated from 1,2-dichlorobenzene at a concentration of 40 mg mL^{-1}, and 1.15% for the SSQ-1 solar cell, spin coated from a blend of chloroform and chlorobenzene. It was noted during optimization that the series resistance decreased significantly, indicating that the initial high resistance stemmed from both the solar cell architecture and the intrinsic resistance of the SQ:$PC_{71}BM$ layer. It was also discovered that employing solvents with slower evaporation rates could enhance short-circuit current by extending the drying time of the film, thus allowing more time for the squaraines to crystallize. Atomic force microscopy and grazing incidence wide-angle X-ray scattering, provided more detailed insights into charge extraction within the bulk heterojunction layer comprising SQ5c, SSQ-1, and ASQ-2.

The ASQ series demonstrated ideal properties for photovoltaic applications yet resulted in the least efficient devices, primarily due to the presence of large aggregates measuring approximately 500 nm, which, compared to the more homogeneous bulk-heterojunction of SQ5c and SSQ-1, limited photocurrent generation owing to a significantly reduced interface surface area and poor charge carrier separation.

Scattering profile analyses indicated that SSQ-1 possessed greater crystallinity than SQ5c, leading to higher extracted currents; this difference could be attributed to the aromatic rings present at the ends of the SQ5c molecule, which resulted in a less planar conformation that hindered crystallization.

In recombination studies, light intensity measurements of SSQ-1 solar cells with donor-to-acceptor ratios ranging from 1:x (x = 2–6) were used to establish the optimum ratio. This is critical as maximizing the incorporation of squaraine material into the thin film is essential while avoiding the onset of squaraine aggregation that could diminish optical properties and impede charge carrier separation.

Additionally, SQ5c and SSQ-1 solar cells were analyzed in a light intensity study, where the correlation of open-circuit voltage with light intensity provided ideality factors for both devices, confirming that the enhanced crystallinity of SSQ-1 improves charge carrier mobility within the donor layer. This enhanced mobility compensates for the differences in charge carrier mobility between the donor and acceptor layers within the bulk heterojunction, resulting in improved device efficiencies.

Squaraines exhibit substantial potential for the creation of easily synthesized, environmentally friendly, and photochemically stable solar cells. Looking ahead, reducing series resistance is likely to significantly boost both short-circuit current and fill factor, thereby exponentially increasing overall efficiency. The properties of an ideal photovoltaic organic semiconductor include optical bandgaps below 1.8 eV, deep HOMO levels, good solubility in common organic solvents, high charge carrier mobility, and the ability to form an optimal bulk heterojunction with the acceptor material to minimize energy losses. It is worth noting that this research utilized a widely recognized electron acceptor known for its compatibility with various electron donors, suggesting that all compounds synthesized in this study may yield better devices when paired with a more suitable electron acceptor.

References

1. Peter LM (1988) Semiconductor electrochemistry. https://doi.org/10.1016/0013-4686(88)80055-0
2. Harris LA, Wilson RH (1978) Semiconductors for photoelectrolysis. Annu Rev Mater Sci 8:99–134. https://doi.org/10.1146/annurev.ms.08.080178.000531
3. Brédas JL, Calbert JP, Da Silva Filho DA, Cornil J (2002) Organic semiconductors: a theoretical characterization of the basic parameters governing charge transport. Proc Natl Acad Sci USA 99:5804–5809. https://doi.org/10.1073/pnas.092143399
4. Brütting W (2012) Physics of organic semiconductors. https://doi.org/10.1002/9783527654949
5. Gotovac V (1982) Odabrana pogljavlja fizike kondenziranog stanja, 1st edn. Faculty of Technology Split
6. Li S (2006) Semiconductor physical electronics
7. Grundmann M (2016) Graduate texts in Physics. In: The Physics of semiconductors. http://www.springer.com/series/8431
8. Natali D (2020) Fundamentals of organic electronic devices, LTD. https://doi.org/10.1016/B978-0-12-818890-3.00001-1
9. Kitai A (2011) Principles of solar cells, LEDs and diodes: the role of the PN junction. https://doi.org/10.1002/9781119974543
10. Morab S, Sundaram MM, Pivrikas A (2023) Review on charge carrier transport in inorganic and organic semiconductors. Coatings 13. https://doi.org/10.3390/coatings13091657
11. Bässler H, Köhler A (2012) Charge transport in organic semiconductors. Top Curr Chem 312:1–65. https://doi.org/10.1007/128_2011_218
12. Dambhare MV, Butey B, Moharil SV (2021) Solar photovoltaic technology: a review of different types of solar cells and its future trends. J Phys Conf Ser 1913. https://doi.org/10.1088/1742-6596/1913/1/012053

13. Parida B, Iniyan S, Goic R (2011) A review of solar photovoltaic technologies. Renew Sustain Energy Rev 15:1625–1636. https://doi.org/10.1016/j.rser.2010.11.032
14. Brabec CJ, Gowrisanker S, Halls JJM, Laird D, Jia S, Williams SP (2010) Polymer-fullerene bulk-heterojunction solar cells. Adv Mater 22:3839–3856. https://doi.org/10.1002/adma.200903697
15. Rafique S, Abdullah SM, Sulaiman K, Iwamoto M (2018) Fundamentals of bulk heterojunction organic solar cells: an overview of stability/degradation issues and strategies for improvement. Renew Sustain Energy Rev 84:43–53. https://doi.org/10.1016/j.rser.2017.12.008
16. Nian, Kan Y, Gao K, Zhang M, Li N, Zhou G, Jo SB, Shi X, Lin F, Rong Q, Liu F, Zhou G, Jen AKY (2020) Approaching 16% efficiency in all-small-molecule organic solar cells based on ternary strategy with a highly crystalline acceptor. Joule 4:2223–2236. https://doi.org/10.1016/j.joule.2020.08.011
17. Zhang Y, Li X, Dai T, Xu D, Xi J, Chen X (2019) Charge transport and extraction of PTB7:PC_{71}BM organic solar cells: effect of film thickness and thermal-annealing. RSC Adv 9:24895–24903. https://doi.org/10.1039/c9ra02877c
18. Varma PCR, Namboothiry MAG (2016) Squaraine based solution processed inverted bulk heterojunction solar cells processed in air. Phys Chem Chem Phys 18:3438–3443. https://doi.org/10.1039/c5cp05080d
19. Scharber MC, Sariciftci NS (2013) Efficiency of bulk-heterojunction organic solar cells. Prog Polym Sci 38:1929–1940. https://doi.org/10.1016/j.progpolymsci.2013.05.001
20. Viterisi A, Montcada NF, Kumar CV, Gispert-Guirado F, Martin E, Escudero E, Palomares E (2014) Unambiguous determination of molecular packing in crystalline donor domains of small molecule solution processed solar cell devices using routine x-ray diffraction techniques. J Mater Chem A 2:3536–3542. https://doi.org/10.1039/c3ta13116e
21. Pfeiffer M, Leo K, Zhou X, Huang JS, Hofmann M, Werner A, Blochwitz-Nimoth J (2003) Doped organic semiconductors: physics and application in light emitting diodes. Org Electron 4:89–103. https://doi.org/10.1016/j.orgel.2003.08.004
22. Pakravesh F, Izadyar M, Arkan F (2021) Effect of electron donor and acceptor on the photovoltaic properties of organic dyes for efficient dye-sensitized solar cells. Phys B Condens Matter 609:412815. https://doi.org/10.1016/j.physb.2021.412815
23. Hou J, Guo X (2013) Active layer materials for organic solar cells. Green Energy Technol 128:17–42. https://doi.org/10.1007/978-1-4471-4823-4_2
24. Ravishankar S, Gharibzadeh S, Roldán-Carmona C, Grancini G, Lee Y, Ralaiarisoa M, Asiri AM, Koch N, Bisquert J, Nazeeruddin MK (2018) Influence of charge transport layers on open-circuit voltage and hysteresis in perovskite solar cells. Joule 2:788–798. https://doi.org/10.1016/j.joule.2018.02.013
25. Susarova DK, Akkuratov AV, Kukharenko AI, Cholakh SO, Kurmaev EZ, Troshin PA (2017) ITO modification for efficient inverted organic solar cells. Langmuir 33:10118–10124. https://doi.org/10.1021/acs.langmuir.7b01106
26. Yu X, Yu X, Zhang J, Zhang D, Ni J, Cai H, Zhang D, Zhao Y (2015) Efficient inverted polymer solar cells based on surface modified FTO transparent electrodes. Sol Energy Mater Sol Cells 136:142–147. https://doi.org/10.1016/j.solmat.2015.01.001
27. Mabbott GA (1983) An introduction to cyclic voltammetry. J Chem Educ 60:697–702. https://doi.org/10.1021/ed060p697
28. Giessibl FJ (2003) Advances in atomic force microscopy. Rev Mod Phys 75:949–983. https://doi.org/10.1103/RevModPhys.75.949
29. He J, Jo YJ, Sun X, Qiao W, Ok J, Kim T, Li Z (2021) Squaraine dyes for photovoltaic and biomedical applications. Adv Funct Mater 31:1–35. https://doi.org/10.1002/adfm.202008201
30. Wang S, Hall L, Diev VV, Haiges R, Wei G, Xiao X, Djurovich PI, Forrest SR, Thompson ME (2011) N, N-Di aryl anilinosquaraines and their application to organic photovoltaics. Chem Mater 23:4789–4798. https://doi.org/10.1021/cm2020803
31. Beverina L, Salice P (2010) Squaraine compounds: tailored design and synthesis towards a variety of material science applications. Eur J Org Chem 1207–1225. https://doi.org/10.1002/ejoc.200901297

32. Kaczmarek-Kedziera A, Kędziera D (2016) Molecular aspects of squaraine dyes aggregation and its influence on spectroscopic properties. Theor Chem Acc 135:87–100. https://doi.org/10.1007/s00214-016-1971-0
33. Hu L, Yan Z, Xu H (2013) Advances in synthesis and application of near-infrared absorbing squaraine dyes. RSC Adv 3:7667–7676. https://doi.org/10.1039/c3ra23048a
34. Ilina K, MacCuaig WM, Laramie M, Jeouty JN, McNally LR, Henary M (2020) Squaraine dyes: molecular design for different applications and remaining challenges. Bioconjug Chem 31:194–213. https://doi.org/10.1021/acs.bioconjchem.9b00482
35. Wei G, Wang S, Sun K, Thompson ME, Forrest SR (2011) Solvent-annealed crystalline squaraine: PC 70BM (1:6) solar cells. Adv Energy Mater 1:184–187. https://doi.org/10.1002/aenm.201100045
36. Ajayaghosh A (2005) Chemistry of squaraine-derived materials: near-IR dyes, low band gap systems, and cation sensors. Acc Chem Res 38:449–459. https://doi.org/10.1021/ar0401000
37. Völker SF, Uemura S, Limpinsel M, Mingebach M, Deibel C, Dyakonov V, Lambert C (2010) Polymeric squaraine dyes as electron donors in bulk heterojunction solar cells. Macromol Chem Phys 211:1098–1108. https://doi.org/10.1002/macp.200900670
38. Park J, Barolo C, Sauvage F, Barbero N, Benzi C, Quagliotto P, Coluccia S, Di Censo D, Grätzel M, Nazeeruddin MK, Viscardi G (2012) Symmetric vs. asymmetric squaraines as photosensitisers in mesoscopic injection solar cells: a structure-property relationship study. Chem Commun 48:2782–2784. https://doi.org/10.1039/c2cc17187b
39. Khopkar S, Shankarling G (2019) Synthesis, photophysical properties and applications of NIR absorbing unsymmetrical squaraines: a review. Dye Pigment 170:107645. https://doi.org/10.1016/j.dyepig.2019.107645
40. Merritt VY, Hovel HJ (1976) Organic solar cells of hydroxy squarylium. Appl Phys Lett 29:414–415. https://doi.org/10.1063/1.89101
41. Silvestri F, Irwin MD, Beverina L, Facchetti A, Pagani GA, Marks TJ (2008) Efficient squaraine-based solution processable bulk-heterojunction solar cells. J Am Chem Soc 130:17640–17641. https://doi.org/10.1021/ja8067879
42. Yang D, Sasabe H, Jiao Y, Zhuang T, Huang Y, Pu X, Sano T, Lu Z, Kido J (2016) An effective π-extended squaraine for solution-processed organic solar cells with high efficiency. J Mater Chem A 4:18931–18941. https://doi.org/10.1039/c6ta08684e
43. Yang D, Sano T, Sasabe H, Yang L, Ohisa S, Chen Y, Huang Y, Kido J (2018) Colorful squaraines dyes for efficient solution-processed all small-molecule semitransparent organic solar cells. ACS Appl Mater Interfaces 10:26465–26472. https://doi.org/10.1021/acsami.8b08878
44. Mikroyannidis JA, Tsagkournos DV, Sharma SS, Sharma GD (2011) Synthesis of a broadly absorbing modified PCBM and application as electron acceptor with poly(3-Hexylthiophene) as electron donor in efficient bulk heterojunction solar cells. J Phys Chem C 115:7806–7816. https://doi.org/10.1021/jp201337u
45. Guo Q, Guo Q, Geng Y, Tang A, Zhang M, Du M, Sun X, Zhou E (2021) Recent advances in PM6:Y6-based organic solar cells. Mater Chem Front 5:3257–3280. https://doi.org/10.1039/d1qm00060h
46. Shen CA, Bialas D, Hecht M, Stepanenko V, Sugiyasu K, Würthner F (2021) Polymorphism in squaraine dye aggregates by self-assembly pathway differentiation: panchromatic tubular dye nanorods versus J-aggregate nanosheets. Angew Chem Int Ed 60:11949–11958. https://doi.org/10.1002/anie.202102183
47. Scheunemann D, Kolloge O, Wilken S, Mack M, Parisi J, Schulz M, Lützen A, Schiek M (2017) Revealing the recombination dynamics in squaraine-based bulk heterojunction solar cells. Appl Phys Lett 111. https://doi.org/10.1063/1.4996080
48. Chen Y, Zhu W, Wu J, Huang Y, Facchetti A, Marks TJ (2019) Recent advances in squaraine dyes for bulk-heterojunction organic solar cells. Org Photon Photovolt 6:1–16. https://doi.org/10.1515/oph-2019-0001
49. Ávila-Rovelo NR, Ruiz-Carretero A (2020) Recent progress in hydrogen-bonded π-conjugated systems displaying J-type aggregates. Org Mater 2:47–63. https://doi.org/10.1055/s-0040-1708502

50. Li M, Igbari F, Wang ZK, Liao LS (2020) Indoor thin-film photovoltaics: progress and challenges. Adv Energy Mater 10:1–25. https://doi.org/10.1002/aenm.202000641
51. Anitha B, Vijith KP, Alexander A, Srivastava V, Namboothiry MAG (2020) Understanding the poor fill factor of solution-processed squaraine based solar cells in terms of charge carrier dynamics probed: via impedance and transient spectroscopy. J Mater Chem C 8:14748–14756. https://doi.org/10.1039/d0tc03012k
52. Koster LJA, Mihailetchi VD, Xie H, Blom PWM (2005) Origin of the light intensity dependence of the short-circuit current of polymer/fullerene solar cells. Appl Phys Lett 87:1–3. https://doi.org/10.1063/1.2130396
53. Menke SM, Ran NA, Bazan GC, Friend RH (2018) Understanding energy loss in organic solar cells: toward a new efficiency regime. Joule 2:25–35. https://doi.org/10.1016/j.joule.2017.09.020

Application of Silicon-Based Thin Films in High-Efficiency Silicon Solar Cells for Carbon Neutrality

Zhengping Li, Dong Ding, Sheng Ma, Xunming Deng, and Wenzhong Shen

Abstract High-efficiency and low-cost photovoltaic (PV) technology is considered a top choice toward achieving net-zero carbon dioxide emissions - carbon neutrality. In this Chapter, latest technologies in crystalline silicon (c-Si) PV and the underlined physics and device principles are reviewed. A critical step to achieve high efficiency in c-Si PV is the effective passivation of carrier recombination centres (e.g., dangling bonds) at the surfaces and interfaces, often using silicon-based thin film layers. Based on different approaches to accomplish the passivation, various device structures, such as silicon heterojunction (SHJ) and tunnel oxide passivated contact (TOPCon), are developed and optimized. Other devices structures, including passivated emitter and rear contact (PERC), interdigitated back contact (IBC), combinations of these structures such as TBC, HBC, THBC, and tandem device using perovskite top cell on top of a c-Si bottom cell, are also reviewed and discussed.

1 Introduction

Carbon neutrality is a state of net-zero carbon dioxide emissions. The application of renewable energies can make an important contribution to carbon neutrality. Among renewable energies, photovoltaic (PV) solar energy is a significant answer to the challenges. PV refers to the direct energy conversion of solar radiation into electricity. Solar cells are the key devices for realizing energy conversion. At present, crystalline silicon (c-Si) solar cells are the leading products with a market share of more than 95%. Among c-Si solar cells, tunnel oxide passivated contact (TOPCon) solar cells

Z. Li · D. Ding · S. Ma · W. Shen (✉)
Institute of Solar Energy, and Key Laboratory of Artificial Structures and Quantum Control (Ministry of Education), School of Physics and Astronomy, Shanghai Jiao Tong University, Shanghai, China
e-mail: wzshen@sjtu.edu.cn
URL: https://www.physics.sjtu.edu.cn/jsml/shenwenzhong.html

X. Deng
Department of Physics and Astronomy, The University of Toledo, Toledo, OH 43606, USA

1America Solar, LLC, Toledo, OH 43606, USA

D. Shi (ed.), *Energy Materials Science and Engineering: From New Devices to AI Power Systems*, Green Energy and Technology, https://doi.org/10.1007/978-3-031-88550-1_3

and silicon heterojunction (SHJ, also known as HJT or HIT) solar cells currently have the greatest market shares, and both employ silicon-based thin films as their critical passivation layers to achieve high efficiency.

In this chapter, carbon neutrality and PV are introduced firstly. Then an outline of basic knowledge of semiconductors, p–n junction, solar cells is given. Afterwards, the concept of passivating contact cells is provided. Subsequently, the intrinsic hydrogenated amorphous silicon (i-a-Si:H) thin films and doped silicon films, which constitute the passivated contact stack (i-a-Si:H/doped Si) in SHJ solar cells, are introduced in detail. The stack of silicon suboxide (SiO_x) film and doped polycrystalline silicon (poly-Si) film (SiO_x/poly-Si) forming another passivated contact in TOPCon c-Si cell, is also discussed. At the end, the latest developments in some of the advanced c-Si solar cells—TOPCon/heterojunction hybrid back-contact cell and perovskite/c-Si tandem cell—are briefly introduced. It is anticipated that the readers can learn about the passivated contact c-Si solar cells and the related Si-based thin films in this chapter.

2 Photovoltaic Playing a Dominant Role Towards Achieving Carbon Neutrality

2.1 Current Situation and Trend of Photovoltaic Power Generation

Carbon neutrality refers to a state in which the greenhouse gases emitted into the atmosphere are equivalent to those absorbed from it. In the field of carbon management, carbon emissions generally encompass all greenhouse gas emissions; consequently, carbon neutrality typically refers to the neutralization of all greenhouse gases. To achieve carbon neutrality, two primary approaches exist: reducing carbon emissions and increasing carbon absorption. Reducing carbon emissions necessitates the phasing out of coal-fired, oil-fired, and gas-fired power plants. Currently, 30% of the world's fossil energy is utilized for electricity generation, and eliminating this usage would reduce global carbon emissions by 30%. The phase-out of fossil-fueled power plants must be accompanied by a significant effort to replace them with PV, wind, hydroelectric, and nuclear power, which produce negligible carbon emissions. This transition will subsequently be followed by the gradual conversion of equipment currently utilizing fossil fuels to generate electricity, such as replacing fuel-powered vehicles with electric ones and coal-fired boilers with electric alternatives. Increased carbon absorption is predominantly achieved through carbon capture and storage facilities or via afforestation to augment forest carbon sinks. Forest carbon sinks currently represent the more economically viable method of carbon sequestration and fulfill the dual function of climate change mitigation and adaptation.

Achieving the goal of global carbon neutrality requires vigorous promotion of transition to clean energy. When clean energy consumption accounts for more than

80% of total future energy consumption, solar, wind, hydro, nuclear, hydrogen, and biomass will play significant roles. However, the question arises as to which energy source will play the biggest role in this transition?

The answer is, unequivocally, solar energy.

Solar energy is ubiquitous, and its total available quantity is substantial. It is a sustainable, clean energy source with zero emissions and zero pollution that can be accessed at a relatively low cost. Fundamentally, both traditional fossil energy and newer energy forms such as wind, hydropower, and biomass are derivatives of solar energy, albeit in different forms. Solar energy refers to the direct conversion of solar radiation through photoelectric, photothermal, and photochemical processes. Notably, photoelectric conversion, specifically solar PV power generation, is considered the most promising energy technology.

The PV power generation offers the following advantages:

(1) The most advanced technical principle. Fossil energy requires combustion to convert to heat energy, which is then transformed into electricity; wind and hydropower convert the kinetic energy of wind and water into mechanical energy of the turbines, which is subsequently converted to electricity. In contrast, solar PV power generation directly converts photon energy into electric energy.
(2) Abundant total quantity. The Earth's surface receives daily solar energy equivalent to the energy produced by 100 million barrels of oil. Moreover, the solar energy received by the Earth in a single hour is sufficient to meet human energy consumption for an entire year.
(3) Wide distribution. All regions of the world receive varying degrees of solar irradiation, with certain areas such as North Africa, the Middle East, and the southwestern United States receiving particularly high levels of annual solar irradiation, reaching 10,080, 9720, and 10,512 MJ/m^2, respectively.
(4) Abundant raw material. Silicon, the primary raw material, is highly abundant, constituting 26.4% of the Earth's crust, second only to oxygen.
(5) Broad range of applications. Solar PV power generation can be applied in numerous fields, including aerospace, transportation, agriculture, construction, military operations, and urban lighting. Even remote desert and island locations can benefit from illumination provided by PV power generation.

Solar PV power generation plays an increasingly significant role in our effort to achieve a global carbon emission plateau and carbon neutrality. Among all energy generation and consumption methods, using traditional fossil energy sources such as oil and coal emits 778–960 g of carbon dioxide per kWh of electricity generated, while the carbon emissions from clean energy power generation are all below 45 g/kWh, of which the carbon emissions from solar PV power generation are only 30 g/kWh, less than 1/30 of the carbon emissions from coal-fired power generation. From Fig. 1a [1], it can be noted that the global total power generation in 2014 and 2023 is ~25,000 and ~30,000 TWh, respectively, with an average growth rate of only 2.5% during the 10-year period; whereas solar PV power generation has a compound annual growth rate as high as 50–60%, and its share in the structure of the global total power generation has increased from less than 1% a decade ago to 5.5% in 2023. Figure 1b illustrates the cost trend for solar PV and conventional thermal power generation [2]. In 2023, the global average levelized cost of electricity (LCOE) of solar PV is \$0.044/kWh, which is beginning to be lower than the LCOE of conventional thermal power generation, which is \$0.05/kWh, and it will demonstrate a very strong competitiveness in the future of the global energy transition. Figure 1c, d are charts from the International Energy Agency's (IEA) "Renewable Energy Report" released in 2024 [3], covering the current status and future trends of PV power generation. As evident from the charts, more than 5,500 GW of renewable energy capacity will be added globally between 2024 and 2030, almost triple the increase between 2017 and 2023, which translates to approximately 800 GW per year, of which solar PV's share of the total renewable energy capacity will also continue to remain at 60–70% and above. In 2023, China, Europe and the US account for approximately 32%, 20% and 10%, respectively, of the world's total installed renewable energy capacity. While China has the largest installed capacity of new renewable energy, India's growth rate is among the fastest among major economies. Notably, solar PV is expected to triple to more than 16% of the total global power generation mix in 2030 from 2023's 5.5% levels and is expected to further increase its share to 30–40% in 2050.

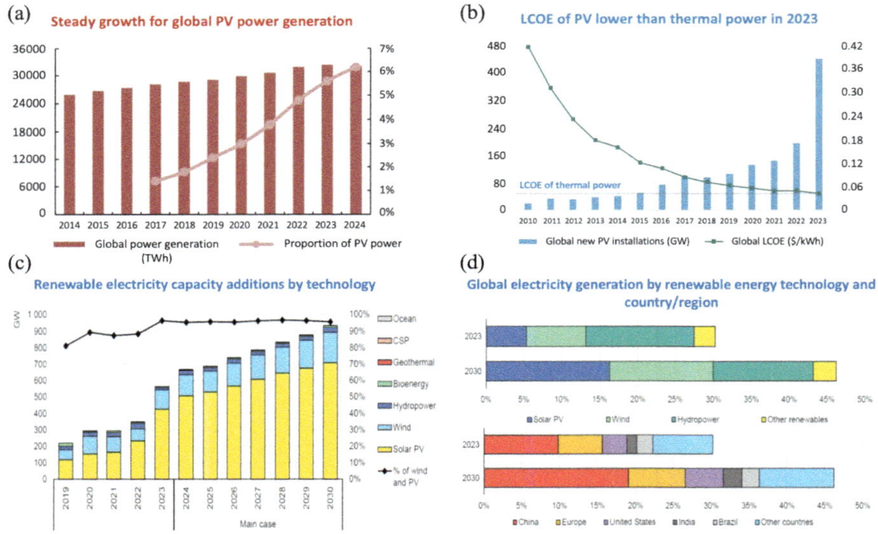

Fig. 1 **a** Global electricity generation and share of PV energy [1]. **b** Trends in global new PV installations and LCOE [2]. **c** Trends in multiple renewable energy generating capacities [3]. **d** Characteristics of the global renewable energy mix and country/region distributions in 2023 (actual) and 2030 (forecast) [3]

2.2 Latest Technologies and Development for c-Si Solar Cells

The primary component of solar PV power generation is the solar cell device. The development of new processes and cell structures aims primarily to enhance cell efficiency by mitigating optical and electrical losses in the PV conversion process. Advanced solar cell technology encompasses numerous aspects, including polycrystalline silicon ingot technology, slicing technology, screen-printed electrode, metal pastes, novel passivation materials, innovative cell structures, and new technologies for modules and power stations, which have significantly reduced the cost of PV power generation. Among the various solar cell products, crystalline silicon (c-Si) solar cells have maintained a market share exceeding 90% for an extended period. The diffusion-doped p–n junction plays a crucial role in the generation and separation of photogenerated carriers in the cell. In the following, the advanced technology and development of solar cells are summarized, focusing on the p–n junction from three perspectives, primarily pertaining to c-Si solar cells.

(1) c-Si solar cells with diffusion-doped p–n junction (homojunction)

In the past few decades, significant progress in c-Si solar cells has come from the technology improving surface passivation. The passivated emitter and rear cell (PERC) structure was developed in 1989 to mitigate the carrier recombination caused by the direct contact between the metal Al and c-Si substrate on the backside of the cell [4]. As illustrated in Fig. 2a, the primary structural features of PERC c-Si solar cells utilize p-type silicon wafers as the substrates, with pyramidal texturing on the front surface and a polished structure on the back surface [5]. High-purity oxygen is passed through methyl chloroform to grow high-quality SiO_2 as the passivation film. The selective emitter structure is prepared using a photolithographic mask method, forming a heavily doped region with a sheet resistance of 20 Ω/sq in the metal–semiconductor contact region, and a lightly doped region with a sheet resistance of 250 Ω/sq. In addition to the localized contact of rear Al and c-Si substrate, the PERC c-Si solar cell employs an Al/SiN_x/AlO_x stack films as the back surface passivation layer and back reflector, which is capable of achieving 97% back reflectance in long wavelengths exceeding 1000 nm. To further enhance the efficiency of PERC cells, researchers proposed passivated emitter and rear locally diffused (PERL) solar cells in 1990 [6], as depicted in Fig. 2b. Its structural improvement primarily involves the preparation of the backside localized contact by diffusing phosphorus sources. The liquid diffusion method has less impact on the lifetime of the c-Si wafer body compared to the solid-state source and spin-coating method for preparing the localized heavy doping. When combined with a front-side Al_2O_3 passivation film, the short-circuit current density (J_{SC}) can be increased by 1.9 mA/cm^2 compared to PERC c-Si solar cells. The tunnel oxide passivated contact (TOPCon) c-Si solar cell in Fig. 2c, completely eliminates direct contact between the back-side metal electrodes and the c-Si wafers [7]. By growing a rear SiO_2 layer with a thickness of only 1–2 nm and subsequently depositing a doped polycrystalline silicon (poly-Si) layer, the two layers together form a passivated contact structure with carrier-selective

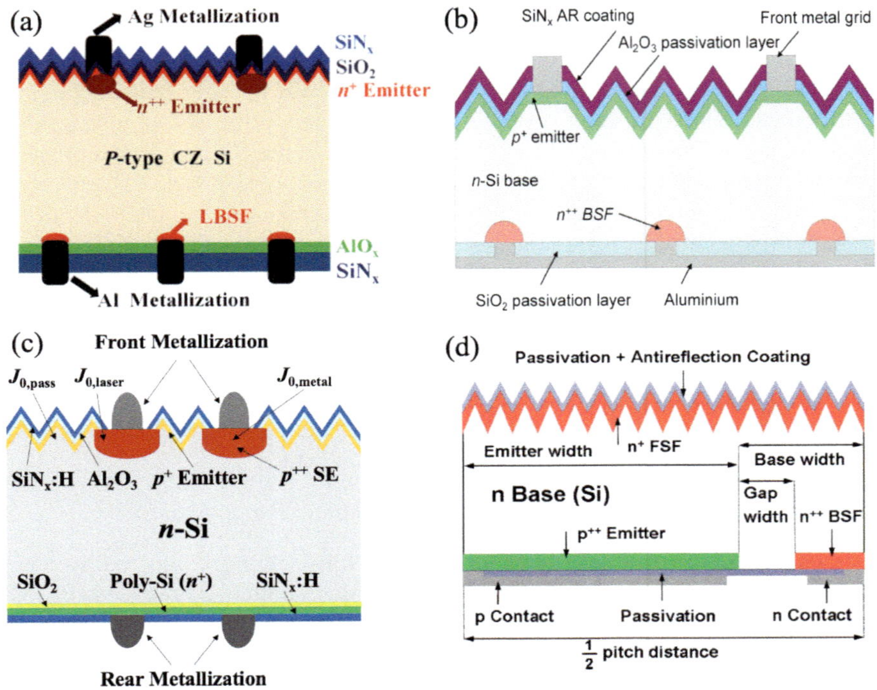

Fig. 2 Schematic diagrams of **a** PERC [5], **b** PERL [6], **c** TOPCon [7], and **d** IBC [8] c-Si solar cell structures

transport capability. The metal grid lines on the front side tend to obstruct part of the incident light. The interdigitated back contact (IBC) c-Si solar cell shown in Fig. 2d relocates all the positive and negative electrodes to the back side [8], allowing the entire frontal area to receive solar irradiation, thereby maximizing the collection of photogenerated carriers.

(2) c-Si solar cells without diffusion-doped p–n junction (heterojunction)

Carrier separation is typically considered to be accomplished by the built-in electric field of the cell device, i.e., diffusion-doped p–n junction. However, it has been demonstrated that carrier separation fundamentally relies on the existence of carrier-selective contacts, and the built-in electric field is merely the result of the formation of a certain type of carrier-selective contacts, which theoretically permits the construction of a cell device without a built-in electric field. Silicon heterojunction (SHJ) solar cells, incorporating phosphorus-doped (P-doped) and boron-doped (B-doped) amorphous silicon (a-Si:H) film, can be approximated as a c-Si solar cell without diffusion-doped p–n junction due to the deposition temperature being less than 250 °C, which impedes the doping element from forming an effective diffusion depth. As illustrated in Fig. 3a [9], the SHJ c-Si solar cell structure is relatively simple and symmetrically distributed. The front side comprises P-doped amorphous silicon and intrinsic a-Si:H (n-a-Si:H/i-a-Si:H/c-Si) to form a field passivation structure, while the back side consists of B-doped amorphous silicon and intrinsic a-Si:H (p-a-Si:H/i-a-Si:H/c-Si) to form the emitter structure. Owing to the superior passivation performance of a-Si:H, SHJ solar cells can be approximated as a diffusion-free p–n junction with the open-circuit voltage (V_{OC}) of SHJ solar cells frequently exceeding 750 mV. Figure 3b depicts the energy band structure of SHJ solar cells for carrier-selective transport.

In addition to heterojunctions formed by doped silicon thin film materials and c-Si substrates that can be utilized as selective contacts, the implementation of certain heterojunctions composed of materials with lower (<4.0 eV) or higher (>5.2 eV) work functions with c-Si wafer also induces band bending on the silicon surface, which facilitates the function of selective contacts. This type of cell is referred to as a dopant-free solar cell, and its structure is schematically represented in Fig. 3c [10]. Materials with electron selective contact function primarily include metal oxides such as TiO_x, MgO_x and metal halides such as LiF_x, MgF_x, which are generally termed electron transport layer (ETL) materials. Materials with hole selective contact function predominantly include metal oxides such as MoO_x, WO_x, VO_x, as well as organic semiconductors such as PEDOT:PSS, which are typically designated as hole transport layer (HTL) materials. By substituting the c-Si substrate of a dopant-free

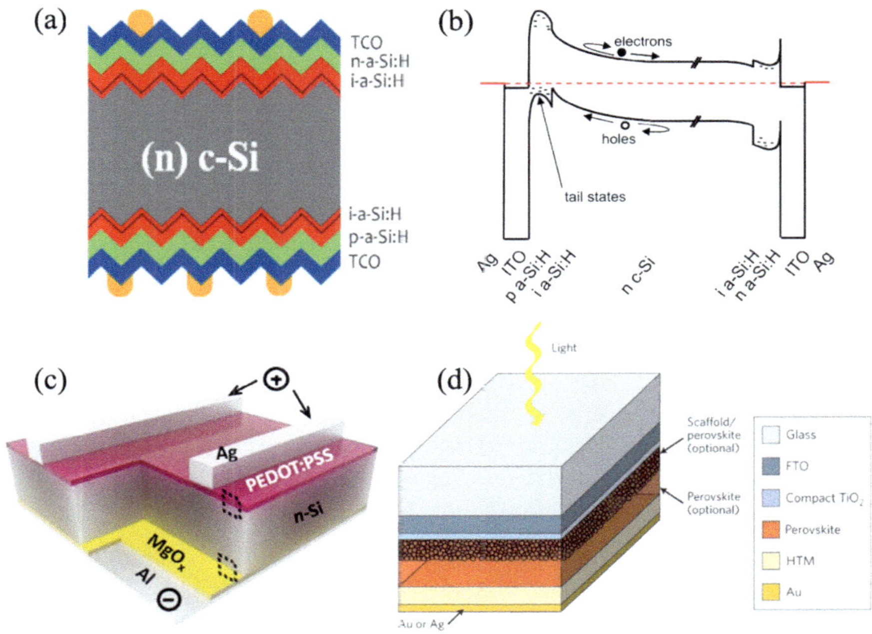

Fig. 3 **a** Schematic structure of SHJ [9] solar cell and **b** energy band structure; **c** Dopant-free [10] and **d** perovskite [11] solar cell structures

solar cell with a perovskite material, a perovskite solar cell is formed, as illustrated in Fig. 3d [11]. Given that the energy bands of perovskite materials are tunable from 1.15 to 3.06 eV, this range allows perovskite materials to exhibit excellent PV conversion performance and broad applicability to sunlight in solar energy. Researches have indicated that the limiting efficiency of single-junction perovskite solar cells is 33%, which significantly surpasses the limiting efficiency of 29.4% for monocrystalline silicon solar cells [12].

(3) Tandem solar cells

To absorb the solar spectrum effectively, semiconductor materials with two different bandgaps are stacked in series on top of each other for effective spectral absorption. Typically, the top wide bandgap material absorbs high-energy photons, while the bottom narrow bandgap c-Si material absorbs low-energy photons. This configuration achieves a broader wavelength absorption region, mitigates carrier heat loss due to energy mismatch, and enables the device to surpass the single junction's Shockley–Queisser limit, thereby realizing high energy conversion efficiency. Compound materials such as GaAs, CdTe, and $CuInSe_2$ can serve as top sub-cells in tandem solar cells. More interestingly, perovskite/c-Si tandem solar cells have been studied since 2015, and their conversion efficiencies have demonstrated remarkable growth from an initial 13.7% [13] to the current 34.6% [14]. Under the irradiation of 1 standard sun, the theoretical conversion efficiencies of the two-junction, three-junction, and infinite junction are 43%, 50%, and 65%, respectively [12, 15–18], assuming 100% radiative compositing and considering solar radiation and electroluminescence. Notably, the theoretically predicted efficiency of the infinite junction cell is 86.8% under a highly concentrated solar spectrum [19, 20].

High efficiency and low cost always are the main subjects of the PV industry. Undoubtedly, higher efficiency solar cells are important for realizing carbon neutrality.

3 Fundamentals of Solar Cells

The photoelectric effect, the conversion of solar radiation to electric energy, is the key point of a solar cell operation. The PV effect of semiconductive materials is the foundation of solar cells. Here, some basic knowledge, including the p–n junction of semiconductors and the working principle of solar cells, is introduced.

3.1 Basis of Semiconductor and p–n Junction

Solid materials can be classified as superconductors, conductors, semiconductors and insulators according to their electrical conductivity. Conductors are materials with good conductivity, of which the electrical resistivity is below 10^{-5} Ω m. The metals such as the gold, silver, copper, iron, aluminium, are conductors. Insulators are materials with very weak conductivity or even non-conducting, of which the electrical resistivity is higher than 10^{8} Ω m. Such materials like rubber, plastic, wood, and glass are insulators. The electrical conductivity of semiconductors is between those of conductors and insulators, of which the electrical resistivity lies in 10^{-5}–10^{8} Ω m. Elementary substances such as Si (silicon), Ge (germanium), and chemical compounds such as GaAs (gallium arsenide), CdTe (cadmium telluride), are typical semiconductors.

The electroconductibility of solid materials can be interpreted by the band theory. The electrical properties of semiconductors are determined by their electronic band structure. In an insulator, the valence band (the band of electrons that are tightly bound to the atoms) is filled, and the conduction band (the band of electrons that are free to move) is empty. In a conductor, the conduction band is partially filled. In a semiconductor, the conduction band is empty, and the valence band is filled, but there is a small energy gap (E_g) between the two bands.

When a semiconductor is exposed to light or heat, electrons can be excited from the valence band to the conduction band. These electrons are then free to move, and the semiconductor becomes more conductive. This phenomenon is known as intrinsic conduction. Semiconductors can also be doped with impurities to increase their conductivity. Donor impurities add electrons to the semiconductor, while acceptor impurities remove electrons. Doped semiconductors are used to create transistors, diodes, and other electronic devices. Figure 4 is the atomic structure diagram of doping semiconductors.

A p–n junction is a semiconductor interface formed between p-type and n-type materials. This junction is a fundamental building block of many electronic devices, such as diodes, transistors, and integrated circuits. In p-type material, the majority carriers are holes (positive charge carriers), while in n-type material, the majority carriers are electrons (negative charge carriers). At the p–n junction, the differing types of charge carriers combine, resulting in a depletion region where no free carriers exist. This depletion region acts as an insulator, preventing current flow in one direction while allowing it in the other, making the p–n junction diode a critical component for controlling current flow in electronic circuits. The schematic diagram of forming

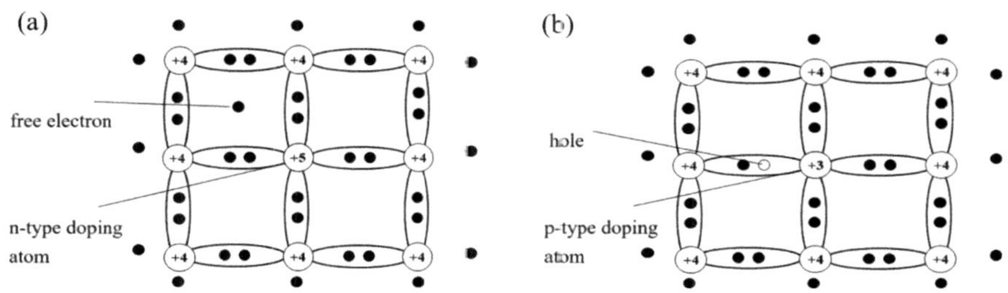

Fig. 4 Atomic structure diagram of doping semiconductors. **a** n-type, **b** p-type

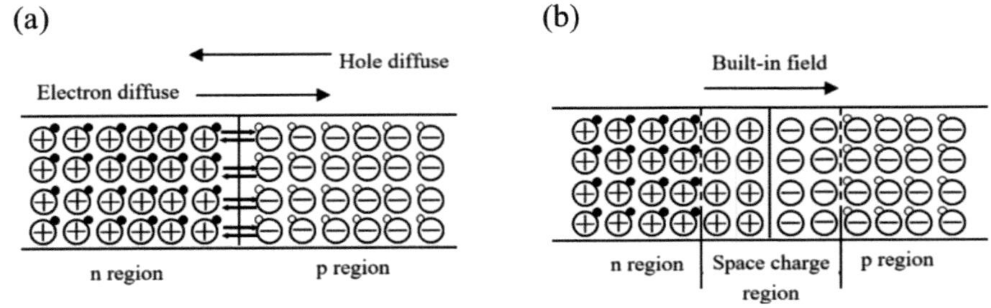

Fig. 5 Schematic diagram of forming a p–n junction. **a** Carriers diffuse before p–n junction formation, **b** space charge region and built-in field after p–n junction formation

a p–n junction can be found in Fig. 5. When a voltage is applied across the junction, the depletion region can either expand or contract, influencing the flow of current. The behavior of the p–n junction can be manipulated to create various electronic components, such as light-emitting diodes (LEDs), solar cells, and bipolar junction transistors (BJTs), making it an essential element in modern electronics. More about semiconductors and p–n junction can be found in book by Sze et al. [21].

3.2 *Photovoltaic and Crystalline Silicon Solar Cells*

The PV effect, which is the generation of voltage when a device is exposed to light, was discovered by Becquerel [22] in 1839, in a junction formed between an electrode and an electrolyte. Since then, there have been reports of similar effects on different solid-state devices. It was not until 1954 that the solar cell received much-increased interest, initiated by the works of Chapin et al. on single-crystal silicon cells [23]. To date, solar cells have been made with many other semiconductors, using various device configurations, and employing single-crystal, polycrystal, and amorphous thin-film structures.

Solar cells are semiconductor devices that convert light energy into electrical energy. The basic principle behind the operation of a solar cell is the PV effect. When light strikes a semiconductor material, it can excite electrons in the material, causing them to move freely. This movement of electrons creates an electric current, which can be used to power electrical devices. So far, more than 100 kinds of solar cells have been found. The most common type of solar cell is the silicon-based solar cell because silicon is a semiconductor material that is relatively inexpensive and easy to process.

The typical structure of a p-type silicon solar cell is in Fig. 6a. The substrate material is a silicon wafer with a normal thickness of 100–200 μm. The processes for making a crystalline silicon solar cell are shown in Fig. 6b and are described below.

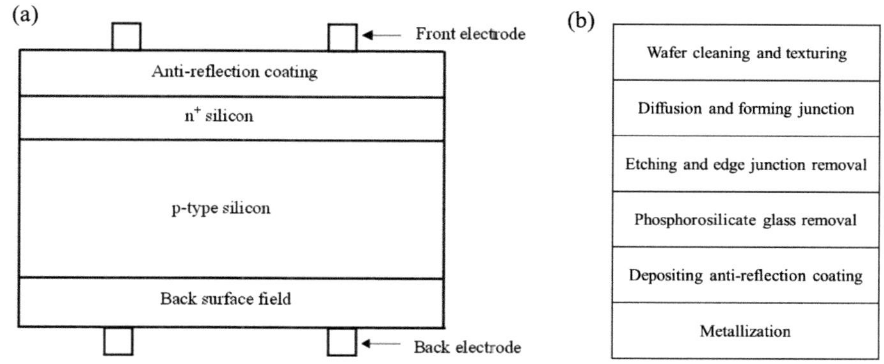

Fig. 6 **a** Schematic structure diagram of a BSF silicon solar cell, **b** processes for a c-Si solar cell fabrication

(1) Wafer cleaning and texturing. The wafer firstly undergoes a wet-chemical etching procedure, e.g. in alkaline potassium hydroxide (KOH) solution, to remove saw damage from the as-cut wafer, and to form light-trapping texture structures (like the pyramids) to enhance anti-reflection properties.
(2) Diffusion and forming junction. The cleaned wafer is put into a diffusion furnace by aerating phosphorus oxychloride ($POCl_3$) to form n^+ silicon layer at a high temperature like 800 °C and above. After that, the p–n junction is formed.
(3) Etching and edge junction removal. The next step is to etch and remove the p–n junction at the edges of the diffused silicon wafer.
(4) PSG removal. A layer of phosphorated silica, called phosphosilicate glass (PSG), has been formed on the surface of diffused silicon wafer. The PSG can be removed by hydrofluoric acid (HF) etching.
(5) Depositing ARC. To enhance the light trapping, a layer of anti-reflection coating (ARC) is deposited by method of plasma-enhanced chemical vapor deposition (PECVD). The ARC is usually hydrogenated silicon nitride (SiN_x:H).
(6) Metallization. Front electrodes are screen printed by using silver (Ag) paste on the front of the silicon wafer. While on the back of the silicon wafer, firstly aluminum (Al) pastes are screen printed to form back surface field (Al-BSF), then back electrodes are screen printed by using Ag paste. After screen printing, the wafers are fired to form an ohmic contact between Ag and Si. Hereto, a silicon solar cell has been prepared. The next is testing and sorting.

3.3 Current–Voltage Characteristics of Solar Cells

A solar cell is assumed to have ideal diode current–voltage (I–V) characteristics. The equivalent circuit is shown in Fig. 7a, where a constant-current source of photocurrent is in parallel with the junction. The source I_{ph} results from the excitation of excess

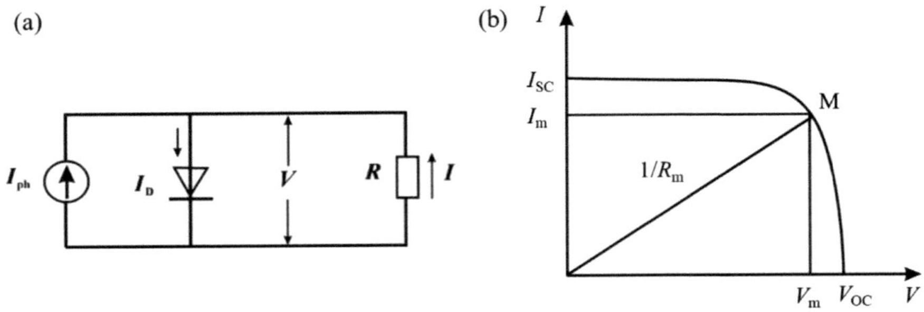

Fig. 7 **a** Ideal equivalent circuit of solar cell under illumination, **b** I–V characteristics of a solar cell under illumination

carriers by solar radiation, I_D is dark current which means the current passing through p–n junction with no illumination, and R is the load resistance.

The total I–V characteristics of such a device under illumination is simply a summation of the dark current and the photocurrent, given as

$$I = I_D - I_{ph} = I_0\left(\exp\frac{qV}{nkT} - 1\right) - I_{ph} \tag{1}$$

where I_0 is the diode saturation current, q is the electron charge, V is the terminal voltage of the equivalent diode, n is the ideality factor and is generally between 1–2, k is the Boltzmann constant, T is temperature in degree Kelvin.

When the p–n junction is in an open circuit, i.e., the load R is infinity and the current I is zero. From Eq. (1) it can be obtained the open-circuit voltage (V_{OC}) by setting $I = 0$, given as

$$V_{OC} = \frac{nkT}{q}\ln\left(\frac{I_{ph}}{I_0} + 1\right) \tag{2}$$

When the p–n junction is in short-circuit, i.e., the R, V, and I_D are all equal to 0, it can be obtained the short-circuit current (I_{SC}). I_{SC} can be derived from Eq. (1) as.

$$I_{SC} = I = I_{ph} \tag{3}$$

V_{OC} and I_{SC} are the key parameters to characterize a solar cell, which increase with the increase of solar radiation. The V_{OC} and I_{SC} can be obtained by measuring the I–V curve of a solar cell under standard test conditions (STC), i.e., solar spectral irradiance distribution of AM1.5G, irradiance level of 1,000 W/m^2, and cell temperature of 25 °C. Like in Fig. 7a, a variable resistance R is connected between positive and negative ends of a cell. Under definite irradiance level and temperature, the currents passing through R and voltages at both ends of R are tested, along with the value of R varying from zero (i.e., short-circuit) to infinity (i.e., open-circuit). Then a plot of current–voltage can be obtained. Figure 7b is the schematic diagram of the I–V curve. Any point on the curve is a working point. The connecting line between the original point and the working point is called load line, and the reciprocal of the slope of load line is the value of R. The corresponding values of abscissa and ordinate are the working voltage V and working current I, respectively. The output power of the working point is $P = IV$. By properly choosing a load to R_m in which the area of rectangle is largest, a point M called maximum power point (MPP) can be obtained. I_m, V_m, and R_m, are the optimum operating current, optimum operating voltage, optimum load, respectively. P_m ($= I_m V_m$) is also defined as the maximum power output.

Fill factor (FF) is another key parameter for characterizing solar cells. FF measures the sharpness of the I–V curve and is defined as

$$FF = I_m V_m / I_{SC} V_{OC} \tag{4}$$

The most important aspect of a solar cell is the power conversion efficiency (PCE, η) which is the ratio of maximum power output to the incident power. The η can be described as

$$\begin{aligned} \eta &= \frac{P_m}{P_{in}A} \times 100\% = \frac{I_m V_m}{P_{in}A} \times 100\% = \frac{FF \times V_{OC} \times I_{SC}}{P_{in}A} \times 100\% \\ &= \frac{FF \times V_{OC} \times J_{SC}}{P_{in}} \times 100\% \end{aligned} \tag{5}$$

where P_{in} is the incident power per unit area, A is the area of the solar cell, J_{SC} is the short-circuit current density which is the I_{SC} per unit area.

The basic parameters of solar cells, like I_{SC}, V_{OC}, FF, η, and P_m can be obtained from the I–V curve. Optical properties and electrical properties are optimized to improve the efficiency of a solar cell. The steadily increasing bulk carrier lifetimes of c-Si wafers for application to commercial c-Si solar cells make recombination at the cell surfaces and at the contacts the major fundamental limitation in c-Si cells. The surface passivation based on dielectric layers such as silicon oxide, silicon nitride, aluminum oxide and stacks thereof, has led to the development of passivated emitter and rear cells (PERCs) [24–26]. The passivation based on carrier-selective passivation layers which provide an effective passivation of non-contacted as well as contacted areas of a c-Si solar cell, increases the efficiency potential of c-Si solar cells significantly [27]. The so-called "Tunnel Oxide Passivated Contact (TOPCon)" solar cells which are passivated by a stack of silicon oxide and polycrystalline silicon (poly-Si) [28], is a type of carrier-selective passivation solar cells. Other excellent carrier-selective contacts based on hydrogenated amorphous silicon (a-Si:H) layers are well known as HIT (heterojunction with intrinsic thin layer) [29, 30], which have led to the current record efficiency for c-Si solar cells of 27.3% [31]. In the next sections, the silicon thin films in TOPCon and silicon heterojunction (SHJ) solar cells are discussed.

4 Passivating Contact Solar Cell Technology

In conventional c-Si homojunction solar cells, electron and hole-selective contacts are realized by implementing n-type and p-type doping on both sides of the absorber layer of the c-Si substrate. The direct contact between metal and silicon is inevitably utilized in the localization where carriers need to be exported out of the cell device, and the contact exhibits significant carrier recombination current (J_0) in the range of 10^3–10^5 fA/cm^2, which constitutes the primary limitation to cell efficiency improvement [32, 33]. To mitigate the adverse effects of Schottky contact in the metal region, the concept of passivated contact is introduced. Passivated contact structures comprise of thin film materials such as poly-Si, a-Si:H, silicon oxide, and transition metal oxides that are inserted between the metal electrode and the c-Si substrate. A passivation contact is a structure that facilitates the passage of one type of carrier while impeding the transport of another type of carrier, as evidenced by its differential conductivity for different carrier types.

Passivating contact solar cells generally include SHJ and TOPCon c-Si solar cells, as well as other novel cell structures. Figure 8a illustrates the basic principle of passivation contact for the distribution of the density of defect states (D_{it}) and fixed charge density (Q_f) at the interface of thin film materials [34]. It is observed that a-Si:H and SiO_2 exhibit lower D_{it} values, both below 10^{11}/cm^2 eV, which enables superior chemical passivation; while Q_f serves as the source of field passivation, a layer of positively charged fixed charge exists within a 2 nm range on the SiO_2 side of the c-Si/SiO_2 interface, which is insensitive to the position of the Fermi energy level, and the charge density generally falls within the range of 10^{11}–10^{12} cm^{-2}. If Q_f is reduced from 2×10^{12} to 1×10^{12} cm^{-2}, the corresponding effective surface recombination rate increases by a factor of 8. The value of Q_f is dependent on the surface structure and the preparation process: the fixed charge in the SiO_2 film prepared on different c-Si surfaces varies such that Q_f (100) < Q_f (111); the higher the doping concentration in the silicon near the surface of the SiO_2 layer, the larger the value of Q_f. The a-Si:H deposited by PECVD equipment can be employed for SHJ c-Si solar cells. Also, the a-Si:H can be converted to poly-Si using high-temperature annealing, which constitutes the key material for the preparation of TOPCon c-Si solar cells.

Figure 8b illustrates the fundamental principle of passivated contact with recombination current (J_0) and contact resistivity (ρ_c) distributions [35]. Both P-doped amorphous silicon n-a-Si:H/i-a-Si:H and B-doped amorphous silicon p-a-Si:H/i-a-Si:H achieve a remarkably low J_0 of less than 5 fA/cm^2, whereas P-doped poly-Si/SiO_2 stack can attain a notably low ρ_c of only 0.1 mΩ cm^2. The corresponding J_0 and ρ_c distributions for various combinations of hole-selective and electron-selective contact layer materials are depicted in Fig. 8c [35]. As a result of the optimization of surface passivation and metal electrode/c-Si contact, both SHJ and TOPCon solar cells demonstrate substantially increased efficiency limits, to 28.5% for SHJ solar cells and 28.7% for TOPCon solar cells, which are not significantly different from one another.

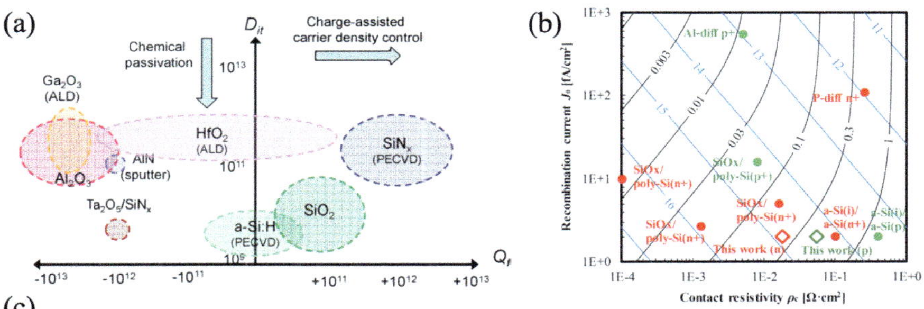

(c)

	Electron selective		P-diffused n+		a-Si:H(i)/a-Si:H(n)		Thermal/PECVD SiO_x/poly-Si(n+)		Thermal/LPCVD SiO_x/poly-Si(n+)		Chemical/LPCVD SiO_x/poly-Si(n+)	
Hole selective			$J_{0,e}$	$\rho_{c,e}$	$J_{0,e}$	$\rho_{c,e}$	$J_{0,e}$	$\rho_{c,e}$	$J_{0,e}$	$\rho_{c,e}$	$J_{0,e}$	$\rho_{c,e}$
			109	0.26	2	0.017	5	0.016	2.7	0.0013	10	0.0001
Al-doped p+	$J_{0,h}$	550	11.7	56.5%	12.9	34.6%	12.8	23.4%	12.9	8.5%	13.0	1.2%
	$\rho_{c,h}$	0.005	3.5%	24.5	1.2%	27.0	1.2%	26.9	1.2%	27.1	1.2%	27.1
a-Si:H(i)/a-Si:H(p)	$J_{0,h}$	2	11.9	45.6%	14.0	24.4%	14.0	14.9%	14.6	5.5%	14.6	0.8%
	$\rho_{c,h}$	0.055	97.9%	24.9	43.9%	28.5	27.6%	28.6	26.2%	28.9	26.0%	28.9
Chemical/PECVD SiO_x/poly-Si(p+)	$J_{0,h}$	16	11.9	46.7%	14.0	23.0%	13.8	15.5%	14.2	5.7%	14.2	0.8%
	$\rho_{c,h}$	0.008	21.4%	24.9	5.9%	28.5	6.1%	28.4	5.8%	28.7	5.7%	28.7

Boxes with blue background: combined selectivity greater than 14
Boxes with pink background: limiting efficiency greater than 28%
Red dotted line: SHJ solar cell
Black dotted line: PERC solar cell
Blue dotted line: p-type TOPCon solar cell
Green dotted line: n-type TOPCon solar cell
Orange dotted line: Bi-facial TOPCon solar cell

Fig. 8 Selective passivation contact fundamentals: **a** interfacial defect states (D_{it}) and fixed charge (Q_{f}) distribution [34], **b** recombination current (J_0) and contact resistivity (ρ_{c}) distribution for different tunneling materials [35]; **c** effect of different passivation contact materials on the cell efficiency limit [35]

5 Silicon Thin Films in Silicon Heterojunction Solar Cells

5.1 SHJ Solar Cells

The design of silicon heterojunction solar cell is based on an emitter and BSF that are produced by low-temperature growth of ultra-thin layers of a-Si:H on both sides of a thin c-Si wafer which is less than 200 μm, where electrons and holes are photogenerated. A schematic structure diagram of a silicon heterojunction solar cell is in Fig. 9a. It can be seen that there are three kinds of silicon thin films, i.e., intrinsic amorphous silicon (i-a-Si:H), p-type amorphous silicon (p-a-Si:H), and n-type amorphous silicon (n-a-Si:H). The SHJ solar cells use c-Si wafers for both carrier transport and absorption, and amorphous and/or nanocrystalline thin silicon layers for passivation and junction formation. The top electrode is comprised of a transparent conductive oxide (TCO) layer in combination with a metal grid. The amorphous/crystalline silicon heterojunction solar cell combines silicon wafer and thin film technologies. Actually, the idea of making solar cells from silicon heterojunction is a rather old one which was published in 1974 by Fuhs et al. [36]. However, it turned out that to realize $V_{OC} > 700$ mV inherent to the heterojunction concept, it is mandatory to include an additional, very thin (of the order of 10 nm) undoped a-Si:H buffer layer between the wafer and doped a-Si:H layers (emitter or BSF). The reason is that the defect density in a-Si:H increases strongly with doping, which leads to high interface defect density at the a-Si:H/c-Si junction, resulting in enhanced recombination and a lower V_{OC}. This finding is the essence of "HIT" concept which is to insert an intrinsic a-Si:H layer between c-Si wafer and doped a-Si:H [29, 30].

In recent years, the structure of SHJ has been developed into a back junction (BJ), i.e., the emitter on the back of the cell. At the same time, the doped layers are nanocrystallized, i.e., n-type hydrogenated nanocrystalline silicon oxide film (n-nc-SiO_x:H) as window layer and p-type hydrogenated nanocrystalline silicon film (p-nc-Si:H) as the emitter. The schematic structure of BJ-SHJ solar cell can be found in Fig. 9b [37]. Figure 9c shows the *I–V* curve of a BJ-SHJ cell with 26.81% efficiency certificated by the Institute für Solarenergieforschung (ISFH) [37]. There are four major manufacturing processes for SHJ solar cells. The c-Si wafer surface is randomly textured firstly. Secondly, the intrinsic a-Si:H (passivation layers) and doped silicon thin layers (emitter and BSF) are formed by depositing a-Si:H and/or nanocrystalline silicon layers (nc-Si:H) at a temperature below 250 °C in a PECVD

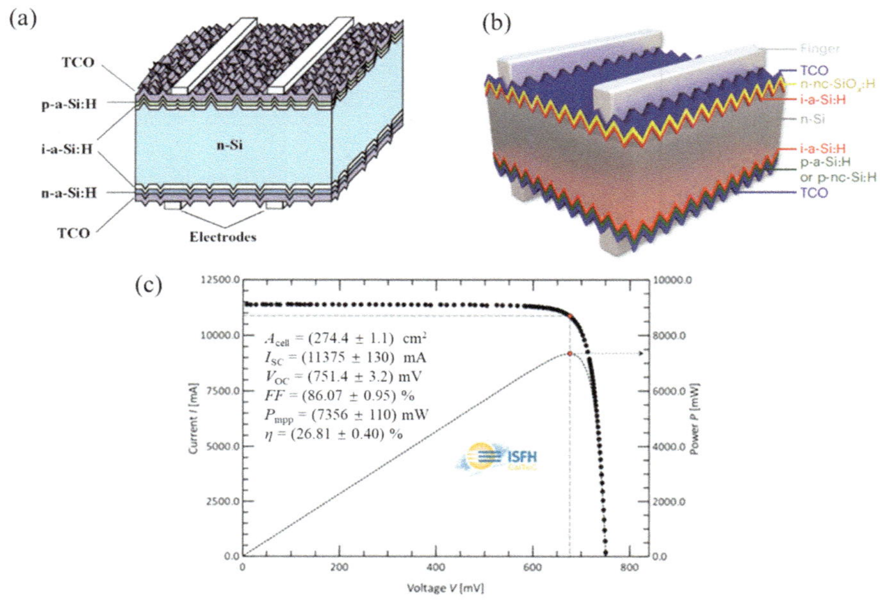

Fig. 9 **a** Schematic structure diagram of SHJ solar cell [30], **b** schematic structure diagram of BJ-SHJ solar cell [37], **c** *I–V* curve of a BJ-SHJ cell with 26.81% certified efficiency [37]

process. Thirdly, TCO layers are deposited by physical vapor depositing (PVD). Fourthly, silver pastes are screen printed and cured to form grids.

SHJ solar cells have several advantages in comparison to those of homojunction c-Si cells. The structure of SHJ solar cell is symmetric, and the process temperature of SHJ solar cell is typically less than 250 °C, while those of homojunction c-Si solar cells are typically higher than 800 °C. The wafer bowing can be suppressed due to the symmetric structure and low-temperature processes, which enables the use of thinner wafers for cost reduction. The V_{OC} of SHJ solar cell is higher than conventional c-Si cell because of excellent surface passivation. SHJ solar cells also exhibit a smaller drop in performance with increasing temperature in comparison with conventional c-Si cell, which enables SHJ solar cell to generate more power, especially on hot days. The processes of SHJ solar cells include only four major steps, which are simpler and shorter than those of homojunction c-Si cell. However, the processes of SHJ solar cell are totally different from those of homojunction c-Si cells, causing it to require new capital investment for SHJ cell manufacturing.

5.2 *Preparation Methods of Silicon Thin Films*

Methods of depositing thin films are generally classified as two types, (1) physical deposition, such as evaporation and sputtering, and (2) chemical deposition, like the chemical vapor deposition (CVD). Normally, chemical deposition is applied to obtain the hydrogenated Si thin films. PECVD and hot-wire CVD (HWCVD) are usually used in SHJ processes for depositing Si thin films.

Deposition of Si thin layers, whether they are intended to be used as passivation, emitter or BSF layer, or for material characterization, is usually performed in a parallel plate, mostly capacitively coupled, PECVD system. Figure 10a is the schematic diagram of a PECVD system. It includes a plasma excitation source, matcher, loading system for substrate, heater, vacuum system, gas delivery system and pumping system. In this technique, a plasma occurs during the decomposition of the gaseous precursors, which can be silane (SiH_4), hydrogen (H_2), methane (CH_4), and doping gases such as phosphine (PH_3) and diborane (B_2H_6). During the decomposition, inelastic collision between highly-energetic electrons and the gaseous precursor atoms result in a dissociation into atomic and ionic species [38]. The pathways for the chemical reaction of SiH_4 and its plasma products occurring during the operation of the PECVD system can be found in the literature [39].

The RF (radio frequency) PECVD system operates at a fixed frequency of 13.56 MHz or is driven by a very-high-frequency (VHF) generator, at frequencies ranging up to 110 MHz. Changing the excitation frequency does not necessarily lead to higher deposition rates, but possible changes in the microstructure are likely. The other main factors that affect the deposition rate and structure of resulting Si thin films include plasma excitation power, electrode interspacing, deposition pressure, gas concentration, gas flow, and deposition pressure.

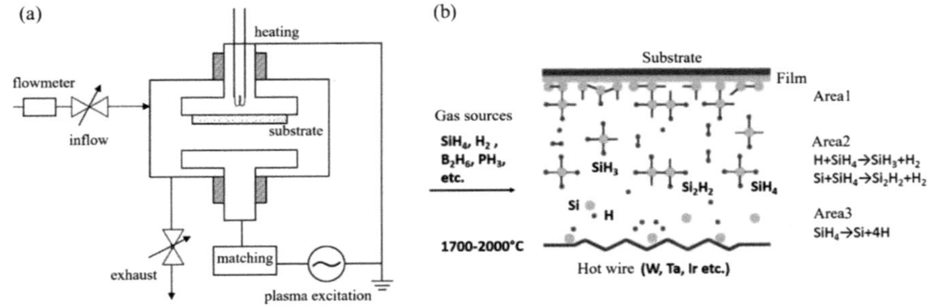

Fig. 10 **a** Schematic diagram of PECVD system, **b** schematic diagram of SiH_4 chemical reactions in HWCVD system [40]

HWCVD is also called catalytic CVD. It is believed that a hot-filament, normally a metal filament (like W, Ta, Ir), acts as a catalytic to decompose the process gas. Figure 10b is the schematics of the SiH_4 chemical reactions in HWCVD for Si thin films [40]. There are three regions in the sketch. The hottest region (Area 3) is near the filament. Typical filament temperatures used in HWCVD roughly vary between 1700 and 2000 °C. Silane decomposes into Si and H when the molecule moves close to the hot wire. In Area 2, the temperature gradient region is between the filament and the substrate. Here, atomic Si and H react with SiH_4 in the gas phase. The majority radials in this region are Si_2H_2 resulting from Si and SiH_4 reaction and some of SiH_3 from H and SiH_4 reaction. The last region (Area 1) is at the substrate surface, where the temperature is constant. Temperature at the surface, to a large degree, determines the structure of the film. Surface reactions and the subsurface release of hydrogen take place at the substrate, after which the silicon matrix relaxes, and the layer is formed. HWCVD is quite different from thermal CVD in that a hot surface absorbs the radicals and deposits the film. There is no film deposition on the hot filament.

HWCVD has several potential advantages compared to PECVD methods. First of all, there are no limitations to the scalability of this technique due to the finite wavelength effect, which is the case in PECVD methods. Secondary, no ions are bombarding the growing film, which could lead to detrimental film properties or damage the film substrates, Thirdly, the substrate holder is not part of the decomposition mechanism, which facilitates the use of moving substrates. Finally, HWCVD has a higher gas utilization compared to PECVD techniques and higher deposition rates can be achieved. Two important drawbacks of the technique are (i) filament ageing and (ii) limited independent control of the substrate temperature due to radiation from the filaments [41].

5.3 Intrinsic Amorphous Silicon Films in SHJ Solar Cells

Figure 11 is the atomic structure sketches of crystalline silicon (c-Si), amorphous (a-Si,) and hydrogenated (a-Si:H) [42]. The well-defined band structure of c-Si principally stems from their lattice periodicity. The diamond like crystal structure results from the formation of strongly directional covalent hybrid bond orbitals. To create a surface, bonds need to be broken, thus the so-called dangling bond (DB) are created. The DBs of c-Si can only exist at the dislocation line (Fig. 11a). For the a-Si, the DBs are formed because of the non-perfect network structure (Fig. 11b), i.e., short-range order while long-range disorder of Si atoms. In Fig. 8c, the DBs are passivated because of the incorporation of H atom in a-Si:H.

On the surface of c-Si, an atom gives rise to a half-saturated DB. The electronic structure of DB gives rise to a surface state within the bandgap [43], which is called recombination center. To resolve surface recombination through DBs, a layer of intrinsic hydrogenated amorphous silicon film is inserted between doped a-Si:H and c-Si to passivate the interface in SHJ solar cell. The physical passivation mechanism of i-a-Si:H stems from chemical surface passivation by hydrogen, similar to defect passivation in the a-Si:H bulk. Intrinsic amorphous silicon films have been known already for some decades to yield good c-Si surface passivation [44].

The DB density changes from 10^{15} to 10^{8} cm^{-2} after i-a-Si:H passivation, thus the silicon wafer gets a longer effective carrier lifetime (τ_{eff}) [45], which may result in a large V_{OC} of SHJ solar cell. This is the reason that the structure of i-a-Si:H film is vital to the V_{OC} and efficiency of SHJ solar cell. Fourier transform infrared absorption (FTIR) spectrum can be used to analyze the structure of a-Si:H film. In the past, the i-a-Si:H was one layer, resulting in the highest efficiency of 25.1% SHJ solar cells [45]. Nowadays, the i-a-Si:H is bilayers or multilayers. Figure 12a, b are the FTIR spectra of i-a-Si:H bilayers, i.e., the buffer layer and normal i-a-Si:H layer [9]. The FTIR spectra are fitted with two Gaussian-distributed absorption peaks centered at wavenumbers of 2000 cm^{-1} and 2100 cm^{-1}, representing monohydride (Si–H) bond and dihydride (Si–H_2) bond stretching modes, respectively [46]. The relative intensities of Si–H and Si–H_2 absorption for buffer (Fig. 12a) and normal

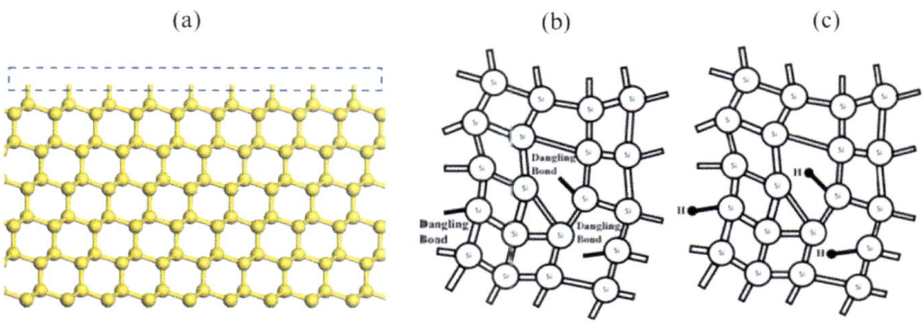

Fig. 11 Structure sketches [42] of **a** c-Si, **b** a-Si, and **c** a-Si:H

i-a-Si:H (Fig. 12b) layers are significantly different. It is observed that there is a greater contribution of Si–H_2 mode in the overall absorption for the buffer layer, implying a higher void density and more disordered hydrogen incorporation within the film [47]. The microstructure factor R^*, is defined as

$$R^* = \frac{I_{2100}}{I_{2000} + I_{2100}} \tag{6}$$

where I_{2100} and I_{2100} are the integral intensity of peak 2000 and 2100 cm^{-1}. The hydrogen in a-Si:H exists mostly in Si–H_2 form if R^* is large, which means that the a-Si:H is porous and has higher void density and richer H content. The a-Si:H film with large R^* is beneficial to passivation of a-Si:H/c-Si interface, while it is not good for the performance of SHJ solar cell. The buffer with high R^* is in a thickness of 0.5–1.0 nm. Then a normal i-a-Si:H with lower R^* which is relatively dense, with lower void density and lower H content, is deposited on the buffer. Furthermore, the normal i-a-Si:H can be divided into two layers, i_1-a-Si:H with moderate R^* and i_2-a-Si:H with lowest R^*. The FTIR spectra and fitness of the triple-layer of i-a-Si:H

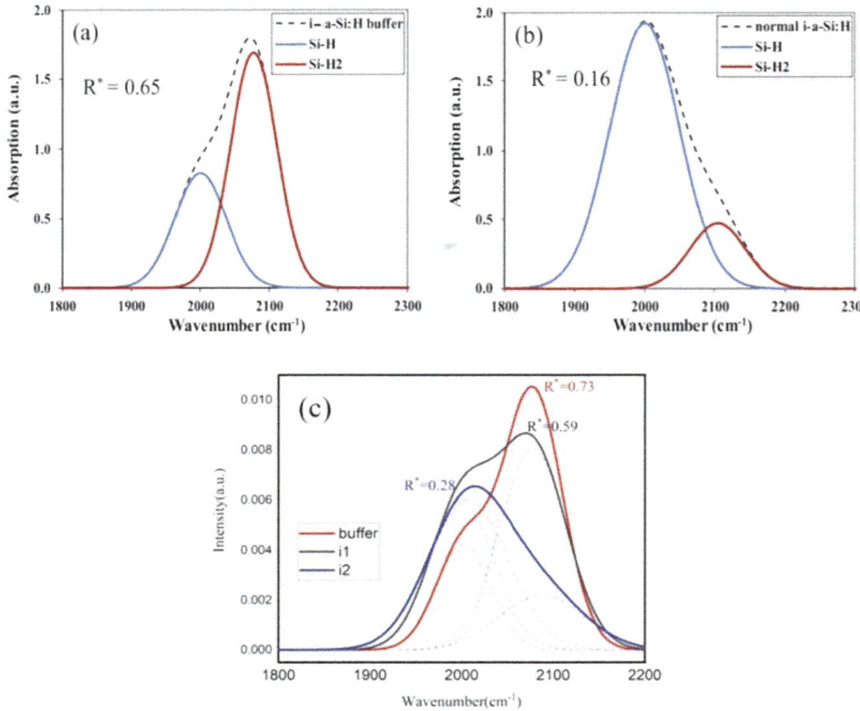

Fig. 12 FTIR spectra fitted by two Gaussian-distributed absorption peaks for **a** buffer, **b** normal i-a-Si:H [9], and **c** triple-layer of i-a-Si:H [48]

[48] is in Fig. 12c. The V_{OC} of around 750 mV for SHJ solar cells can be achieved [37, 48] by using a bilayer or triple-layer structure for i-a-Si:H at the a-Si:H/c-Si interface, which demonstrates the excellent passivation using i-a-Si:H.

A necessary condition for the improvement of i-a-Si:H/c-Si interface passivation is that the interface is atomically sharp, i.e., abrupt, which means that no epitaxial film should be present at the interface [43]. Controlling the thickness of the buffer layer in a range of 0.5–1.0 nm can lead to an abrupt interface. An i-SiO_x:H buffer layer instead of i-a-Si:H buffer can suppress epitaxy [37, 49]. By using a two-stage composite gradient layer of oxygen-containing amorphous silicon sublayer (a-SiO_x:H) and i-a-Si:H, i.e., i-a-SiO_x:H/i-a-Si:H stack, an extremely high V_{OC} of 761 mV was obtained [49].

In brief, intrinsic hydrogenated amorphous film is a key point for highly efficient SHJ solar cells. The i-a-Si:H near the c-Si interface should be porous and with high H content to ensure excellent passivation. A bilayer or triple-layer of i-a-Si:H is necessary to realize the passivation, and epitaxy growth of i-a-Si:H should be avoided.

5.4 Doped Silicon Films in SHJ Solar Cells

To fabricate heterojunction devices, doped films are required to form the emitter and BSF (or front surface field, FSF). Doped a-Si:H layers are usually deposited in similar plasma systems as the intrinsic layers, where for p-type layers either trimethylboron (TMB) or B_2H_6 is mixed in the SiH_4 gas flow, and for n-type films, PH_3 is used. These dopant gases are generally strongly diluted in H_2.

Doping occurs by the incorporation of substitutional impurities. For any semiconductor, intentional doping turns out to form defects; this is a fundamental nature of a semiconductor. It is generally considered that Si–H bond rupture results in defect formation. The formation of defects can be expressed as [50]

$$\mathrm{Si{-}H} \leftrightarrow \mathrm{Si{-}H{-}Si} + \mathrm{Si_{DB}} \tag{7}$$

where Si_{DB} means dangling bond.

Although doped films principally produce a field effect at the interface with c-Si wafer, their electronic passivation properties are often found to be inferior to those of intrinsic films, because the defects formation can lead to reduced passivation at the a-Si:H/c-Si interface. An example of the difference in passivation quality between intrinsic and doped a-Si:H films can be seen in Fig. 13a [51]. In Fig. 13a, τ_{eff} of samples with layers of device-relevant thicknesses deposited in the processing sequence are shown. The structures of samples are i-a-Si:H/c-Si/i-a-Si:H, n-a-Si:H/i-a-Si:H/c-Si/i-a-Si:H/p-a-Si:H, and n-a-Si:H/c-Si/p-a-Si:H, respectively. All results represent films in their as-deposited state (without any postdeposition annealing), deposited on random-pyramid textured c-Si wafers of 200 μm thickness. Despite higher defect densities in progressively thinner films, minority carrier lifetimes as high as 7 ms (at an excess carrier density of 10^{15} cm^{-3}) were obtained with intrinsic films as thin as 15 nm. The excess carrier density of the sample under 1-sun illumination at V_{OC} conditions is marked by an open circle in the figure. The implied-V_{OC} (iV_{OC}) is defined as the energetic distance between the electron and hole quasi-Fermi levels. An implied-V_{OC} value of 738 mV is obtained at 1 sun for the intrinsic layers shown. Doped films deposited directly on c-Si wafer surfaces provide much poorer passivation. Figure 13a shows a lifetime curve for a sample with 15 nm thin p- and n-type layers deposited on the wafer surfaces. At an excess carrier density of 10^{15} cm^{-3}, a carrier lifetime of less than 0.1 ms is obtained. Under 1-sun illumination, this corresponds to an implied-V_{OC} value of only 613 mV. Similarly, SHJ devices with doped films deposited directly either on n- or p-type c-Si surfaces were limited by their low V_{OC} values. Such drastic passivation loss is almost certainly related to doping-related defect generation in the amorphous host matrix. This effect is most severe for p-type films, but can also play a role in n-type films. Note that this effect is not so much caused by the presence of dopant atoms in the a-Si:H material, but rather due to the shift of the Fermi level away from midgap [52]. Such a shift can significantly lower the formation energy of native defects that counteract intentional doping. Increased doping may thus lead to higher defect densities, which ultimately pin the Fermi level.

Due to such defect formation, it is challenging to simultaneously fulfill both the surface passivation and doping requirements. For this reason, a few-nanometer-thick intrinsic layer is typically inserted between the c-Si surface and the doped a-Si:H films for device fabrication, as was first demonstrated by Sanyo [29]. The benefit of inserting an intrinsic layer underneath the doped layers is clearly demonstrated by the data in Fig. 13a for a-Si:H stacks with total thicknesses of 25 nm. Again, on one wafer side a p-a-Si:H layer was deposited, while the other side received an n-type film, making this asymmetric structure a SHJ solar cell precursor. At an excess

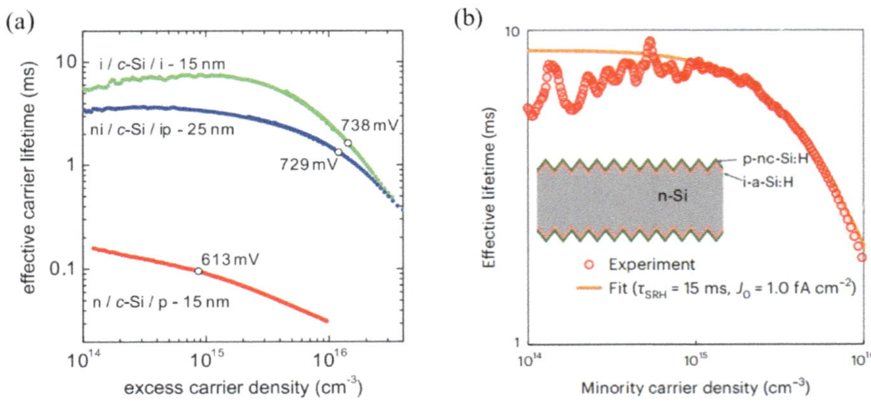

Fig. 13 **a** Effective carrier lifetime and implied-V_{OC} of passivated c-Si wafers consisting of device-relevant a-Si:H stacks [51]. **b** Effective carrier lifetime and structure of p-nc-Si:H/i-a-Si:H/n-Si [37]

carrier density of 10^{15} cm^{-3}, a carrier lifetime higher than 3 ms is obtained. Under 1-sun illumination, this yields an implied-V_{OC} value of 729 mV. The slightly lower passivation quality compared to the case of intrinsic a-Si:H films without doped overlayers may be explained either by defect formation in the intrinsic layer induced by the p-type overlayer, or by probing of the electron wavefunction through the ultra-thin i-layers into the defective doped overlayers.

However, the conductivity of doped a-Si:H is not good enough to transport carriers. The replacement of the p-a-Si:H layer with a p-nc-Si:H layer improves the passivation quality and reduces ρ_c values at both the p layer/i-a-Si:H/n-Si contact and the TCO/p layer interface. The implementation of p-nc-Si:H together with matched TCO leads to a dramatic reduction in the contact resistivity at the rear side, resulting in an efficiency increase in BJ-SHJ cell [37]. Figure 13b is a symmetric structure used to characterize J_0 for the structure based on p-nc-Si:H. The transient mode of a Sinton measurement was used for testing minority carrier lifetime, and the results are indicated by red circles. Tuning the parameters of Shockley–Read–Hall (SRH) recombination (τ_{SRH}, reflecting the quality of silicon wafer) and J_0 (representing the quality of surface passivation), a fitted total J_0 of 1.0 fA/cm^2 is obtained. Considering the symmetry of the structure, it should be 0.5 fA/cm^2 for the J_0 of the p-nc-Si:H/i-a-Si:H/n-Si.

The defect formation because of doping is unavoidable, but its negative impact to the passivation is limited if i-a-Si:H is inserted between c-Si and doped layers. Therefore, the topic for doped a-Si:H is to optimize its optical and electrical properties, not the passivation. However, the lifetime of minority carriers generated in the a-Si:H layers—particularly in the doped layers—is very short so light absorption in these layers is mostly parasitic [51]. This is not a problem at the rear of SHJ cell since the c-Si wafer absorbs all visible light, but light absorbed in the a-Si:H stack at the front of the cell leads to losses of short-circuit current density J_{SC}. The absorption in a-Si:H films in SHJ solar cell is the main factor that results in lower J_{SC} in comparison with TOPCon cell. To minimize the effect of absorption of a-Si:H, the simplest way is to decrease the thickness of a-Si:H layers. However, the a-Si:H layers should be sufficiently thick to achieve passivation and desired field effect.

To improve the J_{SC} of a-Si:H/c-Si heterojunction solar cells further, it is preferable to employ Si-based film as window layer that has larger optical bandgap than a-Si:H to suppress light absorption or to reduce recombination. The doped Si-based films with wide-gap, such as hydrogenated amorphous silicon carbide (a-SiC:H), hydrogenated amorphous silicon-oxygen alloy (a-SiO$_x$:H), and the nanocryatallized Si-based films like nc-Si:H, nc-SiO$_x$:H, nc-SiC:H, are all studied.

Investigating the BJ-SHJ structure (see Fig. 9b) which is the mainstream of SHJ nowadays, n-nc-SiO$_x$:H is employed as a window layer. The external quantum efficiency (EQE) of n-nc-SiO$_x$:H and n-a-Si:H are shown in Fig. 14a. It can be seen that the parasitic absorption of n-nc-SiO$_x$:H is less than n-a-Si:H, as illustrated by the improved short-wavelength response. The higher EQE at short-wavelength increases the J_{SC}. Similarly, replacing the p-a-Si:H by p-nc-Si:H, the long-wavelength response is enhanced (Fig. 14b), which is beneficial to J_{SC} improvement. In addition to the improved J_{SC}, the application of doped nc-Si:H or nc-SiO$_x$:H is also related, to a

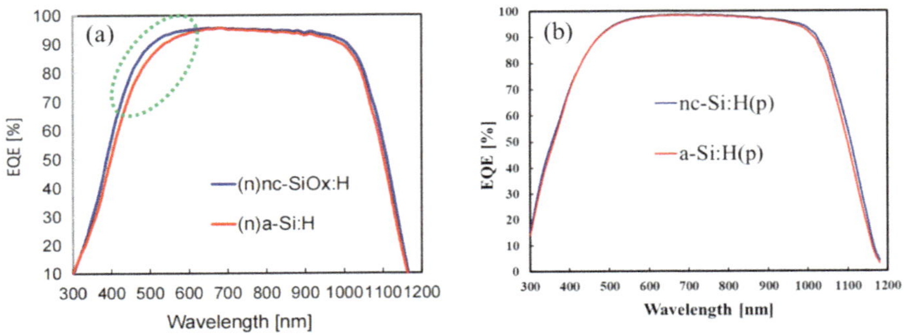

Fig. 14 **a** EQE of n-nc-SiO_x:H versus n-a-Si:H, **b** EQE of p-nc-Si:H versus p-a-Si:H [48]

certain extent, to the achieved high *FF* in SHJ solar cells [53]. The nc-Si:H or nc-SiO_x:H has a higher doping efficiency than a-Si:H, and the dark conductivity can be higher by more than two orders of magnitude than that of doped a-Si:H. The higher doping efficiency also leads to lower series resistances and an adequate nc-Si:H (or nc-SiO_x:H)/TCO electronic contact. The carrier transport properties can be improved because of better conductivity, lower series resistances, and lower contact resistivity with TCO, and hence a higher *FF* of the SHJ devices.

The application of triple-layer i-a-Si:H as passivation layer, n-nc-SiO_x:H as the window layer, p-nc-Si:H as emitter, plus the improved performance of TCO layers and fine screen-printing grids, results in SHJ solar cell with 26.81% efficiency [37]. In 2024, Trina Solar reported an efficiency as high as 27.08% based on n-type SHJ solar cell with fully passivated contact [54]. This is the first time a solar cell efficiency exceeding 27% is achieved using bifacial passivated contact. Besides the excellent passivation of i-a-Si:H films, the key points for improvement of optical and electrical properties of SHJ solar cell include: (1) the optimization of back passivated contact, (2) deposition of doped nanocrystalline silicon by PECVD generated by multi-frequency RF (including VHF) to control the structure and growth rate of doped nc-Si:H, and (3) ultra-fine screen-printed grids.

5.5 Development of HBC Solar Cells

SHJ solar cells have achieved high efficiency due to the application of i-a-Si:H as excellent passivation layers. However, the lower J_{SC} in comparison to TOPCon cell is a drawback for further efficiency improvement. Silicon solar cells usually have grids on both sides so that they are front- and back-contact cells. The grids on the sunny side obstruct light, thus reducing energy input. Placing both terminals on the shaded side creates the so-called back-contact (BC) solar cells that are potentially more efficient. In fact, solar cells with IBC grids, in which the grids are interdigitated on the back side, were introduced in the 1970s [55]. SunPower, established in 1985,

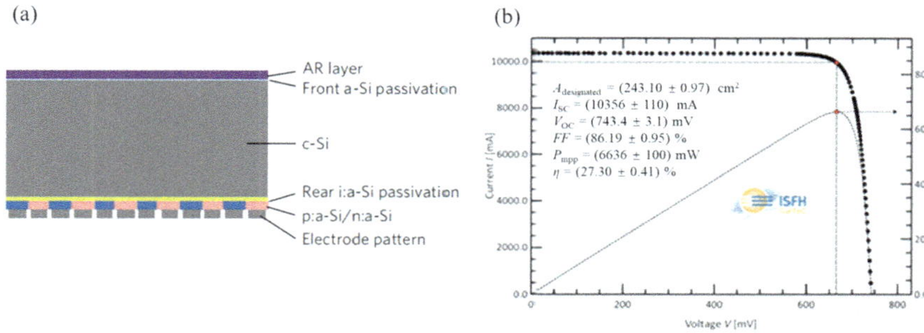

Fig. 15 **a** Schematic structure diagram of HBC solar cell [59], **b** *I*–*V* and power–voltage (*P*–*V*) curves of the 27.3% efficiency HBC cell [31]

became a pioneer in commercializing IBC technology. The IBC structure can also be used together with the SHJ structure to form a heterojunction back-contact (HBC) cell structure which was first developed by Lu et al. in 2007 [56]. Subsequently, companies like Sharp [57], Panasonic [58] and Kaneka [59] have improved HBC technology. In 2017, Kaneka achieved a record 26.7% power conversion efficiency for Si solar cells with the HBC structure [58]. The schematic structure diagram of HBC solar cells can be seen in Fig. 15a [59].

A big challenge in making HBC solar cells is the patterning on the back side, which usually introduces process complexity and power output losses. Although potential solutions have been explored, such as photolithography, inkjet printing, in situ shadow masking and laser patterning, they face technological or cost issues that are still inadequately addressed in mass production. Laser patterning is considered the most economical method for fabricating back-contact solar cells [31]. Pulsed picosecond lasers operating at different wavelengths were used to create the back-contact patterns. Using laser patterning, accompanied with the i-a-Si:H passivation and dense passivating contact technologies, a silicon solar cell with HBC structure has attained an efficiency of 27.3% (V_{OC} = 743.4 mV, J_{SC} = 42.60 mA/cm^2, *FF* = 86.19%) [31]. The *I*–*V* curve of the HBC cell with a 27.3% record efficiency is shown in Fig. 15b.

The low-temperature processes of SHJ and HBC solar cells have a low carbon footprint. Additionally, using laser patterning enhances HBC's performance. The remaining issues are manufacturability and lowering of capital expenditure, to make HBC a prominent terawatt-scale technology for mass production.

6 Silicon Films in TOPCon Solar Cells

The combination of doped poly-Si and tunneling oxide layers has been a focal point of extensive research in both laboratory and industrial settings in recent years. The tunneling oxide plays a crucial role in passivating the c-Si surface, while the doped poly-Si layer selectively allows the passage of a single type of charge carrier. This concept dates back to the 1980s but gained significant momentum in 2013 when it was incorporated into c-Si solar cells, leading to the development of TOPCon solar cells by Fraunhofer ISE [28]. TOPCon solar cells feature a hybrid architecture on an n-type wafer, incorporating a p^+ emitter and an AlO_x/SiN_x bilayer on the front surface. The rear surface consists of a 1–3 nm thin interfacial SiO_x layer and an n^+ poly-Si layer (Fig. 16a). This design is compatible with the established fabrication processes for PERC solar cells, requiring only additional steps to form the interfacial SiO_x layer, deposit poly-Si, and perform high-temperature annealing and hydrogenation (Fig. 16b). Unlike the PERC architecture with partial-area contacts, the full-area contacts on the rear surface of TOPCon solar cells facilitate one-dimensional vertical transport of charge carriers. The current state-of-the-art stack of SiO_x/n^+ poly-Si contacts is achieved through various deposition and doping techniques, with reported J_0 values below 3 fA/cm^2 and ρ_c below 2 mΩ cm^2 [60]. The excellent passivation contact can primarily be attributed to the synergistic effects of chemical passivation of defects at the c-Si surface and field-effect passivation provided by the doped poly-Si layer. The ultrathin interfacial SiO_x layer plays a key role in saturating dangling bonds on the c-Si surface, thereby reducing the density of interface defects and recombination losses. Simultaneously, field-effect passivation is achieved through high-temperature annealing and crystallization of doped amorphous silicon, which activates the dopant atoms and allows them to diffuse from the poly-Si surface into the c-Si absorber. This results in the formation of a shallow junction at the poly-Si/c-Si interface, creating a strong band bending between the interfaces [61] (Fig. 16c). The SiO_x/n^+ poly-Si stack forms a potential barrier that prevents the migration of minority carriers toward the tunneling oxide, thereby ensuring carrier-selective contact [62]. The superior performance of TOPCon solar cells has driven the exploration of optimal process routes and re-parametrization for the deposition of each functional layer, highlighting the complexity and precision required in fabricating high-efficiency devices [63]. Due to the excellent passivation of SiO_x/n^+ poly-Si stack, combined with the optimization of cell processes, a TOPCon cell with the currently highest efficiency of 26.58% is achieved, and the I–V curve can be found in Fig. 16d.

6.1 Ultrathin Interfacial SiO_x Film

The interfacial SiO_x layer plays a crucial role in TOPCon solar cells, acting as a barrier between the poly-Si layer and the c-Si substrate while also reducing carrier recombination. The quality of interfacial oxides is dependent on the method of preparation. Thermally grown oxides outperform photo-oxidized (UV/O_3) oxides in terms of passivation, even at higher annealing temperatures up to 1000 °C [64]. However, high-temperature thermal oxidation will reduce bulk lifetime, particularly in lower-quality silicon wafers, and the associated high thermal budget may limit its industrial scalability. Wet-chemical oxidation methods, such as nitric acid oxidation (NAOS) and ozone-gas oxidation (wet O_3), present more feasible low-temperature

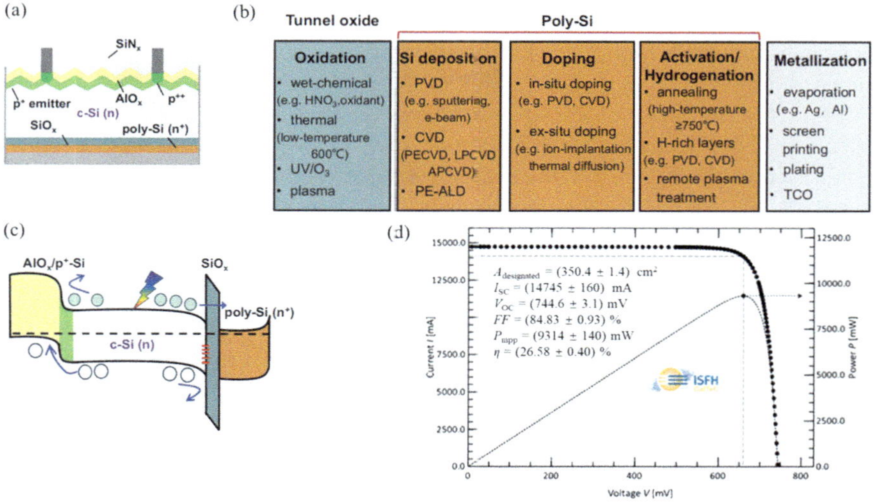

Fig. 16 **a** Structural diagram of a typical n-TOPCon solar cell, showing the layer stack on the front and rear surfaces. **b** Main fabrication steps for TOPCon solar cells, including the deposition of the interfacial SiO_x layer, poly-Si layer, high-temperature annealing, and hydrogenation [63]. **c** Schematic energy band diagram illustrating the band alignment at the poly-Si/c-Si interface and the potential barrier formed by the SiO_x/n^+ poly-Si stack [63]. **d** I–V curve of a TOPCon cell with 26.58% efficiency

alternatives for growing interfacial oxide layers [65]. These methods have achieved good passivation quality, with iV_{OC} values exceeding 730 mV and J_0 values below 10 fA/cm^2. However, they may face compatibility challenges with in-situ doped thin film silicon growth processes. Plasma-assisted nitrous oxide oxidation (PANO), coupled with PECVD technology, shows promising industrial potential, with iV_{OC} values of 747 mV and 742 mV for n-Si and p-Si, respectively [66, 67]. Additionally, plasma-enhanced atomic layer deposition (PEALD) has been developed for growing interfacial SiO_x, offering precise thickness control down to 2 nm. This method has demonstrated exceptional passivation quality for n-TOPCon cells, with a high iV_{OC} of 759 mV, remarkably low J_0 values of 2.8 fA/cm^2, and a correspondingly low ρ_c value of 0.6 mΩ cm^2 [68].

Several fundamental aspects of the interfacial oxide layer remain the subject of ongoing debate, particularly its thickness, stoichiometry, and charge transport mechanisms. Some researchers have proposed that carrier transport through pinholes might also contribute to the overall transport of the oxide [69]. It is well-established that the thickness of the SiO_x layer plays a crucial role in determining the transport mechanism. For ultra-thin SiO_x layers (<1.7 nm), traditional tunneling models can explain the charge carrier transport [70]. However, tunneling alone cannot account for the low junction resistance observed in thicker SiO_x layers (>2 nm). Interestingly, both tunneling and pinhole-assisted transport mechanisms have been shown to coexist in thin SiO_x layers containing pinholes [71]. While both mechanisms are applicable to some extent, their relative contributions depend on the specific arrangement of the junctions. Numerical investigations revealed that direct pinhole transport occurs alongside tunneling charge carrier transport at the poly-Si/SiO_x/Si junctions. As shown in Fig. 17 [72], the morphology of the oxide layer is significantly affected by the annealing process. Transmission electron microscopy (TEM) and energy-dispersive x-ray spectroscopy (EDS) were used to assess film quality and detect pinhole formation. Figure 17f illustrates the formation mechanism of pinholes during the annealing step, where the poly-Si and SiO_x films, along with the Si substrate, expand due to the increasing annealing temperature. The poly-Si and Si substrate expand more than the SiO_x film due to their higher thermal expansion coefficients. Tensile stress leads to concave distortion, eventually causing SiO_x fracture. This fracture, which relieves accumulated tension, forms pinholes. This process is reversed during the cooling phase, where the SiO_x layer contracts less than the poly-Si and Si substrate, leading to increased tensile stress in the SiO_x layer and facilitating further crack propagation, thereby increasing pinhole formation [73].

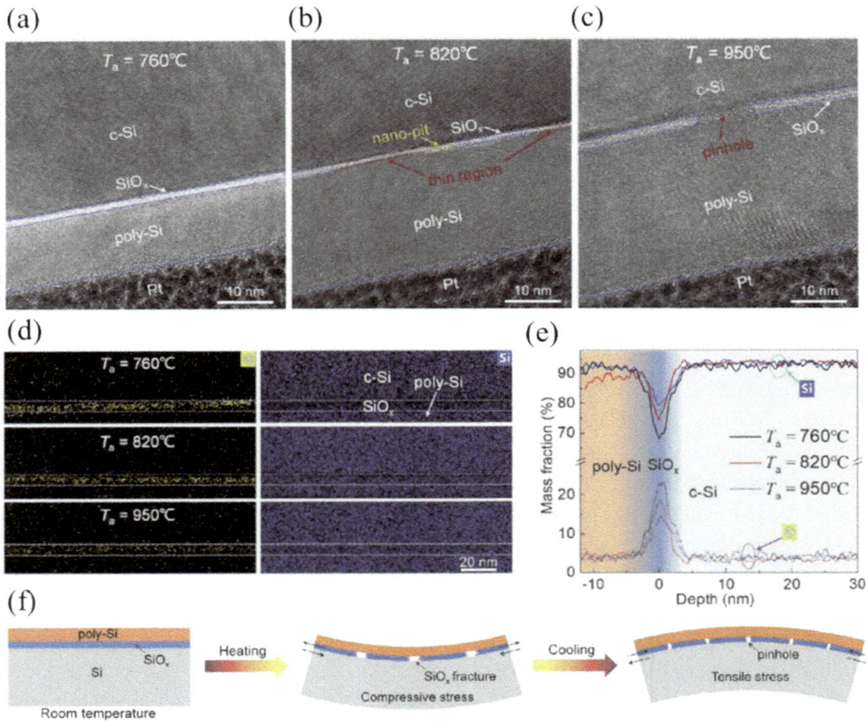

Fig. 17 Appearance of pinholes and their formation mechanism [72]. **a–c** Transmission electron microscope (TEM) images of TOPCon structures after high-temperature annealing at T_a = **a** 760 °C, **b** 820 °C, and **c** 950 °C. **d** Local energy dispersive spectroscopy (EDS) profiles. **e** Variation of the mass fractions of Si and O elements with depth. **f** Schematic diagrams illustrating the mechanism of pinhole formation

6.2 Doped Poly-Si Film

The heavily doped poly-Si layer in TOPCon cells serves several crucial functions, including: (1) Contact: the doping concentration of the poly-Si layer plays a significant role in determining the selectivity of the contact. By adjusting the doping levels, the electrical properties of the contact can be optimized, enhancing carrier collection and reducing contact resistance; (2) Barrier against metallization and diffusion: the poly-Si layer acts as a protective barrier, preventing the destruction of the SiO_x passivation layer during metallization and diffusion processes. This helps to preserve the integrity of the passivation, which is essential for reducing recombination losses and maintaining high efficiency; (3) Gettering: the heavily doped poly-Si layer improves the bulk quality of the silicon wafer by acting as a gettering layer. This process helps trap and neutralize harmful impurities, thereby enhancing the overall material quality and reducing the defect density, which can otherwise degrade cell performance; (4) Field effect: the doping in the poly-Si layer induces a field effect that generates a charge-rich region on the silicon surface. This charge-rich layer can repel one type of carrier (either electrons or holes), which contributes to reducing recombination at the surface and improving the overall device efficiency.

In the preparation of the poly-Si layer for the TOPCon solar cells, the dominant approach in industrial applications is low-pressure CVD (LPCVD) of poly-Si, where silane is thermally decomposed in low-pressure quartz tubes. While LPCVD provides high-quality poly-Si films, its non-directional nature necessitates the removal of the poly-Si layer from one side of the wafer, which can detract from the homogeneity and consistency of the process [74]. In contrast, PECVD is a directional deposition method that relies on the plasma-assisted decomposition of precursor gases at low temperatures, which offers several benefits: faster deposition rates, the ability for in situ doping, manageable wrap-around areas that can be easily removed, and the use of graphite boats that can be periodically cleaned without the need for consumables like quartz tubes. Numerous studies have shown that PECVD can achieve comparable performance to LPCVD [75], making PECVD a viable and competitive route for industrial TOPCon cell manufacturing. Dopant incorporation into silicon layers for TOPCon cells can be achieved through various methods. The simplest in situ doping techniques involve introducing controlled rates of dopants, such as PH_3 for n-type or B_2H_6/BCl_3 for p-type, alongside silane during LPCVD or PECVD deposition. However, while this method allows for direct doping during deposition, it is limited by slower deposition rates for doped silicon compared to undoped silicon, and the lateral homogeneity of the doped films is less than ideal. Thermal diffusion, particularly using $POCl_3$ for n-type doping, is the most widely used ex-situ technique in industrial settings. It allows for the simultaneous recrystallization of the silicon layer and activation of the dopants, eliminating the need for a separate high-temperature process. Extensive research has examined how PECVD deposition parameters, such as temperature and deposition power, influence the structural and electrical properties of poly-Si films. For example, higher deposition temperatures reduce Si–H bond concentrations, resulting in denser films that are more conducive to crystallization during subsequent thermal dopant diffusion processes [76]. However, increasing the thickness of poly-Si films can lead to higher parasitic absorption losses, with a decrease in J_{SC} by approximately 0.15 mA/cm^2 for every 50 nm increase in poly-Si thickness [74].

Doped poly-Si typically exhibits a higher absorption coefficient than c-Si, and its performance is influenced by the doping concentration. Specifically, higher doping levels (above 10^{20} cm^{-3}) can lead to dopant diffusion into the c-Si absorber, which may increase the J_0 and negatively affect the passivation quality [77]. Additionally, high-temperature annealing is often necessary to activate the dopants and promote recrystallization of the poly-Si layer, which is vital for achieving the desired electrical and optical properties. To further improve the optical properties and reduce parasitic absorption, elements like carbon (C), oxygen (O), or nitrogen (N) can be introduced into the poly-Si film. This modification effectively increases the optical band gap of poly-Si, helping to mitigate absorption losses while improving its passivation performance [78]. However, the doping and recrystallization processes generally require specialized, high-cost equipment, and the need for stringent safety protocols due to the involvement of high temperatures and specialty gases. Furthermore, deposition conditions such as gas flow rates, annealing temperatures, and substrate temperatures play a crucial role in determining the quality of the SiO_x/poly-Si junction, doping distribution, and crystallization of the poly-Si layer [79].

6.3 *Development of TOPCon Technology*

Recent advancements in TOPCon solar cell efficiency have been remarkable, with laboratory-scale efficiencies reaching 25.8% and 26.1% for n-TOPCon and p-TOPCon cells, respectively [80]. Industrial-scale cells have also surpassed 25% efficiency since 2023, with a world record for n-TOPCon cells at 26.58% created by Trina Solar in 2024 [81]. However, efficiency gaps primarily arise from the front surface, particularly the boron-doped p$^+$ emitter region. Quantitative analysis shows that this area contributes a cumulative loss of 1.19 mW/cm^2, accounting for 63% of the total losses. Optimizing the front surface contact structure is therefore critical for enhancing the performance of TOPCon cells. Figure 18 illustrates the proprietary laser-enhanced contact optimization (LECO) technology, which has been successfully integrated into the pilot production line since 2019 and subsequently scaled up for mass production in 2023. The LECO process, as illustrated in Fig. 18a, involves applying a reverse bias to an already metalized solar cell [82]. A laser scans the cell surface, creating a high density of electron–hole pairs. These pairs are then directed through tiny contact areas between the silicon and the metal surface, resulting in the formation of unique contact structures named current-fired contact (CFC) as shown in Fig. 18b [83]. Fellmeth et al. [84] applied the LECO technology on TOPCon solar cells, and the results reveal that LECO improves the metal–semiconductor contact resistivity ρ_c on the boron emitter and the n-TOPCon side from an underfired state of 2.9 and 14.1 mΩ cm^2 to an enhanced state of 1.8 and 2.9 mΩ cm^2. Ultimately,

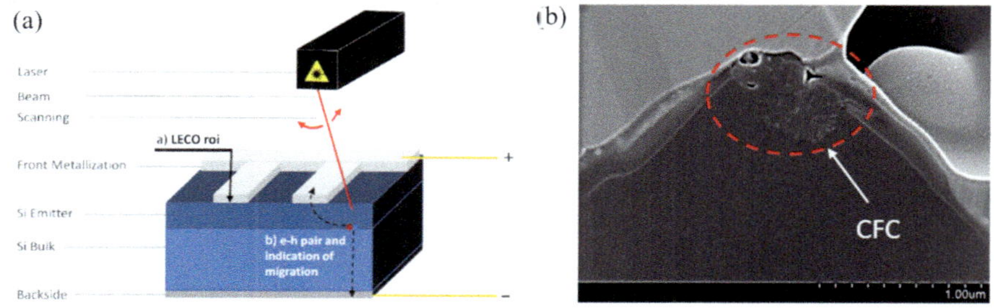

Fig. 18 **a** Schematic diagram of LECO working principle [83]. **b** SEM image of current fired contact (CFC) [84]

LECO improved TOPCon solar cells conversion efficiency by 0.6% compared with the untreated reference.

Other key strategies include adopting a front-surface selective emitter (SE) design, a double-sided TOPCon architecture, and a TOPCon back-contact design (TBC) to minimize efficiency losses [63] (Fig. 19). The contact resistance at the front emitter/metal interface is significant, but it can be substantially reduced using a front-surface SE design. Increasing the doping concentration in the metal contact area helps lower the ρ_c and optimize the J_0 of the front emitter. Another promising option is using a SiO_x/poly-Si passivating contact structure on the front surface, which could push the efficiency to as high as 28.7% [27]. However, careful consideration must be given to the effects of boron doping and the resulting defects. Additionally, the poly-Si layer on the front surface introduces parasitic absorption, requiring limited passivating contact to the front finger area, which increases process complexity. A combined TOPCon and IBC design, the TBC design, is another approach that can further enhance J_{SC} by eliminating the front-contact metal. However, developing metallization techniques compatible with boron-doped poly-Si contacts on the front surface while minimizing costly Ag consumption is a notable challenge. Electroplating technology offers a promising solution by reducing front-surface grid shading, improving cell efficiency, and cutting costs [85]. The choice of metallization scheme is crucial and must align with the doping level and thickness of the poly-Si layer, allowing for adjustments based on the selected metallization techniques and specific requirements.

7 Further Development of Silicon Solar Cells

As the development of TOPCon and SHJ solar cells continues, the highest certified efficiency of single-junction c-Si solar cells has approached their Shockley–Queisser limits of 29.4% [12]. At present, TOPCon is shifted to TBC, while SHJ is shifted to HBC. One trend is that TBC and HBC are merging together as HTBC toward achieving higher efficiency and technological feasibility. On the other hand, to break the efficiency limits of single-junction c-Si solar cells and enrich the application

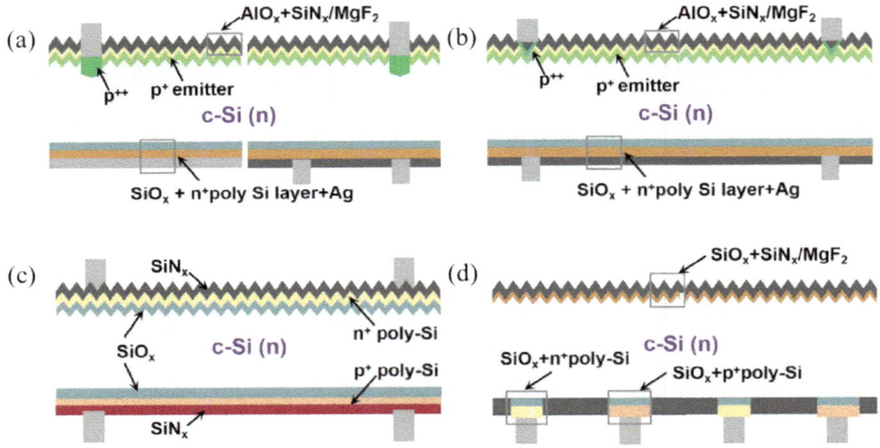

Fig. 19 Development trends of TOPCon solar cells [63], including: **a** optimization of typical n-TOPCon, **b** front-surface SE design, **c** double-sided TOPCon architecture, and **d** TBC design

aspects of PV devices, multi-junction solar cells with a much higher efficiency limit have been intensively studied, of which perovskite/silicon tandem solar cells (TSCs) have attracted special attention because of their potential advantages in efficiency enhancement and cost reduction. In this section, the development trend of silicon solar cells for the above-mentioned two approaches will be discussed.

7.1 TOPCon/Heterojunction Hybrid BC Solar Cells

The i-a-Si:H and doped silicon films in SHJ and HBC solar cells are usually prepared by PECVD. However, the costly use of parallel-plate PECVD limits the spreading and expansion of SHJ and HBC solar cells. Nowadays, TOPCon solar cells with bifacial passivated contact (application of SiO_x/poly-Si on the front side, i.e., poly-finger technology) and with BC electrodes (TBC) are developed. The poly-finger and TBC employ boron-doped polycrystalline silicon (poly-Si(p^+)) as a passivated contacting layer. The preparation of poly-Si(p^+) is not easy, and the passivation of poly-Si(p^+) is not effective. While the i-a-Si:H/p-a-Si:H or i-a-Si:H/p-nc-Si:H in SHJ and HBC solar cells have showed good passivation and technological feasibility, a structure of merging HBC and TBC, which combine p-nc-Si:H or p-a-Si:H with SiO_x/poly-Si(n^+) as p-contact and n-contact, respectively, is developed and called TOPCon/heterojunction hybrid BC (THBC or HTBC) solar cell. A schematic structure diagram of THBC solar cell is shown in Fig. 20a. On the front, the FSF and AlO_x act as passivation layers, while the SiN_x:H is antireflection coating. On the back, c-Si/SiO_x/

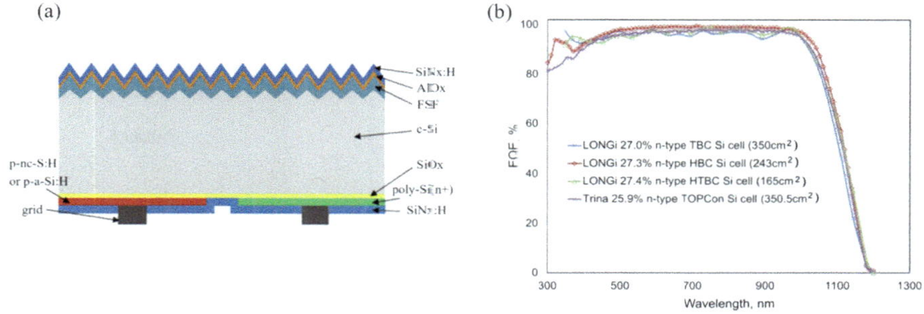

Fig. 20 **a** Schematic structure diagram of THBC solar cell. **b** EQE of THBC cell versus TOPCon, TBC and HBC cells [86]

p-nc-Si:H (or p-a-Si:H) formed the heterojunction and p-type passivating contact, while the SiO_x/poly-Si(n^+) formed the n-type passivating contact.

The THBC solar cell gives a limiting efficiency of 28.9% [35], which shows potential in the future. An efficiency of 27.4% ($V_{OC} = 745.6$ mV, $J_{SC} = 42.35$ mA/cm^2, $FF = 86.7\%$) is reported for a large-area (182×91 mm^2) n-type THBC cell, with rear heterojunction p-type contacts and rear TOPCon n-type contacts, fabricated by LONGi Solar and measured by ISFH [86]. The cell, establishing a new record for silicon, has both polarity contacts on the rear surface, restricting loss by the absence of contacts on the front illuminated surface. Figure 20b illustrates the EQE of a THBC cell compared with those of TOPCon, TBC and HBC cells.

7.2 *Perovskite/Silicon Tandem Solar Cells*

There are many kinds of solar cells. Crystalline silicon solar cells are undoubtedly the dominant ones in the present PV market. Recently, perovskite (PSK) solar cells have attracted much attention, the efficiency of which has been improved from 3.8% [87] to over 26% [88] in the past few years. The combination of perovskite and c-Si, i.e., PSK/c-Si tandem solar cell (TSC) is the most recognized by researchers. The efficiency of PSK/c-Si TSC has soared from 13.7% [13] initially to 33.9% [89] and 34.6% [14] during the past 10 years. According to some studies, the Si bottom-cell with 1.12 eV band-gap stacked with the 1.68 eV PSK top-cell can ensure the optimal absorption of solar spectrum and avoid the thermal relaxation due to the mismatch of band-gap. The theoretical efficiency limit of the full-spectrum matched two-junction PSK/c-Si TSC can exceed 43% [16]. The higher efficiency limit of PSK/c-Si TSC, together with the advantages of the well developed c-Si solar cells technologies, make the TSC widely regarded as one of the most promising future PV technology approaches.

There are currently five options reported in the literature for the preparation of PSK on silicon substrates, including low-cost spin-coating, large-area blade-coating

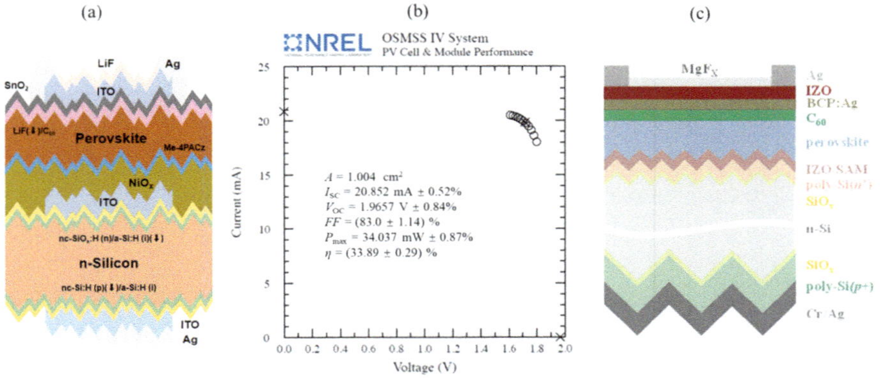

Fig. 21 **a** Schematic structure diagram of PSK/SHJ tandem solar cell, **b** *I*–*V* curve of a PSK/c-Si TSC with 33.89% efficiency [89], **c** schematic diagram of PSK/TOPCon tandem solar cell [91]

and slot-die coating, facile physical stacking, and excellent conformal co-evaporation + solution. It is worth noting that while co-evaporation + solution is suitable for substrates with various roughness, the complex two-step approach can result in high cost and low performance of PSK/c-Si TSC. In order to obtain PSK films with large area, excellent quality and robust stability, it is essential to explore new simple one-step solution schemes for conformal structures, but this process still requires much further studies [90].

Initially, the PSK/c-Si TSC used an SHJ solar cell as bottom-cell. The schematic structure of PSK/SHJ TSC is shown in Fig. 21a. The *I*–*V* curve of a PSK/c-Si TSC with 33.89% efficiency certified by the National Renewable Energy Laboratory (NREL) is shown in Fig. 21b. The highest efficiency of PSK/SHJ TSC is 34.6% at present [14]. TOPCon solar cell is also a candidate as bottom-cell for PSK/c-Si TSC. The schematic structure of PSK/TOPCon TSC is shown in Fig. 21c [91]. The highest efficiency of PSK/TOPCon TSC is 33.84% currently [92].

Despite the breakthroughs in cell efficiency, large-scale manufacturing issues, stability issues, and bifacial properties, further research and development is still needed for the adoption of two-terminal (2T) PSK/c-Si TSC in large-scale mass production.

8 Summary

The usage of renewable energies will promote the progress of worldwide carbon neutrality, and PV energy is one of the most important renewable energies. Solar cells are the key devices to realize the conversion of solar energy into electric power. Among the different types of solar cells, c-Si solar cells are dominant, which currently occupy a market share of more than 95%. Surface passivation is the key point for producing high-efficiency c-Si solar cells. SHJ and TOPCon are the main passivating contact c-Si solar cells.

For SHJ, the key features are that intrinsic hydrogenated amorphous Si films are inserted between c-Si and doped Si layers and the i-a-Si:H/doped Si layers act as passivated contact stack. The i-a-Si H shows excellent passivation capability. The p-type doped Si layer, either p-a-Si:H or p-nc-Si:H, acts as the emitter. For BJ-SHJ, the n-type doped Si layer, either n-a-Si:H or n-nc-SiO_x:H, acts as the window and front surface field layer. Carefully depositing i-a-Si:H, p-a-Si:H or p-nc-Si:H, n-a-Si:H or n-nc-SiO_x:H are crucial for achieving high-efficiency SHJ cells. In the future, optimal performance of i-a-Si:H and doped Si thin films is beneficial to SHJ and promotes carbon neutrality as the market share of SHJ solar cells increases.

For TOPCon, the stack of interfacial oxide SiO_x and doped polycrystalline films (SiO_x/poly-Si) forms as the passivating contact layers. The excellent passivation contact can primarily be attributed to the synergistic effects of chemical passivation of defects at the c-Si surface and field-effect passivation provided by the doped poly-Si layer. The ultrathin interfacial SiO_x layer plays a key role in saturating dangling bonds at the c-Si surface, thereby reducing the density of interface defects and recombination losses. TOPCon solar cells have been the mainstream device of the present PV industry.

It is clear that Si-based thin films are vital for the high efficiency of c-Si passivating contact solar cells. These high-efficiency solar cells generate more PV power. Widespread use of PV power helps to achieve carbon neutrality in the near future.

Acknowledgements Co-Authors ZL, DD, SM, and WS acknowledge the support of the Major State Basic Research Development Program of China (2022YFB4200101), National Natural Science Foundation of China (11834011 and 11974242), Shanghai New Energy Technology Research and Development Project of 24DZ3000900, and Inner Mongolia Science and Technology Project (Grant No. 2022JBGS0036). Co-Author XD acknowledges support from the University of Toledo and 1America Solar, LLC.

Exercises for this Chapter

1. What proportions of installed photovoltaic and wind power capacity will be in achieving the carbon neutrality, analyzed in terms of electricity generation?

Answer: The average annual growth rate of global electricity generation is 2.5%, while the average annual growth rates of photovoltaic and wind power generation in the past fifteen years are 28% and 14%, respectively. In 2023, the global total electricity generation was about 30 trillion kWh, with photovoltaic and wind power generation contributing 1.6 trillion kWh and 2.3 trillion kWh, respectively. It is anticipated that the annual growth rate of global total electricity generation will remain constant, and global total electricity generation will achieve carbon neutrality by 2060, reaching 74.7 trillion kWh. At fifteen-year intervals, the photovoltaic power generation is projected to reach 11.4, 31.5, and 41.6 trillion kWh in 2038, 2053, and 2060, respectively. Consequently, the share of photovoltaic power generation in total power generation is estimated to be 56% when global carbon neutrality is achieved. Similarly, the wind power generation is projected to reach 6.3, 11.3, and 13 trillion kWh in 2038, 2053, and 2060, respectively. The share of wind power in total electricity generation is estimated to be 17% when carbon neutrality is achieved.

2. China's annual electricity consumption is currently 10 trillion kWh. Given the present photovoltaic module efficiency of 20%, what area of photovoltaic power plants would need to be installed in the desert regions of western China to meet this demand?

Answer: The standard intensity of solar irradiation on the Earth's surface is 1000 W/ m^2. Considering the photovoltaic module's photoelectric conversion efficiency of 20%, the power generated by a 1 m^2 module per hour is 0.2 kWh. Disregarding the effects of nocturnal inactivity and diurnal fluctuations in light intensity, and assuming that solar irradiation in the desert area of western China occurs for 7–8 h per day at standard intensity for 365 days annually, a 1 m^2 module installed in this region would generate ~500 kWh of electricity per year. To meet the 10 trillion kWh electricity demand, the installation of ~20 billion m^2 of modules, or 20,000 km^2 of photovoltaic power plants, would be required in the western desert region.

3. What are the primary factors to be considered in designing a high-performance solar cell?

Answer: The design of a high-performance solar cell necessitates consideration of two principal factors: optical and electrical performance. Optical performance primarily involves the reduction of reflection, transmission, and parasitic absorption losses of incident light, thereby maximizing sunlight absorption by the crystalline silicon substrate. This can be optimized through the pyramid texturing, passivation film material selection, layer quantity, refractive index, and thickness. Electrical properties are predominantly associated with minimizing the recombination loss of photogenerated carriers to enhance carrier collection by the metal electrodes. These properties can be optimized through interface chemical passivation, field effect passivation, and selective contact implementation.

References

1. Fortes P, Simoes SG, Amorim F et al (2022) How sensitive is a carbon-neutral power sector to climate change? The interplay between hydro, solar and wind for Portugal. Energy 239:122106
2. Gorjian S, Sharon H, Ebadi H et al (2021) Recent technical advancements, economics and environmental impacts of floating photovoltaic solar energy conversion systems. J Clean Prod 278:124285
3. International Energy Agency (IEA) Renewables 2024, analysis and forecast to 2030. https://www.iea.org/reports/renewables-2024. Accessed 18 Dec 2024
4. Blakers AW, Wang A, Milne AM et al (1989) 22.8% efficient silicon solar cell. Appl Phys Lett 55:1363–1365

5. Tang HB, Ma S, Lv Y et al (2020) Optimization of rear surface roughness and metal grid design in industrial bifacial PERC solar cells. Sol Energy Mater Sol Cells 216:110712
6. Benick J, Hoex B, van de Sanden MCM et al (2008) High efficiency n-type Si solar cells on Al_2O_3-passivated boron emitters. Appl Phys Lett 92:253504
7. Ding D, Du ZR, Shen WZ et al (2023) Laser doping selective emitter with thin borosilicate glass layer for n-type TOPCon c-Si solar cells. Sol Energy Mater Sol Cells 253:112230
8. Reichel C, Granek F, Hermle M et al (2011) Investigation of electrical shading effects in back-contacted back-junction silicon solar cells using the two-dimensional charge collection probability and the reciprocity theorem. J Appl Phys 109:024507
9. Ru X, Qu M, Wang J et al (2020) 25.11% efficiency silicon heterojunction solar cell with low deposition rate intrinsic amorphous silicon buffer layers. Sol Energy Mater Sol Cells 215:110643
10. Lin H, Ding D, Wang ZL et al (2018) Realization of interdigitated back contact silicon solar cells by using dopant-free heterocontacts for both polarities. Nano Energy 50:777–784
11. Green MA, Ho-Baillie A, Snaith HJ et al (2014) The emergence of perovskite solar cells. Nat Photon 8:506–514
12. Shockley W, Queisser HJ (1961) Detailed balance limit of efficiency of p-n junction solar cells. J Appl Phys 32:510–519
13. Mailoa J, Bailie C, Johlin E et al (2015) A 2-terminal perovskite/silicon multijunction solar cell enabled by a silicon tunnel junction. Appl Phys Lett 106:121105
14. LONGi sets a new efficiency world record of 34.6% for silicon-perovskite tandem solar cells. https://www.longi.com/cn/news/2024-snec-silicon-perovskite-tandem-solar-cells-new-world-efficiency/. Accessed 6 Jan 2025
15. Leijtens T, Bush KA, Prasanna R et al (2018) Opportunities and challenges for tandem solar cells using metal halide perovskite semiconductors. Nat Energy 3:828–838
16. Yu Z, Leilaeioun M, Holman Z (2016) Selecting tandem partners for silicon solar cells. Nat Energy 1:16137
17. Wang J, Li J, Zhou Y et al (2021) Tuning an electrode work function using organometallic complexes in inverted perovskite solar cells. J Am Chem Soc 143:7759–7768
18. Bremner SP, Levy MY, Honsberg CB (2008) Analysis of tandem solar cell efficiencies under AM1.5G spectrum using a rapid flux calculation method. Prog Photovolt Res Appl 16:225–233
19. Araújo GL, Martí A (1994) Absolute limiting efficiencies for photovoltaic energy conversion. Sol Energy Mater Sol Cells 33:213–240
20. Yamaguchi M, Dimroth F, Geisz JF et al (2021) Multi-junction solar cells paving the way for super high-efficiency. J Appl Phys 129:240901
21. Sze SM, Ng KK (2007) Physics of semiconductor devices, 3rd edn. Wiley, New Jersey
22. Becquerel E (1839) On electric effects under the influence of solar radiation. Compt Rend 9:561
23. Chapin DM, Fuller CS, Pearson GL (1954) A new silicon p-n junction photocell for converting solar radiation into electrical power. J Appl Phys 25:676–677
24. Green MA, Blakers AW, Zhao J et al (1990) Characterization of 23-percent efficiency silicon solar cells. IEEE Trans Electron Dev 37:331–336
25. Hoex B, Heil SBS, Langereis E et al (2006) Ultralow surface recombination of c-Si substrates passivated by plasma-assisted atomic layer deposited Al_2O_3. Appl Phys Lett 89:042112
26. Green MA (2015) The passivated emitter and rear cell (PERC): from conception to mass production. Sol Energy Mater Sol Cells 143:190–197
27. Schmidt J, Peibst R, Brendel R (2018) Surface passivation of crystalline silicon solar cells: present and future. Sol Energy Mater Sol Cells 187:39–54
28. Feldmann F, Bivour M, Reichel C et al (2014) Passivated rear contacts for high-efficiency n-type Si solar cells providing high interface passivation quality and excellent transport characteristics. Sol Energy Mater Sol Cells 120:270–274
29. Tanaka M, Taguchi M, Matsuyama T et al (1992) Development of new a-Si/c-Si heterojunction solar cells: ACJ-HIT (artificially constructed junction-heterojunction with intrinsic thin-layer). Jpn J Appl Phys 31:3518–3522

30. Mishima T, Taguchi M, Sakata H et al (2011) Development status of high-efficiency HIT solar cells. Sol Energy Mater Sol Cells 95:18–21
31. Wu H, Ye F, Yang M et al (2024) Silicon heterojunction back contact solar cells by laser patterning. Nature 635:604–609
32. Mader C, Muller J, Eidelloth S (2012) Local rear contacts to silicon solar cells by in-line high-rate evaporation of aluminum. Sol Energy Mater Sol Cells 107:272–282
33. Bullock J, Cuevas A, Samundsett C et al (2015) Simple silicon solar cells featuring an a-Si: H enhanced rear MIS contact. Sol Energy Mater Sol Cells 138:22–25
34. Cuevas A, Allen T, Bullock J et al (2015) Skin care for healthy silicon solar cells. In: IEEE 42nd photovoltaic specialist conference (PVSC), New Orleans, LA, USA
35. Long W, Yin S, Peng FG et al (2021) On the limiting efficiency for silicon heterojunction solar cells. Sol Energy Mater Sol Cells 231:111291
36. Fuhs W, Niemann K, Stuke J (1974) Heterojunctions of amorphous silicón and silicon single crystals. In: Proceedings of the conference on tetrahedrally bound amorphous semiconductors, Yorktown, NY, USA, pp 345–350
37. Lin H, Yang M, Ru X et al (2023) Silicon heterojunction solar cells with up to 26.81% efficiency achieved by electrically optimized nanocrystalline-silicon hole contact layers. Nat Energy 8:789–799
38. Fahrner WR (2013) Amorphous silicon/crystalline silicon heterojunction solar cells. Springer, Berlin Heidelberg, pp 17–19
39. Liebermann MA, Lichtenberg AJ (1994) Principles of plasma and discharges and materials processing. Wiley, New York
40. Wang Q (2009) Hot-wire CVD amorphous Si materials for solar cell application. Thin Solid Films 517:3570–3574
41. Schüttauf JWA (2011) Amorphous and crystalline silicon based heterojunction solar cells. Dissertation, Utrecht University
42. Shen WZ, Gao C, Li ZP (2023) Perovskite/silicon-heterojunction tandem solar cells. Science Press, Beijing (In Chinese)
43. De Wolf S (2012) Intrinsic and doped a-Si:H/c-Si interface passivation. In: van Sark WGJHM, Korte L, Roca F (eds) Physics and technology of amorphous-crystalline heterostructure silicon solar cells. Springer, Berlin Heidelberg
44. Pankove JI, Tarng ML (1979) Amorphous silicon as a passivant for crystalline silicon. Appl Phys Lett 34:156–157
45. Adachi D, Hernández JL, Yamamoto K (2015) Impact of carrier recombination on fill factor for large area heterojunction crystalline silicon solar cell with 25.1% efficiency. Appl Phys Lett 107:233506
46. Tsu DV, Lucovsky G, Davidson BN (1989) Effects of the nearest neighbors and the alloy matrix on SiH stretching vibration in the amorphous SiOr: H ($0 < r < 2$) alloy system. Phys Rev B 40:1795–1805
47. Smets AHM, Kessels WMM, van de Sanden MCM (2003) Vacancies and voids in hydrogenated amorphous silicon. Appl Phys Lett 82:1547–1549
48. Yang M, Ru X, Yin S et al (2022) Progress of high-efficient silicon heterojunction solar cell. In: 18th China solar grade silicon and PV power conference, Taiyuan, China (In Chinese)
49. Li Y, Ru X, Yang M et al (2024) Flexible silicon solar cells with high power-to-weight ratios. Nature 626:105–110
50. van de Walle CG, Street RA (1995) Silicon-hydrogen bonding and hydrogen diffusion in amorphous silicon. Phys Rev B 51:10615–10618
51. De Wolf S, Descoeudres A, Holman ZC et al (2012) High-efficiency silicon heterojunction solar cells: a review. Green 2:7–24
52. De Wolf S, Kondo M (2009) Nature of doped a-Si:H/c-Si interface recombination. J Appl Phys 105:103707
53. Shen WZ, Zhao YX, Liu F (2022) Highlights of mainstream solar cell efficiencies in 2021. Front Energy 16:1–8

54. Trinasolar sets a new efficiency world record of 27.08% for n-type HJT solar cell. http://static.trinasolar.com/cn/resources/newsroom/wed-20241218-2000. Accessed 18 Dec 2024
55. Lammert MD, Schwartz RJ (1977) The interdigitated back contact solar cell: a silicon solar cell for use in concentrated sunlight. IEEE Trans Electron Dev 24(1977):337–342
56. Lu M, Bowden S, Das U et al (2007) Interdigitated back contact silicon heterojunction solar cell and the effect of front surface passivation. Appl Phys Lett 91:063507
57. Nakamura J, Asano N, Hieda T et al (2014) Development of heterojunction back contact Si solar cells. IEEE J Photovolt 4:1491–1495
58. Masuko K, Shigematsu M, Hashiguchi T et al (2014) Achievement of more than 25% conversion efficiency with crystalline silicon heterojunction solar cell. IEEE J Photovolt 4:1433–1435
59. Yoshikawa K, Kawasaki H, Yoshida W et al (2017) Silicon heterojunction solar cell with interdigitated back contacts for a photoconversion efficiency over 26%. Nat Energy 2:17032
60. Nandakumar N, Rodriguez J, Kluge T et al (2019) Approaching 23% with large-area monopoly cells using screen-printed and fired rear passivating contacts fabricated by inline PECVD. Prog Photovolt Res Appl 27:107–112
61. Haug FJ, Ingenito A, Meyer F et al (2019) Contributions to the contact resistivity in fired tunnel-oxide passivating contacts for crystalline silicon solar cells. IEEE J Photovolt 9:1548–1553
62. Richter A, Benick J, Muller R et al (2018) Tunnel oxide passivating electron contacts as full-area rear emitter of high-efficiency p-type silicon solar cells. Prog Photovolt Res Appl 26:579–586
63. Li W, Xu ZY, Yan Y et al (2024) Passivating contacts for crystalline silicon solar cells: an overview of the current advances and future perspectives. Adv Energy Mater 14:2304338
64. Vossen RVD, Feldmann F, Moldovan A et al (2017) Comparative study of differently grown tunnel oxides for p-type passivating contacts. Energy Procedia 124:448–454
65. Moldovan A, Feldmann F, Kaufmann K et al (2015) Tunnel oxide passivated carrier-selective contacts based on ultra-thin SiO_2 layers grown by photo-oxidation or wet-chemical oxidation in ozonized water. In: IEEE 42nd photovoltaic specialist conference (PVSC), New Orleans, LA, USA
66. Huang Y, Liao M, Wang Z et al (2020) Ultrathin silicon oxide prepared by in-line plasma-assisted N_2O oxidation (PANO) and the application for n-type polysilicon passivated contact. Sol Energy Mater Sol Cells 208:110389
67. Rui Z, Zeng YH, Guo XQ et al (2019) On the passivation mechanism of poly-silicon and thin silicon oxide on crystal silicon wafers. Sol Energy 194:18–26
68. Liao B, Wu W, Yeo R et al (2023) Atomic scale controlled tunnel oxide enabled by a novel industrial tube-based PEALD technology with demonstrated commercial TOPCon cell efficiencies> 24%. Prog Photovolt Res Appl 31:220–229
69. Green MA (1978) Effects of pinholes, oxide traps, and surface states on MIS solar cells. Appl Phys Lett 33:178–180
70. Steinkemper H, Feldmann F, Bivour M (2015) Theoretical investigation of carrier-selective contacts featuring tunnel oxides by means of numerical device simulation. Energy Procedia 77:195–201
71. Folchert N, Rienäcker M, Yeo AA et al (2018) Temperature dependent contact resistance of carrier selective Poly-Si on oxide junctions. Sol Energy Mater Sol Cells 185:425–430
72. Zhou J, Su X, Huang Q et al (2022) Recent advancements in poly-Si/SiO_x passivating contacts for high-efficiency silicon solar cells: technology review and perspectives. J Mater Chem A 10:20147
73. Yang Z, Liu Z, Cui M et al (2021) Charge-carrier dynamics for silicon oxide tunneling junctions mediated by local pinholes. Cell Rep Phys Sci 2:100667
74. Fırat M, Radhakrishnan HS, Payo MR et al (2022) Large-area bifacial n-TOPCon solar cells with in situ phosphorus doped LPCVD poly-Si passivating contacts. Sol Energy Mater Sol Cells 236:111544
75. Feldmann F, Steinhauser B, Pernau T et al (2020) Industrial TOPCon solar cells realized by a PECVD tube process. In: 37th European photovoltaic solar energy conference and exhibition, Lisbon, Portugal

76. Chen WH, Truong TN, Nguyen HT et al (2020) Influence of PECVD deposition temperature on phosphorus doped poly-silicon passivating contacts. Sol Energy Mater Sol Cells 206:110348
77. Feldmann F, Muller R, Reichel C et al (2014) Ion implantation into amorphous Si layers to form carrier-selective contacts for Si solar cells. Phys Status Solidi RRL 8:767–770
78. Ingenito A, Nogay G, Stuckelberger J et al (2018) Phosphorous-doped silicon carbide as front-side full-area passivating contact for double-side contacted c-Si solar cells. IEEE J Photovolt 9:346–354
79. Du HJ, Liu ZK, Liu W et al (2023) Concurrently preparing front emitter and rear passivating contact via continuous PECVD deposition plus one-step annealing for high-efficiency tunnel oxide passivating contact solar cells. Sol RRL 7:2201082
80. Zheng PT, Yang J, Wang Z et al (2021) Detailed loss analysis of 24.8% large-area screen-printed n-type solar cell with polysilicon passivating contact. Cell Rep Phys Sci 2:100603
81. Trinasolar unveils i-TOPCon Ultra technology, with cell efficiency of 26.58%. https://www.trinasolar.com/en-mea/resources/newsroom/thu-20241128-2000. Accessed 14 Jan 2025
82. Krassowski E, Jaeckel B, Zeller U et al (2021) Reliability evaluation of PV modules fabricated from treated solar cells by laser enhanced contact optimization (LECO) process. Sol RRL 6:2100537
83. Großer S, Krassowski E, Swatek S et al (2021) Microscale contact formation by laser enhanced contact optimization. IEEE J Photovolt 12:26–30
84. Fellmeth T, Höffler H, Mack S et al (2022) Laser-enhanced contact optimization on iTOPCon solar cells. Prog Photovolt Res Appl 30:1393–1399
85. Grubel B, Cimiotti G, Schmiga C et al (2022) Progress of plated metallization for industrial bifacial TOPCon silicon solar cells. Prog Photovolt Res Appl 30:615–621
86. Green MA, Dunlop ED, Yoshita M et al (2025) Solar cell efficiency tables (Version 65). Prog Photovolt Res Appl 33:3–15
87. Kojima A, Teshima K, Shirai Y et al (2009) Organometal halide perovskites as visible-light sensitizers for photovoltaic cells. J Am Chem Soc 131:6050–6051
88. Yang Y, Chen H, Liu C et al (2024) Amidination of ligands for chemical and field-effect passivation stabilizes perovskite solar cells. Science 386:898–902
89. Liu J, He Y, Ding L et al (2024) Perovskite/silicon tandem solar cells with bilayer interface passivation. Nature 635(2024):596–603
90. Gao C, Du D, Ding D et al (2022) A review on monolithic perovskite/c-Si tandem solar cell: progress, challenges, and opportunities. J Mater Chem A 10:10811–10828
91. Jiang S, Ding Z, Li X et al (2024) Advancing monolithic perovskite/TOPCon tandem solar cells by customizing industrial-level micro-nano structures. Adv Funct Mater 34:2401900
92. Jinkosolar sets a 33.84% efficiency record for perovskite/TOPCon tandem solar cell. https://www.jinkosolar.com/site/newsdetail/2448. Accessed 6 Jan 2025

Perovskite Ferroelectric/Relaxor Materials/Systems for Capacitive Energy Storage

Ivana Panzic and Arijeta Bafti

Abstract This chapter explores the potential of perovskite-based ferroelectric, relaxor, and antiferroelectric materials for high-performance capacitive energy storage. It emphasizes the unique dielectric properties of relaxor ferroelectrics, which exhibit high energy densities due to their low remnant polarization and slim hysteresis loops. Lead-based systems like PZT offer high performance but raise environmental concerns, prompting a shift toward lead-free alternatives. The chapter reviews synthesis methods, structural characteristics, and energy storage performance of both lead-containing and lead-free perovskites such as BNT, BT, NN, BMN, and ATN. Techniques like doping, microstructure engineering, and multilayer design are discussed as means to enhance energy density and efficiency. The role of polar nanoregions (PNRs) and phase transitions in optimizing dielectric response is highlighted. Challenges such as improving breakdown strength, thermal/frequency stability, and reducing leakage currents are analysed. The chapter also outlines promising applications, including actuators, sensors, and biomedical devices. Future directions involve high-entropy designs, advanced characterization methods, and integration of computational modelling. Overall, perovskite relaxors are shown to be highly promising for next-generation energy storage systems.

I. Panzic (✉) · A. Bafti
University of Zagreb, Faculty of Chemical Engineering and Technology, Trg Marka Marulica 19, Zagreb 10000, Croatia
e-mail: ipanzic@fkit.unizg.hr

A. Bafti
e-mail: abafti@fkit.unizg.hr

D. Shi (ed.), *Energy Materials Science and Engineering: From New Devices to AI Power Systems*, Green Energy and Technology, https://doi.org/10.1007/978-3-031-88550-1_4

1 Introduction

1.1 Context

Dielectric capacitors, crucial components in modern electronics and power systems, are constantly being improved to meet the growing demand for higher energy density and efficiency [1, 2]. Among various dielectric materials, perovskite-based relaxor ferroelectrics have emerged as promising candidates due to their unique combination of high dielectric constant and low remnant polarization [3, 4]. This review chapter delves into the research area of perovskite ferroelectric, relaxor and antiferroelectric materials for capacitive energy storage, analyzing their properties, design strategies, and challenges. We will explore both lead-based and lead-free systems, highlighting recent advancements and identifying areas for future research.

1.2 Perovskite Structure and Relaxor Ferroelectricity

Perovskite materials, generally represented by the formula ABO_3, comprise a cubic crystal structure [5]. The A-site cation is typically larger and occupies the corners of the cube, while the smaller B-site cation resides at the center, surrounded by an octahedron of oxygen anions. In principle, many perovskites undergo phase transitions, exhibiting ferroelectric behavior characterized by spontaneous polarization, particularly as a function of heating and/or cooling. This polarization arises due to the displacement of ions within the crystal lattice, typically below a critical temperature called the Curie temperature. Perovskite ferroelectrics, such as barium titanate ($BaTiO_3$) and lead titanate ($PbTiO_3$), are widely studied for their exceptional dielectric, piezoelectric, and pyroelectric properties. These materials find applications in capacitors, sensors, actuators, memory devices, and energy harvesting technologies, making them integral to modern electronic and energy systems.

Relaxor ferroelectrics, a subgroup of ferroelectric materials, exhibit a unique behavior characterized by diffuse phase transitions and frequency-dependent dielectric maxima [6]. Unlike normal ferroelectrics with sharp phase transitions, relaxors show a broad distribution of polar nanoregions (PNRs) with fluctuating polarization vectors [7]. This disordered microstructure is responsible for their superior energy storage properties, particularly their low remnant polarization, which leads to higher energy efficiency [8]. The formation of PNRs is often attributed to compositional disorder, introduced through chemical substitutions at either the A-site or B-site of the perovskite structure [6, 9]. These substitutions create local distortions in the lattice, disrupting long-range ferroelectric order and leading to the formation of PNRs.

Antiferroelectrics are materials characterized by the presence of antiparallel electric dipoles within their crystal structure, resulting in a net-zero spontaneous polarization in their equilibrium state. Unlike ferroelectrics, where dipoles align in the same direction, antiferroelectrics exhibit a periodic arrangement of dipoles pointing in opposite directions. This unique behavior allows them to undergo a reversible phase transition to a ferroelectric state under an applied electric field, accompanied by a characteristic double-loop hysteresis. Antiferroelectrics are of interest for energy storage capacitors, actuators, and electrocaloric cooling applications due to their high energy density, low loss, and tunable properties. Their distinctive polarization switching and high-field behavior make them promising for advanced electronic and energy systems. Common examples include lead zirconate ($PbZrO_3$) and its derivatives.

A typical polarization hysteresis loop for ferroelectrics and relaxors is shown in Fig. 1 (black FE and blue RE).

When a positive electric field surpasses a certain strength known as the coercive field (E_c), the dipole moments align, resulting in a persistent remnant polarization (P_r) even after the field is withdrawn. Conversely, when a negative electric field is applied, the dipole moments switch direction, necessitating a negative coercive field ($-E_c$) for the elimination of net polarization. Upon applying a stronger electric field, the dipole moments complete their switching, leading to the reappearance of net polarization in the ferroelectric material. While aligning the polarization with the applied electric field in the ferroelectric, energy can be stored. Since the appearance of a hysteresis is indicative of a non-linear phenomenon, the standard equation for

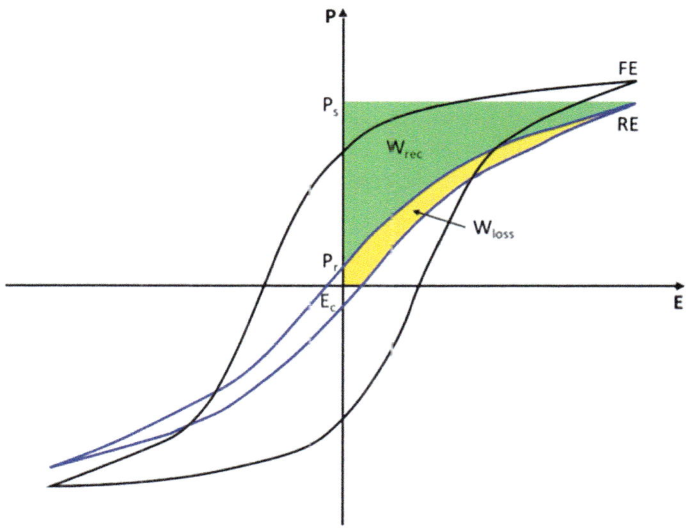

Fig. 1 Schematic polarization versus electric field hysteresis loop (PE-loop) of a classical ferroelectric (FE) and a relaxor material (RE) with the indication of Wrec and Wloss, remnant polarization (P_r) and maximum/saturation polarization (P_s) [10]

calculating the electric energy stored in a capacitor is not valid anymore and changes to

$$W_{rec} = \int_{P_r}^{P_{max}} EdP \tag{1}$$

The region of the Polarization–Electric field (PE) diagram (Fig. 1) that is shaded in green represents the energy density that can be regained (W_{rec}). This recoverable energy density (W_{rec}) can be quantified mathematically as the integral of the applied electric field (E) over the measured polarization (P). This integration occurs within the span between the remnant polarization (P_r) and the maximum polarization (P_{max}).

For energy storage applications, the energy storage efficiency (η) is critical and can be computed from the recoverable energy density and the loss energy density using the following formula:

$$\eta = \frac{W_{rec}}{W_{rec} + W_{loss}} \tag{2}$$

FE exhibits higher permittivity but also higher losses attributed to the hysteretic PE loop, as compared to relaxors (RE). The reduced remnant polarization (P_r) of RE materials, in comparison to ferroelectrics, renders RE materials more suitable for energy storage. This is attributed to their high recoverable energy density resulting from their slim PE loops. The relaxor behavior in ferroelectric perovskites is induced by chemical modifications, wherein the substituted cations contribute to the creation of both polar and nonpolar nanoregions. The presence of nonpolar nanoregions disrupts the long-range ferroelectric order, resulting in slimmer loops and reduced hysteretic loss. This occurs because only regions exhibiting ferroelectricity respond to the applied electric field.

2 Lead-Based Perovskite Relaxors for Energy Storage

2.1 Synthesis Techniques of PZT

Lead zirconate titanate (PZT), a well-known lead-based perovskite, has been extensively studied for its piezoelectric and ferroelectric properties [11]. While PZT exhibits excellent energy storage capabilities, its toxicity has driven research towards lead-free alternatives. However, the understanding of PZT's behavior remains crucial for developing improved lead-free materials. For example, research has shown that mechanically tailoring the microstructure of PZT thick films can induce relaxor ferroelectric behavior, resulting in enhanced energy storage density [12]. This approach, based on high kinetic deposition, simultaneously enhances dielectric breakdown strength and polarization, leading to a record energy density of 124.1 J cm^{-3} [11]. This highlights the potential of microstructure engineering as a strategy to improve energy storage performance.

The discovery of PZT's ferroelectricity spurred the development of numerous synthesis techniques, each offering unique advantages and disadvantages. These methods can be broadly categorized into solid-state methods, liquid-phase methods, and other advanced techniques.

2.1.1 Solid-State Synthesis

Solid-state synthesis is a cost-effective method for producing PZT ceramics [13]. This approach involves mixing and milling the constituent oxides (PbO, ZrO_2, TiO_2) followed by high-temperature sintering. However, a significant challenge with solid-state sintering is the loss of PbO due to vaporization at high temperatures [14]. Variations in compaction pressure during the process significantly impact the resultant piezoelectric and dielectric properties, with an optimal pressure identified to improve these properties and optimize sintering temperature [14]. The development of alternative methods to mitigate PbO loss and reduce sintering temperatures is a crucial area of ongoing research. The solid-state reaction route is also employed in the synthesis of lead-based antiferroelectric perovskite materials like PZ and PZT, with structural changes observed at varying concentrations [15].

2.1.2 Liquid-Phase Synthesis

Liquid-phase methods, including sol–gel, co-precipitation, and hydrothermal methods, offer advantages in terms of producing PZT in nano-ceramic powder forms [16]. Sol–gel methods, for example, are widely used for synthesizing PZT thin films due to their low densification temperatures and ease of application [17, 18]. The sol–gel technique allows for the fabrication of both thin and thick films without requiring costly physical deposition equipment. However, the piezoelectric properties of PZT sol–gel-derived films are often lower than those of bulk materials [17, 18]. Hydrothermal synthesis is another liquid-phase method that allows for the preparation of PZT nanoceramic powders [16, 19]. This method involves dissolving the precursor materials in a high-temperature, high-pressure aqueous solution, followed by precipitation and crystallization of PZT nanoparticles. The formation mechanisms in hydrothermal synthesis can involve dissolution/precipitation or in-situ processes. Parameters such as temperature, basicity (pH), time, and drying methods significantly influence the characteristics of the resulting PZT nanoceramic powders [19]. Freezing drying methods are particularly effective in obtaining PZT powders with better dispersion and symmetrical granularity [19]. The modified sol–gel method using ethylene glycol as a solvent has been explored for the synthesis of Nd^{3+}-doped PZT nanopowder, showing a higher heating temperature compared to undoped PZT [20]. Co-precipitation is yet another liquid-phase method used to prepare PZT nanoceramics [16]. This involves the simultaneous precipitation of the metal ions in solution using a precipitating agent. The resulting precipitate is then calcined and sintered to obtain the desired PZT phase. The control over particle size and morphology is a key aspect of this method.

2.1.3 Advanced Techniques

Beyond the traditional methods, several advanced techniques are employed for PZT synthesis. These include spray pyrolysis, pulsed laser deposition (PLD), chemical solution deposition (CSD), and magnetron sputtering [11]. Spray pyrolysis involves spraying a precursor solution into a hot furnace, leading to the formation of PZT particles. PLD uses a pulsed laser to ablate a target material, depositing PZT thin films on a substrate. CSD involves the deposition of a solution onto a substrate, followed by heat treatment to form the PZT film. Magnetron sputtering is a physical vapor deposition technique used to deposit high-quality PZT thin films over large areas, offering high feasibility and controllability [20]. The relationship between synthesis parameters and the resulting film properties is crucial in these advanced techniques. Direct ink writing (DIW), combined with furnace sintering and resin impregnation, is used to fabricate PZT piezoelectric composites for transducers. This method allows for the creation of complex 3D structures. A gravity-driven sintering (GDS) process has also been developed to fabricate curved PZT ceramics, offering advantages in terms of scalability and cost-effectiveness [13]. A bio-template synthesis method using polysaccharide and bacteriophage templates has been reported for the fabrication of PZT nanowires, achieving high piezoelectric constants [15]. Wet-etch patterning of sol–gel PZT films is used for microelectromechanical systems (MEMS) applications, enabling the fabrication of PZT films with thicknesses ranging from 2 to 10 microns [16].

2.2 Properties of PZT

2.2.1 Piezoelectric Properties

PZT's exceptional properties are the foundation of its widespread applications. These properties are highly tunable depending on the synthesis method, composition (Zr/Ti ratio), and doping.

The piezoelectric effect, the ability to generate an electric charge in response to mechanical stress or vice versa, is central to PZT's utility [12, 17]. The piezoelectric coefficients (d33, d31, etc.) quantify the magnitude of this effect. The piezoelectric properties of PZT can be enhanced by aligning the crystal grains along specific orientations, a technique known as texturing [18]. This texturing process, using $Ba(Zr,Ti)O_3$ microplatelet templates, results in superior piezoelectric coefficients (d33 of 760 $pC\,N^{-1}$, g33 of 100 $mV\,m\,N^{-1}$, and k33 of 0.85) and Curie temperatures of 360 °C [18]. The effect of compaction pressure on the piezoelectric properties of PZT prepared via high planetary milling has also been investigated. The incorporation of ZrO_2 nanoparticles into PZT nanocomposites leads to improved mechanical and piezoelectric properties, with optimized properties obtained at specific ZrO_2 concentrations [17]. Subcoercive degradation, resulting from repeated electrical loading, leads to a progressive decrease in PZT functionality and affects local piezoelectric properties and domain structure [19]. The thickness dependence of the piezoelectric properties at the nanoscale has also been studied in polycrystalline PZT thin films revealing a self-polarization effect that increases with film thickness. The piezoelectric properties of PZT aerogel/PVDF composite coatings and sintered sheets, designed with a biomimetic shell structure, exhibit high piezoelectric coefficients (d33 of 435.15 pC N^{-1} and d31 of −144.55 pC N^{-1}) and excellent electromechanical coupling properties [21].

2.2.2 Dielectric Properties

PZT exhibits high dielectric constants, making it suitable for capacitor applications. The dielectric constant decreases with increasing temperature [22]. The ac conductivity in PZT is influenced by the ionization state of charge carriers, being singly ionized in the ferroelectric region and doubly ionized in the paraelectric region. The dielectric properties of PZT are also influenced by doping, such as Gd doping, resulting in systematic changes in dielectric constant, phase transition temperature, and diffusivity [23]. The influence of defect engineering on dielectric and piezoelectric properties has been investigated, achieving exceptionally well-balanced performances through multiscale coordination among defect dipole, domain wall density, and grain boundary [12]. The dielectric properties of screen-printed PZT thick films on silicon electrodes have also been studied, yielding relative permittivity values around 970 at 1 kHz [24]. The liquid-phase flash sintering (LPFS) of PZT results in a significantly higher dielectric constant, particularly at higher frequencies, with lower dielectric loss compared to conventional sintering. Domain wall motion contributes significantly to the extrinsic component of the dielectric response in PZT films, with the contribution increasing with film thickness and grain size.

2.2.3 Ferroelectric Properties

PZT's ferroelectric properties, characterized by spontaneous polarization that can be switched by an applied electric field, are essential for many applications [19]. The remanent polarization (P_r) and coercive field (E_c) are key parameters characterizing ferroelectric behavior. Gd doping in PZT enhances the remnant polarization [23]. The effect of subcoercive electrical cycling on local ferroelectric properties has been studied using piezoresponse force microscopy (PFM), revealing a strong depolarization effect in the direction of the applied electric field [19]. The use of PZT in optical memristors is a relatively recent development, offering advantages such as low loss, high precision, large modulation depth, and high stability [12]. In PZT-filled polymer-derived ceramics (PDCs), the crystallization of the PDC matrix is inhibited in the presence of the PZT filler, and the PDC matrix causes a pronounced decomposition of the filler compared to pure PZT [24].

2.3 Applications of PZT

2.3.1 Microelectromechanical Systems (MEMS) and Nanoelectromechanical Systems (NEMS)

The unique combination of piezoelectric, dielectric, and ferroelectric properties makes PZT highly versatile, leading to its applications in a wide range of fields.

PZT's excellent piezoelectric properties make it ideal for MEMS and NEMS devices [25]. Its use in these applications ranges from actuators and sensors to transducers and resonators. The ability to fabricate PZT thin films with controlled properties is crucial for MEMS and NEMS applications [16, 25]. Wet-etch patterning of PZT thick films is particularly relevant for MEMS applications. The development of lead-free alternatives for MEMS applications is driven by environmental concerns [26].

2.3.2 Sensors and Actuators

PZT-based sensors and actuators are used extensively in various applications, including automotive systems, micro-propulsion systems, and biomedical devices [27]. The high strain capabilities of PZT are particularly beneficial for actuator applications. A novel PLZT sensor with a platinum wire implant has been developed for biomedical applications, specifically for detecting cardiac micropulses on human skin [28]. The use of PZT transducers in structural health monitoring (SHM) is also gaining traction, providing a reliable method for damage diagnosis in reinforced concrete structures [29]. EMI (Electromechanical Impedance) technique combined with GCN (Graph Convolutional Networks) is used for real-time monitoring of multiple bolt looseness [30].

2.3.3 Energy Harvesting

PZT's ability to convert mechanical energy into electrical energy makes it a promising material for energy harvesting applications [7, 8, 15]. The development of efficient energy harvesters is crucial for powering self-powered devices and reducing reliance on batteries [31]. PZT nanowires, synthesized using biotemplated methods, have shown promise in generating significant power [15]. The use of PZT in ferroelectric energy harvesters is also being explored, offering advantages in terms of flexibility and potentially higher energy harvesting efficiency [31].

2.3.4 Other Applications

PZT finds applications in diverse areas including ultrasound transducers, ceramic capacitors [12], transistors, ferroelectric memories [16], displays, resonators, and chips [17]. The development of PZT-based optical memristors represents a significant advancement in the fusion of photonics and electronics, offering potential applications in neuromorphic computing and artificial intelligence [28]. PZT is also being explored as a potential anode material for lithium-ion batteries, offering high initial discharge capacity and stable cycling performance [12]. The use of PZT in PZT/diamond heterostructures is gaining attention for acoustic devices and microelectromechanical systems, operating in the GHz frequency range. However, challenges associated with the direct integration of PZT on the diamond, including the formation of undesired pyrochlore phase and thermal expansion coefficient mismatch, need to be addressed [30]. In a different application, PZT is used as a detector for hypervelocity particles, enabling the identification of impact positions. The use of PZT in ferroelectric photovoltaics is also a promising area of research, offering potential advantages over traditional semiconductor solar cells [31]. Finally, the application of PZT in 3D printing is an active area of research, aiming to achieve defect-free dense piezoelectric ceramics [32].

3 Lead-Free Alternatives

3.1 Evolution of the Need for the Lead-Free Alternatives

The toxicity of lead in PZT has driven significant research into lead-free piezoelectric materials [33–40]. Several lead-free systems, such as BNT (bismuth sodium titanate), KNN (potassium sodium niobate), BKT (barium potassium tantalate), and BZT-BCT (barium zirconate titanate-barium calcium titanate), have been explored. However, achieving comparable performance to PZT remains a challenge. High-strain lead-free piezo ceramics are needed for various sensor and actuator applications [41]. While some lead-free systems deliver non-linear strains exceeding those of PZT, their overall performance and reliability in practical applications are still under development [42]. Texturing techniques, similar to those used for PZT, are being applied to lead-free systems like $BaTiO_3$, KNN, and BNT to enhance their piezoelectric properties [44]. A life cycle assessment comparing KNN and PZT reveals that KNN has lower environmental and health impacts, but improvements in KNN processing are needed to reduce energy consumption [45]. Multi-length engineering of KNN films has led to the development of lead-free piezoelectric acoustic sensors with high sensitivity, demonstrating the potential of lead-free alternatives in practical applications [46].

3.2 *Barium Titanate (BT)-Based Relaxors*

BT is another widely studied perovskite material known for its high dielectric constant. However, its relatively high remnant polarization limits its energy storage efficiency. The addition of $BiYb_{0.9}Sc_{0.1}O_3$ to BT-based ceramics has been shown to enhance energy storage properties by inducing relaxor behavior and improving the microstructure [9]. This resulted in a recoverable energy density of 1.24 J cm^{-3} at 100 kV cm^{-1} [9]. Furthermore, $BaTiO_3$-based relaxors modified with $(Sr,Bi)TiO_3$-$Bi(Mg_{0.5}Ti_{0.5})O_3$ have been shown to exhibit a superparaelectric state, leading to improved energy storage capabilities [39]. A recoverable energy density of ~1.12 J cm^{-3} and efficiency of ~94% were achieved at 170 kV cm^{-1} (Fig. 2) [39].

3.2.1 Synthesis Methods for BT-Based Relaxors

The synthesis of BT-based relaxor ferroelectrics is crucial for controlling the material's microstructure and ultimately its properties. Several methods have been employed, each with its own advantages and disadvantages. Conventional solid-state reaction methods, involving high-temperature sintering of precursor oxides or carbonates, are widely used due to their simplicity and scalability [48–50]. However, this approach often leads to larger grain sizes and can result in inhomogeneous compositions. To achieve finer grain sizes and better compositional control, wet chemical methods such as sol–gel [51], hydrothermal [52], and co-precipitation techniques [53] have been explored. These methods allow for better control over stoichiometry and particle size distribution, leading to improved dielectric properties. More recently, advanced techniques like electrospray synthesis have emerged, offering further control over particle morphology and size. The choice of synthesis method significantly impacts the final properties of the BT-based relaxor, necessitating careful consideration of the desired application.

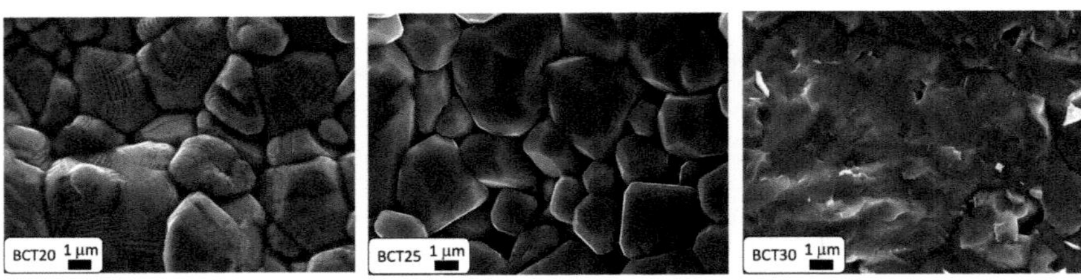

Fig. 2 SEM images of BCT ceramics with different amounts of Ca, w% 20, 25, 30 [47]

3.2.2 Influence of Doping on BT Relaxors

Doping BT with various cations is a common strategy to engineer its properties and induce relaxor behavior. Substituting Ti^{4+} with other ions of similar valence, such as Zr^{4+} [54], Sn^{4+} [55], or Nb^{5-} [56] can introduce disorder into the crystal lattice, leading to a diffuse phase transition. The extent of this disorder and the resulting relaxor characteristics depend on the type and concentration of the dopant. For example, Zr doping in $Ba(Zr,Ti)O_3$ leads to the formation of Ti-rich and Zr-rich polar regions, resulting in two distinct types of relaxor behavior [54]. Similarly, Nb doping in BT can modify the conductivity, dielectric constant, and loss relaxation phenomena, particularly in the paraelectric phase at high dopant concentrations [56]. Furthermore, A-site substitution, such as with Na^+ and La^{3+} [57], can also modify the dielectric properties and induce relaxor behavior. The interplay between dopant type, concentration, and synthesis method significantly influences the final properties of the BT-based relaxor, providing a pathway to tailor the material for specific applications.

3.2.3 Structure–Property Relationships in BT-Based Relaxors

The dielectric properties of BT-based relaxors are intimately linked to their microstructure and chemical composition. The presence of polar nanoregions (PNRs), which are dynamic clusters of dipoles with correlated orientations, is a key characteristic of relaxor ferroelectrics [58]. The size, distribution, and dynamics of these PNRs govern the dielectric response, which can be characterized by parameters such as the freezing temperature (T_f) and activation energy (E_a) [58]. The grain size also plays a significant role; smaller grain sizes generally lead to higher dielectric constants due to increased grain boundary effects [56]. Moreover, the degree of compositional disorder introduced by doping affects the distribution and dynamics of PNRs, directly influencing the dielectric properties. Understanding these structure–property relationships is vital for designing BT-based relaxors with optimized dielectric properties for various applications.

3.2.4 Impact of Microstructure on Dielectric Properties

The microstructure of BT-based relaxors, including grain size, grain boundary characteristics, and the presence of secondary phases, strongly influences dielectric behavior. The grain size affects the dielectric constant, with smaller grain sizes often leading to enhanced permittivity due to increased grain boundary contributions [56]. Moreover, the presence of grain boundaries and interfaces can create internal barriers to charge transport, affecting the dielectric loss and conductivity. The morphology of the BT particles also plays a role; for instance, the use of BT nanoparticles in polymer composites can lead to improved dielectric properties [53]. The distribution and orientation of BT particles within the polymer matrix are crucial factors in determining the composite's dielectric behavior [54]. Furthermore, the presence of secondary phases or impurities can significantly alter the dielectric properties, highlighting the importance of careful synthesis and processing techniques.

3.2.5 Applications of BT-Based Relaxors in Energy Storage and Biomedical Area

The unique dielectric properties of BT-based relaxors make them attractive for various applications. Their high dielectric constants and relatively low losses make them suitable for high-capacitance applications, such as multilayer ceramic capacitors (MLCCs) [48]. The ability to tune the dielectric properties through doping and processing offers flexibility for optimizing the material for specific requirements [54, 55]. Furthermore, the relaxor behavior, with its frequency-dependent dielectric response, can be exploited in tunable microwave devices [52]. The combination of high dielectric constant and mechanical flexibility in BT-polymer composites opens possibilities for flexible electronics and dielectric elastomer applications [52]. In addition, the piezoelectric properties of BT-based materials find applications in energy harvesting [48, 52], sensors [58], and actuators [57]. However, challenges remain in improving the energy storage density and breakdown strength of BT-based relaxors to meet the demands of high-power applications.

BT-based relaxors have shown promise in energy storage applications, particularly in dielectric capacitors, due to their ability to store a significant amount of energy [48]. The energy storage density (W) is determined by the dielectric constant (ε), breakdown strength (E_b), and charge–discharge efficiency (η) [52]. While the incorporation of BT nanoparticles can improve ε and W, it often reduces E_b and η [52]. The use of ultralow loadings of BT nanoparticles in a polymer matrix can lead to a simultaneous enhancement of both ε and E_b, resulting in significant improvements in energy density. The addition of other materials, such as boron nitride nanosheets, can further enhance the energy storage properties by improving E_b and η [25]. However, further research is needed to overcome limitations such as high dielectric loss and low breakdown strength to realize the full potential of BT-based relaxors in high-power energy storage applications.

The biocompatibility of BT and its potential for use in biomedical applications has also been explored [59–62]. Surface functionalization of BT nanoparticles with biocompatible materials is crucial for improving their dispersibility in aqueous media and reducing cytotoxicity [61]. Studies have shown that BT can enhance the radiopacity of bio-ceramics, making them suitable for imaging applications [59]. The piezoelectric properties of BT could also be exploited in bone tissue engineering, potentially stimulating bone regeneration [59]. Furthermore, BT-based nanocomposites have been investigated for use as contrast agents in computed tomography (CT) imaging [61]. However, further research is needed to fully assess the long-term biocompatibility and efficacy of BT-based materials in biomedical applications.

3.2.6 Research Gaps and Future Directions

Despite significant progress, several research gaps remain in the field of BT-based relaxors. A deeper understanding of the fundamental mechanisms underlying relaxor behavior, including the role of PNRs and defects, is needed to further optimize material properties. Developing novel synthesis methods to achieve precise control over microstructure and composition is crucial for tailoring materials for specific applications. Improving the breakdown strength and energy storage density of BT-based relaxors is essential for their wider adoption in high-power applications. Furthermore, exploring new dopants and combinations of dopants to enhance the desired properties is a promising area of research. The integration of BT-based relaxors into complex device architectures requires further investigation, particularly in the area of flexible electronics and biomedical applications. Finally, more research is needed to understand and mitigate the potential environmental and health impacts of BT-based materials, especially those containing lead, as many relaxor systems are lead-based. Addressing these research gaps will pave the way for the development of advanced BT-based relaxor materials with enhanced performance and expanded applications.

3.3 Bismuth Sodium Titanate (BNT)-Based Relaxors

3.3.1 Synthesis and Characterization Techniques

BNT-based relaxors are particularly attractive due to their high Curie temperature and relatively high polarization. However, their large remnant polarization often limits their energy storage efficiency [63–65]. Researchers have explored various strategies to improve BNT-based relaxors, including chemical modifications and microstructure engineering. For instance, the incorporation of $Sr(Zn_{1/3}Nb_{2/3})O_3$ (SZN) into BNT ceramics has been shown to create coexisting ferroelectric and relaxor phases at the nanoscale, leading to enhanced energy storage performance. This resulted in a recoverable energy density of 5.0 J cm^{-3} with an efficiency of 82% at 370 kV cm^{-1} [66]. Similarly, modifying BNT with $Bi(Mg_{0.5}Zr_{0.5})O_3$ has been shown to optimize energy storage properties by retaining high polarization while reducing remnant polarization and enhancing breakdown strength [67–69]. This resulted in a recoverable energy density of 8.3 J/cm^3 [67]. The work of Liu et al. demonstrates a chemical framework for designing linear-like relaxors, resulting in recoverable energy densities as high as 17.3 J cm^{-3} with 90% efficiency [70]. This framework categorizes ions based on their radius and ferroelectric activity, allowing for the rational design of high-performance materials.

Several methods are employed to synthesize BNT-based ceramics, each impacting the resulting microstructure and properties. The conventional solid-state reaction method is widely used, offering a relatively straightforward approach to producing bulk ceramics [71, 72]. However, this method often results in inhomogeneous microstructures and compositional variations. To address these issues, researchers have explored alternative techniques such as sol–gel auto-combustion [72], hydrothermal synthesis [73], and the more recently employed microwave sintering [74], which offer better control over particle size and morphology. The choice of synthesis method significantly affects the grain size, density, and ultimately, the dielectric and piezoelectric properties of the resulting material. Further refinement of these techniques, such as hot-pressed sintering [75], has been shown to improve the density and breakdown strength, leading to enhanced energy storage performance.

Characterization techniques are crucial for understanding the structure–property relationships in BNT-based relaxors. X-ray diffraction (XRD) is routinely used to determine the crystal structure and phase purity [71–75]. Rietveld refinement of XRD data allows for a more detailed analysis of the crystal structure and the quantification of secondary phases. Scanning electron microscopy (SEM) provides information on the microstructure, grain size, and morphology [71–75], while transmission electron microscopy (TEM) offers higher resolution imaging, allowing for the visualization of PNRs [69]. Piezoresponse force microscopy (PFM) is a powerful technique for studying the nanoscale domain structure and dynamics of relaxors [72], providing insights into the origin of their unique properties. Raman spectroscopy can be used to probe the local structure and vibrational modes, providing information on the phase transitions and the degree of disorder [73].

3.3.2 Compositional Modification and Defect Engineering

The properties of BNT-based relaxors can be significantly tuned by compositional modification and defect engineering. The addition of other perovskite materials, such as $BaTiO_3$ (BT) [10, 76–82], $SrTiO_3$ (ST) [83], significantly affects the phase transitions, dielectric constant, and piezoelectric properties. These modifications often aim to create a morphotropic phase boundary (MPB) [10, 83], which is associated with enhanced piezoelectric activity. The introduction of rare-earth ions [79] or other dopants such as Mn [82] and Nb [84] can further modify the properties by influencing the defect chemistry and the formation of PNRs. For example, the introduction of La^{3+} [79] has been shown to enhance the antiferroelectric-like relaxor ferroelectric characteristics, boosting energy storage performance. Similarly, the addition of BBN ($BaBi_2Nb_2O_9$) [80] enhances the relaxation behavior and improves breakdown strength. The impact of defect engineering on the strain response is of significant interest. Studies by Wang et al. [66] review the generation of strain in BNT-based materials, focusing on the role of defect dipoles and their coupling with ferroelectric spontaneous polarization. The authors highlight the importance of understanding the asymmetric effect caused by this coupling, which can influence the strain characteristics. Different doping strategies can lead to varying degrees of defect concentration and distribution, leading to different levels of strain response.

3.3.3 Dielectric and Ferroelectric Properties

BNT-based relaxors exhibit unique dielectric and ferroelectric properties, largely due to their relaxor behavior. The dielectric constant is typically high and exhibits a broad peak with temperature, unlike the sharp Curie transition observed in normal ferroelectrics [10]. The frequency dependence of the dielectric constant is also a characteristic feature of relaxors. The ferroelectric hysteresis loops are often slim, indicating a lower remnant polarization compared to normal ferroelectrics [10]. The temperature dependence of the dielectric and ferroelectric properties reflects the complex phase transitions that occur in these materials, which include rhombohedral, orthorhombic, and tetragonal phases as shown by Kobald et al. in Fig. 3 [10]. The depolarization temperature (T_d) marks a significant transition, where the material loses its macroscopic ferroelectric order. The position of T_d relative to room temperature significantly influences the material's performance in various applications. The presence of PNRs plays a crucial role in shaping the dielectric and ferroelectric behavior, as these nanoscale regions exhibit fluctuating polarization that contributes to the relaxor characteristics. Several studies have focused on enhancing the energy storage performance of BNT-based relaxors [76–80]. Li et al. [65] reported a high recoverable energy density of 5.7 J/cm^3 and energy storage efficiency of 86.4% under a moderate electric field of 390 kV cm^{-1} in BFN-modified BNT-ST ceramics. Zhang et al. [71] achieved a recoverable energy density of 8.58 J/cm^3 and an efficiency of 94.5% through defect engineering, demonstrating the potential for significant performance enhancement. The combination of relaxor and antiferroelectric properties has also been explored as a strategy to improve energy storage performance, leading to higher energy density and efficiency. However, challenges remain in optimizing the balance between energy density, efficiency, and breakdown strength.

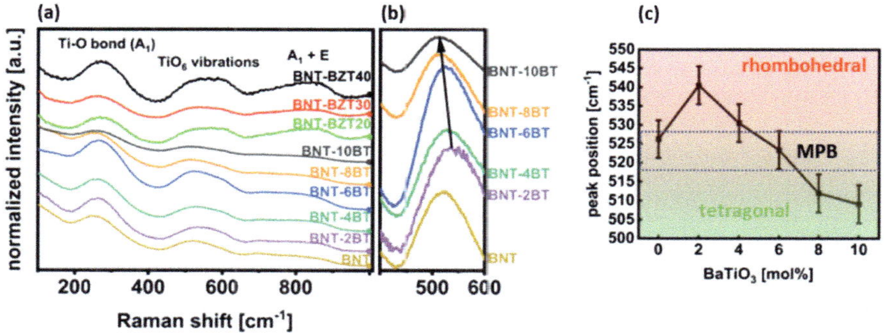

Fig. 3 **a** Raman spectra of BNT-100xBT (with $x = 0$, 0.02, 0.04, 0.06, 0.08, and 0.10) and BNT-6BZT100y (with $y = 0$, 0.2, 0.3, 0.4) thin films at room temperature, **b** zoomed-in area of the peak-position around 540 cm^{-1} that marks the morphotropic phase transition from rhombohedral to tetragonal of the BNT-100xBT thin films, and **c** corresponding peak-positions of the rhombohedral and tetragonal peaks of the BNT-100xBT thin films [10]

3.3.4 Piezoelectric and Electrostrictive Properties

BNT-based relaxors exhibit significant piezoelectric and electrostrictive properties, making them attractive for actuator applications [80]. The large strain response is often attributed to the field-induced transition from the relaxor to the ferroelectric state. This transition is accompanied by significant domain switching and changes in the microstructure [10], leading to a considerable strain. However, challenges remain in achieving a high strain with low hysteresis [76], which is crucial for precise actuator applications. The introduction of dopants and compositional modifications can effectively tune the strain response and hysteresis [76–80], often aiming to optimize the properties near the relaxor phase boundary. The electrostrictive coefficient (Q) is another important parameter, and achieving ultrahigh Q values is a significant goal in the development of high-performance electrostrictive materials [77].

3.3.5 Pyroelectric Properties

The significant polarization fluctuation near the depolarization temperature (T_d) in BNT-based relaxors makes them suitable for pyroelectric applications [77]. Wen et al. [69] demonstrated a high pyroelectric coefficient in Li^+, Sr^{2+}, and Ta^{5+} co-doped samples, highlighting the potential for tailoring the pyroelectric performance through composition engineering. The temperature-dependent evolution of PNRs plays a critical role in determining the pyroelectric response, and manipulating this evolution through compositional modifications is a key strategy for optimizing the pyroelectric coefficient [69]. The achievement of high pyroelectric coefficients at room temperature is particularly important for practical applications.

3.3.6 Applications

BNT-based relaxors are finding applications in various fields, owing to their unique combination of properties. Their high energy storage capacity makes them promising candidates for dielectric capacitors in advanced pulsed power systems [10, 77, 80, 82, 83]. Their large strain response and piezoelectric properties are suitable for piezoelectric actuators in precision drive systems [76, 81], offering a lead-free alternative to traditional PZT-based actuators. The significant pyroelectric response makes them suitable for pyroelectric sensors and energy harvesting devices [71]. Furthermore, research is ongoing to explore their potential in other applications, such as force-electric conversion [82] and flexible multilayer film capacitors (Fig. 4) [10].

3.3.7 Research Gaps and Future Directions

Despite significant progress, several research gaps remain in the field of BNT-based relaxors. A deeper understanding of the atomic-scale structure–property relationships

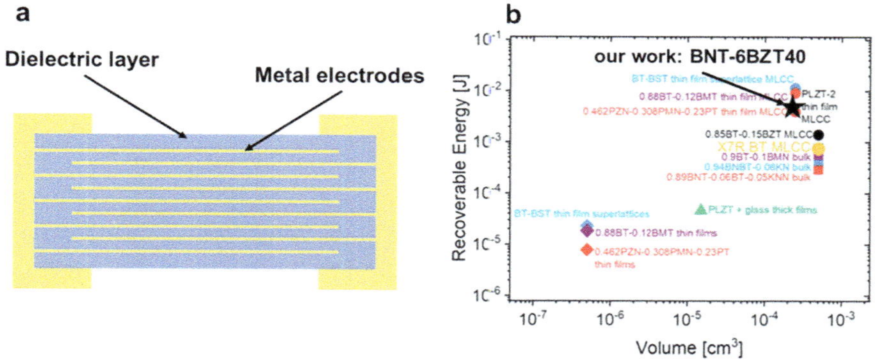

Fig. 4 **a** Scheme of the virtual device model of a BNT-6BZT40 thin film capacitor structure. **b** Recoverable energy density at the maximum applied electric field for some selected relaxor compositions in bulk, thick-film multilayers, and thin-film multilayers. Values for MTFCs were extrapolated from the respective J_r values of single-layer thin films assuming an MLCC consisting of 500 layers with 1 μm thickness and 0.5 × 1 mm^2 area (the best composition BNT-6BZT40 of this work is marked with a star) [10]

is crucial for rational design and optimization of materials. Further investigation into the role of PNRs in various properties, including energy storage, piezoelectricity, and pyroelectricity, is needed. The development of advanced characterization techniques to probe the dynamic behavior of PNRs is also important. Improving the breakdown strength while maintaining high polarization remains a significant challenge. Exploring novel compositional modifications and defect engineering strategies to overcome these limitations is essential. The development of scalable and cost-effective synthesis methods for high-quality BNT-based relaxors is also crucial for their widespread commercialization. Finally, more research is needed to explore the potential of BNT-based materials in emerging applications, such as flexible electronics and energy harvesting.

3.4 Bismuth Magnesium Niobate (BMN)-Based Relaxors

3.4.1 Synthesis and Fabrication Methods

Bismuth magnesium niobate (BMN), with its chemical formula $Bi_{1.5}MgNb_{1.5}O_7$, has emerged as a promising material in the field of relaxor ferroelectrics. Relaxors are a class of ferroelectric materials characterized by a diffuse phase transition, meaning that their dielectric permittivity exhibits a broad peak rather than a sharp peak at the Curie temperature, as seen in normal ferroelectrics [86]. This diffuse transition is attributed to the presence of polar nano-regions (PNRs) with varying sizes and orientations, which interact weakly with each other [87]. The unique properties of

BMN-based relaxors, including their high dielectric constant, tunability, and low dielectric loss, make them attractive candidates for various applications, particularly in electric field tunable transparent capacitors [86], microwave dielectric devices [87], and energy storage applications [88]. This review will comprehensively examine the research on BMN-based relaxors, focusing on their synthesis, characterization, and potential applications.

Several methods have been employed for the synthesis and fabrication of BMN-based materials, each with its own advantages and limitations. One common approach is the solid-state reaction method, where stoichiometric amounts of Bi_2O_3, MgO, and Nb_2O_5 are mixed, calcined, and sintered at high temperatures [86–90]. This method is relatively simple and cost-effective, but it can lead to inhomogeneous mixing and the formation of secondary phases. To address these issues, researchers have explored alternative techniques such as sol–gel [87], pulsed laser deposition (PLD) [87], and radio frequency magnetron sputtering [89]. PLD and sputtering allow for the deposition of thin films with precise control over thickness and composition, making them suitable for applications requiring specific microstructures. The choice of synthesis method significantly influences the final properties of the BMN material, including its crystallinity, microstructure, and dielectric performance. For instance, post-annealing conditions, particularly oxygen partial pressure, play a crucial role in determining the crystalline quality of sputtered BMN thin films [90].

3.4.2 Structural and Microstructural Characterization

The crystal structure of BMN is typically described as a pyrochlore structure, which is a complex oxide structure with a general formula $A_2B_2O_7$ [86–88, 90]. However, deviations from the ideal pyrochlore structure, such as cation disorder and oxygen vacancies, are often observed, depending on the synthesis method and processing conditions [90–95]. X-ray diffraction (XRD) is a primary technique for characterizing the crystal structure and phase purity of BMN materials [91]. XRD patterns can reveal the presence of secondary phases, determine the crystallite size, and assess the degree of crystallinity. Scanning electron microscopy (SEM) provides information about the microstructure, including grain size, morphology, and grain boundary characteristics [90]. Transmission electron microscopy (TEM) offers higher resolution imaging and can reveal nanoscale features such as the presence of PNRs [92]. Other characterization techniques, such as Raman spectroscopy [92] and neutron diffraction [93], can provide complementary information about the local structure and atomic arrangements within the BMN material (Fig. 5). The interplay between the synthesis method, microstructure, and resulting properties is complex and requires a multi-technique approach for comprehensive understanding. For instance, the size and distribution of grains significantly affect the dielectric and electrical properties of BMN-based materials, as observed in studies of BMN thin films with varying thicknesses [94].

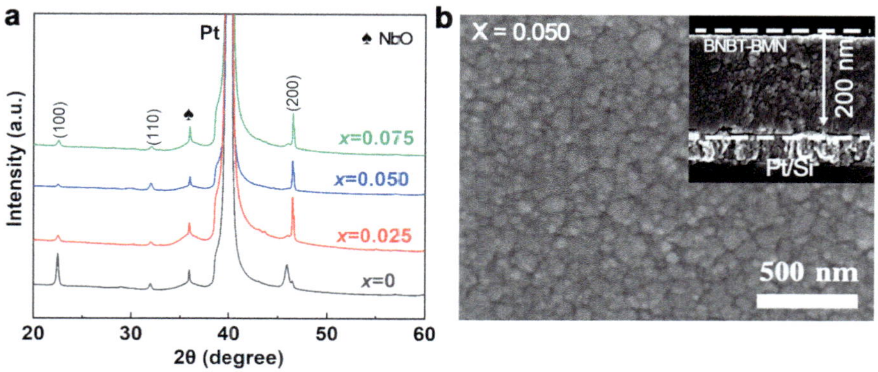

Fig. 5 **a** XRD patterns of the BNBT-xBMN thin films with different compositions, and **b** SEM image of the sample with $x = 0.050$ [95]

3.4.3 Dielectric Properties

BMN-based relaxors exhibit distinctive dielectric properties, characterized by a high dielectric constant (ε), low dielectric loss (tan δ), and significant tunability [87]. The dielectric constant is a measure of a material's ability to store electrical energy, while the dielectric loss represents the energy dissipated as heat during polarization reversal. Tunability refers to the ability of the dielectric constant to change under the application of an electric field. The high dielectric constant of BMN is attributed to the presence of PNRs, which contribute to a large polarization response [88]. The low dielectric loss is desirable for many applications, as it minimizes energy dissipation. The tunability of BMN, observed in both bulk ceramics and thin films [89, 91], arises from the field-induced reorientation of PNRs. The magnitude of tunability depends on various factors, including the composition, microstructure, and temperature. Studies on BMN-based thin film capacitors have shown tunability values as high as 28% at 1.3 MV cm^{-1} [90], while bulk ceramics exhibit lower tunability. The temperature dependence of dielectric properties is also a crucial consideration. The dielectric constant of BMN typically decreases with increasing temperature, but the rate of decrease depends on the composition and microstructure [90]. The frequency dependence of dielectric properties is another important characteristic of relaxors. The dielectric constant and dielectric loss exhibit dispersion, meaning that their values change with frequency. This dispersion is related to the dynamics of PNRs and their response to the applied electric field.

3.4.4 Energy Storage Properties

The energy storage performance of BMN-based materials has been investigated extensively due to their potential applications in energy storage devices [85, 86,

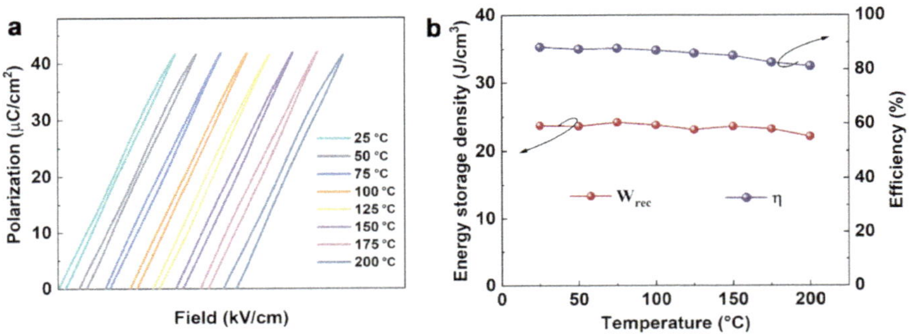

Fig. 6 **a** P–E loops of $x = 0.050$ simple under 1000 kV/cm subjected to different temperatures and **b** the corresponding calculated W_{rec} and η [95]

88]. The energy storage density (W) is a key parameter that determines the amount of energy that can be stored in a capacitor, while the efficiency (η) represents the ratio of recoverable energy to the total energy stored. Several studies have demonstrated that BMN additions to other lead-free dielectric materials, such as $BaTiO_3$ [87], and $NaNbO_3$ [89], can significantly enhance their energy storage properties. The incorporation of BMN can lead to the formation of relaxor ferroelectrics, which exhibit lower remnant polarization (P_r) and higher breakdown strength (BDS), resulting in improved energy storage density and efficiency [91]. However, the energy storage properties of BMN-based materials are sensitive to temperature, frequency, and electric field. Studies have shown that energy storage density and efficiency can decrease with increasing temperature or frequency, highlighting the need for further research to improve the thermal and frequency stability of these materials [93]. Further optimization of the composition and microstructure of BMN-based materials is needed to achieve even higher energy storage performance (Fig. 6).

3.4.5 Piezoelectric and Ferroelectric Properties

While BMN is primarily known for its dielectric properties, its piezoelectric and ferroelectric characteristics are also of interest [94]. Piezoelectricity refers to the ability of a material to generate an electric charge in response to mechanical stress, while ferroelectricity involves the spontaneous polarization of the material that can be switched by an applied electric field. The piezoelectric and ferroelectric properties of BMN are often intertwined with its relaxor behavior. The presence of PNRs contributes to a diffuse phase transition, leading to a broad range of temperatures over which the material exhibits ferroelectric or relaxor characteristics [94]. Studies have shown that the addition of BMN to other ferroelectric materials, such as $BaTiO_3$ and $PbTiO_3$, can significantly influence their piezoelectric and ferroelectric properties [95, 96]. For example, the addition of BMN to $BaTiO_3$-$BiFeO_3$ ceramics has led to enhanced piezoelectric coefficients and large effective piezoelectric coefficients

[90]. The element ratio in the doping component BMN significantly affects the strain, ferroelectric, piezoelectric, and dielectric properties of the host matrix. The presence or absence of specific elements like Nb and Mg also influences the ergodicity of the system, impacting the strain response and piezoelectric coefficient. These findings highlight the tunability of the properties of BMN-based materials through compositional engineering. However, more research is needed to fully understand the underlying mechanisms governing the piezoelectric and ferroelectric properties of BMN-based materials and to optimize them for specific applications.

3.4.6 Applications

The unique combination of high dielectric constant, low loss, tunability, and energy storage capabilities of BMN-based materials makes them promising candidates for various applications. In the field of optoelectronics, BMN thin films have shown potential for use in electric field tunable transparent capacitors [90]. Their high transparency in the visible light range and superior tunable dielectric properties are advantageous for applications requiring both optical and electrical functionality. In microwave technology, BMN-based ceramics have been investigated for use in dielectric voltage-controlled microwave devices [92]. Their high dielectric constant and low dielectric loss are beneficial for achieving high performance in these devices. In energy storage applications, BMN-based materials offer the potential for high energy density and efficiency [86–88]. They are being explored as lead-free alternatives to traditional lead-based energy storage materials, addressing environmental concerns. Further research is focused on enhancing the thermal and frequency stability of BMN-based materials to improve their suitability for practical applications. Moreover, the piezoelectric and ferroelectric properties of BMN-based materials make them potentially useful for actuators and sensors [94]. The ability to tailor their properties through compositional and microstructural engineering opens up possibilities for developing advanced devices with enhanced performance.

3.4.7 Research Gaps and Future Directions

Despite significant progress in the research of BMN-based relaxors, several research gaps remain. Further investigations are needed to fully understand the relationships between synthesis methods, microstructure, and resulting properties. More detailed studies are required to elucidate the mechanisms responsible for the diffuse phase transition and the dynamics of PNRs in BMN-based materials. Advanced characterization techniques, such as advanced microscopy and spectroscopy, are crucial for gaining deeper insights into the nanoscale structure and local atomic arrangements. Improving the thermal and frequency stability of BMN-based materials is essential for their widespread application in energy storage and other devices. Exploring new synthesis methods and processing techniques could lead to materials with enhanced properties and improved performance. Furthermore, research on BMN-based composites and hybrid materials could open up new avenues for tailoring the properties of these materials for specific applications. The development of predictive models based on theoretical calculations and simulations could facilitate the design and optimization of BMN-based materials with desired properties. Finally, exploring the potential applications of BMN-based materials in emerging technologies, such as flexible electronics and wearable sensors, is a promising area for future research.

3.5 *Sodium Niobate (NN)-Based Relaxors Antiferroelectrics*

3.5.1 Synthesis and Processing of NN-Based Relaxors

NN-based ceramics exhibit antiferroelectric behavior, characterized by a high energy density due to field-induced phase transitions. However, the large hysteresis associated with these transitions often leads to low energy efficiency. Introducing $Sr(Fe_{0.5}Ta_{0.5})O_3$ into the NN matrix creates relaxor antiferroelectric ceramics, resulting in a slimmer hysteresis loop and improved energy storage performance [96]. A recoverable energy density of 11.5 J cm^{-3} and efficiency of 86.2% was achieved in this system [96].

The synthesis of NN-based relaxors presents several challenges, primarily associated with the volatility of alkali elements (Na, K) at high temperatures [96–99]. Conventional solid-state reaction methods often result in compositional inhomogeneity and low density. To address these limitations, researchers have explored alternative synthesis techniques, including sol–gel methods [97], hydrothermal synthesis [98], and microwave sintering [99]. The sol–gel method offers advantages in controlling stoichiometry and achieving fine particle sizes, while hydrothermal synthesis allows for the fabrication of morphology-controlled particles, such as plate-like structures. Microwave sintering has shown promise in reducing sintering temperatures and times, improving densification, and enhancing properties. Furthermore, a novel approach using a water-soluble precursor and a modified Pechini method has been developed to synthesize KNN powders at moderate temperatures, mitigating the volatility issue [100]. The use of sintering aids, such as CuO [100] and ZnO/MnO_2 [101, 102], has also been investigated to enhance densification and improve the overall properties of the ceramics. However, achieving phase purity and chemical homogeneity remains a significant challenge, especially in complex compositions [103–105].

3.5.2 Structure–Property Relationships in NN-Based Relaxors

The dielectric and piezoelectric properties of NN-based relaxors are strongly influenced by their crystal structure and microstructure [101]. NN itself exhibits a complex sequence of phase transitions, with multiple ferroelectric and antiferroelectric phases coexisting at different temperatures [104]. This unique relaxor behavior, characterized by frequency dispersion at the transition between different ferroelectric phases, differs significantly from conventional relaxor behavior observed at the paraelectric-ferroelectric phase transition [104]. The introduction of dopants or the formation of solid solutions can further modify the phase transitions and enhance the desired properties. For instance, doping with tantalum (T_a) enhances resistivity and energy efficiency in sodium niobate-based relaxors, leading to improved energy storage performance [105]. The development of a new phase boundary (NPB) in KNN-based ceramics has also been reported to significantly improve piezoelectric properties and temperature stability [104]. The NPB is believed to be a region of phase coexistence, allowing for enhanced polarization rotation and reduced hysteresis. However, a complete understanding of the underlying mechanisms governing the NPB and its impact on properties requires further investigation.

The grain size and morphology also play a crucial role in determining the properties of NN-based relaxors. Fine-grained microstructures with controlled grain orientation (texturing) are often desirable for enhancing piezoelectric properties [98, 101]. However, achieving high-density, fine-grained microstructures remains a challenge, particularly for KNN-based ceramics. The use of specific processing techniques, such as pressure-assisted sintering and spark plasma sintering, can be effective in enhancing the density of KNN ceramics.

3.5.3 Applications of NN-Based Relaxors

The unique combination of piezoelectric and dielectric properties of NN-based relaxors makes them suitable for a wide range of applications. Their most prominent application is as lead-free alternatives to PZT in piezoelectric devices, such as actuators, sensors, and transducers [95–99]. The superior piezoelectric properties achieved in some KNN-based ceramics, particularly those with high d_{33} values and improved temperature stability [101], are driving their adoption in high-power applications, including ultrasonic motors and resonators [105].

In addition to piezoelectric applications, NN-based relaxors are also being explored for energy storage applications [102, 104]. Their antiferroelectric or relaxor characteristics offer the potential for high energy density and efficiency. Strategies to enhance energy storage performance include doping with various elements to modify the polarization–electric field (P–E) loops and improve breakdown strength. The development of NN-based antiferroelectric ceramics with high recoverable energy density and efficiency is a significant area of ongoing research [99–105].

3.5.4 Challenges and Future Directions

Despite significant progress, several challenges remain in the development and application of NN-based relaxors. These include:

- Improving the densification and microstructure control of KNN-based ceramics: Achieving high density and fine-grained microstructures with controlled grain orientation remains a key challenge, which might be resolved by novel processing techniques and sintering aids.
- Understanding the fundamental mechanisms governing the NPB and other phase boundaries: A deeper understanding of the structure–property relationships at phase boundaries is essential for optimizing the piezoelectric and dielectric properties of NN-based relaxors.
- Developing high-quality single crystals of complex compositions: The growth of large, high-quality single crystals of NN-based materials remains challenging and is crucial for fundamental studies and high-performance applications.
- Enhancing the temperature stability of piezoelectric and dielectric properties: Improving the temperature stability of the properties is crucial for many applications, which might be resolved by better microstructure control.
- Reducing the leakage current density in thin films: Controlling stoichiometry and reducing leakage current are important for the successful application of KNN-based thin films in MEMS devices.
- Exploring new applications: The unique properties of NN-based relaxors offer potential for applications beyond traditional piezoelectric and dielectric applications, which warrants further research.

Future research should focus on addressing these challenges through innovative synthesis techniques, advanced characterization methods, and a deeper understanding of structure–property relationships. The combination of experimental and theoretical approaches, including advanced computational modeling, will be crucial for further advancements in this field. The development of new compositions and processing techniques, potentially incorporating novel dopants and nanostructuring strategies, holds immense potential for creating high-performance NN-based relaxors for a wide range of applications. The ongoing research in this area promises to lead to the development of environmentally friendly and high-performance materials for various technological applications.

3.6 Silver Tantalate Niobate (ATN)-Based Relaxors Antiferroelectrics

3.6.1 Synthesis Methods

Silver tantalate niobate (ATN), a lead-free perovskite material, has garnered significant attention for its potential applications in high-frequency devices [106] and energy storage [107–109]. Its relaxor and antiferroelectric characteristics offer unique opportunities for tunable microwave devices and high-power capacitors. This review synthesizes current research on ATN-based relaxors and antiferroelectrics, examining their synthesis methods, structural properties, and electrical performance while highlighting research gaps and future directions.

Several methods have been employed to synthesize ATN materials, each with its own advantages and limitations. The screen printing technique has been used to create ATN thick films for high-frequency applications [109]. This method allows for the creation of large-area films, but control over the microstructure and composition may be limited. Pulsed laser deposition (PLD) offers greater control over film thickness and crystallinity, making it a preferred technique for thin-film applications [110]. However, PLD can be more expensive and complex than screen printing. A wet-chemical method, employing a citric solution of Nb^{5+} and Ta^{5+}, has also been successfully used to synthesize nano-sized ATN powder with a grain size of approximately 30–40 nm [111]. This approach can lead to uniform and small particle sizes, beneficial for certain applications, but achieving high purity may require careful control of reaction conditions. Other methods, such as reactive DC magnetron sputtering, are also being explored to produce ATN thin films with customized chemical composition and microstructure [112]. The choice of synthesis method depends on the desired application and the required level of control over the material's properties.

3.6.2 Structural Properties and Phase Transitions

ATN exhibits complex structural behavior, including phase transitions that significantly influence its electrical properties. The M1–M2 phase transition in ATN thick films exhibits a ferroelectric relaxor-like character [106], with a tunability of 18% in the transition range. This relaxor behavior is attributed to the presence of polar nanoregions (PNRs) [7], where B-site ions deviate from their ideal positions, inducing weak ferroelectricity. The coexistence of ferroelectric and antiferroelectric coupling, predominantly related to B-site and A-site cations respectively, has been observed in related lead-based ABO_3-type relaxors [108], suggesting a frustrated ferrielectric state. Studies using transmission electron microscopy (TEM) have revealed the existence of ferroelastic and ferrielectric domains in $AgNbO_3$ single crystals [108], indicating a polar ferrielectric phase at room temperature. The effect of tensile strain on epitaxial ATN films has also been investigated, showing an expansion of the tetragonal and orthorhombic phase field regions and changes in octahedral tilting [109]. Further research is needed to fully elucidate the relationship between the microstructure, phase transitions, and the relaxor and antiferroelectric behavior in ATN.

3.6.3 Electrical Properties

The electrical properties of ATN are strongly influenced by its composition, microstructure, and temperature. At low frequencies and in a temperature range from −150 to 120 °C, the permittivity, losses, and tunability of ATN thick films have been measured [106, 109]. Compared to barium strontium titanate (BST), the permittivity of ATN is relatively low, for example, $\varepsilon = 180$ for $AgTa_{0.1}Nb_{0.9}O_3$ at room temperature [107]. High-field leakage current remains a challenge in ATN thin films [2], but the introduction of Mn^{4+} doping can effectively reduce this leakage [109]. Dielectric relaxation studies on ATN thin film varactors have shown frequency dispersion in the range of 1 kHz to 1 MHz [110], with tunability and K-factor values varying depending on the substrate material. The slow and weak relaxation of capacitance upon voltage switching suggests potential for applications requiring stable dielectric properties [110].

3.6.4 Energy Storage Applications

ATN-based materials, particularly those with antiferroelectric characteristics, are promising candidates for energy storage applications [106–112]. The incorporation of Ta into $AgNbO_3$ enhances antiferroelectricity, leading to increased energy storage density [112]. For example, Ta-modified $AgNbO_3$ ceramics have achieved a recoverable energy density of 4.2 J cm^{-3} and high thermal stability [112]. $NaNbO_3$-based relaxor antiferroelectric (AFE) ceramics have shown even higher energy storage densities, reaching 12.2 J cm^{-3} with acceptable energy efficiency (69%) at 68 kV/mm [110]. The synergistic effect of the orthorhombic AFE R phase and relaxor characteristics contributes to this high performance. Epitaxial lead-free thin films with coexisting ferroelectric and antiferroelectric phases have shown even more impressive energy storage densities (154 J cm^{-3}) [109], highlighting the potential of ATN-based materials for next-generation energy storage devices. However, challenges remain in improving energy efficiency and achieving high breakdown strength, especially in thin films.

3.6.5 Comparison with Other Materials

The electrical properties of ATN are often compared to those of other ferroelectric and antiferroelectric materials. Compared to BST, ATN exhibits lower permittivity [113], but it offers advantages in terms of tunability and potential for lead-free applications. The energy storage performance of ATN-based materials is often compared to that of lead-based antiferroelectrics, such as $Pb(Zr,Ti)O_3$ [35], demonstrating comparable or superior energy storage density in lead-free alternatives. The use of relaxor ferroelectrics and antiferroelectrics as fillers in composite dielectrics is being explored for high-voltage insulation systems [107, 113]. Studies have also shown that combining relaxor and antiferroelectric properties in BNT-based ceramics leads to improved energy storage performance [43], suggesting that further exploration of composite materials and phase engineering could lead to even better performance in ATN-based systems. The comparison with other materials underscores the importance of considering multiple factors, such as permittivity, tunability, loss tangent, energy storage density, and efficiency, when selecting a material for a specific application.

3.6.6 Limitations and Future Research Directions

Despite the significant progress in understanding and developing ATN-based materials, several limitations and research gaps remain. High-field leakage current in thin films needs to be addressed to fully exploit their potential for high-power applications. Further research is needed to elucidate the complex relationship between microstructure, phase transitions, and electrical properties, particularly the role of PNRs in the relaxor behavior [111, 112, 114]. The optimization of processing parameters for different synthesis methods is crucial for achieving high-quality materials with consistent properties. The development of robust theoretical models that accurately predict the electrical behavior of ATN under various conditions is essential for guiding material design and optimization. Exploring novel doping strategies and composite structures could lead to further improvements in energy storage density and efficiency. Finally, a comprehensive understanding of the long-term reliability and stability of ATN-based devices under operating conditions is necessary for their practical implementation.

4 Design Strategies for Enhanced Energy Storage

Several strategies have been employed to enhance the energy storage performance of perovskite relaxors. These include:

- **Chemical Modification**: Substituting ions at the A-site or B-site of the perovskite structure can introduce compositional disorder, leading to the formation of PNRs and improved relaxor behavior. This approach has been successfully employed in BNT, BT, and NN-based systems, as discussed above.
- **Microstructure Engineering**: Controlling the grain size and morphology of the ceramic can significantly affect energy storage properties. For example, reducing the grain size can increase the breakdown strength, while creating a specific grain boundary structure can enhance polarization.
- **Doping**: Introducing dopants can modify the electronic structure and defect chemistry of the material, influencing its dielectric and ferroelectric properties. This can lead to improved breakdown strength, reduced leakage current, and enhanced relaxor behavior.
- **Multilayer Structures**: Fabricating multilayer structures using different perovskite compositions can combine the advantages of each layer, leading to superior energy storage performance. For example, combining relaxor and paraelectric layers can enhance both polarization and breakdown strength.
- **High-Entropy Design**: Utilizing high-entropy perovskites, characterized by multiple cations occupying each lattice site, can result in enhanced energy storage performance due to the creation of local polymorphic distortions and ultrasmall polar nanoregions. This approach has led to giant recoverable energy densities in lead-free relaxors.

- **Phase Separation**: Controlling the phase separation can lead to dual-phase structures with synergistic effects, improving energy density and efficiency. This method creates interactive relaxor phases with diversified nanoscale polar structures and heterogeneous grain boundaries.

5 Challenges and Future Directions

Research on PZT continues to evolve, focusing on several key areas. These include developing new synthesis techniques to control material properties at the nanoscale, improving the understanding of domain wall dynamics and its influence on PZT's functional properties, exploring new applications of PZT in advanced technologies like optical memristors and energy harvesting, and developing high-performance lead-free alternatives to address environmental concerns. The development of novel processing methods such as liquid-phase flash sintering [25] and gravity-driven sintering [14] demonstrates the ongoing efforts to improve the efficiency and sustainability of PZT fabrication. Furthermore, the integration of computational methods and advanced characterization techniques is crucial for understanding the complex relationships between microstructure, processing, and properties of PZT materials [45]. The ongoing research on PZT and the development of lead-free alternatives promise to further expand the applications of these versatile materials in various technological domains. The combination of experimental and computational approaches will continue to be essential in unlocking the full potential of PZT and its lead-free counterparts [45]. The development of high-performance, environmentally friendly piezoelectric materials remains a priority, driving innovation in materials science and engineering [42]. The ultimate goal is to create sustainable and highly efficient piezoelectric devices for applications ranging from energy harvesting to advanced sensing technologies [42]. The exploration of PZT's potential in novel applications, coupled with the continued development of lead-free alternatives, ensures its ongoing relevance in the realm of materials science and technology. Further research is needed to fully understand the complex interplay between processing parameters, microstructure, and the resultant properties of PZT and its lead-free counterparts [46]. The development of more efficient and sustainable methods for material extraction and processing is also crucial for minimizing the environmental impact of both PZT and lead-free alternatives [44]. The field of piezoelectric materials is dynamic, with ongoing advancements shaping the future of numerous technological applications.

Despite significant advancements, several challenges remain in the development of high-performance perovskite relaxors for energy storage:

- **Breakdown Strength**: Improving the breakdown strength remains a significant challenge, as it directly limits the achievable energy density. Strategies to enhance breakdown strength include microstructure engineering, doping, and the use of multilayer structures.

- **Leakage Current**: High leakage current can significantly reduce energy efficiency. Strategies to reduce leakage current include defect engineering and improving the quality of interfaces in multilayer structures.
- **Thermal Stability**: Maintaining stable energy storage performance over a wide temperature range is crucial for many applications, which might be addressed by raising the understanding of the thermal stability of lead-free relaxors.
- **Frequency Stability**: The energy storage properties of relaxors can be frequency-dependent. Developing materials with stable performance over a broad frequency range is essential for practical applications.
- **Lead-Free Materials**: While significant progress has been made in developing lead-free relaxors, further research is needed to achieve performance comparable to that of lead-based PZT.

Future research directions include exploring novel perovskite compositions, developing advanced processing techniques, and utilizing computational modeling to guide the design of high-performance materials. A deeper understanding of the relationship between microstructure, defect chemistry, and energy storage properties is also crucial. Furthermore, the development of more robust theoretical models that accurately predict the energy storage performance of perovskite relaxors is needed. This would accelerate the development and optimization of these materials for practical applications.

6 Conclusion

Perovskite ferroelectric, relaxor, and antiferroelectric materials have attracted significant attention for high-performance capacitive energy storage applications due to their excellent dielectric properties, high polarization, and fast charge–discharge capabilities. These materials exhibit high-energy storage densities, making them suitable for advanced electronic and power applications. Both lead-based and lead-free perovskite systems have demonstrated promising recoverable energy densities exceeding 10 J cm^{-3}. However, key challenges remain in optimizing their performance, particularly in enhancing breakdown strength, reducing leakage current, and improving thermal and frequency stability. To address these limitations, researchers are actively exploring various strategies, including chemical modifications, microstructure engineering, and the incorporation of dopants to tailor the dielectric and ferroelectric properties. The use of nanostructuring and domain engineering techniques has also shown potential in improving energy storage efficiency by minimizing energy losses and enhancing polarization response. Additionally, the exploration of high-entropy perovskite compositions is emerging as a promising avenue to achieve improved material stability and tenability. Another crucial aspect is the integration of experimental and computational approaches to accelerate material discovery and optimization. Density functional theory (DFT) calculations, phase-field simulations, and machine learning-based predictive modeling are increasingly being used to guide the design of next-generation perovskite relaxors with enhanced energy storage performance. Moreover, the development of novel processing techniques, such as spark plasma sintering and thin-film fabrication methods, can further improve material quality and reliability.

Overall, perovskite-based ferroelectric, relaxor, and antiferroelectric materials hold great promise for capacitive energy storage. Continued advancements in material design, processing methods, and theoretical modeling will be essential in unlocking their full potential for next-generation energy storage applications.

References

1. Li D, Zeng X, Li Z, Shen Z, Hao H, Luo W, Wang X, Song F, Wang Z, Li Y (2021) Progress and perspectives in dielectric energy storage ceramics. J Adv Ceram 10:675–703. https://doi.org/10.1007/s40145-021-0500-3
2. Sekhar KC, Silva JPB, Kamakshi K, Dastan D, Annapureddy V, Pereira M, Sekhar KC (2022) Are lead-free relaxor ferroelectric materials the most promising candidates for energy storage capacitors? Prog Mater Sci 132:101046. https://doi.org/10.1016/j.pmatsci.2022.101046
3. Veerapandiyan V, Benes F, Gindel T, Deluca M (2020) Strategies to improve the energy storage properties of perovskite lead-free relaxor ferroelectrics: a review. Materials 13:5742. https://doi.org/10.3390/ma13245742
4. Liu Y, Lin Y, Haibibu A, Xu W, Zhou Y, Li L, Kim SH, Wang Q (2021) Relaxor ferroelectric polymers: insight into high electrical energy storage properties from a molecular perspective. Small Sci 1:2000061. https://doi.org/10.1002/SMSC.202000061
5. Prasatkhetragarn A, Sareein T, Triamnak N, Yimnirun R (2022) Dielectric and ferroelectric properties of modified-$BaTiO_3$ lead-free ceramics prepared by solid solution method. Ferroelectrics 586:224–241. https://doi.org/10.1080/00150193.2021.2014274
6. Veerapandiyan VK, Popov MN, Mayer F, Spitaler J, Svirskas S, Kalendra V, Lins J, Canu G et al (2021) Origin of relaxor behavior in barium titanate-based lead-free perovskites. Adv Electron Mater 8:2100812. https://doi.org/10.1002/aelm.202100812
7. Cao L, Yuan Y, Meng X, Li E, Tang B (2022) Ferroelectric-relaxor crossover and energy storage properties in $Sr_2NaNb_5O_{15}$-based tungsten bronze ceramics. ACS Appl Mater Interfaces 14:9318–9329. https://doi.org/10.1021/acsami.1c23673
8. Deka B, Cho KH (2021) $BiFeO_3$-based relaxor ferroelectrics for energy storage: progress and prospects. Materials 14:7188. https://doi.org/10.3390/ma14237188
9. Shen Z, Wang X, Luo B, Li L (2015) $BaTiO_3$–$BiYbO_3$ perovskite materials for energy storage applications. J Mater Chem A 3:18146–18153. https://doi.org/10.1039/C5TA03614C
10. Kobald H, Kobald AM, Panzic I, Deluca M (2025) Enhanced energy storage in relaxor (1-x)$Bi_{0.5}Na_{0.5}TiO_3$-xBaZryTi1-yO_3 thin films by morphotropic phase boundary engineering. Commun Mater 6:12. https://doi.org/10.1038/s43246-024-00730-x
11. Peddigari M, Wang B, Wang R, Yoon WH, Jang J, Lee H, Song K, Hwang GT, Wang K, Hou Y et al (2023) Giant energy density via mechanically tailored relaxor ferroelectric behavior of PZT thick film. Adv Mater 35:2302554. https://doi.org/10.1002/adma.202302554
12. Ma Y, Song J, Wang X, Liu Y, Zhou J (2021) Synthesis, microstructure and properties of magnetron sputtered lead zirconate titanate (PZT) thin film coatings. Coatings 11:944. https://doi.org/10.3390/coatings11080944
13. Yao S, Liu S, Wang B, Hong Y, Zhang C, Lim CW, Zhang G, Yang Z (2021) A gravity-driven sintering method to fabricate geometrically complex compact piezoceramics. Nat Commun 12:6066. https://doi.org/10.1038/s41467-021-26373-x

14. Li J (2022) Microstructure and piezoelectric properties of lead zirconate titanate nanocomposites reinforced with in-situ formed ZrO_2 nanoparticles. Materials 15:1389. https://doi.org/10.3390/ma15041389
15. Qiao H, Jones JL, Balke N (2024) Effect of subcoercive degradation on the local piezoelectric properties in lead zirconate titanate ceramics. J Am Ceram Soc 108:e20277. https://doi.org/10.1111/jace.20277
16. Long X, Wei X, Qiu Y, Jiang M, Chen Z, Song Y, Bi L, Wang J, Wang S, Liao J (2021) Lead zirconate titanate aerogel piezoelectric composite designed with a biomimetic shell structure for underwater acoustic transducers. Chem Commun 57:9764–9767. https://doi.org/10.1039/d1cc03037j
17. Li Z, Thong HC, Zhang YF, Xu Z, Zhou Z, Liu YX, Cheng YYS, Wang SH, Zhao C, Chen F, Bi K, Han B, Wang K (2020) Defect engineering in lead zirconate titanate ferroelectric ceramic for enhanced electromechanical transducer efficiency. Adv Funct Mater 31:2005012. https://doi.org/10.1002/adfm.202005012
18. Arya K, Singh RP, Chakrabarti T (2024) Novel liquid-phase flash sintering of lead zirconate titanate piezoceramics. J Am Ceram Soc 107:8007–8022. https://doi.org/10.1111/jace.20075
19. Eichhorn F, Kellermann S, Betke U, Fey T (2020) Phase evolution, filler-matrix interactions, and piezoelectric properties in lead zirconate titanate (PZT)-filled polymer-derived ceramics (PDCs). Materials 13:1520. https://doi.org/10.3390/ma13071520
20. Bhattarai MK, Shweta S, Choudhary S, Meyer HM, Thapaliya B, Weiner B, Katiyar RS, Morell G (2024) Exploring lead zirconate titanate, the potential advancement as an anode for Li-ion batteries. ACS Omega 9:19219–19226. https://doi.org/10.1021/acsomega.4c00090
21. Song L, Glinsek S, Defay E (2021) Toward low-temperature processing of lead zirconate titanate thin films: advances, strategies, and applications. Appl Phys Rev 8:041315. https://doi.org/10.1063/5.0054004
22. Li J, Qu W, Daniels J, Wu H, Liu L, Wu J, Wang M, Checchia S, Yang S et al (2023) Lead zirconate titanate ceramics with aligned crystallite grains. Science 380:87–93. https://doi.org/10.1126/science.adf6161
23. Tiwari B, Babu T, Choudhary RNP (2021) Piezoelectric lead zirconate titanate as an energy material: a review study. Mater Today Proc 43:407–412. https://doi.org/10.1016/j.matpr.2020.11.692
24. Huang Y, Zhang L, Jing R, Tang M, Alikin D, Shur V, Wei X, Jin L (2023) Lead zirconate titanate-based ceramics with high piezoelectricity and broad usage temperature range. Chem Eng J 477:147192. https://doi.org/10.1016/j.cej.2023.147192
25. Li Z, Thong HC, Zhang YF, Xu Z, Zhou Z, Liu YX, Cheng YYS, Wang SH, Zhao C, Chen F, Bi K, Han B, Wang K (2021) Defect engineering in lead zirconate titanate ferroelectric ceramic for enhanced electromechanical transducer efficiency. Adv Funct Mater 31:2005012. https://doi.org/10.1002/adfm.202005012
26. Hidayat D, Taufik M, Setianto S (2022) One-step synthesis of lead zirconate titanate particles using a solid-state reaction method. Heliyon 8(3):e09125. https://doi.org/10.1016/j.heliyon.2022.e09125
27. Liu CL, Du Q, Zhou H, Chen S, Wu JM, Zhang G, Shi YS (2023) 3D printing of lead zirconate titanate piezoelectric ceramics via digital light processing (DLP). Ceram Int 49(17):28492–28499
28. Atanova AV, Zhigalina OM, Khmelenin DN et al (2021) Microstructure analysis of porous lead zirconate–titanate films. J Am Ceram Soc 105:639–652. https://doi.org/10.1111/jace.18064
29. Shi Y, He R, Zhang B, Zhong Z (2024) Revisiting the phase diagram and piezoelectricity of lead zirconate titanate from first principles. Phys Rev B 109:174104. https://doi.org/10.1103/PhysRevB.109.174104
30. Akkopru-Akgun B, Bayer TJM, Tsuji K, Randall CA, Lanagan MT, Trolier-McKinstry S (2021) The influence of Mn doping on the leakage current mechanisms and resistance degradation behavior in lead zirconate titanate films. Acta Mater 208:116680. https://doi.org/10.1016/j.actamat.2021.116680

31. Godard N, Grysan P, Defay E, Glinšek S (2021) Growth of {100}-oriented lead zirconate titanate thin films mediated by a safe solvent. J Mater Chem C 9:281–287. https://doi.org/10.1039/D0TC04066E
32. Gupta S, Wang D, Randall CA, Trolier-McKinstry S (2021) Comparison of different sintering aids in cold sinter-assisted densification of lead zirconate titanate. J Am Ceram Soc 104:5479–5488. https://doi.org/10.1111/jace.17934
33. Wang H, Jiang X, Liu X, Yang R, Yang Y, Zheng Q, Kwok K, Lin D (2019) An effective approach to achieve high energy storage density and efficiency in BNT-based ceramics by doping $AgNbO_3$. Dalton Trans 48:17864–17873. https://doi.org/10.1039/c9dt03654g
34. Pan T, Zhang J, Che D, Wang Z, Wang J, Wang J, Wang Y (2023) Improved capacitive energy storage in sodium niobate-based relaxor antiferroelectric ceramics. Appl Phys Lett 122:072902. https://doi.org/10.1063/5.0134282
35. Zhu W, Shen Z, Deng W, Li K, Luo W, Song F, Zeng X, Wang Z, Li Y (2023) A review: (Bi, Na)TiO_3 (BNT)-based energy storage ceramics. J Mater 10:86–123. https://doi.org/10.1016/j.jmat.2023.05.002
36. Jiang S, Zhang J, Karpinsky DV, Li J, Liu L, Wang Y (2024) Optimization in the energy storage properties of $Bi_{0.5}Na_{0.5}TiO_3$-$SrTiO_3$ ceramics by introducing $Bi(Mg_{0.5}Zr_{0.5})O_3$. J Am Ceram Soc 107:4846–4853. https://doi.org/10.1111/jace.19774
37. Liu H, Sun Z, Zhang J, Luo H, Zhang Y, Sanson A, Hinterstein M, Liu L, Neuefeind J, Chen J (2024) Chemical framework to design linear-like relaxors toward capacitive energy storage. J Am Chem Soc 146:3498–3507. https://doi.org/10.1021/jacs.3c13405
38. Jain A, Wang Y, Wang N, Li Y, Wang F (2020) Synergetic effect of $BiYb_{0.9}Sc_{0.1}O_3$ substitution on the energy storage performance of $Ba_{0.9}Sr_{0.1}Ti_{0.9}Zr_{0.1}O_3$ ferroelectric ceramic. J Phys D Appl Phys 53:235301. https://doi.org/10.1088/1361-6463/ab7bb1
39. Yoo IR, Choi SH, Park JY, Kim MS, Yadav AK, Cho KH (2024) A superparaelectric state in relaxor ferroelectric (Sr, Bi)TiO_3-Bi(Mg, Ti)O_3-modified BaT_iO_3 ceramics to achieve high energy storage performance. Materials 17:20426. https://doi.org/10.3390/ma17020426
40. Guan Z, Wang J, Pan T, Zhang J, Wang J, Wang Y (2023) Improved electric breakdown strength and energy storage performances in $La(Mg_{2/3}Nb_{1/3})O_3$ and MnO_2-modified $BiFeO_3$-$SrTiO_3$ ceramics. Inorg Chem 62:1234–1239. https://doi.org/10.1021/acs.inorgchem.2c03824
41. Chen J, Zhao P, Si F, Zhang S, Fang Z, Tang B (2023) Dielectric, energy storage, and charge-discharge properties of Yb-modified $Sr_{0.7}Bi_{0.2}TiO_3$ relaxor ferroelectric ceramic. J Am Ceram Soc 107:2504–2516. https://doi.org/10.1111/jace.19600
42. Nguyen MD, Houwman E, Birkhlzer Y, Vu HN, Koster G, Rijnders G (2024) Toward design rules for multilayer ferroelectric energy storage capacitors: a study based on lead-free and relaxor-ferroelectric/paraelectric multilayer devices. Adv Mater 36:2402070. https://doi.org/10.1002/adma.202402070
43. Chen L, Deng S, Liu H, Wu J, Qi H, Chen J (2022) Giant energy-storage density with ultrahigh efficiency in lead-free relaxors via high-entropy design. Nat Commun 13:2347. https://doi.org/10.1038/s41467-022-30821-7
44. Sun Z et al (2023) Superior capacitive energy-storage performance in Pb-free relaxors with a simple chemical composition. J Am Chem Soc 145:2495–2502. https://doi.org/10.1021/jacs.2c12200
45. Bajpai PK (2012) Dielectric relaxation phenomena in some lead and non-lead based ferroelectric relaxor materials: recent advances. Sci Technol 189:233–240. https://doi.org/10.4028/www.scientific.net/SSP.189.233
46. Liu Y-X, Zhou J, Jiang Y, Li C-B-W, Li C, Lu J-T, Xu Z, Yao F-Z, Nan H, Wang D, Xu L et al (2024) Multi-length engineering of (K, Na)NbO_3 films for lead-free piezoelectric acoustic sensors with high sensitivity. Adv Funct Mater 34:2312699. https://doi.org/10.1002/adfm.202312699
47. Feliksik K, Makowska J, Bartkowska JA, Pikula T, Panek R, Strczewska O, Adamczyk-Habrajska M (2024) Dielectric, electric and pyroelectric properties of $Ba_{1-x}Ca_xTiO_3$ ceramics. Materials 17:6040

48. Wang M (2024) Preparation of barium titanate-based ferroelectric ceramics by solid-phase sintering and its energy storage performance. Sci Adv Mater 46:1021–1028. https://doi.org/10.1166/sam.2024.4608
49. Jiang B, Iocozzia J, Zhao L, Zhang H, Harn YW, Chen Y, Lin Z (2019) Barium titanate at the nanoscale: controlled synthesis and dielectric and ferroelectric properties. Chem Soc Rev 48:3812–3829. https://doi.org/10.1039/c8cs00583d
50. Zamperlin N, Bottacini A, Callone E, Pegoretti A, Fontana M, Dir S (2022) Barium titanate functionalization with organosilanes: effect on particle compatibility and permittivity in nanocomposites. Molecules 27:6499. https://doi.org/10.3390/molecules27196499
51. Haronin V, Yang Z, Grigalaitis R, Calisir I, Banys J, Hall D (2024) Broadband dielectric spectroscopy of Nb-doped 0.7$BiFeO_3$–0.3$BaTiO_3$ ceramics. J Phys Commun 8:1154. https://doi.org/10.1088/2399-6528/ad55a2
52. Liu J, Qu S, Chen Z, Ni J, Jia Z, Yin X, Ma Y, Li T (2024) Significantly enhanced electrostatic energy storage performances of polyetherimide nanocomposites with ultralow loadings of barium titanate nanoparticles. Appl Phys Lett 124:152903. https://doi.org/10.1063/5.0190323
53. Okhay O, Tkach A (2023) Current achievements in flexible piezoelectric nanogenerators based on barium titanate. Nanomaterials 13:988. https://doi.org/10.3390/nano13060988
54. Reda M, Ateia EE, El-Dek SI, Arman MM (2024) New insights into optical properties and applications of Zr-doped $BaTiO_3$. Appl Phys A 130:240. https://doi.org/10.1007/s00339-024-07381-2
55. Wu B, Zheng H, Wu Y-Q, Huang Z, Thong H-C, Tao H, Ma J, Zhao C, Xu Z, Liu Y-X, Xing Z, Liang N, Yao F-Z, Wu C-F, Wang K, Han B (2024) Origin of ultrahigh-performance barium titanate-based piezoelectrics: stannum-induced intrinsic and extrinsic contributions. Nat Commun 15:7700. https://doi.org/10.1038/s41467-024-52031-z
56. Wu T-B, Lin J-N (1994) Transition of compensating defect mode in niobium-doped barium titanate. J Am Chem Soc 77:759–764. https://doi.org/10.1111/j.1151-2916.1994.tb05362.x
57. Jain A, Panwar AK (2020) Synergetic effect of rare-earths doping on the microstructural and electrical properties of Sr and Ca co-doped $BaTiO_3$ nanoparticles. Ceram Int 46:10270–10278. https://doi.org/10.1016/j.ceramint.2020.01.020
58. Li B, Zheng T, Wei X, Zhong S, Li Y, Wu J (2022) Constructing relaxor/ferroelectric pseudo-composite to reveal the domain role in electrostrain of bismuth ferrite-barium titanate based ceramics. ACS Appl Mater Interfaces 14:12345–12353. https://doi.org/10.1021/acsami.2c03817
59. Choi Y, Hwang Y, Yu M-K, Lee K, Min K (2022) Effects of barium titanate on the dielectric constant, radiopacity, and biological properties of tricalcium silicate-based bioceramics. Dent Mater J 41:682–690. https://doi.org/10 4012/dmj.2022-069
60. Putra DF, Aji BB, Ningsih HS, Wu T-W, Nakanishi A, Moriga T, Shih S (2023) Preparation and characterization of freeze-dried-tricalcium phosphate/barium titanate/collagen composite scaffolds for bone tissue engineering in orthopedic applications. Ceramics 6:132. https://doi.org/10.3390/ceramics6040132
61. Sood A, Singhmr R, Sahoo S, Lee D, Kim C, Kumar A, Han S (2023) Physicochemical, electrochemical, and biological characterization of field-assisted gold nanocluster-coated barium titanate nanoparticles for biomedical applications. J Mater Chem B 12:525–539. https://doi.org/10.1039/d3tb01928d
62. Zate TT, Abdurrahmanolu C, Esposito V, Haugen AB (2025) Textured lead-free piezoelectric ceramics: a review of template effects. Materials 18:477. https://doi.org/10.3390/ma18030477
63. Li Y, Lu G, Zhao Y, Zhao R, Zhao J, Hao J, Bai W, Li P, Li W (2024) Ferroelectric and relaxor-ferroelectric phases coexisting boosts energy storage performance in $(Bi_{0.5}Na_{0.5})TiO_3$-based ceramics. Molecules 29:3187. https://doi.org/10.3390/molecules29133187
64. Zhang DQ, Li QI, Hou N, Zhao Y, Cai CF (2008) Recent progresses of sodium bismuth titanate based lead-free piezoelectric ceramics In: IEEE symposium on piezoelectricity, acoustic waves, and device applications, pp 493–498.https://doi.org/10.1109/SPAWDA.2008.4775838
65. Li Z, Yang Q, Wang C, Zhang Y, Wang Z, Gao B, Li Z, Wang Z, Yan X, Tao A, Wang D, Niu Y (2023) A brief review of sodium bismuth titanate-based lead-free materials for

energy storage: solid solution modification, metal/metallic oxide doping, defect engineering and process optimizing. Crystals 13:295. https://doi.org/10.3390/cryst13020295
66. Jiang S, Zhang J, Luo H, Liu L, Li J, Karpinsky DV, Wang J, Wang Y (2024) Enhanced comprehensive energy storage properties in $Bi_{0.5}Na_{0.5}TiO_3$-based relaxor ferroelectric ceramics under a moderate electric field. Appl Phys Lett 124:251901. https://doi.org/10.1063/5.0217327
67. Huiping Y, Xuefan Z, Haojie F, Xiaoyun Z, Hang L, Dou Z (2022) Field-induced strain property of lead-free ferroelectric ceramics based on sodium bismuth titanate. J Inorg Mater 37:603–610. https://doi.org/10.15541/jim20210453
68. Akshayveer Buroni FC, Melnik, Roderiguez-Tembleque L, Sez A (2024) Phase-field thermo-electromechanical modelling of lead-free BNT-based piezoelectric materials. https://doi.org/10.48550/arXiv.2410.04548
69. Wen L, Wu X, Yin J, Zhang Y, Yang D, Wu J (2024) Realization of the giant pyroelectric response via modulated polar structures. Small 20:2307326. https://doi.org/10.1002/smll.202307326
70. Zhang MH, Cheon Thong H, Lu YX, Sun W, Li JF, Wang K (2017) (K, Na)NbO_3-based lead-free piezoelectric materials: an encounter with scanning probe microscopy. J Kor Ceram Soc 54:261–271. https://doi.org/10.4191/kcers.2017.54.4.10
71. Zhang L, Jing R, Huang Y, Hu Q, Aikin DO, Shur VY, Gao J, Wei X, Zhang L, Liu G, Yan Y, Jin L (2022) Enhanced antiferroelectric-like relaxor ferroelectric characteristic boosting energy storage performance of $(Bi_{0.5}Na_{0.5})TiO_3$-based ceramics via defect engineering. J Materiomics 8:527–536. https://doi.org/10.1016/j.jmat.2022.01.007
72. Sehga K (2016) Synthesis and characterization of Ba doped bismuth sodium titanate (BNT). http://hdl.handle.net/10266/4387
73. Preeti M, Saurav M, Gaur A (2023) Dielectric and transport properties of Mn-modified bismuth sodium strontium titanate composite. Ferroelectrics 617:43–50. https://doi.org/10.1080/00150193.2023.2271132
74. Reshetnikova EA, Lisnevskaya IV, Zalyubovskaya EA, Butova V, Soldatov A (2020) The effect of hydrothermal synthesis parameters on the formation of sodium bismuth titanate. Comments Inorg Chem 40:314–326. https://doi.org/10.1080/02603594.2020.1813728
75. Azam F, Rafiq MA, Ahmed F, Moqbool A, Fayyaz O, Imran Z, Habib MS, Shakoor R (2024) Microwave-assisted fabrication and characterization of carbon fiber-sodium bismuth titanate composites. Crystals 14:798. https://doi.org/10.3390/cryst14090798
76. Liu C, Xu J, Zheng T, Qian H, Tao Z, Jiang X, Lyu C, Liu Y, Lyu Y (2023) Decoding the atomic-scale structural origin of large strain performance in BNT-6BT based relaxor ferroelectrics. J Phys Chem Lett 14:6234–3240. https://doi.org/10.1021/acs.jpclett.3c01318
77. Bai W, Wang L, Zhao X, Zheng P, Wen F, Li L, Zhai J, Ji Z (2019) Tailoring frequency-insensitive large field-induced strain and energy storage properties in $(Ba_{0.85}Ca_{0.15})(Zr_{0.1}Ti_{0.9})O_3$-modified $(Bi_{0.5}Na_{0.5})TiO_3$ lead-free ceramics. Dalton Trans 48:10160–10173. https://doi.org/10.1039/c9dt01738k
78. Tang X, Hu Z, Kova V, Yang B, Smith GC, Yan H (2023) Energy storage properties of samarium-doped bismuth sodium titanate-based lead-free ceramics. Chem Eng J 473:145363. https://doi.org/10.1016/j.cej.2023.145363
79. Li Y, Rao R, Wang Y, Du H, Shi J, Liu X (2023) Effect of chemical inhomogeneity on the dielectric and impedance behaviors of bismuth sodium titanate based relaxors. ECS J Solid State Sci Technol 12:013005. https://doi.org/10.1149/2162-8777/acb28e
80. Bin C, Hou X, Liao L, Liu Y, Yang H, Liu Y, Wangn J (2023) Improved energy storage performance in rare-earth modified lead-free BNT-based relaxor ferroelectric ceramics. Appl Phys Lett 123:012901. https://doi.org/10.1063/5.0158219
81. Sun H, Wu X, Peng D, Kwok K (2017) Room-temperature large and reversible modulation of photoluminescence by in situ electric field in ergodic relaxor ferroelectrics. ACS Appl Mater Interfaces 9:34042–34049. https://doi.org/10.1021/acsami.7b09354
82. Mao P, Guo Y, Wang T, He L, Li W, Liu Z, Xie B, Guo K, Shu L, Gao J (2024) Boosting high electric breakdown strength for excellent energy storage performance in $Bi_{0.5}Na_{0.5}TiO_3$-based

lead-free ceramics via a high entropy strategy. ACS Appl Mater Interfaces 17:1485–1498. https://doi.org/10.1021/acsami.4c15339
83. Li H, Gao J, Li M, Zhang Q, Zhang Y (2023) Optimized strain response in $(Co_{0.5}Nb_{0.5})^{4+}$-doped $76Bi_{0.5}Na_{0.5}TiO_3$–$24SrTiO_3$ relaxors. Coatings 13:1331. https://doi.org/10.3390/coatings13081331
84. Lin Y, Nie H, Xie M, Cao F, Peng P, Wang G (2023) Ultrahigh polarization $Bi_{0.5}Na_{0.5}TiO_3$-based relaxor ceramics for force-electric conversion. Appl Phys Lett 123:083903. https://doi.org/10.1063/5.0153889
85. Zhang H, Krynski M, Fortes AD, Saunders TG, Palma M, Hao Y, Krok F, Yan H, Abrahams I (2024) Origin of polarization in bismuth sodium titanate-based ceramics. J Am Chem Soc 146:5569–5579. https://doi.org/10.1021/jacs.3c13927
86. Amu D, Jesse NO, Zhang C, Hussain S, Issaka E, Otoo SLeumas, Agyemang MF, Okoroafo E (2022) Dielectric properties and relaxor behavior of (1-x)$Ba(Zr_{0.15}Ti_{0.85})O_3$ - xBa(Mg(1/3)Nb(2/3))O_3 ceramics for capacitor applications. Mater Chem Phys 287:126328. https://doi.org/10.1016/j.matchemphys.2022.126328
87. Koroleva M, Ishchenko A, Vlasov M, Krasnov AG, Istomina E, Shein I, Weinstein I, Piir IV (2022) Structural, optical, luminescence, and electrical properties of Eu/Li- and Eu/Na-codoped magnesium bismuth niobate pyrochlores. Inorg Chem 61:9295–9307. https://doi.org/10.1021/acs.inorgchem.2c01037
88. Sahoo S, Badapanda T, Kumar D, Rout SK, Mohanty S, Ray J, Tripathy SN (2023) Structural, relaxor behavior, and energy storage performance of $BaTiO_3$–$Bi(Mg_{2/3}Nb_{1/3})O_3$ solid solutions for potential MLCC applications. Ceram Int 49:37700–37711. https://doi.org/10.1016/j.ceramint.2023.09.096
89. Dasin NAM, Tan KB, Khaw CC, Zainal Z, Chen SK (2017) Subsolidus solution and electrical properties of Sr-substituted bismuth magnesium niobate pyrochlores. Ceram Int 43:10183–10191. https://doi.org/10.1016/j.ceramint.2017.05.043
90. Kim SS, Kim WJ (2005) Electrical properties of sol-gel derived pyrochlore-type bismuth magnesium niobate $Bi_2(Mg_{1/3}Nb_{2/3})_2O_7$ thin films. J Cryst Growth 281:432–439. https://doi.org/10.1016/j.jcrysgro.2005.04.033
91. Bobic JD, Vijatovic Petrovic MM, Stojanovic BD (2018) Review of the most common relaxor ferroelectrics and their applications. In: Stojanovic BD (ed) Metal oxides, magnetic, ferroelectric, and multiferroic metal oxides. Elsevier. https://doi.org/10.1016/B978-0-12-811180-2.00011-6
92. Gao L, Jiang S, Li R, Li B, Li Y (2012) Structure and dielectric properties of sputtered bismuth magnesium niobate thin films. Thin Solid Films 520:6295–6298. https://doi.org/10.1016/j.tsf.2012.06.035
93. Tan PY, Tan KB, Khaw CC, Murthy HCA, Balachandran R, Chen SK, Lee OJ, Chan KY, Lu M (2024) Electrical circuit modeling for the relaxor response of bismuth magnesium tantalate pyrochlore. J Sci Adv Mater Dev 9:100715. https://doi.org/10.1016/j.jsamd.2024.100715
94. Tan PY, Tan KB, Khaw CC, Zainal Z, Chen SK, Chon MP (2014) Phase equilibria and dielectric properties of $Bi_{3+(5/2)x}Mg_{2-x}Nb_{3-(3/2)x}O_{14-x}$ cubic pyrochlores. Ceram Int 40:4237–4246. https://doi.org/10.1016/j.ceramint.2013.08.087
95. Wu J, Zhang T, Gao X, Ning L, Hu Y, Lou X, Liu Y, Sun N, Li Y (2024) Excellent energy storage and photovoltaic performances in $Bi_{0.45}Na_{0.45}Ba_{0.1}TiO_3$-based lead-free ferroelectricity thin film. Ceramics 7:1043–1052. https://doi.org/10.3390/ceramics7030068
96. Jin L, Pang J, Luo W, Lan Y, Du H, Yang S, Li F, Tian Y, Wei X, Xu Z, Guo D, Gao F (2019) Phase transition behavior and high electrostrictive strains in $Bi(Li_{0.5}Nb_{0.5})O_3$-doped lead magnesium niobate-based solid solutions. J Alloys Compd 806:206–214. https://doi.org/10.1016/j.jallcom.2019.07.227
97. Park S, Jeong HY, Kim S, Peddigari M, Hwang GT, Moon GD, Roh JW, Min Y (2025) Strategic design of emerging (K, Na)NbO_3-based perovskites for high-performance piezocatalysis and photo-piezocatalysis. Nanoscale Adv Article. https://doi.org/10.1039/d4nr04415k
98. Malič B, Koruza J, Hreščak J, Bernard J, Wang K, Fisher J, Benčan A (2015) Sintering of lead-free piezoelectric sodium potassium niobate ceramics. Materials 8:8117–8146. https://doi.org/10.3390/ma8125449

99. Yang L, Kong X, Cheng Z, Zhang S (2020) Enhanced energy storage performance of sodium niobate-based relaxor dielectrics by ramp-to-spike sintering profile. ACS Appl Mater Interfaces 12:32834–32841. https://doi.org/10.1021/acsami.0c08737
100. Hussain F, Khesro A, Lu Z, Alotaibi N, Mohamad AA, Wang G, Wang D, Zhou D (2020) Acceptor and donor dopants in potassium sodium niobate-based ceramics. Front Mater 7. https://doi.org/10.3389/fmats.2020.00160
101. Yang L, Kong X, Li Q, Lin Y, Zhang S, Nan C (2022) Excellent energy storage properties achieved in sodium niobate-based relaxor ceramics through doping tantalum. ACS Appl Mater Interfaces 14:32218–32226. https://doi.org/10.1021/acsami.2c05205
102. Wang X, Wang X, Huan Y, Li C, Ouyang J, Wei T (2022) A combined optimization strategy for improvement of comprehensive energy storage performance in sodium niobate-based antiferroelectric ceramics. ACS Appl Mater Interfaces 14:9330–9339. https://doi.org/10.1021/acsami.1c23914
103. Yang L, Kong X, Cheng Z, Zhang S (2021) Enhanced energy density and electric cycling reliability via MnO_2 modification in sodium niobate-based relaxor dielectric capacitors. J Mater Res 36:1214–1222. https://doi.org/10.1557/s43578-020-00085-2
104. Wang X, Sun X, Yang Y, Hao X, Lv X, Zhang Y, Ma Y, Zhang X, Wu H, Wu J (2024) Modified ReEntrant-like (K, Na)NbO_3-based relaxors with superior electrostrain properties. Adv Funct Mater 34:2406154. https://doi.org/10.1002/adfm.202406154
105. Soheli SN, Lu Z, Sun D, Shyha I (2024) Lead-free $NaNbO_3$-Based ceramics for electrostatic energy storage capacitors. Ceramics 7:712–734. https://doi.org/10.3390/ceramics7020047
106. Liu X, Zhai Y, Xie X, Zhou R, Li X, Liu S (2020) Relaxation dielectric enhancement of barium strontium titanate ceramics by trace doping sodium niobate. Mat Technol 37:2049–2055. https://doi.org/10.1080/10667857.2020.1821454
107. Li G, Liu HY, Zhao L, Gao J, Li J, Zhu J (2022) Local [001]c off-centering nanoregions in silver niobate. J Appl Phys 131:124107. https://doi.org/10.1063/5.0084467
108. Zhao L, Liu Q, Gao J, Zhang S, Li J (2017) Lead-free antiferroelectric silver niobate tantalate with high energy storage performance. Adv Mater 29:1701824. https://doi.org/10.1002/adma.201701824
109. Krylov AS, Krylova SN, Vtyuri AN, Roginskii EM, Li J, Tian Y, Wei X (2023) Temperature phase transitions in silver niobate and lithium tantalate-modified silver niobate ceramics. Phys Chem Chem Phys 25:11410–11417. https://doi.org/10.1039/D3CP00607G
110. Koh JH, Khartsev SI, Grishin A (2000) Ferroelectric silver niobate-tantalate thin films. Appl Phys Lett 77:4416–4418. https://doi.org/10.1063/1.1334655
111. Tian Y, Jin L, Hu Q, Yu K, Zhuang Y, Viola G, Abrahams I, Xu Z, Wei X, Yan H (2019) Phase transitions in tantalum-modified silver niobate ceramics for high power energy storage. J Mater Chem A 7:834–842. https://doi.org/10.1039/C8TA10075F
112. Telli MB, Trolier-McKinstry S, Woodward DI, Reaney IM (2007) Chemical solution deposited silver tantalate niobate, $Ag_x(Ta_{0.5}Nb_{0.5})O_{3-y}$, thin films on (111)Pt/Ti/SiO_2/(100)Si substrates. J Sol-Gel Sci Technol 42:407–414. https://doi.org/10.1007/s10971-006-0204-8
113. Kölbl L, Kobald AM, Griesser T, Munnik F, Mitterer C (2024) Sputter deposited silver niobate thin films: pathway towards phase purity. Thin Solid 804:140505. https://doi.org/10.1016/j.tsf.2024.140505
114. Kölbl L, Mehrabi M, Griesser T, Munnik F, Mitterer C (2025) Lead-free dielectric thin films: synthesis of $Ag(Nb_{1-x}Ta_x)O_3$ via reactive dc magnetron sputtering. J Vac Sci Technol A 43:013103. https://doi.org/10.1116/6.0004162

Energy Conversion, Generation, and Storage

Advancing Lithium/Sulfur (Li/S) Batteries

An L. Phan

Abstract This chapter aims to provide a comprehensive foundation for understanding lithium/sulfur (Li/S) batteries and their current research. It begins with an introduction to their fundamentals, followed by an overview of the current state of Li/S battery development and progress. Two major operation mechanisms of sulfur-based cathodes (i.e., solid-liquid-solid and full-solid mechanisms) will be discussed and compared. Key challenges hindering commercial viability of Li/S batteries will be analyzed, along with potential solutions to address these issues. By the end of this chapter, readers will have a clear understanding of the field's landscape and the essential knowledge to embark on research in Li/S battery technology.

1 Introduction to Lithium/Sulfur (Li/S) Batteries

The rapid growth of the electric vehicle (EV) market is driving the demand for batteries with low cost and high energy density (that is, energy provided per mass unit, Wh/kg). While lithium-ion batteries (LIBs) have been the backbone of current energy storage technologies, they are approaching their theoretical energy density limit. This is a critical concern because energy density not only determines how long a battery can last after each charge (which translates directly into the driving range of an EV), but also impacts the overall cost of energy storage, as cost per kilowatt-hour ($/kWh) depends inversely on stored energy. In addition, high-energy LIBs intrinsically rely on expensive and geopolitically constrained transition metals like cobalt (Co) and nickel (Ni). Their continuously rising costs and potential environmental impact exacerbate the economic and sustainability challenges of LIB, well explaining the increasing interest in the development of battery chemistries that utilize Earth-abundant, inexpensive materials while delivering similar or higher energy density.

Among many so-called "beyond lithium-ion" technologies, lithium/sulfur (Li/S) batteries stand out for their high theoretical energy density and low material costs.

A. L. Phan (✉)
Pacific Northwest National Laboratory, Richland, WA, USA
e-mail: an.phan@pnnl.gov

D. Shi (ed.), *Energy Materials Science and Engineering: From New Devices to AI Power Systems*, Green Energy and Technology, https://doi.org/10.1007/978-3-031-88550-1_5

On the material level, sulfur can store more electrochemical charge than conventional LIB cathodes. Its complete redox conversion involves two moles of electrons per one mole of sulfur atoms, as shown in the half-cell reaction (2), corresponding to a theoretical specific capacity of 1675 mAh/g_{sulfur}. As will be discussed later, reaction (2) typically proceeds through multiple steps, with an average voltage of around 2.15 V versus Li^+/Li. With lithium metal as the anode (half-cell reaction (1), specific capacity 3860 mAh/g_{Li}), Li/S cells can deliver an average voltage of 2.15 V and a theoretical energy density of around 2500 Wh/kg based on the total weight of active materials (i.e., cathode + anode), far exceeding those of graphite/ $LiNi_{0.8}Co_{0.1}Mn_{0.1}O_2$ (NMC811) (around 450−500 Wh/kg) and Li/NMC811 (around 700−750 Wh/kg). This significant advantage in theoretical energy density serves as an important incentive for Li/S battery research.

Besides, sulfur is readily accessible and inexpensive. In January 2024, the United States Geological Survey (USGS) reported that the average cost of sulfur was $100 per metric ton, significantly lower than nickel's $22,000 per metric ton. They also stated that sulfur can be sourced as a byproduct from the petrochemical industry and its reserves are sufficient for the foreseeable future [1]. In brief, Li/S batteries, with their high energy density, cost-effectiveness, and sustainability, are becoming an important research focus in the field of Energy Materials Science and Engineering.

This chapter aims to provide a comprehensive foundation for understanding Li/ S batteries and their current research. It begins with an introduction to their fundamentals, followed by an overview of the current state of development and progress in the field. Key challenges hindering commercial viability will be analyzed, along with potential solutions to address these issues. By the end of this chapter, readers will have a clear understanding of the field's landscape and the essential knowledge to embark on research in Li/S battery technology.

2 Fundamental Concepts

The concept of Li/S battery was first introduced in the 1960s [2], initially as a primary (non-rechargeable) battery. During (discharge) operation, the lithium anode undergoes oxidation, with each lithium atom producing one Li^+ ion and one electron (Eq. (1), forward direction). The electrons flow through and supply electrochemical energy to the external circuit, while the Li^+ ions travel through the electrolyte. Both eventually arrive at the cathode, where they react with sulfur to form lithium sulfide (Li_2S) (Eq. (2), forward direction). This process was later demonstrated to be reversible to some extent by Peled et al. [3, 4], laying the foundation for the development of rechargeable Li/S batteries.

$$Li \rightleftharpoons Li^+ + e \tag{1}$$

$$\text{Lithium capacity} = \frac{1mol_e \times \frac{26800\,mAh}{mol_e}}{1\,mol_{Li} \times \frac{6.94\,g_{Li}}{mol_{Li}}} = 3860\ mAh/g_{Li}$$

$$S + 2e + 2Li^{+} \rightleftharpoons Li_2S \tag{2}$$

$$\text{Sulfur capacity} = \frac{2\,mol_e \times \frac{26800\,mAh}{mol_e}}{1\,mol_S \times \frac{32\,g_S}{mol_S}} = 1675\ mAh/g_S$$

Despite the long history and immense potential for delivering high energy density at low cost, research on Li/S batteries has only gained significant momentum in the past two decades. This is attributable to the major challenges in achieving reasonable long-term reversibility for sulfur cathode and lithium metal anode:

- *The insulating nature of sulfur and its discharge products.* This is the primary problem from the cathode side because it limits electron transport and consequently, the kinetics of sulfur redox conversion (reaction (2)) throughout the cathode at all states of charge. It is common for the active materials in sulfur cathodes to be underutilized due to kinetic limitations, resulting in a significantly lower capacity than the theoretical value of 1675 mAh/g_{sulfur}. As will be discussed in Sect. 4, achieving and sustaining reasonable electrochemical reactivity for sulfur cathodes generally requires substantial engineering efforts, often accompanied by trade-offs.
- *The shuttling effect and other polysulfides-related issues.* Although generally depicted by Eq. (2), the sulfur-Li_2S conversion might proceed with the formation of various lithium polysulfide (LiPS) intermediates, such as Li_2S_4, Li_2S_6, Li_2S_8. The high nucleophilicity of LiPSs renders highly electrophilic electrolyte solvents, including conventional organic carbonates, unsuitable for Li/S batteries involving LiPSs. Their use can result in severe irreversible loss of active material and/or electrode passivation within the first few cycles [5, 6].

  When less reactive solvents, such as ethers, are used, direct chemical reactions between the electrolyte and LiPSs can be avoided. However, LiPSs can still dissolve into the electrolyte, introducing new challenges. The most well-known challenge is the polysulfide shuttling effect, which involves (1) the dissolution of sulfur active species (in the form of LiPSs) into the electrolyte, (2) the uncontrollable migration and diffusion of LiPSs between cathode and anode, (3) the reduction of LiPSs into lower-order (i.e., shorter-chain) LiPSs at the lithium metal anode, and (4) the (re)oxidation of LiPSs into higher-order (i.e., longer-chain) LiPSs at the cathode during charge. This repetitive behavior leads to self-discharge and compromises energy efficiency. Furthermore, during shuttling, sulfur can undergo uncontrolled reduction, forming insoluble species such as Li_2S_2 and Li_2S. These reactions are generally irreversible because the insulating nature of Li_2S_2 and Li_2S limits their electrochemical activity unless they are well dispersed in a conductive matrix–a condition not met in this case–ultimately leading to significant loss of active materials (both sulfur and lithium) and severe passivation of electrode surfaces upon long-term cycling. In retrospect, the high reactivity of lithium metal anode with LiPSs is a key factor that drives and perpetuates the shuttling effect. As a result, lithium protection has become a major approach to tackling this issue [7].

With lithium protection measures, the cycling stability of Li/S cells improves significantly, yet remains far from requirements for practical applications [8]. The reason is that even without shuttling, the dissolution of LiPSs is problematic by itself. Firstly, the repeated dissolution and precipitation of sulfur species facilitates unfavorable sulfur redistribution and changes in the cathode structure, eventually leading to cathode passivation. A more serious issue that is often overlooked is that LiPS dissolution continuously alters the electrolyte properties, such as conductivity, viscosity, and wettability on electrodes, thus introducing substantial instability during the operation of Li/S cells. This issue becomes particularly pressing when electrolyte amounts need to be limited in the interest of energy density [9].

That said, not all electrolytes dissolve LiPSs. Examples of electrolytes with limited or no LiPS solubility include solid-state electrolytes [10], high-concentration electrolytes [11], and localized high-concentration electrolytes (LHCEs) [12]. The preference for electrolytes with certain LiPS solubility is simply a strategy to improve sulfur redox kinetics (involving liquid phase increases the contact area, plus LiPSs can serve as redox mediators for sulfur conversion, as further discussed in Sect. 4.1). Therefore, LiPS-related problems may be viewed as a consequence of the poor conductivity of sulfur and sulfides, rather than an inherent flaw in Li/S batteries. By addressing the core conductivity issue through alternative strategies, these problems can be effectively eliminated [13–15].

- *Problems of lithium metal anode.* The exceptionally low redox potential of lithium ($E^0_{Li^+/Li} = -3.04$ V versus standard hydrogen electrode) makes it ideal for high-energy battery applications, but also incredibly difficult to manage. Its high reactivity with most known electrolytes leads to complex and problematic behaviors at the lithium-electrolyte interface, including (1) the formation of lithium dendrites and non-uniform, porous lithium morphology upon recharging, (2) thick layers of solid-electrolyte interphase (SEI)[1] on lithium surfaces, particularly around dendrites; and (3) "dead" lithium, i.e., isolated lithium particles trapped in the tangled network of SEI-coated dendrites and broken SEI debris, thus become electronically isolated and electrochemically inactive [16]. The formation of "dead" lithium and SEI depletes the active lithium inventory while the latter also consumes the electrolyte, both contributing to rapid performance degradation. It is noted that due to high surface area, dendrites and "dead" lithium show elevated chemical reactivity[2] compared to bulk lithium metal, causing the degradation rate to further accelerate once they are formed. While adding excess lithium and electrolyte to

[1] The SEI is composed of electrolyte decomposition products. It is supposed to only conduct ions and to insulate electrons.

[2] Chemical reactivity is different from electrochemical reactivity. In the context of batteries, chemical reactivity is often associated with the propensity for (unwanted) side reactions.

the cell can compensate for these losses, such measures lower material utilization and reduce overall cell energy density (discussed further in Sect. 5.1). Beyond performance, the aggressive reactivity of dendrites and "dead" lithium also pose significant safety risks, especially considering that the sharp dendrites can puncture the separator to directly contact and chemically react with the cathode materials. The resulting heat release may escalate uncontrollably, leading to thermal runaway or more severe outcomes. In fact, it was only recently that significant progress in stabilizing lithium metal began to emerge, despite its longer history than all intercalation anodes widely used in commercial LIBs [17]. And even with these advancements, lithium reversibility, characterized by average lithium plating/stripping Coulombic efficiency (Li CE, as referred to in this chapter), remains generally limited to 99.5% or below, corresponding to an average lithium inventory loss rate of 0.5% or more per cycle. For comparison, the average loss rate for graphite anode is at least five times smaller (i.e., average Coulombic efficiency exceeding 99.9%). It is noteworthy that the abovementioned issues of lithium metal anode tend to aggravate in the presence of LiPSs [18].

3 Defining Specific Capacity in Li/S Battery Research

The specific capacity of sulfur-based materials is a cornerstone in Li/S battery research, serving as one of the primary motivators for research activity and advancements in this field. However, it can often be perplexing, especially for newcomers, as many studies tend to report specific capacity values without specifying the mass basis used for normalization. This lack of standardization obstructs fair comparisons between various sulfur-based materials and complicates the interpretation of specific capacity in cell-level contexts. Addressing this issue is crucial for advancing the field.

The specific capacity confusion stems from the fact that, in most sulfur-based cathodes, the active material is not pure sulfur but rather a sulfur-containing composite or compound. In other words, the active material itself already contains an "internal" conductive part, such as carbons or conducting polymers, that acts as the host for active sulfur species. This design helps compensate for the poor electronic conductivity and complex redox behavior of sulfur compared to conventional cathode materials. It is important to note that the "internal" conductive host is different from and, in most cases, does not replace the "external" conductive agent (usually carbon black) used for electrode formulation (Fig. 1).

Aligned with its intricate compositional structure, a typical sulfur-based cathode can be characterized by three distinct specific capacity values: C_{sulfur}, $C_{active\ material}$, and $C_{cathode}$, corresponding to three mass bases shown in Fig. 1: m_{sulfur}, $m_{active\ material}$, and $m_{cathode}$. While each specific capacity value has its own physical meaning and significance, as outlined in Eqs. (3)[3], (4), and (5), they are often not distinguished in

[3] In theory, each sulfur atom can exchange two electrons, corresponding to a theoretical specific capacity of 1675 mAh/g_{sulfur}, as shown in Eq. (2). In practice, however, full conversion is rarely achieved due to kinetic restrictions, resulting in a smaller practical C_{sulfur}. The ratio between practical

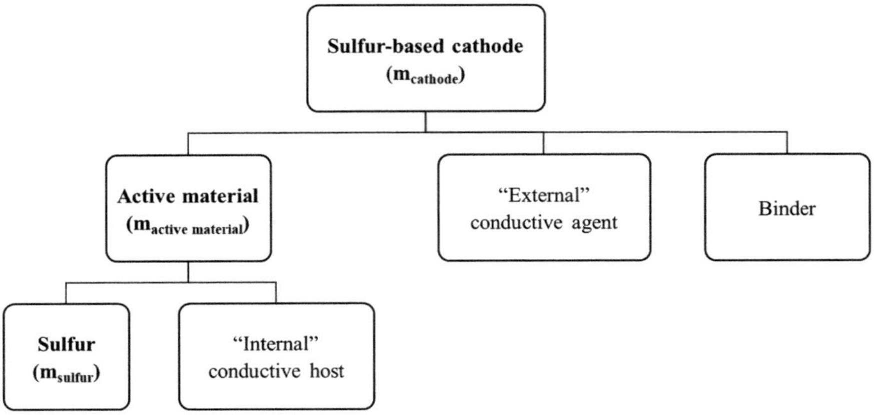

Fig. 1 Compositional structure of a typical sulfur-based cathode

the literature, being uniformly referred to as "specific capacity" and reported in the same unit, mAh/g. This vagueness frequently leads to misunderstandings and unfair comparisons. Therefore, it is highly recommended to clearly specify the mass basis whenever a specific capacity value is reported (e.g., 800 mAh/g_{sulfur} or 800 $mAh/g_{active\ material}$, rather than 800 mAh/g).

$$C_{sulfur} = 1675\, mAh/g_{sulfur} \times sulfur\ utilization \quad (3)$$

$$C_{active\ material} = C_{sulfur} \times sulfur\ content\ in\ active\ material \quad (4)$$

$$C_{cathode} = C_{active\ material} \times active\ material\ content\ in\ cathode \quad (5)$$

4 Current Progress

As highlighted earlier, the poor conductivity leading to limited electrochemical activity and low material utilization of sulfur and its discharge products is the primary challenge for Li/S batteries. Addressing this issue has been the central focus of the field. The underlying principle is straightforward: incorporating sulfur into conductive hosts. However, following substantial progress in the Li/S field, particularly during the intensive research phase starting in the early 2010s, its practical implementation is highly diverse. A wide range of conductive materials with distinct structures and properties have been explored as sulfur hosts, leading to notable differences

value C_{sulfur} and theoretical value 1675 mAh/g_{sulfur}, termed sulfur utilization and ranging from 0–100%, is a good indicator of sulfur redox kinetics.

among sulfur-based cathodes in terms of sulfur's allotropy, interactions with the hosts, dispersion level, reactivity and, ultimately, its redox behaviors. To date, sulfur-based cathodes can be broadly classified into two categories based on their predominant conversion pathway: those preferentially undergoing liquid-assisted conversion and those favoring full-solid conversion.

4.1 Liquid-Assisted Sulfur Redox

Liquid-assisted sulfur redox (also known as the solid–liquid-solid mechanism) has a long history and remains a key area of focus in the field of Li/S batteries. In this mechanism, sulfur discharge occurs in two steps, each involving a phase transition, as illustrated in Fig. 2a. The first step corresponds to the first voltage plateau at around 2.3 V, where solid sulfur transitions into solubilized high-order LiPSs with the general chemical formula Li_2S_x (where $x \geq 4$). The second step corresponds to the second plateau at around 2.1 V, which represents the transfer of sulfur in the liquid electrolyte (as high-order LiPSs) back to the solid phase (as insoluble Li_2S and/or Li_2S_2).

For this mechanism to function, an electrolyte with sufficient LiPS solubility (and negligible reactivity towards LiPSs) is required. It has been argued that without the stabilizing effect of solubilization, LiPSs would not be formed as intermediates and sulfur redox would not follow this two-plateau mechanism [19]. In good agreement, the voltage of the first plateau has been observed to be sensitive to the electrolyte solution and its solvation energy for LiPSs [20, 21].

4.1.1 Motivation and Advantages

The advantage of this liquid-assisted sulfur redox mechanism lies in its ability to loosen the strict requirements for the conductivity and sulfur dispersion within sulfur-based cathodes, as the involvement of soluble LiPS intermediates partially addresses the insulating issues of sulfur and its discharge products. First, liquid-phase LiPSs exhibit superior mobility and reactivity compared to solid sulfur and insoluble Li_2S/Li_2S_2. Inherently, LiPSs-involved conversions encounter less kinetic hindrance than direct conversions.

Second, this mechanism avoids the direct formation and aggregative deposition of insulating solid products (i.e., Li_2S/Li_2S_2 during discharge and sulfur during charge) on the reacting particle surface, which would otherwise block Li^+ transport from the electrolyte and prematurely terminate the reactions. Instead, the formation of soluble LiPSs ensures the continuous exposure of fresh, unreacted surfaces, thereby maximizing the cathode-electrolyte contact area.

Third, LiPSs can act as "carriers" to facilitate electron transport throughout the cathode, thanks to their ease of disproportionation reactions. For this reason, LiPSs are often considered inherent redox mediators for sulfur/sulfides conversion. Due

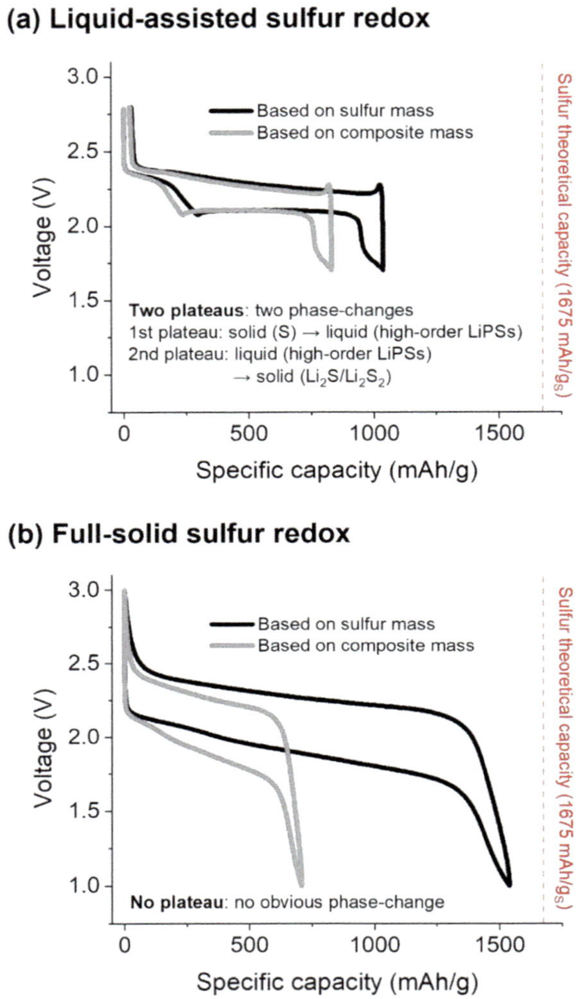

Fig. 2 Typical voltage profiles of two main sulfur conversion pathways. **a** Liquid-assisted mechanism and **b** Full-solid mechanism

to the extensive contact area between liquid-phase LiPSs and active species in the cathode, this indirect electron delivery significantly outperforms the direct electron conduction across the interface between sulfur particles and conductive hosts, which can become ineffective if, for instance, the diffusion distance (i.e., the particle size) is large. By simultaneously enhancing the mobility of all three components needed for the (de)lithiation of sulfur (i.e., active sulfur species, Li^+ ions, and electrons), this mechanism enables adequate kinetics even under non-ideal cathode conductivity, such as when suboptimal sulfur dispersion and relatively large sulfur/sulfides particles (i.e., larger than 2–3 nm) are present.

An important practical implication of this mechanism's tolerance for low cathode conductivity is its capacity to accommodate materials with very high sulfur content (>60 wt.% in the composite/active material or >50 wt.% in the entire electrode), which are typically characterized by suboptimal sulfur dispersion and larger sulfur particle sizes (i.e., larger than a few nm). In addition, this tolerance reduces the need for stringent control over material structure and properties (relative to those required for full-solid sulfur redox), enabling greater flexibility in material design and less demanding preparation processes. In fact, as demonstrated in early works [22, 23], this mechanism allows relatively reversible sulfur redox even with pure sulfur as the active material, although sulfur utilization and cycle life may be inferior.

4.1.2 Challenges and Mitigation Strategies

The most well-recognized challenge of the liquid-assisted sulfur redox mechanism is the shuttling effect as well as the side reactions between soluble LiPSs and Li metal anode. As discussed in Sect. 2, these phenomena can lead to self-discharge, low energy efficiency, irreversible active material loss, and electrode passivation. The active material design represents another conspicuous challenge for this direction. Even with the assistance of soluble LiPSs as redox mediators, a decent degree of sulfur dispersion and particle size optimization is still required to achieve desirable sulfur utilization values. First, reaction kinetics is proportional to the reactive contact area, irrespective of the reaction pathway. Besides, at moderate levels of sulfur utilization (theoretically around 50%), the concentration and redox-mediating potency of long-chain LiPSs would vanish, as a significant portion of sulfur has already been converted into insoluble Li_2S_2 and/or Li_2S.

Owing to extensive past research efforts, these two aforementioned issues are generally under reasonable control in state-of-the-art liquid-assisted Li/S cells. One significant breakthrough is the concept of sulfur nanoconfinement, introduced by Nazar's group in 2009 [24]. They demonstrated that infiltrating sulfur into CMK-3, a mesoporous carbon material with pore diameter on the lower end (around 3 nm), realizes nanostructured sulfur with enhanced contact area and favorable kinetics. At the same time, this approach facilitates the entrapment of LiPSs within the pores of the carbon matrix, thereby alleviating the shuttling effect. Inspired by this pioneering work, a wide variety of host materials, ranging from various carbon types and conductive polymers to inorganic frameworks and hybrid composites, has been explored [25]. Another widely adopted solution to date is the use of $LiNO_3$ as an interphase-forming electrolyte additive to suppress LiPS shuttling [26]. Its proven effectiveness and practicality have subsequently driven the advancement of other protective interphase designs [8, 27]. Recently, the critical role of electrolytes has garnered increasing attention. Innovative electrolyte design concepts, such as high-donor electrolytes and weakly solvating electrolytes, have been proposed, aiming to identify a sweet spot for solvation power and LiPS solubility that balances high sulfur content, favorable redox kinetics, and minimal disruption from LiPSs [28]. Other remarkable approaches in this direction include the strategic integration of

LiPS binding sites/electrocatalysts [29, 30] and separator modification [31]. These strategies proved effective in enhancing the cyclability of sulfur-based cathodes and lithium metal anode at the electrode level. However, achieving high-energy, long-life Li/S batteries requires further advancements to address cell-level challenges under practical conditions: high active material loading with limited electrolyte and lithium inventories.

One formidable but frequently overlooked problem of liquid-assisted sulfur conversion is the electrolyte contamination by soluble LiPSs, which causes continuous changes in the physicochemical properties of the electrolyte (e.g., wettability, viscosity, conductivity), thereby introducing significant instabilities, especially kinetic-wise, into the battery system during cycling. While this effect appears harmless at common lab-scale test conditions (i.e., low active material loading and flooded electrolyte), it can become increasingly severe with decreasing electrolyte/sulfur (E/S) ratio, as higher LiPS concentration can accumulate. In extreme cases, LiPS precipitation may happen, further impairing electron/ion transport [32, 33]. The most daunting aspect of this issue is its intrinsic connection to the mechanism: preventing the dissolution of LiPSs in any manner inevitably triggers trade-offs in redox kinetics. Maintaining the original, supposedly good sulfur dispersion within the active material and the cathode over repeating sulfur dissolution–precipitation cycles is another difficult problem. Sulfur dislocation during cycling has been reported [34], even when localized high-concentration electrolytes (LHCEs) with low LiPS solubility are used [35, 36].

Apart from the cathode-side and cross-contamination issues, the liquid-assisted mechanism indirectly affects the cyclability of Li metal anode by limiting electrolyte choices. For example, the lithium bis(fluorosulfonyl)imide (LiFSI) salt, well-known for its excellent SEI-forming capability on Li metal, is often excluded due to its tendency to undergo reductive decomposition at relatively high potentials and/or chemical reactions with LiPSs. This behavior can lead to cathode-electrolyte interphase (CEI) formation, typically of low quality, on sulfur-based cathodes, inactivate LiPSs and undermine redox kinetics. For this reason, sulfur-based cathodes following this mechanism generally require electrolyte salts that remain stable within their operating potential range (1.7–2.8 V versus Li^+/Li), such as lithium triflate (LiOTf) and lithium bis(trifluoromethanesulfonyl)imide (LiTFSI). The necessary exclusion of LiFSI presents a significant drawback in terms of Li reversibility, as most practical electrolytes with high Li plating/stripping Coulombic efficiency (Li CE) rely on LiFSI as an important SEI-forming component. For reference purpose, many LiFSI-based electrolytes deliver average Li CE values of 99.5% and above [37]. Meanwhile, electrolytes developed for liquid-assisted sulfur redox, even with the presence of $LiNO_3$ and other SEI-forming additives, barely reaches Li CE of 99.0% [8, 38–40]. It should be noted that a Li CE of 99.0% represents a 1.0% loss of the cycled Li per cycle, which doubles the loss rate associated with a Li CE of 99.5%.

4.2 *Full-Solid Sulfur Redox*

A straightforward solution for LiPSs-related problems is to adopt the full-solid sulfur redox mechanism (also known as the solid–solid mechanism), which involves direct conversions between insoluble active species, namely sulfur, Li_2S, and possibly Li_2S_2. This mechanism operates in a solid-solution-like fashion, with sulfur redox reactions occurring explicitly in one solid phase, consistent with the fact that distinct voltage plateaus, indicators of phase transitions, are absent from the charge–discharge curves, as shown in Fig. 2b.

In contrast to the liquid-assisted mechanism, directing sulfur redox toward the full-solid mechanism requires electrolytes with low LiPS solubility and/or a strong tendency to form LiPS-blocking CEIs. Examples include carbonate electrolytes, LiFSI-based LHCEs, LiFSI-based weakly-solvating electrolytes, and $LiNO_3$-containing electrolytes [41]. Several past studies have even shown that rather than the cathode identity, the preference for a specific sulfur conversion mechanism is primarily influenced by the electrolyte composition and the cycling voltage protocol because they govern LiPSs solvability and CEI chemistry [42–44], two factors that directly impact LiPS stability and likelihood of formation. However, it is worth highlighting that only active materials with properly dispersed and sized active-sulfur particles/moieties, preferably on the molecular/atomic scale, can achieve reasonable redox kinetics through the full-solid conversion pathway.

4.2.1 Motivation and Advantages

The main motivation for Li/S batteries with full-solid sulfur redox is the void of LiPSs and their associated issues, including (1) the shuttling effect and Li corrosion, (2) the uncontrollable changes in cathode structure and electrolyte properties during cycling, and (3) the paradoxical constraints on electrolyte design, as discussed earlier in Sect. 4.1.2. While the shuttling- and Li corrosion-free behavior is the most frequently cited benefit [45, 46], it can also be achieved in liquid-assisted sulfur conversions with the presence of appropriate electrolyte additives (e.g., $LiNO_3$) and thus, is clearly not enough to justify the growing interest in this more demanding full-solid mechanism.

The unique competitive advantage of LiPSs-free operation is the superior stability and greater flexibility in terms of cell design. To begin with, obviating continuous sulfur dissolution–precipitation cycles effectively mitigates sulfur redistribution and aggregation within the cathode active material, thereby reducing the risk of cathode passivation. Furthermore, materials designed for full-solid sulfur redox are often engineered with special protection for atomic/molecular-level sulfur dispersion, typically through micropore confinement or chemical bonding [47]. These complementary factors greatly promote the material-level performance (i.e., the performance of the material itself, without cell-level constraints such as limited electrolyte and Li inventory). In fact, full-solid sulfur-based materials consistently exhibit the ability to sustain high sulfur utilization (over 80%) for thousands of cycles [48–50]. Notably, outstanding utilization of up to 98% has also been demonstrated [51].

More importantly, LiPSs-free operation facilitates efficient cell design, enabling significant enhancement in cell-level merits, namely energy density and cycle life. In particular, the absence of LiPSs greatly alleviates concerns associated with implementing Li/S systems under a reasonably lean electrolyte amount (i.e., comparable to the commonly used values in Li batteries, such as 3 $g_{electrolyte}$/Ah), including (1) LiPSs saturation hindering further dissolution, thus leading to redox kinetic penalties, and (2) high LiPSs concentration exacerbating Li corrosion and disruptions in electrolyte transport and wetting behavior. Furthermore, the elimination of LiPS compatibility criterion enables greater flexibility in electrolyte design, facilitating the discovery of electrolytes with good stability on both sulfur-based cathodes and Li metal anode. As discussed in Section 5, electrolyte amount and Li CE, rather than cathode-specific capacity, are among the key factors to unlocking the full potential of Li/S batteries due to their critical influence on cell-level energy density and cycle life. With the LiPSs-free feature, full-solid Li/S batteries offer unparalleled benefits and opportunities from a practical perspective.

4.2.2 Challenges and Mitigation Strategies

The most fundamental challenge for full-solid sulfur conversion is the design of cathode material. The obviation of liquid assistance necessitates a demanding technical challenge to enable sufficient redox kinetics: establishing and stabilizing subnanostructured sulfur. To date, only two approaches are known to realize this high standard, both of which require strict design criteria, accounting for the limited number of relevant designs reported in the literature.

The first approach involves spatially confining active sulfur particles in a microporous carbon matrix. In general, carbons with only sub-nanoscale pores are preferred [52], because smaller pores enhance sulfur utilization within each pore, while a higher density of suitably small pores across the material improves macroscopic sulfur utilization. However, those with slightly larger pores may also be tolerated. Examples include mesoporous carbon with relatively small pores (i.e., pore diameter close to the micro-mesopore boundary), such as CMK-3 [53], and mixed micro-mesoporous carbon, such as Ketjen Black [54, 55]. It should be mentioned that although these options allow higher sulfur content, the loss in sulfur utilization and rate capability caused by less ideal sulfur-carbon contact may lead to a net negative impact on the overall performance. Nevertheless, this approach still generally delivers better long-term sulfur utilization (up to over 70% [52]) than those achievable through liquid-assisted sulfur conversions, especially at high active material loadings and lean electrolyte conditions [42, 54, 56].

In the second approach, active sulfur species, typically consisting of fewer than a few sulfur atoms, are chemically anchored to the conductive agent. This strategy maximizes sulfur contact with the conductive surface for optimal electrochemical reactivity while minimizing sulfur-sulfur interactions, thereby reducing the risk of

Table 1 Comparison between typical materials for liquid-assisted and full-solid sulfur conversions

	Full-solid mechanism	Liquid-assisted mechanism
Average discharge potential	1.85 V versus Li^+/Li	2.15 V versus Li^+/Li
Material-level specific capacity	700–750 mAh/$g_{active\ material}$	>800 mAh/$g_{active\ material}$
Sulfur content	40–50 wt.%	>60 wt.%

sulfur aggregation. As chemical bonds provide more robust protection for sub-nanostructured sulfur than physical interactions, this approach often demonstrates superior effectiveness in enhancing sulfur dispersion and utilization, especially over extended cycling, compared to the first approach–sulfur confinement in microporous carbon. Sulfurized polyacrylonitrile (SPAN), a material with sulfur covalently bonded to the pyrolyzed polyacrylonitrile backbone, is an exemplar of this concept. Since its first report by Wang et al. in [14], SPAN has been well-known for its impressive sulfur utilization and cycle life. While most sulfur-based cathode materials, including microporous carbon–sulfur composites, struggle to achieve 70% sulfur utilization, typical SPAN can maintain over 80% utilization for hundreds, or even thousands of cycles [15], even at particularly high active material loading [57, 58].

Regardless of the approach, full-solid conversions of materials with exclusively sub-nanostructured sulfur generally exhibit lower average discharge potential and reduced material-level specific capacity due to their lower sulfur content (Table 1). While the lower discharge potential is often regarded as an intrinsic and unchangeable limitation of full-solid conversions, a hopeful prospect arises if sulfur is chemically bonded to the conductive agent, as described in the second approach. The chemical bonds may act as an intermediary, transferring electronic effects from the backbone to the active sulfur species and enabling the possibility of raising the sulfur redox potential through appropriate backbone modifications, although developing these new materials may require considerable effort [15]. Notably, similar structures to SPAN have demonstrated favorable response in redox potential to the introduction of electron-withdrawing groups [59]. Meanwhile, enhancing the sulfur content and material-level capacity is generally considered a more achievable goal and has consistently been the primary focus in the field of full-solid sulfur conversion; however, success has been limited. Beyond the inherent challenges of full-solid conversions, such as their particular sensitivity to sulfur dispersion and sulfur volume change during (de)lithiation,[4] the subpar progress observed with many materials of this type, including SPAN, can also be attributed to the lack of clear and effective design principles, stemming from the insufficient understanding of the material's (electro)chemistry [15].

---

[4] The conversion from pure sulfur to pure Li_2S (i.e., 100% sulfur utilization) requires enough empty space to withstand its 80% expansion in the active material volume, restricting the amount of sulfur that can be accommodated by the host materials. Full-solid sulfur conversions are more vulnerable to this limitation than liquid-assisted counterparts because they take place in a more confined space and do not allow sulfur redistribution. Another important factor is the reliance of full-solid mechanism on the CEI, which is prone to damages from large, continuous electrode volume changes.

It is worth reiterating that although full-solid sulfur conversions seem to exhibit inferior performance and suffer great challenges for improvements at the material level, their excellent cycling stability and ability to operate under demanding yet practically relevant conditions (e.g., high active material loading and limited electrolyte reservoir) remain a formidable advantage at the cell level. In fact, Li/SPAN chemistry has been demonstrated to afford a cell-level energy density of over 750 Wh/kg [60], exceeding most reported values for Li/S batteries of any kind [61].

5 Building Better Li/S Batteries

Regardless of the sulfur conversion mechanism, specific capacity and other material-level attributes are not the final goal of Li/S battery research. Ultimately, it is the cell-level outputs, such as cell-level energy density and cycle life,[5] that hold the actual significance. Therefore, in addition to gaining insights into individual cell materials, it is important to understand Li/S batteries from a cell-level perspective to establish reasonable research directions for their advancement. The models, analyses and interpretations presented in this section are adopted and modified from those originally developed for Li/SPAN batteries [15, 62].

5.1 Energy Density

Equations (6) and (7) describe the dependence of Li/S cell-level energy density E (Wh kg^{-1}) on cell parameters.

$$E = V \times C \tag{6}$$

$$\begin{aligned}\frac{1}{C} &= \frac{1}{C_{active} \times R_{active}} + \frac{R_{N/P}}{C_{Li}} + ew_E + \frac{1}{aC} \\ &\times \left[aw_S + \left(\frac{1}{2} + \frac{1}{2n} \right) \times aw_{CC} + \frac{2}{n} \times aw_P \right]\end{aligned} \tag{7}$$

where C is the cell specific capacity (mAh g^{-1}) and accordingly, 1/C is total cell equivalent weight (g/mAh); V is the average cell discharge voltage (V); C_{active} and C_{Li} are the specific capacities of the sulfur-based active material and Li (mAh/g); R_{active} is the weight percentage of sulfur-based active material in cathode; $R_{N/P}$ is the ratio of capacity between negative (Li) and positive electrodes; ew_E is the equivalent

[5] Cell-level cycle life refers to the cycle life of a practically designed cell (reasonably high active material loadings, limited electrolyte amount, limited Li inventory, etc.) for optimal energy density, in contrast to the material-level cycle life, which is concerned solely with the cyclability of the cathode material.

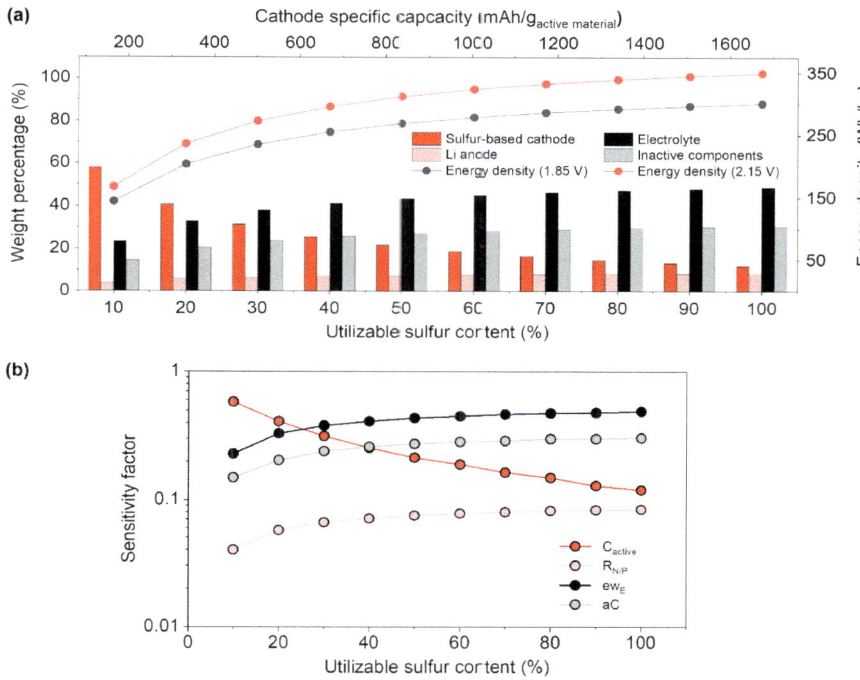

Fig. 3 Analysis of Li/S cell-level energy density **a** Cell composition and energy density as a function of utilizable sulfur content in the active material (i.e., the product of sulfur content in the material and sulfur utilization). Energy density calculations were conducted at two average voltage values, 1.85 V and 2.15 V, to represent Li/S cells with full-solid and liquid-assisted sulfur redox, respectively **b** Sensitivity factors of the key cell-design parameters to the cell-level energy density. The sensitivity factor of a parameter X is defined as the ratio between the percentage change in energy density and the percentage change in X given that the changes are infinitesimally small and other parameters remain constant (also known as the partial elasticity coefficient of energy density with respect to X). All calculations are based on Eqs. (6) and (7) with inputs for a standard Li/S cell [15]: $R_{active} = 0.8$; $R_{N/P} = 2$ (i.e., one excess of lithium); $C_{Li} = 3860$ mAh/g_{Li}; $ew_E = 3$ g_E/Ah; $aC = 5$ mAh/cm^2; $aw_S = 1.3$ mg/cm^2; $aw_{CC} = 13.7$ mg/cm^2 (C-coated Al + Cu); $aw_P = 12.4$ mg/cm^2; $n = 25$

weight of electrolyte (g/mAh); aC is the effective areal capacity of each layer (mAh/cm^2); aw_S, aw_{CC}, and aw_P are the areal weights of separator, current collectors, and cell package (g/cm^2); n is the number of layers (i.e., cathode–anode pairs) in the cell. The right-hand side of Eq. (7) consists of four terms, corresponding sequentially (left to right) to the contributions of the cathode, anode, electrolyte, and inactive components to the overall cell weight. Their normalized values (in percentage) at various materials' utilizable sulfur content[6] are given in Fig. 3a.

[6] Utilizable sulfur content of an active material is defined as the product of its sulfur content and sulfur utilization.

Figure 3a clearly demonstrates that under the common lean-electrolyte condition of 3 g_E/Ah and at meaningful specific capacity (above 700 mAh/$g_{active\ material}$, corresponding to over 40% utilizable sulfur content), the order of weight fraction within the cell follows: electrolyte > inactive components > sulfur-based cathode > Li anode. Additionally, the improvement in energy density in response to increasing specific capacity (or utilizable sulfur content) of the cathode active material progressively slows and becomes particularly modest in the range of interest (i.e., above 700 mAh/$g_{active\ material}$), regardless of the sulfur redox mechanism and average cell voltage. In good agreement, the sensitivity factors of these parameters with regard to cell-level energy density (Fig. 3b) rigorously follow the same trend. This finding contradicts the common belief within the Li/S battery field, which was unquestioningly adopted from the LIB field, that the cathode-specific capacity is the bottleneck in achieving higher cell-level energy density. However, it should be emphasized that sulfur-based cathodes exhibit at least three times the specific capacity of conventional metal oxide and polyanion cathodes (typically below 220 mAh/$g_{active\ material}$), justifying the significant reduction in their contribution to the total cell weight and in their impact on the cell-level energy density, compared to the case of LIBs.

From the analysis, electrolytes appear to be the most effective target for weight cutback to improve energy density, as it is the heaviest cell component, with the highest sensitivity factor, even in the current cell design with lean electrolyte condition of 3 g_E/Ah. As an example, at a utilizable sulfur content of 40% (i.e., specific capacity of around 700 mAh/$g_{active\ material}$), the sensitivity factor of electrolyte is 0.41, implying that a 33% improvement from 3 g_E/Ah to 2 g_E/Ah would result in a roughly 15% improvement in the cell-level energy density [62, 63]. At the same time, it is essential to maintain the electrolyte volume at an appropriate level to ensure sufficient wettability and ionic conductivity throughout the cell. In most cases, the initial electrolyte loading also needs to account for the electrolyte consumption due to side reactions during long cycles.

The intuitive solution to simultaneously minimize electrolyte weight for high energy density and maximize electrolyte volume in favor of cyclability is to utilize electrolytes with low gravimetric density. Meanwhile, the overall electrolyte stability against reductive forces, such as Li metal and some lithiated sulfur-based cathodes, needs to be enhanced to limit irreversible losses. Together, these criteria imply advocacy for less-fluorinated solvents and highlight the importance of carefully formulating such solvents–along with proper choices of other components (e.g., salts, additives)–to ensure electrolytes with favorable (electro)chemical behavior, solvation structure, and interphase chemistry. Also from this perspective, full-solid sulfur redox appears to be a more straightforward approach for optimal cell design, because the scenario can become highly complicated if liquid-assisted sulfur redox is employed. The presence and continuous fluctuations in the concentration of soluble lithium polysulfide (LiPS) species during cycling severely aggravate all the abovementioned issues (i.e., wetting, transport, and side reaction). On the other hand, the 16% higher discharge voltage associated with liquid-assisted sulfur redox presents a counterbalancing advantage. An optimal balance between all factors remains to be figured out.

Following electrolytes, inactive materials also exhibit a relatively significant influence on energy density, especially as the specific capacity of the sulfur-based material gets larger. Another key motivation for reducing inactive weight is that the manipulation can be accomplished through well-defined strategies. The direct approach, minimizing the mass of inactive components, appears challenging, as current collectors and other inactive materials have been extensively optimized over the past decades for LIBs. Unlike active materials and electrolytes, the design of inactive components remains relatively unaffected by variations in cell chemistry. A common alternative approach is increasing active material loading, as it allows higher cell capacity and energy while still requiring the same absolute inactive weight (i.e., the inactive weight needed to support a specific cell capacity/energy is reduced). Interestingly, a crucial determinant of this approach's success is the electrolyte, particularly the nature of the interphases. For Li metal anode and some sulfur-based cathodes, the formation of stable, inorganic-rich interphases with uniform coverage is essential to maintaining cyclability, given their substantial volume fluctuations during cycling. As active material loading increases, these dimensional changes become more severe at the electrode scale, making the role of effective electrolyte and interphase engineering even more critical.

5.2 Cycle Life

Along with high energy density, a prolonged cycle life is a critical requirement in facilitating the adoption of Li/S battery technology. For Li/S batteries with full-solid sulfur redox, represented by Li/SPAN systems, Li plating/stripping Coulombic efficiency (Li CE) is widely acknowledged as the primary factor limiting cycle life. In alignment, Fig. 4a and b demonstrate cycle life extension with increasing Li CE in both flooded and various lean electrolyte conditions. Interestingly, Fig. 4b indicates that unless the electrolyte weight reaches 4 g_E/Ah and above, which is unsuitable for high-energy applications, the cycle life is insensitive to the N/P ratio. This observation is rather counterintuitive: since Li CE represents Li reversibility, a common interpretation of the high Li CE requirement is the need to decelerate Li consumption during cycling. However, although rarely noted, Li CE also controls the electrolyte consumption rate, because the electrolyte is the other participant in the side reactions that cause Li inventory loss. The fact that, within the relevant range (electrolyte weight $\leq$ 3 g_E/Ah), cycle life is dependent on Li CE and electrolyte weight but does not increase with the addition of excess Li metal/Li inventory proves that electrolyte consumption, rather than Li consumption, is the limiting factor for cycle life in practical high-energy cells. Considering that (1) increasing Li CE is already challenging and will get even more challenging as Li CE rises, (2) a rarely achieved Li CE value of 99.9% is needed to realize practically meaningful cycle life (500 cycles) at 3 g_E/Ah, and (3) further reduction in electrolyte amount is highly desirable [15], the finding that electrolyte consumption controls the cycle life is a good news. It opens up new possibilities to enhance cycle life, because unlike Li

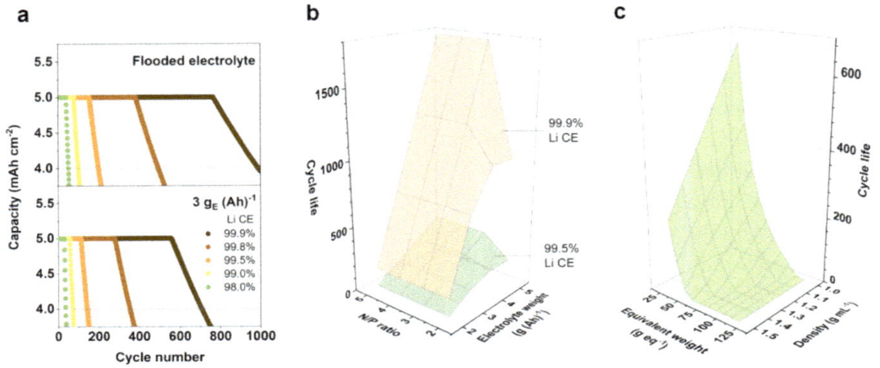

Fig. 4 Analysis of Li/SPAN cycle life. **a** Predicted cycling stability of Li/SPAN cell at various Li CE, with and without electrolyte reservoir limitation. **b** Effect of cell design on cycle life. **c** Effect of side reaction chemistry on cycle life at N/P = 2, $ew_E = 2$ g/Ah, Li CE = 99.9%. Figure is adapted from reference [15]. Copyright 2024, Elsevier

consumption, which solely responds to Li CE, electrolyte consumption can be also controlled through the parameters of the side reaction chemistry, such as density and equivalent molecular weight (i.e., molecular weight per electron exchanged) of the reacting electrolyte species, as shown in Fig. 4c.

Simulating the cycle life of Li/S batteries with liquid-assisted sulfur redox is notably more challenging. In addition to the factors previously described, there are two additional loss mechanisms: the intrinsic degradation of the sulfur-based cathodes (for example, due to sulfur aggregation), and the adverse effects of LiPSs on the overall behavior of cell components. These issues have been thoroughly discussed in Sect. 4.1.2. Certainly, achieving a practical cycle life with a liquid-assisted sulfur redox mechanism would require substantial extra effort.

6 Question

1. Given the theoretical data for cathodes and anodes:

	Sulfur	NMC811	Graphite
Specific capacity (mAh/$g_{\text{active material}}$)	1675	200	350
Average voltage (V vs. Li^+/Li)	2.15	3.70	0.10

a. Calculate the theoretical energy density (based on the total weight of the cathode and anode) of Li/S, Li/NMC811, and graphite/NMC811 batteries. Using the results, explain the motivation for Li/S battery research.

b. The practical cell-level energy density of state-of-the-art Li/S, Li/NMC811, and graphite/NMC811 batteries can reach around 450 Wh/kg, 400 Wh/kg, and 300 Wh/kg. Given that Li/S and Li/NMC811 batteries are currently under the research and development (R&D) stage, so they mostly adopt cell design from conventional Li-ion batteries. Compare the practical-to-theoretical energy density ratios for these three battery chemistries and explain the observed order.

2. Explain the polysulfide shuttling effect. Describe its consequences on individual cell components: cathode, anode, electrolyte.
3. A research direction within the Li/S battery field focuses on strongly solvating electrolytes, also known as high-donor electrolytes. These electrolytes utilize solvents with particularly high solubility for lithium polysulfides (LiPSs). What are the motivations for this direction? What are the potential drawbacks?
4. A paper reports a new sulfur-based material with a specific capacity of 1100 mAh/g without specifying the mass basis. The material Thermogravimetric analysis (TGA) indicates that the material contains 70 wt.% sulfur. In the supporting information, the authors report that the sulfur cathode is made of 80 wt.% active material, 10 wt.% carbon, and 10 wt.% binder. What might be the sulfur utilization of the material?
5. What are the two conversion pathways of sulfur-based cathodes? How are they different? What are their advantages and disadvantages?
6. Given the following design parameters for a Li/S cell: $E = 1.85$ V; $C_{active} = 800$ mAh/$g_{active\ material}$; $R_{active} = 0.8$; $R_{N/P} = 2$ (i.e., one excess of lithium); $C_{Li} = 3860$ mAh/g_{Li}; $ew_E = 2.5$ g_E/Ah; $aC = 3$ mAh/cm^2; $aw_S = 1.3$ mg/cm^2; $aw_{CC} = 13.7$ mg/cm^2 (C-coated Al + Cu); $aw_P = 12.4$ mg/cm^2; $n = 25$.

   a. Calculate the cell-level energy density.
   b. Calculate the sensitivity factors for C_{active}, $R_{N/P}$, ew_E, and aC.
   c. Find the maximum value of ew_E and the minimum value of aC to make C_{active} the limiting

References

1. Apodaca LE (2024) Sulfur. U.S. Geological survey: mineral commodity summaries. pp 172–173
2. Danuta H, Juliusz U (1962) Electric dry cells and storage batteries. Google Patents
3. Peled E, Gorenshtein A, Segal M, Sternberg Y (1989) Rechargeable lithium-sulfur battery. J Power Sources 26(3–4):269–271
4. Peled E, Sternberg Y, Gorenshtein A, Lavi Y (1989) Lithium-sulfur battery: evaluation of dioxolane-based electrolytes. J Electrochem Soc 136(6):1621
5. Gao J, Lowe MA, Kiya Y, Abruna HD (2011) Effects of liquid electrolytes on the charge–discharge performance of rechargeable lithium/sulfur batteries: electrochemical and in-situ X-ray absorption spectroscopic studies. J Phys Chem C 115(50):25132–25137

6. Yim T, Park M-S, Yu J-S, Kim KJ, Im KY, Kim J-H, Jeong G, Jo YN, Woo S-G, Kang KS (2013) Effect of chemical reactivity of polysulfide toward carbonate-based electrolyte on the electrochemical performance of Li-S batteries. Electrochim Acta 107:454–460
7. Yan C, Zhang X-Q, Huang J-Q, Liu Q, Zhang Q (2019) Lithium-anode protection in lithium–sulfur batteries. Trends Chem 1(7):693–704
8. Bi C-X, Hou L-P, Li Z, Zhao M, Zhang X-Q, Li B-Q, Zhang Q, Huang J-Q (2023) Protecting lithium metal anodes in lithium–sulfur batteries: a review. Energy Mater Adv 4:0010
9. Zhao M, Li BQ, Peng HJ, Yuan H, Wei JY, Huang JQ (2020) Lithium–sulfur batteries under lean electrolyte conditions: challenges and opportunities. Angew Chem Int Ed 59(31):12636–12652
10. Nagao M, Hayashi A, Tatsumisago M (2011) Sulfur–carbon composite electrode for all-solid-state Li/S battery with Li2S–P2S5 solid electrolyte. Electrochim Acta 56(17):6055–6059
11. Zheng J, Fan X, Ji G, Wang H, Hou S, DeMella KC, Raghavan SR, Wang J, Xu K, Wang C (2018) Manipulating electrolyte and solid electrolyte interphase to enable safe and efficient Li-S batteries. Nano Energy 50:431–440
12. Zheng J, Ji G, Fan X, Chen J, Li Q, Wang H, Yang Y, DeMella KC, Raghavan SR, Wang C (2019) High-fluorinated electrolytes for Li–S batteries. Adv Energy Mater 9(16):1803774
13. Zhang B, Qin X, Li G, Gao X (2010) Enhancement of long stability of sulfur cathode by encapsulating sulfur into micropores of carbon spheres. Energy Environ Sci 3(10):1531–1537
14. Wang J, Yang J, Xie J, Xu N (2002) A novel conductive polymer–sulfur composite cathode material for rechargeable lithium batteries. Adv Mater 14(13–14):963–965
15. Phan AL, Le PM, Wang C (2024) Realizing high-energy and long-life Li/SPAN batteries. Joule 8(6):1601–1618
16. Xu K (2004) Nonaqueous liquid electrolytes for lithium-based rechargeable batteries. Chem Rev 104(10):4303–4418
17. Winter M, Barnett B, Xu K (2018) Before Li ion batteries. Chem Rev 118(23):11433–11456
18. Manthiram A, Chung SH, Zu C (2015) Lithium–sulfur batteries: progress and prospects. Adv Mater 27(12):1980–2006
19. Lin Y, Zheng J, Wang C, Qi Y (2020) The origin of the two-plateaued or one-plateaued open circuit voltage in Li–S batteries. Nano Energy 75:104915
20. Baek M, Shin H, Char K, Choi JW (2020) New high donor electrolyte for lithium–sulfur batteries. Adv Mater 32(52):2005022
21. Lee C-W, Pang Q, Ha S, Cheng L, Han S-D, Zavadil KR, Gallagher KG, Nazar LF, Balasubramanian M (2017) Directing the lithium–sulfur reaction pathway via sparingly solvating electrolytes for high energy density batteries. ACS Cent Sci 3(6):605–613
22. Jin B, Kim J-U, Gu H-B (2003) Electrochemical properties of lithium–sulfur batteries. J Power Sources 117(1–2):148–152
23. Choi Y-J, Kim K-W, Ahn H-J, Ahn J-H (2008) Improvement of cycle property of sulfur electrode for lithium/sulfur battery. J Alloy Compd 449(1–2):313–316
24. Ji X, Lee KT, Nazar LF (2009) A highly ordered nanostructured carbon–sulphur cathode for lithium–sulphur batteries. Nat Mater 8(6):500–506
25. Seh ZW, Sun Y, Zhang Q, Cui Y (2016) Designing high-energy lithium–sulfur batteries. Chem Soc Rev 45(20):5605–5634
26. Aurbach D, Pollak E, Elazari R, Salitra G, Kelley CS, Affinito J (2009) On the surface chemical aspects of very high energy density, rechargeable Li–sulfur batteries. J Electrochem Soc 156(8):A694
27. Huang Y, Guo Y, Niu Q, Pei F, Wang Q, Zhang Y, Du L, Zhang Y, Zhang Y, Zhang Y (2024) Interfaces engineering toward stable lithium-sulfur batteries. Energy & Environ Sci
28. Liu Y, Elias Y, Meng J, Aurbach D, Zou R, Xia D, Pang Q (2021) Electrolyte solutions design for lithium-sulfur batteries. Joule 5(9):2323–2364
29. Zhao M, Peng H-J, Li B-Q, Huang J-Q (2024) Kinetic promoters for sulfur cathodes in lithium-sulfur batteries. Acc Chem Res 57(4):545–557
30. Wang P, Xi B, Xiong S (2024) Insights into the optimization of catalytic active sites in lithium-sulfur batteries. Acc Chem Res 57(15):2093–2104

31. Manthiram A, Fu Y, Chung S-H, Zu C, Su Y-S (2014) Rechargeable lithium–sulfur batteries. Chem Rev 114(23):11751–11787
32. Zhao M, Li B-Q, Zhang X-Q, Huang J-Q, Zhang Q (2020) A perspective toward practical lithium–sulfur batteries. ACS Cent Sci 6(7):1095–1104
33. Shi H, Sun W, Cao J, Han S, Lu G, Ghazi ZA, Zhu X, Lan H, Lv W (2023) Challenges and solutions for lithium-sulfur batteries with lean electrolyte. Adv Func Mater 33(42):2306933
34. Yu X, Pan H, Zhou Y, Northrup P, Xiao J, Bak S, Liu M, Nam K-W, Qu D, Liu J (2015) Direct observation of the redistribution of sulfur and polysufides in Li-S batteries during first cycle by in situ X-Ray fluorescence microscopy. Adv Energy Mater 5(BNL-107699–2015-JA)
35. Tan G, Xu R, Xing Z, Yuan Y, Lu J, Wen J, Liu C, Ma L, Zhan C, Liu Q (2017) Burning lithium in CS2 for high-performing compact Li2S–graphene nanocapsules for Li–S batteries. Nat Energy 2(7):1–10
36. Strubel P, Thieme S, Weller C, Althues H, Kaskel S (2017) Insights into the redistribution of sulfur species during cycling in lithium-sulfur batteries using physisorption methods. Nano Energy 34:437–441
37. Hobold GM, Lopez J, Guo R, Minafra N, Banerjee A, Shirley Meng Y, Shao-Horn Y, Gallant BM (2021) Moving beyond 99.9% Coulombic efficiency for lithium anodes in liquid electrolytes. Nat Energy 6(10):951–960
38. Kim SC, Gao X, Liao S-L, Su H, Chen Y, Zhang W, Greenburg LC, Pan J-A, Zheng X, Ye Y (2024) Solvation-property relationship of lithium-sulphur battery electrolytes. Nat Commun 15(1):1268
39. Kim I, Kim S, Cho H, Jung J, Kwon H, Kim D, Shin Y, Kim HT (2024) Moderately solvating electrolyte with fluorinated cosolvents for lean-electrolyte Li–S batteries. Adv Energy Mater 2403828
40. Li W, Yao H, Yan K, Zheng G, Liang Z, Chiang Y-M, Cui Y (2015) The synergetic effect of lithium polysulfide and lithium nitrate to prevent lithium dendrite growth. Nat Commun 6(1):7436
41. Miao Q, Solan N, Hyun G, Holoubek J, Liu P (2023) Electrolyte engineering for long-life Li-SPAN batteries. ACS Energy Lett 8(11):4818–4830
42. Zhang W, Qiao D, Pan J, Cao Y, Yang H, Ai X (2013) A Li+-conductive microporous carbon–sulfur composite for Li-S batteries. Electrochim Acta 87:497–502
43. Markevich E, Salitra G, Talyosef Y, Chesneau F, Aurbach D (2016) On the mechanism of quasi-solid-state lithiation of sulfur encapsulated in microporous carbons: is the existence of small sulfur molecules necessary? J Electrochem Soc 164(1):A6244
44. Markevich E, Salitra G, Rosenman A, Talyosef Y, Chesneau F, Aurbach D (2015) The effect of a solid electrolyte interphase on the mechanism of operation of lithium–sulfur batteries. J Mater Chem A 3(39):19873–19883
45. Wu X, Zhao Y, Li H, Zhou C, Wang X, Du L (2024) Sulfurized polyacrylonitrile as cathodes for advanced lithium-sulfur batteries: advances in modification strategies. Nanoscale
46. Kang X, He T, Zou R, Niu S, Ma Y, Zhu F, Ran F (2024) Size Effect for inhibiting polysulfides shuttle in lithium-sulfur batteries. Small 20(8):2306503
47. Li X, Yuan L, Liu D, Xiang J, Li Z, Huang Y (2022) Solid/quasi-solid phase conversion of sulfur in lithium–sulfur battery. Small 18(43):2106970
48. Wang W, Cao Z, Elia GA, Wu Y, Wahyudi W, Abou-Hamad E, Emwas A-H, Cavallo L, Li L-J, Ming J (2018) Recognizing the mechanism of sulfurized polyacrylonitrile cathode materials for Li–S batteries and beyond in Al–S batteries. ACS Energy Lett 3(12):2899–2907
49. Liu H, Holoubek J, Zhou H, Chen A, Chang N, Wu Z, Yu S, Yan Q, Xing X, Li Y (2021) Ultrahigh coulombic efficiency electrolyte enables Li‖SPAN batteries with superior cycling performance. Mater Today 42:17–28
50. Yang H, Naveed A, Li Q, Guo C, Chen J, Lei J, Yang J, Nuli Y, Wang J (2018) Lithium sulfur batteries with compatible electrolyte both for stable cathode and dendrite-free anode. Energy Storage Mater 15:299–307
51. Lei J, Chen J, Zhang H, Naveed A, Yang J, Nuli Y Wang J (2020) High molecular weight polyacrylonitrile precursor for S@ pPAN composite cathode materials with high specific capacity for rechargeable lithium batteries. ACS Appl Mater Interfaces 12(30):33702–33709

52. Xin S, Gu L, Zhao N-H, Yin Y-X, Zhou L-J, Guo Y-G, Wan L-J (2012) Smaller sulfur molecules promise better lithium–sulfur batteries. J Am Chem Soc 134(45):18510–18513
53. Li S, Ma J, Zeng Z, Hu W, Zhang W, Cheng S, Xie J (2020) Enhancing the kinetics of lithium–sulfur batteries under solid-state conversion by using tellurium as a eutectic accelerator. J Mater Chem A 8(6):3405–3412
54. Pang Q, Shyamsunder A, Narayanan B, Kwok CY, Curtiss LA, Nazar LF (2018) Tuning the electrolyte network structure to invoke quasi-solid state sulfur conversion and suppress lithium dendrite formation in Li–S batteries. Nat Energy 3(9):783–791
55. Li X, Banis M, Lushington A, Yang X, Sun Q, Zhao Y, Liu C, Li Q, Wang B, Xiao W et al (2018) A high-energy sulfur cathode in carbonate electrolyte by eliminating polysulfides via solid-phase lithium-sulfur transformation. Nat Commun 9(1):4509
56. Li Z, Yuan L, Yi Z, Sun Y, Liu Y, Jiang Y, Shen Y, Xin Y, Zhang Z, Huang Y (2014) Insight into the electrode mechanism in lithium-sulfur batteries with ordered microporous carbon confined sulfur as the cathode. Adv Energy Mater 4(7):1301473
57. Kim J-M, Gao P, Miao Q, Zhao Q, Rahman MM, Chen P, Zhang X, Hu E, Liu P, Zhang J-G (2024) Tailoring solvation solvent in localized high-concentration electrolytes for lithium‖ sulfurized polyacrylonitrile. ACS Appl Mater & Interfaces 16(16):20618−20625
58. Zhang X, Gao P, Wu Z, Engelhard MH, Cao X, Jia H, Xu Y, Liu H, Wang C, Liu J (2022) Pinned electrode/electrolyte interphase and its formation origin for sulfurized polyacrylonitrile cathode in stable lithium batteries. ACS Appl Mater Interfaces 14(46):52046–52057
59. Wang D-Y, Si Y, Li J, Fu Y (2019) Tuning the electrochemical behavior of organodisulfides in rechargeable lithium batteries using N-containing heterocycles. J Mater Chem A 7(13):7423–7429
60. Kakiage K, Yano T, Uehara H, Kakiage M (2024) Ultra-lightweight rechargeable battery with enhanced gravimetric energy densities>750 Wh kg−1 in lithium–sulfur pouch cell. Commun Eng 3(1):177
61. Zhou G, Chen H, Cui Y (2022) Formulating energy density for designing practical lithium–sulfur batteries. Nat Energy 7(4):312–319
62. Phan AL, Nan B, Le PM, Miao Q, Wu Z, Le K, Chen F, Engelhard M, Dan Nguyen T, Han KS (2024) Lightweight electrolyte design for Li/sulfurized polyacrylonitrile (SPAN) batteries. Adv Mater 36(35):2406594
63. Liu T, Li H, Yue J, Feng J, Mao M, Zhu X, Hu YS, Li H, Huang X, Chen L (2021) Ultralight electrolyte for high-energy lithium–sulfur pouch cells. Angew Chem Int Ed 60(32):17547–17555

Sodium-Ion Batteries

Lin Ma

Abstract This chapter discusses sodium-ion batteries (SIBs), a cost-effective, sustainable alternative to lithium-ion batteries, leveraging abundant sodium resources. It covers their operational mechanism, where sodium ions shuttle between positive (e.g., layered oxides, polyanionic compounds, Prussian blue analogs) and negative electrodes (e.g., hard carbon, titanium-based, alloy-based materials). Key electrochemical properties, including voltage, capacity, and cycle life, are detailed, alongside advancements in electrolytes and separators to enhance performance and safety. Safety strategies addressing thermal stability and regulatory needs are discussed. The chapter highlights SIBs' potential for grid storage and low-speed vehicles, driven by ongoing material and safety innovations.

1 General Characteristics

Like lithium-ion batteries, modern sodium-ion (Na-ion) batteries are built from cells that use sodium-based compounds for both the positive and negative electrodes (Fig. 1). During battery operation, sodium ions (Na^+) move back and forth between the two electrodes, which is why they are sometimes called "rocking chair batteries." This rocking motion of sodium ions occurs as the battery charges and discharges. The positive electrode is often made of a layered metal oxide ($NaTMO_2$), such as $NaNi_{1/3}Fe_{1/3}Mn_{1/3}O_2$, attached to an aluminum current collector. The negative electrode typically consists of hard carbon, also supported by an aluminum current collector. As the battery charges and discharges, sodium ions are inserted into or removed from the spaces between atomic layers within the active materials. If Na_xTMO_2 is used as the positive electrode material and hard carbon as the negative electrode material, the electrode and battery reaction equations can be expressed as

L. Ma (✉)
Department of Mechanical Engineering and Engineering Science, College of Engineering, University of North Carolina, Charlotte, NC, USA
e-mail: L.ma@charlotte.edu
URL: https://mees.charlotte.edu/directory/lin-ma/

D. Shi (ed.), *Energy Materials Science and Engineering: From New Devices to AI Power Systems*, Green Energy and Technology, https://doi.org/10.1007/978-3-031-88550-1_6

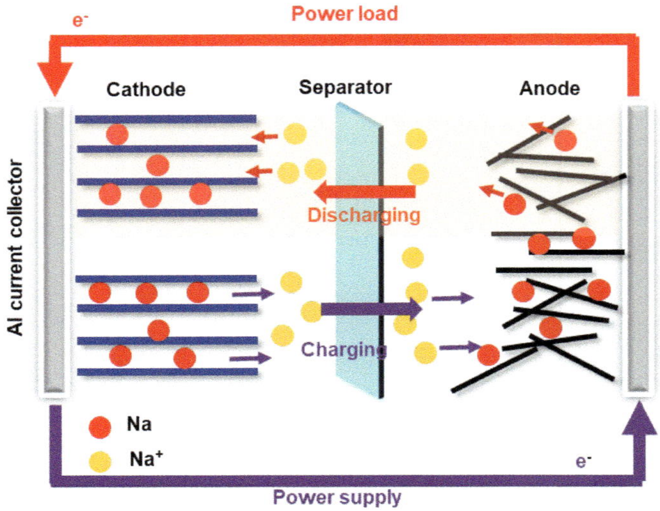

Fig. 1 The schematic of a modern sodium ion cell

$$\text{Positive electrode reaction: } Na_xTMO_2 \rightleftharpoons Na_{x-y}TMO_2 + yNa^+ + ye^-$$

$$\text{Negative electrode reaction: } nC + yNa^+ + ye^- \rightleftharpoons Na_yC_n$$

$$\text{The whole reaction: } Na_xTMO_2 + nC \rightleftharpoons Na_{x-y}TMO_2 + Na_yC_n$$

When describing the electrochemical performance of sodium-ion batteries, various related concepts and terms are typically involved. Here, we will first introduce some commonly used terminology.

- **Open Circuit Voltage (** U_{ocv}**)**: The open circuit voltage refers to the potential difference between the positive and negative electrodes of the battery when no current flows through the external circuit. It is measured in volts (V).
- **Operating Voltage (** U_w**)**: The operating voltage refers to the potential difference between the positive and negative electrodes when current flows through the external circuit. Also known as discharge voltage, it is the actual output voltage of the battery, measured in volts (V). The operating voltage is always lower than the open circuit voltage because the current flowing within the battery must overcome resistance caused by ohmic and polarization resistances.
- **Cutoff Voltage**: The cutoff voltage is the maximum charging voltage or minimum discharging voltage specified for the battery during charging or discharging.
- **Charge/Discharge Rate**: The charge/discharge rate is typically expressed in terms of "C-rate." The C-rate refers to the current required to charge or discharge the battery's capacity within a specified time. For example, a 1C discharge rate

means the discharge current is set so that the battery will be fully discharged in one hour. If the cell capacity is 10 Ah, the discharge current would be 10 A.

- **Internal Resistance**: The internal resistance of a battery refers to the resistance encountered by active ions as they migrate within the battery during operation. This includes both ohmic resistance and polarization resistance. Ohmic resistance is composed of the resistance from the electrode materials, separator, electrolyte, and contact resistance between various parts. Polarization resistance arises from the polarization occurring during electrochemical reactions at the electrodes and includes both electrochemical polarization resistance and concentration polarization resistance.
- **Specific Capacity**: Specific capacity refers to the capacity provided by a battery per unit mass or per unit volume. It is called gravimetric specific capacity (Ah/kg) when expressed per unit mass, or volumetric specific capacity (Ah/L) when expressed per unit volume.
- **Specific Energy**: Specific energy, also known as energy density, refers to the energy output of a battery per unit mass or per unit volume. It is called gravimetric specific energy (Wh/kg) when expressed per unit mass, or volumetric specific energy (Wh/L) when expressed per unit volume.
- **Specific Power**: Also known as power density, specific power refers to the power output of a battery per unit mass or per unit volume, typically expressed in W/kg or W/L. The power density indicates the battery's ability to handle a certain level of operating current, and it is related to the internal resistance of the battery.
- **Constant Current Charging/Discharging**: Constant current charging/discharging refers to the process of charging or discharging a battery at a fixed current. The charging or discharging process typically ends when the set termination voltage is reached.
- **Coulombic Efficiency**: Usually expressed as the percentage of discharge capacity to charge capacity, coulombic efficiency can be used to predict the lifespan of the battery.
- **Cycle Lifetime**: One complete charge and discharge of a battery is referred to as one cycle. Cycle life is defined as the number of cycles a battery can undergo under specific conditions until its discharge capacity reaches a specified value, usually around 80% of its initial capacity.
- **Self-Discharge**: Self-discharge refers to the phenomenon where the capacity of a battery decreases during storage in an open-circuit state under certain conditions. The self-discharge rate is the percentage of capacity reduction per unit of time.
- **State of Charge (SoC)**: SoC indicates the ratio of the remaining capacity of a battery to its initial fully charged capacity after a period of use or after being left unused for a long time, usually expressed as a percentage. An SoC of 100% indicates that the battery is fully charged.
- **Depth of Discharge (DoD)**: Depth of discharge is a measure of the extent of discharge, represented as the percentage of the discharge capacity relative to its total capacity of the battery.

The development of modern sodium-ion batteries has spanned several decades. In 1967, Yao et al. [1] discovered the fast conduction of Na^+ in β-Al_2O_3 (i.e. $Na_2O{\cdot}11Al_2O_3$). A year later, in 1968, Ford Motor Company in the U.S. invented the high-temperature sodium-sulfur (Na-S) battery by using metallic sodium as the negative electrode, elemental sulfur as the positive electrode, and β-Al_2O_3 as the solid-state electrolyte. In 1986, Coetzer [2] replaced sulfur with $NiCl_2$ as the positive electrode material, leading to the invention of the ZEBRA battery. Meanwhile, in 1981, Delmas et al. [3] made a significant contribution to the development of room-temperature sodium-ion batteries by first reporting the electrochemical properties of the layered oxide positive electrode material Na_xCoO_2 at room temperature. During his research on NASICON-structured solid electrolytes, $Na_3M_2(PO_4)_3$ (M = V, Fe, Ti, etc.), Delmas [4] also found that these materials could be used as electrode materials for reversible Na^+ intercalation. In fact, before the successful commercialization of lithium-ion batteries in 1991, some companies in the U.S. and Japan were already exploring room-temperature sodium-ion batteries using Na_xCoO_2 as the positive electrode and a Na–Pb alloy as the negative electrode. However, the average discharge voltage of these batteries was below 3 V, which was less competitive compared to the 3.7 V $LiCoO_2$/C batteries reported at that time, and thus failed to attract significant attention from researchers.

The year 2000 marked a significant turning point in the development of sodium-ion batteries. In 2000, Dahn et al. [5] first synthesized a hard carbon negative electrode material through the pyrolysis of glucose, demonstrating a high specific capacity of 300 mAh/g. To this day, hard carbon remains one of the most promising negative electrode materials for sodium-ion batteries. Additionally, in 2007, Nazar et al. [6] reported the polyanion material Na_2FePO_4F, which exhibited only a 3.7% volume change during sodium-ion intercalation and extraction. Since 2010, as research on lithium-ion battery materials reached maturity, interest in sodium-ion batteries experienced a resurgence, leading to a rapid increase in related studies and publications. A wide variety of positive and negative electrode materials, as well as electrolyte systems, have been reported. The primary positive electrode materials include layered and tunnel-type transition metal oxides, polyanion compounds, Prussian blue analogs, and organic materials. The main negative electrode materials include hard carbon, alloys, and organic carboxylates.

The development of sodium-ion batteries has consistently progressed toward practical application and commercialization. In 2011, Komaba et al. [7] were the first to report the performance of a full $NaNi_{0.5}Mn_{0.5}O_2$/hard carbon battery. That same year, the world's first sodium-ion battery company, Faradion, was founded in the UK. In 2013, Goodenough et al. [8] introduced Prussian white positive electrode materials, which demonstrated high voltage and excellent rate performance. In 2014, Hu et al. [9, 10] discovered the electrochemical activity of the Cu^{3+}/Cu^{2+} redox couple in layered oxides and designed a series of low-cost, high-performance Cu-based positive electrode materials. In 2016, Hu et al. [11] further proposed using low-cost anthracite-based feedstock to produce amorphous hard carbon negative electrode materials for sodium-ion batteries. In 2017, China's first sodium-ion battery research and production company, HiNa Battery Technology Co., Ltd., was established. The company launched the world's first sodium-ion battery-powered low-speed electric vehicle in 2018 and the first 100 kWh sodium-ion battery energy storage station in 2019 [12]. Today, more companies are dedicated to the research and production of sodium-ion batteries, such as Natron Energy, Altris AB, and Tiamat. Simultaneously, cutting-edge research is being conducted to advance sodium-ion battery technology, including efforts to replace organic electrolytes with aqueous electrolytes [13] and solid-state electrolytes [14].

2 Chemistry

2.1 *Positive Electrode Materials*

Overview

Since the late 1970s, when researchers first discovered the reversible extraction and insertion of Na^+ ions in layered oxide Na_xCoO_2, the study of sodium-ion battery-positive electrode materials has gained significant attention. Sodium-ion battery-positive electrode materials primarily include oxides, polyanionic compounds, Prussian blue analogs, and organic materials. Among them, oxides are mainly classified into layered and tunnel structures, while polyanionic compounds include phosphates, fluorophosphates, pyrophosphates, and sulfates. Similar to lithium-ion batteries, nearly all sodium-ion battery positive electrode materials require transition metal (TM) ions capable of reversible valence changes. The redox potential is related to both the type of TM and the structure of the material, and the number of transferable electrons varies. However, simply replacing lithium with sodium in lithium-ion battery electrode materials is not suitable. Therefore, finding appropriate electrode materials for sodium-ion batteries is crucial for their practical application.

Oxides

Oxide-based positive materials are mainly classified into layered oxides and tunnel-structured oxides. Layered oxides are one of the earliest studied intercalation compounds and are characterized by their high energy density and ease of preparation. The general formula for these materials is Na_xTMO_2, where M represents one or more transition metal elements, such as Ti, V, Cr, Mn, Fe, Co, Ni, or Cu, all of which exhibit electrochemical activity. Typically, the TM elements form a MO_6 octahedral structure with surrounding oxygen atoms, creating a TM layer. Sodium ions are located between these TM layers, forming a layered structure with alternating MO_6 polyhedral layers and NaO_6 alkali metal layers.

Delmas et al. [15] classified layered oxides into different structures such as O3, O2, P3, and P2 based on the coordination geometry of sodium ions in the MO_6 polyhedra and the stacking arrangement of oxygen layers. In this classification, the uppercase letters represent the sodium ion coordination geometry (O for octahedral and P for prismatic), while the numbers indicate the number of stacking layers in the smallest repeating unit.

This structural classification is advantageous as it provides a clear description of different layered structures. However, it does not specify the exact space group and atomic position information. In sodium-ion battery layered positive electrode materials, the most common structures are O3 (Fig. 2a) and P2 (Fig. 2b), although the P3 structure can also be obtained under specific synthesis conditions. In the O3 structure, sodium ions occupy a single type of site, forming NaO_6 octahedra that share edges with the transition metal MO_6 octahedra. In contrast, the P2 structure features two types of NaO_6 trigonal prisms. One type shares edges with the MO_6 octahedra above and below it (Na_e, edge-sharing, 2d site), while the other shares faces with the MO_6 octahedra above and below (Na_f, face-sharing, 2b site). In the P2 structure, if both Na_e and Na_f positions are fully occupied, the sodium content can reach 2. However, due to strong Coulombic repulsion, adjacent sites cannot be simultaneously occupied. Generally, the Na_e sites have lower energy relative to the Na_f sites, making them more likely to be occupied. However, the occupancy ratio of Na^+ in these two positions depends on the state of charge/discharge and the choice of TM elements.

Tunnel-structured oxides have more complex structures than layered oxides. For example, $Na_{0.44}MnO_2$ is a typical tunnel-structured oxide. It features large S-shaped tunnels adjacent to smaller hexagonal tunnels (Fig. 3). The large S-shaped tunnel is formed by 12 manganese (Mn) atoms and contains five distinct lattice positions, designated as Mn1, Mn2, Mn3, Mn4, and Mn5. In this structure, Mn1, Mn3, and

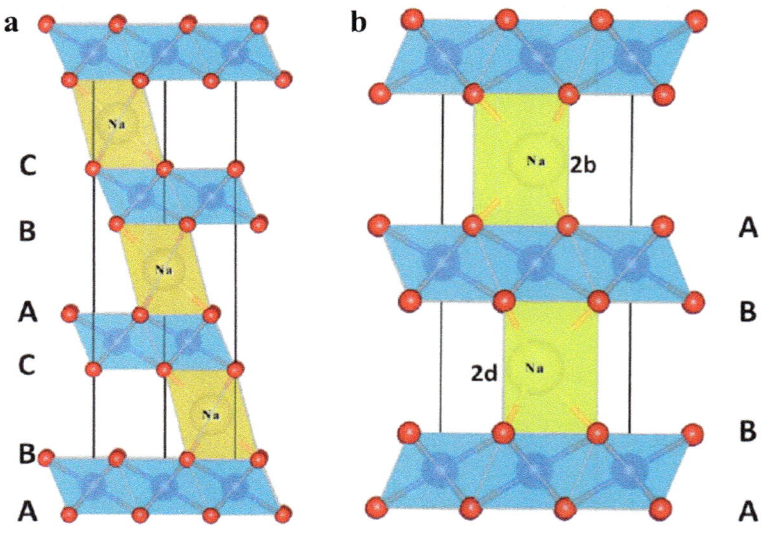

Fig. 2 The stacking types of **a** O3 and **b** P2 phases in Na_xTMO_2. Reproduced from Ref. [16], Copyright 2013 Royal Society of Chemistry

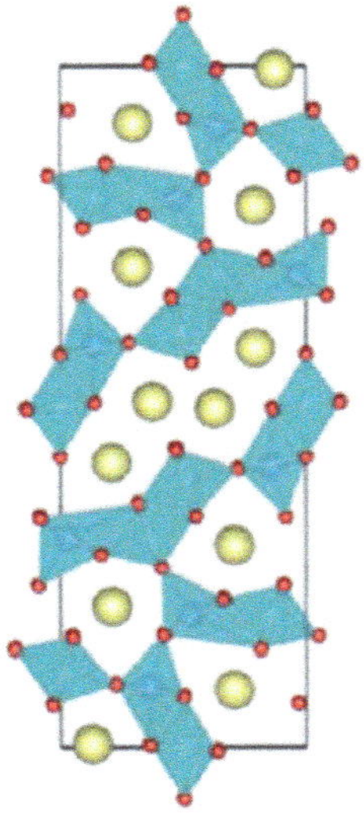

Fig. 3 The crystalline structure of $Na_{0.44}MnO_2$. Reproduced from Ref. [16], Copyright 2013 Royal Society of Chemistry

Mn4 sites are occupied by Mn^{4+} ions, while Mn2 and Mn5 sites are occupied by Mn^{3+} ions, displaying an orderly charge distribution. Within the S-shaped tunnel, there are four rows of sodium ions. The sodium ions near the center of the tunnel occupy the Na3 site, while those near the edges of the tunnel occupy the Na2 site. In the smaller hexagonal tunnels, the sodium ions are found in the Na1 site.

Polyanionic compounds

Polyanion compounds are generally represented as $Na_xM_y(X_aO_b)_zZ_w$, where M can be one or more of the transition metals such as V, Cr, Fe, Mn, Ni, Al, Ti; X represents elements like S, P, Si, W; and Z includes elements such as F. These materials typically feature a framework structure formed by the connection of X polyhedra and M polyhedra through shared edges or corners, with sodium ions distributed in the voids of the framework. Polyanion cathode materials often exhibit excellent structural stability. Additionally, the high electronegativity of F^- and polyhedra like PO_4^{3-}, SO_4^{2-} contributes to a strong inductive effect, which can enhance the operating voltage of the battery.

Phosphate compounds come in various forms, including $Na_3M_2(PO_4)_3$-type phosphates, pyrophosphates, and mixed phosphates. $Na_3M_2(PO_4)_3$-type phosphates feature a three-dimensional framework structure formed by the connection of PO_4 tetrahedra and MO_6 octahedra through shared corners, with sodium ions located in the framework's voids. These materials often exhibit high sodium ion diffusion coefficients. For example, $Na_3V_2(PO_4)_3$ (Fig. 4) [17], which belongs to the $R\overline{3}c$ space group, contains two types of sodium ion environments: Na1 with six-fold coordination and Na2 with eight-fold coordination. The Na2 site, with eight-fold coordination, can undergo extraction and insertion, showing electrochemical activity, while sodium ions at the Na1 site cannot be extracted. Additionally, the PO_4^{3-} anions in

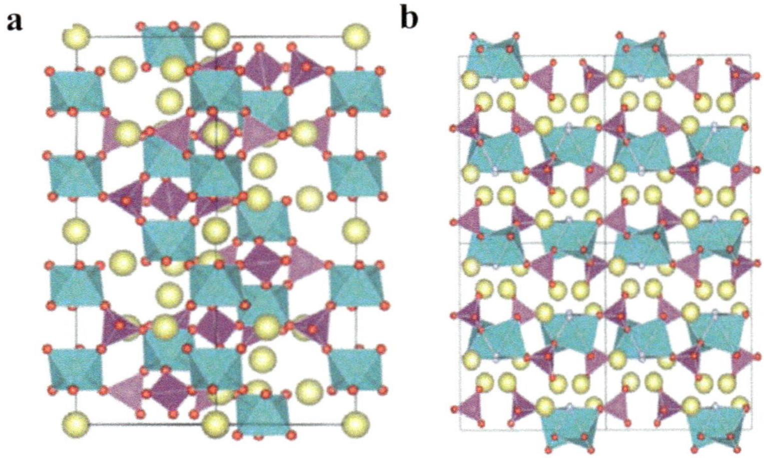

Fig. 4 The crystalline structure of **a** $Na_3V_2(PO_4)_3$ and **b** Na_2FePO_4F. Reproduced from Ref. [16], Copyright 2013 Royal Society of Chemistry

phosphates have an inductive effect that increases the redox potential of transition metals. Mixed polyanionic fluoride phosphate materials are also an important class of positive electrode materials for sodium-ion batteries, such as Na_2FePO_4F (Fig. 4).

Compared to PO_4^{3-}, SO_4^{2-} has higher electronegativity and a stronger inductive effect, resulting in a higher operating voltage for sulfate materials. The $Na_2Fe_2(SO_4)_3$ material [18], for example, has a theoretical specific capacity of 120 mAh/g and an average operating voltage of up to 3.8 V. Other polyanionic compounds, such as silicates and borates, are widely studied due to their diverse crystal structures, abundance of resources, and environmentally friendly nature. Additionally, combining two or more different anions can lead to a variety of materials with different structures and electrochemical properties.

The unique polyanion structural units in polyanionic cathode materials are tightly connected by strong covalent bonds, which isolate the valence electrons of the polyanion groups from the TM ions. Although this isolated electronic structure of the TM ions results in a higher operating voltage for these materials, it also leads to low electronic conductivity. This significantly limits their charge and discharge performance at high rates, posing challenges for practical applications. Therefore, the preparation and modification of these materials primarily focus on enhancing their electronic conductivity. Modification methods mainly include nanostructuring and carbon coating; nanomaterials can increase the solid–liquid contact area and shorten the sodium ion diffusion path, while carbon coating helps improve the surface electronic conductivity of the material and enhance electrical contact between particles. Additionally, the presence of a carbon coating layer can inhibit particle growth, which not only promotes nanostructuring but also mitigates particle agglomeration.

Prussian blue analogs

The chemical formula of Prussian blue analog cathode materials can be represented as $A_xM_1[M_2(CN)_6]_{1-y}\cdot\square_y\cdot nH_2O$ ($0 \leq x \leq 2, 0 \leq y \leq 1$), where A represents alkali metal ions such as Na^+, K^+, etc.; M_1 and M_2 are different coordinated TM ions, such as Mn, Fe, Ni, Cu, Zn, etc.; and $\square$ denotes $[M_2(CN)_6]$ vacancies. Due to the stability of the ferricyanide structure and the simplicity and availability of precursors, research on Prussian blue analogs is often focused on ferricyanide $A_xM_1[Fe(CN)_6]_{1-y}\cdot\square_y\cdot nH_2O$. These materials typically have a face-centered cubic structure, with the metal M and ferricyanide arranged in a three-dimensional framework as Fe-CN-M. The Fe and M ions are located at the vertices of the cubic lattice, the CN groups are positioned along the edges of the cube, and the intercalated ions A^+ and lattice water molecules (H_2O) occupy the spaces within the cubic structure (Fig. 5).

Their main characteristics include:

- The unique electronic structure of the Fe-CN-M framework (with a feedback π-bond between Fe and the C atom, and an inductive effect from different Mn^+ ions on the Fe^{3+}/Fe^{2+} redox pair) ensures a high working potential for the Fe^{3+}/Fe^{2+} redox couple, ranging from 2.7 to 3.8 V versus Na/Na^+.
- By utilizing the M^{3+}/M^{2+} and Fe^{3+}/Fe^{2+} redox pairs, it is possible to achieve the reversible extraction and insertion of up to two Na^+ ions, corresponding to a theoretical specific capacity of 170 mAh/g (e.g., $Na_2Fe[Fe(CN)_6]$).
- The open three-dimensional ion channels facilitate the rapid extraction and insertion of Na^+ ions.
- The Fe-CN coordination has a high stability constant, which helps maintain the stability of the three-dimensional framework structure. This large framework also helps alleviate structural stress during the extraction and insertion of Na^+ ions, contributing to a longer cycle life.

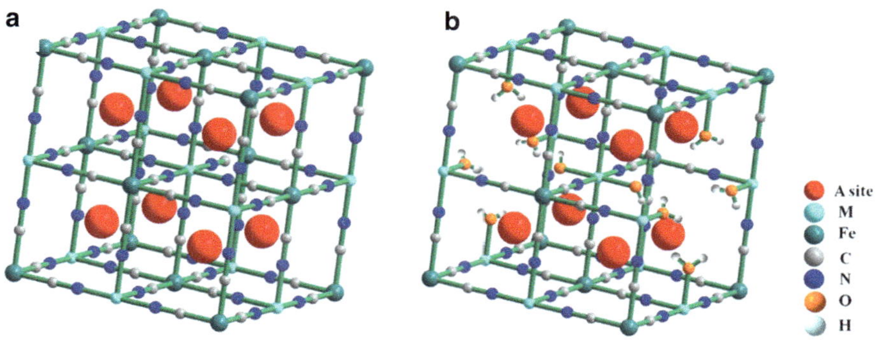

Fig. 5 Scheme of **a** an intact $Na_2M^{II}[Fe^{II}(CN)_6]$ framework without structural defects; **b** an ideally defective $NaM^{II}[Fe^{II}(CN)_6]_{0.75}\cdot\square_{0.25}$ framework with 25% $Fe(CN)_6$ vacancies. Reproduced from Ref. [19], Copyright 2016 American Chemical Society

- Prussian blue analogs have a low solubility product constant, effectively preventing dissolution and loss in aqueous solutions, making them suitable as positive electrode materials for aqueous systems.

Organic materials

Organic compounds, especially polymers, are an important direction in the development of sodium-ion battery electrode materials due to their abundant resources, wide variety, and environmental friendliness. Organic materials can be classified into n-type, p-type, or bipolar-type based on their redox mechanisms (Fig. 6).

N-type organic materials typically undergo a reduction process first, which leads to the formation of a negatively charged state, while simultaneously associating with metal ions and electrons. In contrast, p-type organic materials start from a neutral state and are oxidized during the charging process, requiring the incorporation of anions from the electrolyte (such as PF_6^-) to balance the charge, before undergoing a reversible reduction back to the neutral state. Bipolar-type organic materials, however, can either gain or lose electrons (moving from a half-oxidized state to either a fully reduced or fully oxidized state) within different voltage range. Typically, reversible electrochemical reactions occur in conjugated systems and in groups containing lone pairs of electrons (such as N, O, and S). Conjugated structures facilitate electron transport and charge delocalization, which helps stabilize the molecular structure after electrochemical reactions. In contrast, lone pair electrons or single

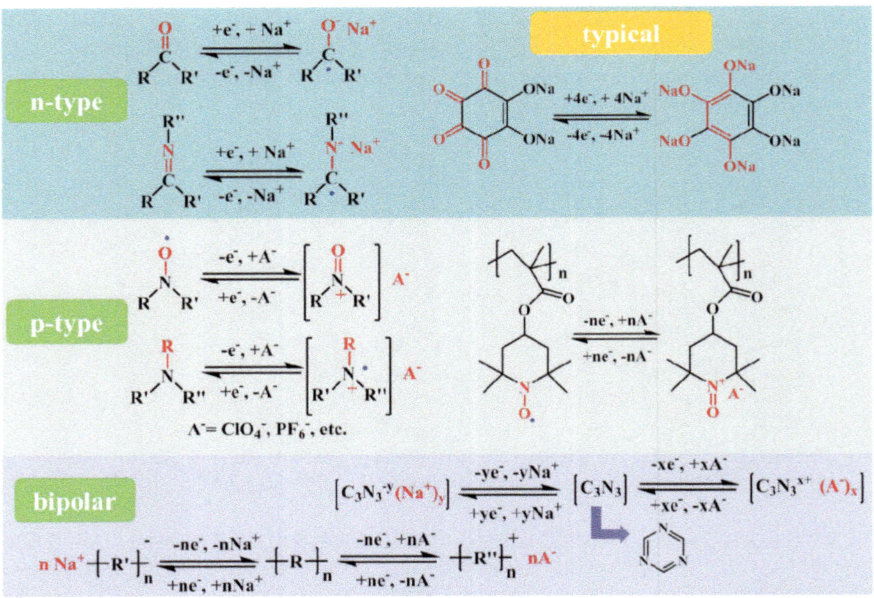

Fig. 6 Categories of energy storage mechanisms of organic electrodes in sodium-ion batteries. Reproduced from Ref. [20], Copyright 2021 Wiley-VCH

electrons generally have higher reactivity. Current research on organic materials is based on these principles and primarily focuses on two main types depending on the active groups: conducting polymers and conjugated carbonyl compounds.

Synthesis methods of positive electrode materials

The most commonly used synthesis method for positive electrode materials is the solid-state reaction method. This method is straightforward to operate, easy to control, has a short process flow, and is suitable for industrial production, allowing researchers to quickly screen and develop high-performance electrode materials. However, samples produced by the solid-state method may not achieve uniformity at the atomic level, which poses certain limitations. Other synthesis methods, such as co-precipitation, sol–gel, hydrothermal, and solvothermal methods, microwave synthesis, spray drying, and ion exchange, can also produce high-performance materials. Currently, the typical industrial approach involves using the co-precipitation method to obtain precursors, followed by solid-state calcination to produce the final positive electrode materials.

- Solid-State Reaction: This is the most traditional and widely used method. It involves mixing solid precursor powders containing sodium, TMs, and other elements in stoichiometric ratios. The mixture is then ground, pressed, and calcined at high temperatures to form the desired crystalline phase. This method is simple and cost-effective, but it often requires high temperatures and long reaction times, and the resulting materials can have large particle sizes and inhomogeneous compositions.
- Sol–Gel Method: In the sol–gel process, metal salts are dissolved in a solution to form a gel. The gel is then dried and calcined to produce the final material. This method allows for better control over the particle size and morphology of the synthesized materials and can result in more homogeneous compositions. However, it usually requires longer synthesis times and involves more steps than solid-state reactions.
- Co-precipitation: This method involves the simultaneous precipitation of metal hydroxides or carbonates from a solution containing metal salts. The precipitates are filtered, dried, and then calcined to obtain the final material. Co-precipitation allows for uniform mixing of the metal ions at a molecular level, which can lead to homogeneous materials with controlled compositions and particle sizes.
- Hydrothermal/Solvothermal Synthesis: In this method, the precursors are dissolved in water (hydrothermal) or an organic solvent (solvothermal) and then heated in a sealed autoclave at high pressure and temperature. This technique can produce materials with unique morphologies, such as nanostructures or single crystals, and allows for better control over the size and shape of the particles. It is especially useful for synthesizing complex oxides and polyanionic compounds.
- Spray Pyrolysis: This method involves spraying a solution of metal salts into a hot furnace, where the droplets evaporate, decompose, and react to form fine particles of the desired material. Spray pyrolysis can produce highly uniform and

fine powders with controlled particle sizes and morphologies. However, it requires specialized equipment and careful control of the process parameters.

2.2 *Negative Electrode Materials*

Historically, advancements in negative electrode materials have significantly accelerated the commercialization of lithium-ion batteries. The first commercialized lithium-ion batteries, introduced by Sony, used petroleum coke as the negative electrode material. By the mid-1990s, most lithium-ion batteries employed graphitic materials as the negative electrode. The adoption of these carbon-based materials helped avoid the safety issues associated with lithium metal anodes. However, graphite is not suitable for storing sodium in carbonate-based electrolytes, making it necessary to identify a cost-effective, high-performance alternative for sodium-ion batteries to enable their widespread commercialization. Sodium metal may not be viable due to safety concerns, such as dendrite formation leading to short circuits. Additionally, sodium metal has a low melting point (~97.7 °C) and high chemical reactivity, which could pose safety risks during battery manufacturing. Recent research has focused on developing alternative negative materials for sodium-ion batteries, including hard carbon [21], organic compounds [22], titanium-based compounds [23], and alloy-based materials [24].

2.2.1 Hard Carbon

Graphite is widely used as a negative electrode material in lithium-ion batteries due to its high specific capacity and excellent reversibility. When fully lithiated to form LiC_6, graphite achieves its theoretical specific capacity of 372 mAh/g. Although sodium and lithium share similar characteristics, the specific capacity of graphite for sodium is significantly limited. In 1958, Asher et al. [25] conducted a study where they reacted sodium vapor with graphite, discovering that only a very small amount of sodium could intercalate into graphite, forming NaC_{64}. Later, in 1988, Ge et al. [26] investigated the electrochemical behavior of sodium intercalation in graphite and identified a plateau at approximately 0.65 V, corresponding to the formation of NaC_{64} (Fig. 7).

To gain a deeper understanding of why sodium cannot intercalate into graphite, Liu et al. [27] conducted extensive density functional theory (DFT) calculations to determine the formation energy (E_f) of M-graphite compounds, where M represents Li, Na, K, Rb, and Cs. They focused on MC_6 and MC_8 stoichiometries, as these are commonly found in non-Na compounds. Their findings showed that the E_f follows the order: Na > Li > K > Rb > Cs, with all metals except Na exhibiting negative E_f values with graphite (Fig. 8), consistent with calculations reported in the literature using different methods. This indicates that Na-graphite compounds with high

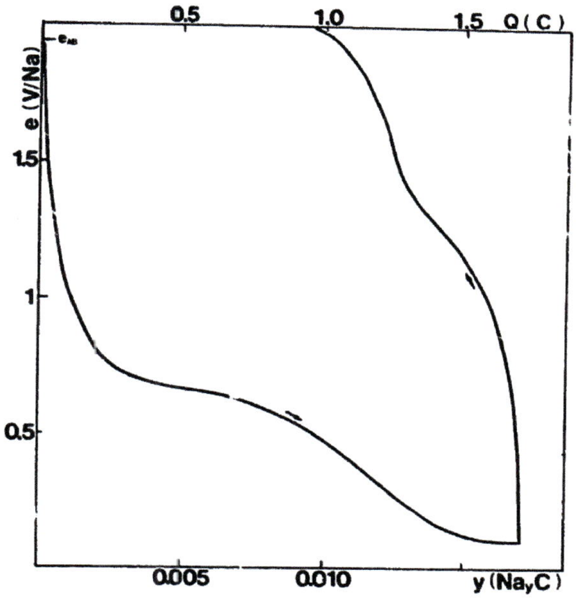

Fig. 7 Variation of graphite electrode potential during reduction and oxidation at constant current. Electrolyte: (PEO, $NaCF_3SO_3$), discharge rate: C/200, temperature: 80 °C. Reproduced from Ref. [26], Copyright 1988 Elsevier

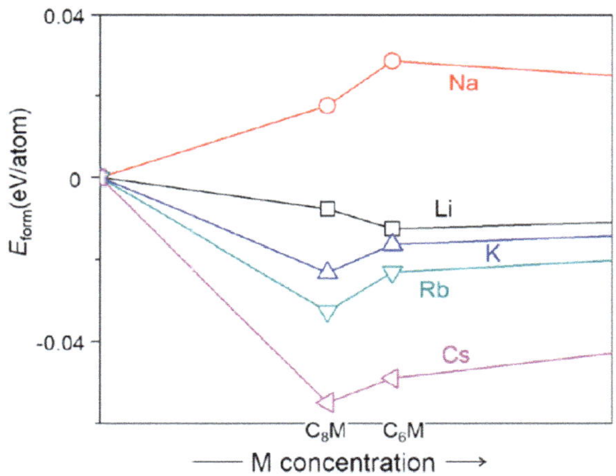

Fig. 8 Calculated formation energies of alkali metal (M)-graphite compounds. Notably, unlike other alkali metals, NaC_6 and NaC_8 exhibit positive formation energies. Reproduced from Ref. [27], Copyright 2016 National Academy of Sciences

sodium content are not thermodynamically stable, unlike the other four M-graphite compounds, which aligns with the experimentally observed low sodium capacity.

In 2000, Dahn et al. [5] were the first to report the use of hard carbon as a negative electrode material, utilizing a glucose precursor. This material demonstrated a high reversible specific capacity of up to ~300 mAh/g within the 0–2 V range. Unlike graphite, which stores sodium (Fig. 9, graphite/Na), the electrochemical curve of hard carbon features a sloping region followed by a plateau region (Fig. 9). This study

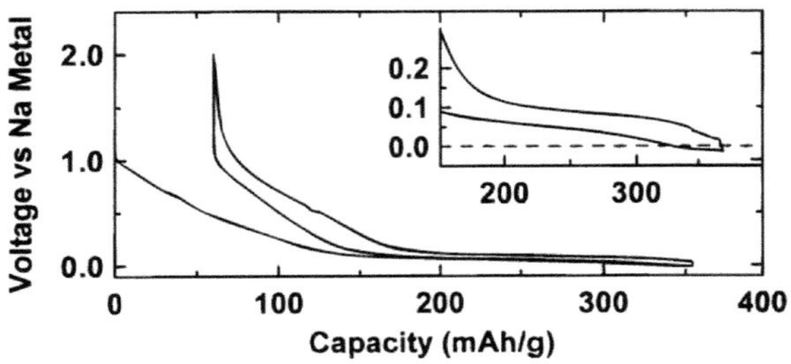

Fig. 9 Sodium potential profiles for glucose pyrolyzed at 1150 °C. The insets provide detailed views of the low-potential region. Reproduced from Ref. [5], Copyright 2000 IOP Science

highlighted the significant potential of hard carbon as a negative material for sodium-ion batteries and spurred extensive subsequent research in the field. The follow-up work primarily focused on understanding the mechanisms of sodium storage in hard carbon and the impact of synthesis conditions on the structure and properties of hard carbon. For a detailed comparison between graphite and hard carbon, readers are referred to the work of Dahn et al. [28], as this topic will not be covered in this chapter.

The mechanisms of sodium storage in HC

In 2000, Dahn et al. [5] introduced the "house of cards" model to explain the mechanisms of sodium storage in hard carbon (Fig. 10). According to this model, small aromatic fragments, roughly 40 Å in lateral extent, are stacked in a somewhat random manner, resembling a house of cards, as depicted in Fig. 10. This random stacking creates small regions where multiple layers (two to three) are parallel to each other, along with nanoscale porosity where layers are randomly oriented. The researchers attribute the sloping potential profile to sodium insertion between these parallel or nearly parallel layers. The insertion potential decreases as the metal content increases because the insertion of metal atoms between layers alters the potential for further insertion, and the turbostratic stacking of parallel sheets leads to a range of insertion-site potentials. Additionally, the metal may interact with hydrogen-terminated sites within the carbon matrix, as previously discussed [29]. Dahn et al. [5] suggest that the low-potential plateaus observed in these hard-carbon materials result from the insertion of metal into the nanoporosity between randomly stacked layers, a process due to adsorption. This insertion occurs at sites with a chemical potential close to that of the metal itself, leading to a potential near 0 V.

Komaba et al. [7] performed ex situ X-ray diffraction (XRD) measurements (Fig. 11) to analyze structural changes during sodium insertion into hard carbon. For the pristine electrode, the broad Bragg peak centered at $2\theta = 23.4°$ corresponds to the parallel stacking of graphene sheets in hard carbon. Typically, hard carbons are highly disordered, resulting in low and broad X-ray diffractive intensity. When the hard-carbon electrode was reduced to 0.2 V, this broad peak shifted to a lower angle, around $2\theta = 21$–$22°$, indicating an expansion of the interlayer spacing between the

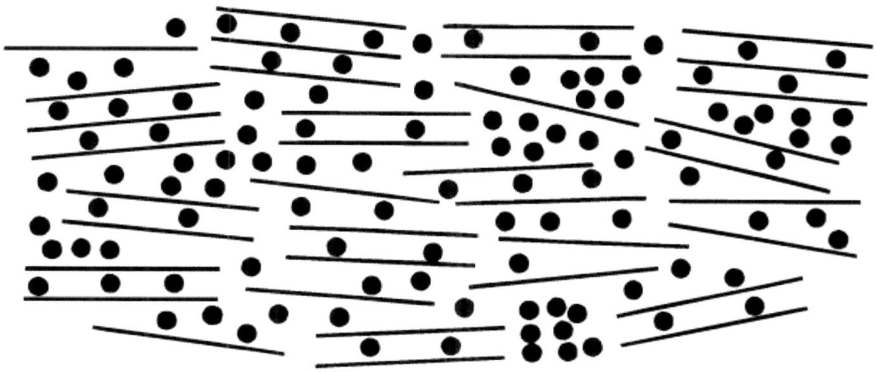

Fig. 10 "House of cards" model for sodium-filled hard carbon. Reproduced from Ref. [5], Copyright 2000 IOP Science

graphene sheets due to sodium insertion. After oxidation to 2.00 V, the peak returned to its original position, demonstrating the reversible nature of sodium removal from the graphene sheets.

In addition, they employed ex situ small-angle X-ray scattering (SAXS) to examine the nanopore structure of the carbon. SAXS is particularly useful for investigating nanopore structures due to variations in electron density between the carbon matrix and the nanopores. When the hard carbon electrode was electrochemically reduced from 0.20 to 0.00 V, the scattering intensity around 0.03–0.07 $Å^{-1}$, attributed to nanopores in the hard carbon with an estimated diameter of approximately 14 Å, decreased reversibly (Fig. 12). This decrease indicates a reduction in the electron density contrast between the carbon matrix and the nanopores, likely due to reversible sodium insertion. The XRD and SAXS results support and are consistent with the "house of cards" model.

New mechanistic insights into sodium-ion storage in hard carbon are emerging in the field. For instance, Bommier et al. [30] propose that the storage mechanism observed in the sloping region of the electrochemical curve is more accurately explained by sodium-ion storage at defect sites rather than intercalation between graphene sheets. Additionally, they hypothesize that the mechanism in the low voltage plateau region is primarily due to intercalation between graphene sheets, with a minor contribution from sodium-ion adsorption on pore surfaces, as illustrated in the graphical representation in Fig. 13.

The effect of synthesis conditions

The synthesis conditions can significantly impact the performance of hard carbon electrode materials. For example, the synthesis temperature affects the voltage profile hysteresis and irreversible capacity: as the synthesis temperature increases, these factors decrease. The reversible capacity remains nearly constant up to around 1100 °C, after which it begins to decline, particularly in the low-voltage region of the voltage profile (Fig. 14).

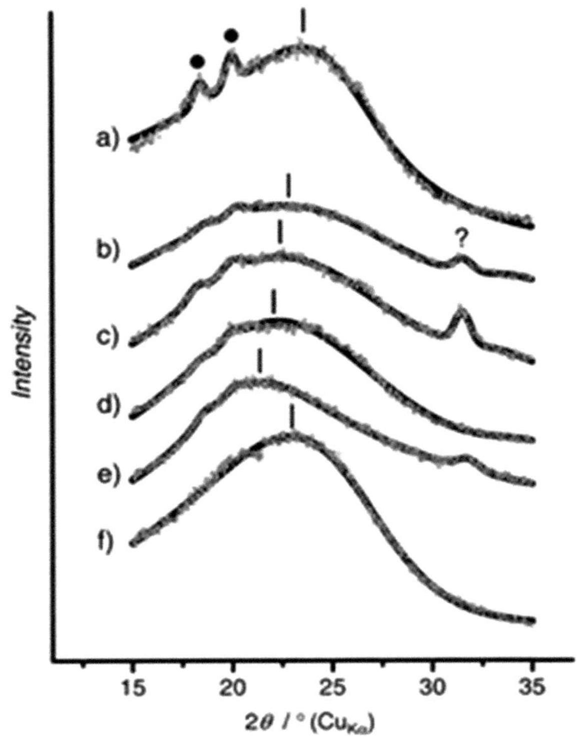

Fig. 11 Ex situ X-ray diffraction (XRD) patterns for hard-carbon electrodes: **a** pristine electrode; **b** reduced to 0.40 V galvanostatically; **c** reduced to 0.20 V; **d** reduced to 0.10 V; **e** reduced to 0.00 V; and **f** oxidized to 2.00 V after reduction to 0.00 V in 1 mol/dm^3 $NaClO_4$ in propylene carbonate (PC) (• PVdF binder, ? unknown). Reproduced from Ref. [7], Copyright 2021 Wiley-VCH

XRD technique reveals that the structural arrangement of graphene sheets within the material shows minimal change when heated between 900 and 1400 °C. The XRD measurements indicate that the arrangement of graphene sheets is largely independent of synthesis temperature (Fig. 15a). SAXS curves (Fig. 15b) demonstrate a continuous increase in nanopore size with higher synthesis temperatures. Thus, the capacity loss cannot be attributed to a loss of nanopore structure. Dahn et al. proposed that nanopore closure explains this capacity loss (Fig. 15c), and this model is supported by CO_2 gas adsorption experiments (Fig. 15d).

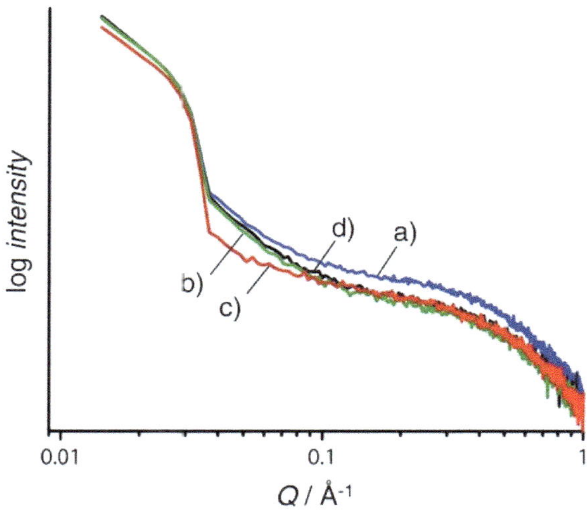

Fig. 12 Ex situ SAXS patterns for hard carbon electrodes: *a* pristine; *b* reduced to 0.20 V galvanostatically; *c* reduced to 0.00 V; and *d* reoxidized to 2.00 V in 1 mol/dm^3 $NaClO_4$ in propylene carbonate (PC). Reproduced from Ref. [7], Copyright 2021 Wiley-VCH

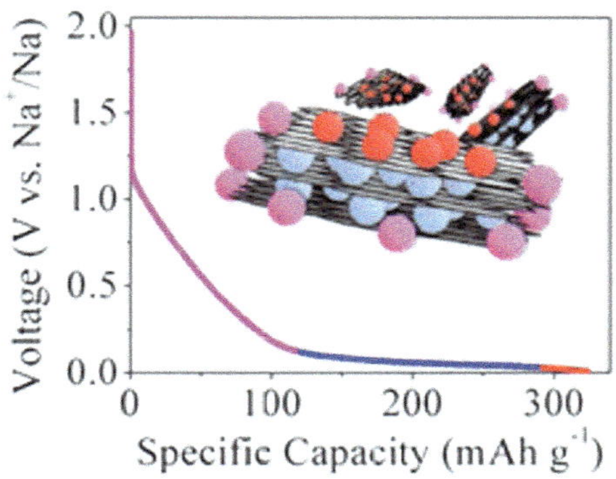

Fig. 13 Potentiogram and schematic of the proposed three-part sodium-ion storage mechanism. Note that the figure is intended as a schematic to represent the three different types of binding sites, rather than an accurate depiction of the hard carbon structure. Reproduced from Ref. [30], Copyright 2015 American Chemical Society

2.2.2 Ti-Based Negative Electrode

Titanium-based negative electrode materials have garnered significant research interest due to the air stability of Ti^{4-} and its ability to exhibit varying sodiation potentials depending on the crystalline structure. In this section, we will explore a few examples of titanium-based negative electrode materials for sodium-ion batteries.

$Na_2Ti_3O_7$

$Na_2Ti_3O_7$ is a highly promising negative electrode material for achieving high energy density in full sodium-ion cells. This is due to its high capacity of 200 mAh g^{-1}

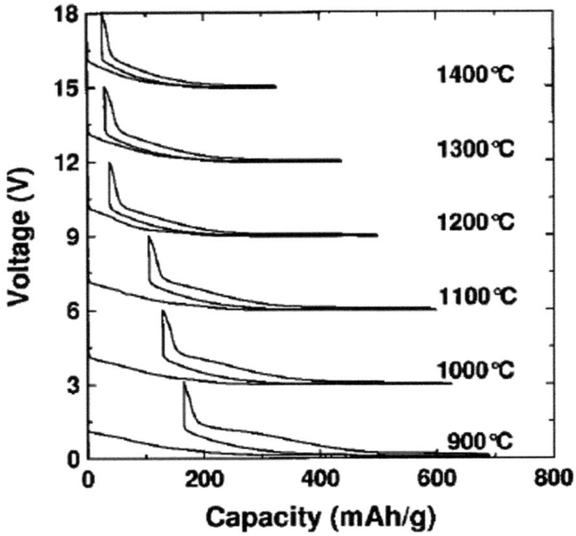

Fig. 14 Voltage profiles of hard carbon prepared by pyrolysis of sucrose in argon gas, with heat treatment temperatures indicated. Reproduced from Ref. [31], Copyright 1999 Elsevier

and its low intercalation voltage of 0.3 V versus Na^+/Na (Fig. 16), which is the lowest among titanium-based materials. However, $Na_2Ti_3O_7$ has poor electronic conductivity, necessitating the addition of a substantial amount of carbon black (over 30%) to enhance conductivity. This addition can reduce the overall energy density and lifespan of the cell.

$NaTi_2(PO_4)_3$

$NaTi_2(PO_4)_3$ (NTP), a NASICON-type material, has recently been investigated for its potential as an electrode in sodium-ion batteries and supercapacitors. NTP is notable for its three-dimensional network made up of TiO_6 octahedra and PO_4 tetrahedra, connected by shared oxygen atoms. This unique arrangement forms large channels along the c-axis, allowing efficient Na^+ ion movement through the structure (Fig. 17) [33]. NTP can accommodate two Na^+ ions per unit cell through a two-phase reaction between $NaTi_2(PO_4)_3$ and $Na_3Ti_2(PO_4)_3$, occurring at around 2.1 V, which provides a theoretical capacity of 133 mAh g^{-1} (Fig. 17). However, the material's low electrical conductivity is a significant drawback for practical applications. To overcome this limitation, several strategies have been employed, such as minimizing particle size to shorten diffusion pathways for electrons and ions and combining NTP with conductive materials to enhance overall conductivity.

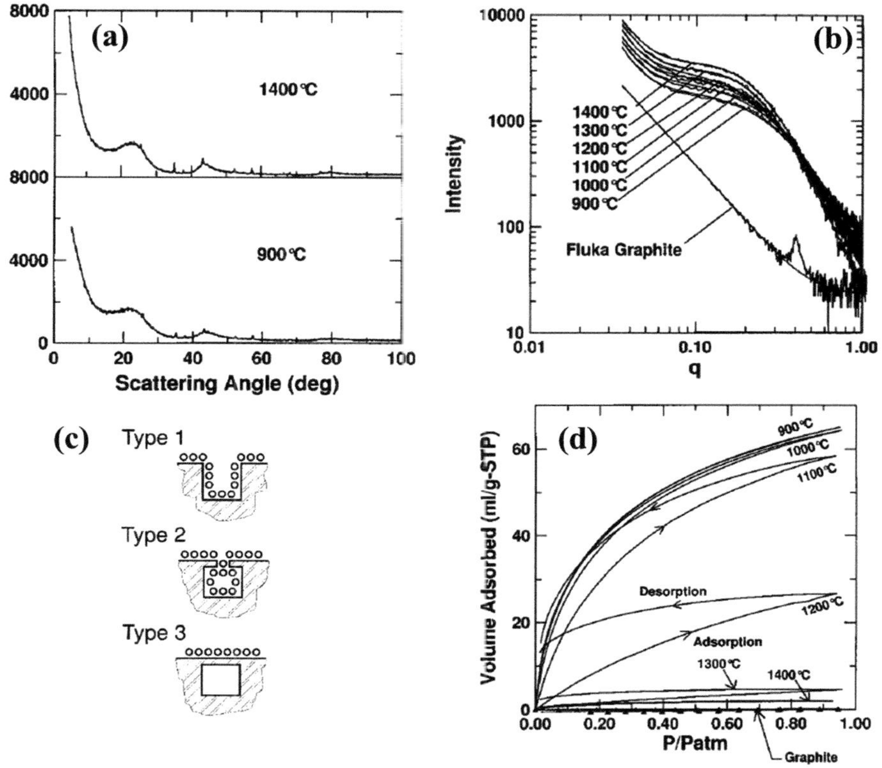

Fig. 15 **a** XRD scattering profiles for samples pyrolyzed at 900 and 1400 °C. **b** SAXS curves for Fluka graphite and samples pyrolyzed between 900 and 1400 °C. **c** Model of nanopore closure showing open nanopores (Type 1), partially closed nanopores (Type 2), and fully closed nanopores (Type 3). **d** CO_2 gas adsorption isotherms for samples pyrolyzed between 900 and 1400 °C, with synthesis temperatures indicated. Reproduced from Ref. [31], Copyright 1999 Elsevier

$Li_4Ti_5O_{12}$

The spinel $Li_4Ti_5O_{12}$ (Fig. 18) is well-known as a "zero-strain" anode for lithium-ion batteries, and it can also be used for sodium storage, showing an average storage voltage of 0.91 V (Fig. 18). The $Li_4Ti_5O_{12}$ electrode offers a reversible capacity of 155 mAh g^{-1} and exhibits excellent cycling stability. DFT calculations predict that the sodium storage process in $Li_4Ti_5O_{12}$ follows a three-phase separation mechanism: $2Li_4Ti_5O_{12} + 6Na^+ + 6e^- \leftrightarrow Li_7Ti_5O_{12} + Na_6LiTi_5O_{12}$.

2.2.3 Organic-Based Negative Electrode

Compared to inorganic materials, organic materials used as negative electrodes offer several advantages: (1) The elements are highly abundant on Earth and are low-cost.

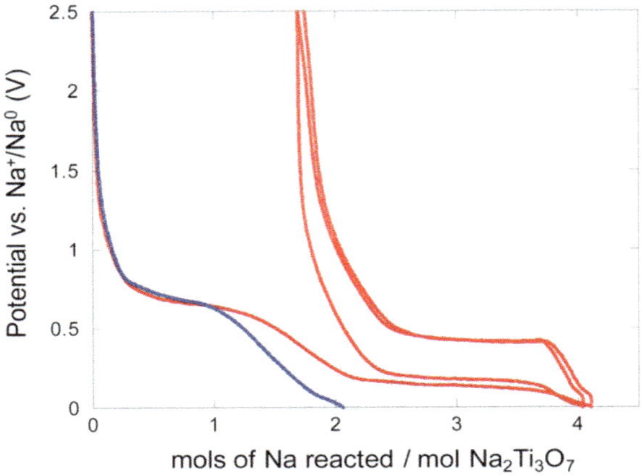

Fig. 16 Voltage versus composition profile for the electrochemical reduction of a blank electrode containing only carbon black (blue curve) compared to a composite electrode containing $Na_2Ti_3O_7$ with 30% carbon black (red curve). The red curve shows the reversible insertion of approximately 2 mol of sodium ions per mole of $Na_2Ti_3O_7$. Reproduced from Ref. [32], Copyright 2011 American Chemical Society

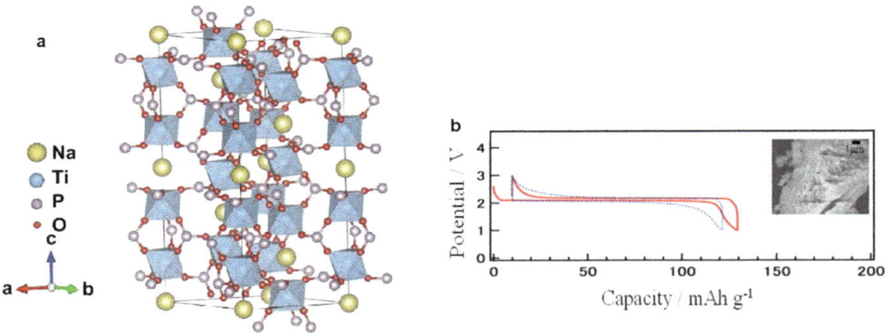

Fig. 17 **a** Crystal structure of $NaTi_2(PO_4)_3$. Reproduced from Ref. [33], Copyright 2019 Royal Society of Chemistry; **b** Galvanostatic cycle profiles of $NaTi_2(PO_4)_3$ at a rate of 0.2 mA cm^{-2} in an organic electrolyte of 1 M $NaClO_4$/EC:DMC (1:1 vol %) together with their FE-SEM images given in the inset. Reproduced from Ref. [34], Copyright 2011 IOP Science

(2) Organic materials have flexible structures, allowing for modifications to their specific capacity and redox potential through the alteration of functional groups. However, they also have several disadvantages: (1) Most organic-based materials exhibit low electronic conductivity, which necessitates the use of a large proportion of carbon black, thereby reducing the cell's energy density. (2) Many organic materials have high solubility in the electrolyte, leading to decreased cycling stability of the cell.

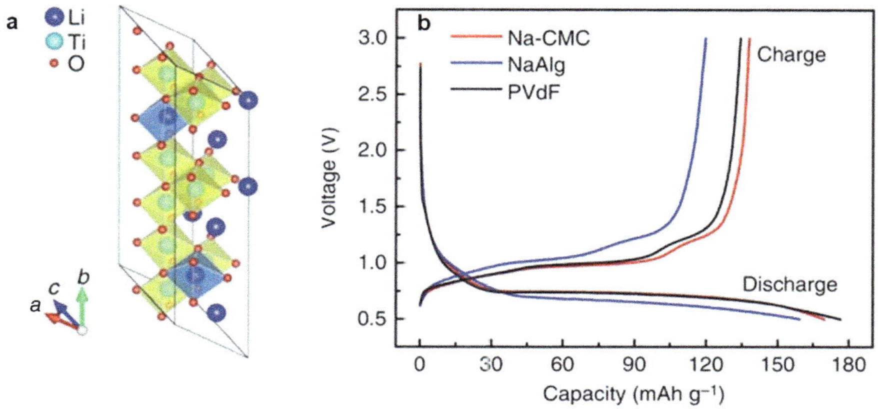

Fig. 18 **a** Crystal structure of $Li_4Ti_5O_{12}$. **b** The $Li_4Ti_5O_{12}$ electrode was tested in a NaFSI/EC electrolyte and cycled at a current rate of C/10. Reproduced from Ref. [35], Copyright 2013 Nature Portfolio

Organic materials studied as negative electrodes for sodium-ion batteries are generally classified into two categories: small organic molecules and organic polymers (Fig. 19). Small organic molecules can be distinguished based on the sites of reversible reactions with sodium ions, including C=O (such as quinones, ketones, carboxylates, anhydrides, and imides derivatives), C=N (such as Schiff bases and pteridine derivatives), and N=N (such as azo derivatives) bond compounds. Organic polymers (doping reaction in Fig. 19) can be grouped into several categories: conjugated conductive polymers, porous organic polymers, organometallic polymers, and organic radical polymers. These different types have all been shown to function effectively as electroactive materials in sodium-ion battery electrodes.

In this discussion, we focus on compounds containing the C=O group as examples of organic-based negative electrodes for sodium-ion batteries. Among these, conjugated carbonyl compounds are of particular interest due to their low cost, ease of synthesis, relatively stable crystal structures, and high specific capacity (>200 mAh/g). Conjugated carbonyl compounds include carboxylate salts and quinones. Carboxylate salts primarily refer to disodium terephthalate and its derivatives, which typically have a voltage plateau below 1 V, making them suitable for use as negative electrodes.

$Na_2C_8H_4O_4$ is the first commercially available organic negative electrode material for sodium-ion batteries, characterized by its $Pbc2_1$ space group and lattice parameters a = 3.5480 Å, b = 10.8160 Å, and c = 18.9943 Å. Hu et al. [37] reported the electrochemical performance of $Na_2C_8H_4O_4$ (Fig. 20) at 0.1C and in the voltage range of 0.1–2 V. This material demonstrates a two-electron reversible redox reaction with a voltage plateau around 0.25 V, a specific capacity of approximately 250 mAh/g, and excellent cycling stability.

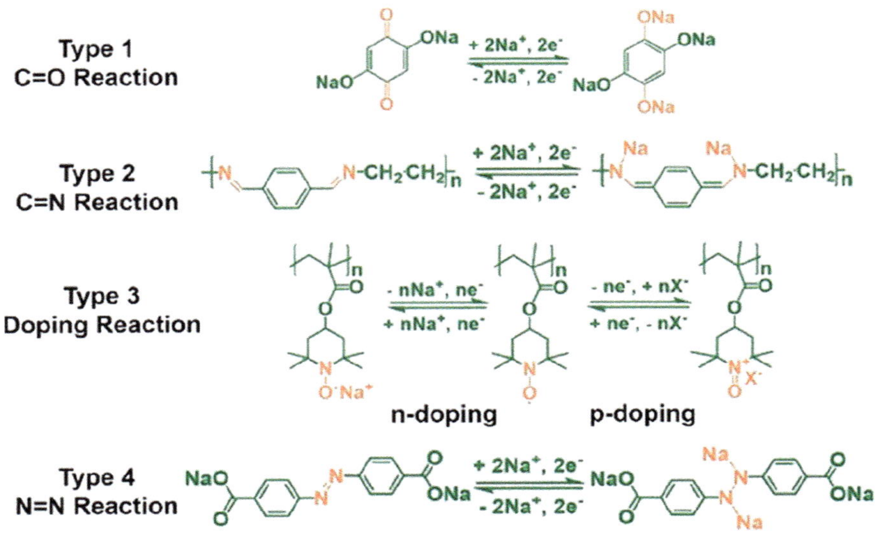

Fig. 19 A schematic illustration depicting the Na-ion storage mechanisms involving C=O, C=N, doping reactions, and N=N functional groups. Reproduced from Ref. [36], Copyright 2023 American Chemical Society

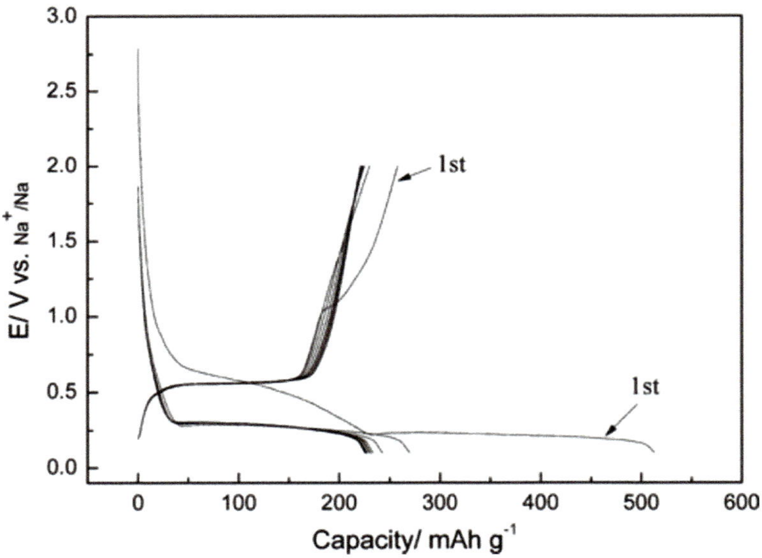

Fig. 20 The first 10 discharge–charge cycles of the $Na_2C_8H_4O_4$/KB composite electrode, recorded within a voltage range of 0.1–2 V. Reproduced from Ref. [37], Copyright 2012 Wiley-VCH

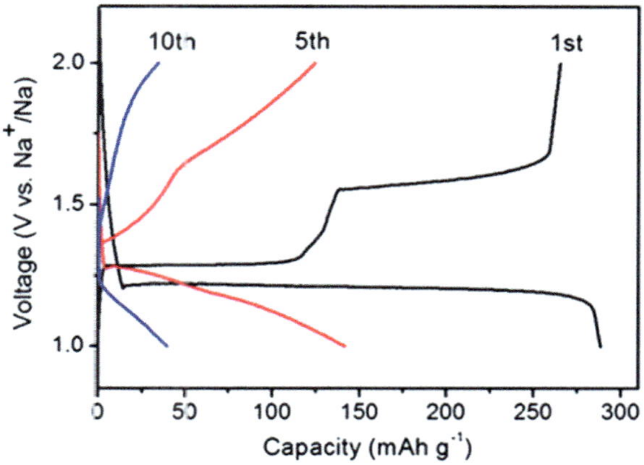

Fig. 21 The discharge/charge curves for the 1st, 5th, and 10th cycles of the control $Na_2C_6H_2O_4$ electrode, measured at a current rate of C/10 (29 mA g^{-1}) within the voltage range of 1.0 to 2.0 V versus Na^+/Na. Reproduced from Ref. [38], Copyright 2015 Royal Society of Chemistry

Typically, quinones have a relatively high specific capacity and a voltage plateau higher than 1 V, which could avoid the formation of a solid electrolyte interphase (SEI). Hu et al. [38] synthesized $Na_2C_6H_2O_4$ at room temperature and reported its electrochemical performance, showing a coulombic efficiency of 91.9% during the first cycle at 0.1C (Fig. 21). Due to its high sodiation potential of around 1 V, side reactions between the electrode and electrolyte are reduced. However, $Na_2C_6H_2O_4$ suffers from poor cycling stability, likely due to the dissolution of the electrode materials into the electrolyte, a common issue among small organic molecules.

To mitigate this problem, two main strategies are typically employed: (1) polymerizing small organic molecules to form more stable structures, and (2) combining these molecules with other carriers, such as carbon nanotubes, to utilize intermolecular forces to suppress the dissolution of small molecules into the electrolyte.

2.2.4 Alloy-Based Negative Electrode Materials

Sodium-ion batteries are attracting growing attention in both research and industrial sectors. However, one of the key challenges hindering their broader adoption is their relatively low volumetric energy density compared to lithium-ion batteries [39, 40]. While there is limited potential for increasing the volumetric capacity of sodium layered-oxide cathode materials, significant improvements can be made in the anode by replacing hard carbon with sodium-alloying compounds.

Sodium can electrochemically and reversibly alloy with several group IV and V elements, such as phosphorus (P), germanium (Ge), tin (Sn), antimony (Sb), lead (Pb), and bismuth (Bi). Alloying materials store sodium through a sequence of reactions that involve breaking and forming bonds, resulting in the formation of various sodiated alloy phases (Na_xM, where M represents P, Ge, Sn, Sb, Pb, or Bi). For practical applications in sodium-ion batteries, the cost and abundance of these elements are critical considerations.

Dahn et al. [24] (Fig. 22) compiled market prices in US dollars per kilogram and per mole, along with the relative atomic abundance compared to silicon, for elements such as nickel (Ni), phosphorus (P), tin (Sn), antimony (Sb), lead (Pb), and bismuth (Bi). Nickel is included in this analysis because, as the primary TM in positive electrode materials of sodium-ion batteries, it is already considered too expensive for sodium-ion batteries to be cost-competitive with lithium iron phosphate ($LiFePO_4$)-based lithium-ion batteries.

In this section, we will focus on phosphorus (P), lead (Pb), and tin (Sn) as examples of alloy-based negative electrode materials. Phosphorus stands out due to its low cost and high abundance, although it suffers from poor electronic conductivity. When fully sodiated, amorphous red phosphorus transforms into a crystalline Na_3P phase, with a theoretical specific capacity of 2596 mAh/g (Fig. 23) [41]. However, Yabuuchi et al. [41] report a significant volume expansion of 292% during the conversion of amorphous red phosphorus to Na_3P.

Lead (Pb) stands out as an affordable alloying element, with a similar abundance to nickel (Ni) but at a significantly lower cost. Its widespread use in lead-acid

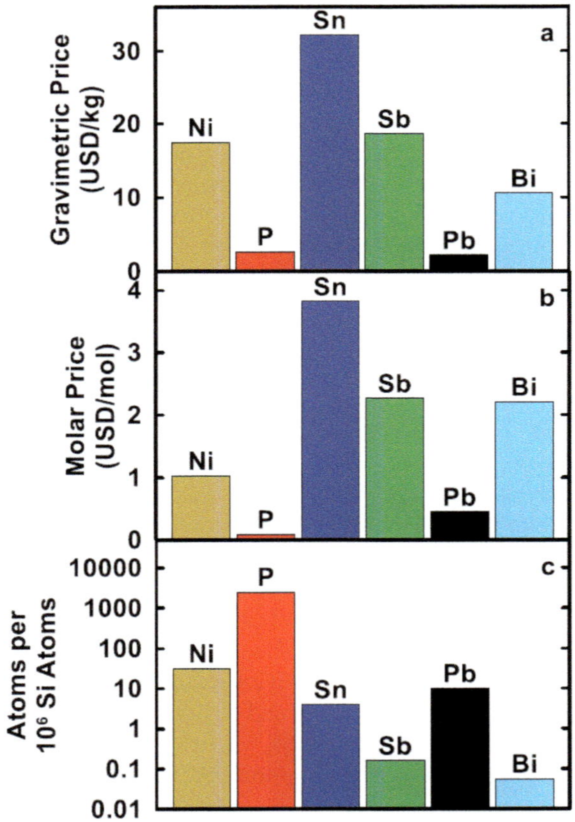

Fig. 22 London Metal Exchange (LME) and Shanghai Metals Market (SMM) prices as of June 18, 2024, for Ni (LME nickel), P (SMM yellow phosphorus), Sn (LME tin), Sb (SMM #1 antimony ingot), Pb (LME lead), and Bi (SMM refined bismuth) in US dollars: **a** per kilogram, and **b** per mole. **c** The atomic abundance of these elements in the Earth's crust relative to silicon (Si). Reproduced from Ref. [24], Copyright 2024 IOP Science

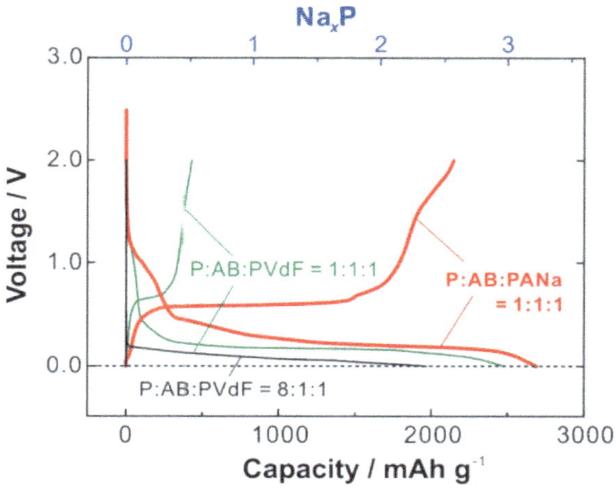

Fig. 23 Galvanostatic charge (reduction) and discharge (oxidation) curves of phosphorus electrodes with varying phosphorus concentrations and different binders (PVdF or PANa) at a current rate of 50 mA/g, within the voltage range of 2.0–0.0 V (vs. Na/Na^+). Reproduced from Ref. [41], Copyright 2014 Wiley-VCH

batteries and its high recycling rate, as reported by the USGS for 2021 [42], make Pb a cost-effective choice compared to other metals like Ni, tin (Sn), copper (Cu), iron (Fe), and aluminum (Al). Pb has also demonstrated good performance in half coin cells [43], requiring only small amounts of binder and carbon additives. On the other hand, Sn, though more costly and less abundant than Pb, has been extensively researched as a sodium alloy material [44, 45]. Recently, UNIGRID Inc. (San Diego, California, USA) shared data on $NaCrO_2$/Sn cylindrical cells, showing almost no capacity loss after 200 charge–discharge cycles [46] (D. H. S. Tan, "Advanced sodium-ion batteries." In 41st International Battery Seminar & Exhibit (2024).). When completely sodiated, both tin (Sn) and lead (Pb) exhibit the same molar composition, $Na_{15}M_4$ (where M = Sn or Pb). The theoretical specific capacities are 847 mAh/g for Sn (Fig. 24) and 485 mAh/g for Pb (Fig. 25).

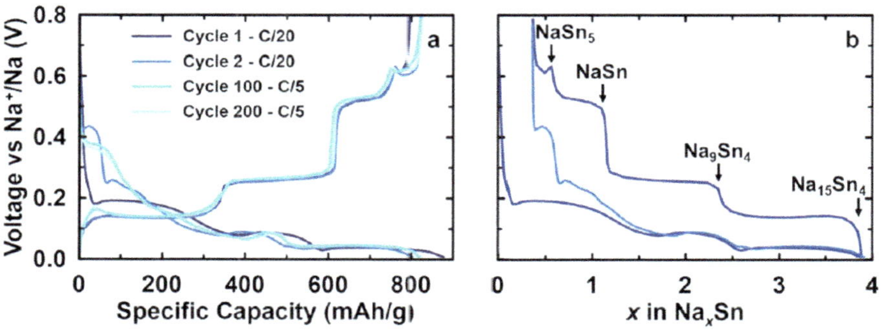

Fig. 24 Voltage profiles for Sn electrodes with PAANa binder versus sodium metal in a 1 M $NaPF_6$ in G1 electrolyte at 30 °C. **a** Voltage versus specific capacity for formation cycles, cycle 100, and cycle 200. **b** Voltage versus sodium content in the Sn electrode for cycles 1 and 2, with Na-Sn phases indicated on the desodiation curve. Reproduced from Ref. [24], Copyright 2024 IOP Science

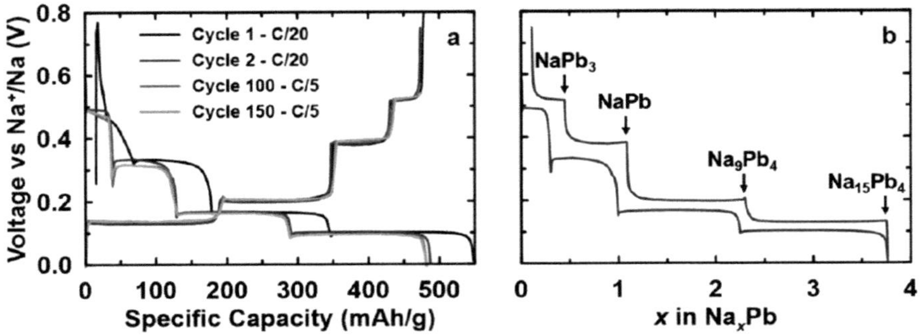

Fig. 25 Voltage profiles for Pb electrodes with PVDF binder versus sodium metal in a 1 M $NaPF_6$ in G1 electrolyte at 30 °C. **a** Voltage versus specific capacity for formation cycles, cycle 100, and cycle 150. **b** Voltage versus sodium content in the Pb electrode for cycle 2, with Na–Pb phases indicated on the desodiation curve. Reproduced from Ref. [24], Copyright 2024 IOP Science

Given that both the unsodiated and sodiated phases are crystalline, the volume expansions are substantial: Sn undergoes a 424% increase, while Pb experiences a 387% increase, according to the ICDD database accessed via Match! Software [24].

Strategies Toward High-Performance Alloy Negative Electrodes

The primary limitation of these alloy materials is that, while they offer impressive volumetric capacities, the substantial volume changes during cycling often result in the fragmentation and detachment of active material particles from the electrode. This expansion and contraction can also undermine the stability of SEI, affecting the long-term reliability and efficiency of the electrode. Several effective strategies have been explored in the literature to mitigate the inherent drawbacks of alloy anodes. One such approach is nanostructuring, which enhances Na^+ flux by increasing surface area through a larger proportion of interfaces in nanostructured alloy anodes. However, caution is needed, as the increased surface area of nanomaterials can lead to more side reactions, significantly reducing cell lifespan. In this section, we will focus on two key strategies: Hybrid Composite Design and the use of Binders and Additives.

Hybrid Composite Design

A hybrid composite design effectively harnesses the synergies between various components to address battery issues related to limitations in electronic and ionic transport rates. This can be achieved through Active/Active Alloys or Active/Inactive Alloys. In an Active/Active Alloy system, one component is an alloy, while the other is an electrochemically active material that does not undergo significant volume expansion. Dahn et al. [24] demonstrated the potential of mixing hard carbon with alloy-based electrodes to enhance volumetric energy density (Fig. 26), offering a practical solution to the challenges posed by alloy materials.

Active elements can be combined with inactive elements to alter voltage profiles, optimize grain structure, and minimize the overall expansion of the alloy during

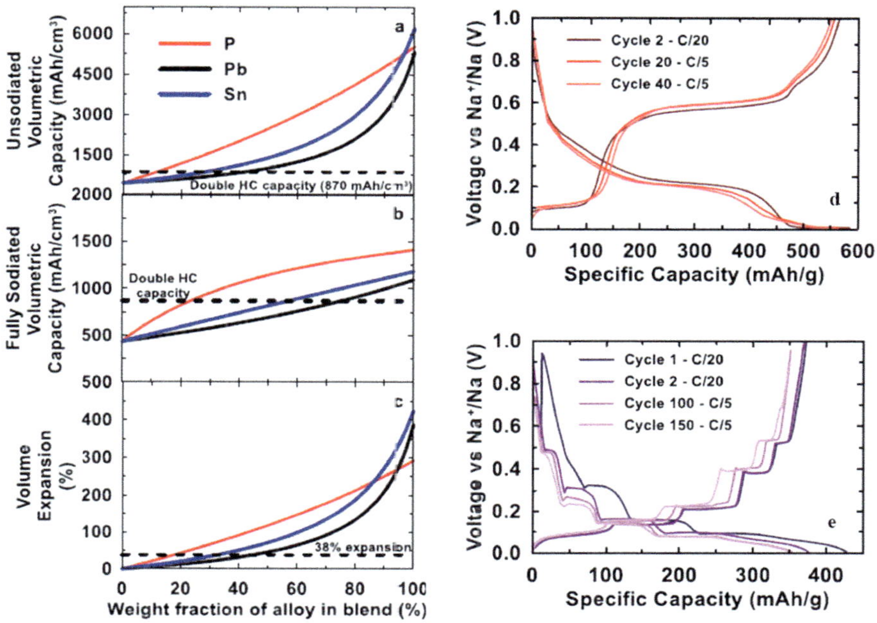

Fig. 26 Calculated **a** unsodiated and **b** fully sodiated volumetric capacity of alloy/hard carbon blends as a function of the weight fraction of phosphorus (P, red), lead (Pb, black), and tin (Sn, blue). The dashed line represents twice the volumetric capacity of hard carbon, set at 870 mAh/cm^3. **c** Calculated volume expansion of alloy-hard carbon blends versus alloy weight fraction, with the dashed line marking 38%, the estimated volume expansion for a 10% silicon-graphite blend in a lithium-ion cell. **d** Voltage versus specific capacity curves at cycles 2, 20, and 40 for P:HC 1:3 electrodes with PAANa binder in 1 M $NaPF_6$ in an EC 1:1 (w/w) electrolyte with 5% FEC by weight at 30 °C. Note that the C-rate increases after cycle 2, from C/20 to C/5, for extended cycling. **e** Voltage versus specific capacity curves for Pb-HC blend electrodes with PVDF binder versus sodium metal in 1 M $NaPF_6$ in G1 electrolyte at 30 °C, shown for formation cycles, cycle 100, and cycle 150. The formation cycles are conducted at C/20, while subsequent cycles are run at C/5. Reproduced from Ref. [24], Copyright 2024 IOP Science

cycling. By embedding the active phase as nanoscale regions within an inactive matrix, the formation of two-phase regions is often prevented, resulting in better cycling durability and improved battery performance [47].

Binders and Additives

The choice of binder [48, 49] and electrolyte [50, 51] plays a critical role in mitigating the volume expansion of sodium alloy anode materials (Fig. 27). Binders with high elasticity and strong adhesion help accommodate the significant volume changes during cycling, reducing mechanical stress and preventing electrode degradation. Meanwhile, optimized electrolytes can form stable SEI, which buffers the expansion and enhances the structural integrity of the anode. Together, the binder and electrolyte

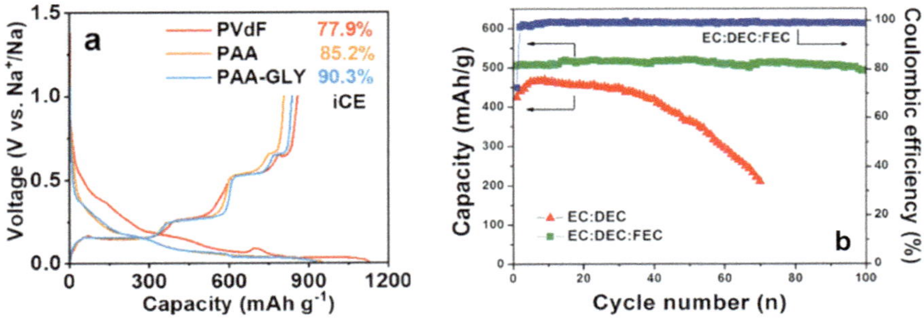

Fig. 27 **a** First discharge/charge voltage profiles of electrodes using different binders at 0.1 A g^{-1}, illustrating the influence of binder choice on electrochemical performance. Reproduced from Ref. [48], Copyright 2023 Wiley-VCH. **b** Cycling performance of the SiC–Sb–C electrode at a cycling rate of 100 mA g^{-1} in FEC-free and FEC-containing electrolytes, highlighting the effect of electrolyte composition on cycling stability. Reproduced from Ref. [51], Copyright 2016 Elsevier

work synergistically to alleviate volume expansion, improving the cycling stability and lifespan of sodium alloy anodes.

2.3 *Electrolytes*

Overview

The electrolyte serves as a bridge connecting the positive and negative electrodes, playing a crucial role in transporting ions between them, and is an essential component of a battery. It significantly affects the battery's rate capability, cycle life, safety, and self-discharge characteristics. Electrolytes in sodium-ion batteries can be classified into liquid and solid types. In this section, we focus on liquid electrolytes. These are primarily composed of solvents, solutes, and additives, with the combined properties of these three components determining the overall characteristics of the electrolyte.

Commonly used sodium-ion electrolytes typically consist of multiple components, with the types and concentrations of these components having a great impact on the operating voltage range, cycle life, and temperature range of sodium-ion batteries. However, the fundamental knowledge of electrolytes currently available is insufficient to fully guide experimental work, meaning that electrolyte formulations are often based heavily on practical experience. Given the similarities in the working mechanisms and electrolyte systems between lithium-ion and sodium-ion batteries, the development of sodium-ion electrolytes can follow and draw from the approaches and experiences established for lithium-ion batteries. However, sodium-ion batteries have many unique characteristics that differentiate them from lithium-ion batteries, making some research conclusions on lithium-ion electrolytes inapplicable to sodium-ion systems. Therefore, fundamental research on sodium-ion electrolytes needs to be further advanced.

Sodium-ion battery electrolytes generally need to meet the following characteristics:

- High Ionic Conductivity: To ensure efficient ion transport within the battery.
- Wide Liquid Temperature Range: To maintain performance across various operating temperatures.
- Good Chemical Stability: To prevent decomposition or undesirable reactions with battery components.
- Excellent Electrochemical Stability: To avoid degradation of the electrolyte during cycling and to ensure a stable operating voltage.
- Good Thermal Stability: To withstand high temperatures without breaking down or losing performance.
- Low Cost: To make the battery more economically viable.
- Environmental Friendliness: To ensure the electrolyte is safe and sustainable for the environment.

This section will focus on the properties of commonly used solvents, sodium salts, and additives in sodium-ion batteries. The aim is to offer guidance and insights for the development of sodium-ion battery electrolytes.

Salt

Sodium salts used in sodium-ion batteries should possess the following characteristics:

- Easy Dissociation: The salt should dissolve readily in the solvent, providing a high working ion concentration.
- Chemical Stability: It should be chemically stable and not corrode current collectors or other non-active components.
- Electrochemical Stability: The dissociated anions should have suitable redox potentials, remaining resistant to oxidation or reduction.
- Thermal Stability: The salt should maintain its properties under high temperatures.
- Environmental Friendliness: It should be environmentally benign, easy to prepare, and cost-effective.

Salts commonly used in sodium-ion batteries are listed in Table 1.

Solvents

Similar to the electrolyte systems in lithium-ion batteries, solvents used in sodium-ion battery electrolytes are typically polar aprotic organic solvents. These solvents have a separation of positive and negative charge centers within their molecules and do not contain highly reactive proton hydrogens. Common solvents in this category include those with polar functional groups such as –C=O, –S=O, –CN, or –C–O–,

Table 1 Salts for Na-ion electrolytes

Name	Formula	Mol. weight (g/mol)	Comments
Sodium perchlorate	$NaClO_4$	122.4	Hygroscopic, can pose safety concerns due to its strong oxidizing nature
Sodium tetrafluoroborate	$NaBF_4$	109.8	Good thermal stability
Sodium hexafluorophosphate	$NaPF_6$	167.9	Most common salt
Sodium trifluoromethanesulfonate	$NaSO_3CF_3$ (NaOTf)	172.1	Can exhibit corrosive behavior towards Al current collectors
Sodium bis(fluorosulfonyl)imide	$Na[(FSO_2)_2N]$ (NaFSI)	203.3	Most common salt
Sodium bis(trifluoromethyl sulfonyl)imide	$Na[(CF_3SO_2)_2N]$ (NaTFSI)	303.1	Can exhibit corrosive behavior towards Al current collectors, especially at high voltages
Sodium-difluoro(oxalato)borate	$NaC_2BF_2O_4$ (NaDFOB)	160	Good thermal stability and SEI forming capability
Sodium bis(oxalato)borate	NaC_4BO_8 (NaBOB)	210	Good thermal stability and SEI forming capability

which are effective at dissolving sodium salts. The choice of solvent may vary slightly depending on the specific requirements of the positive and negative electrode systems.

Organic solvent molecules interact with Na^+ in complex ways, resulting in different solvation structures around Na^+, such as variations in the size, composition, and number of solvent molecules in the solvation shell. These interactions influence Na^+ migration and the reactions occurring at the electrode surface. Sodium-ion battery electrolytes are often composed of mixed solvents, where Na^+ tends to interact preferentially with certain solvent molecules. In such cases, the solvation shell is predominantly formed by these preferred solvents. This preferential solvation behavior can be understood in terms of the donor number, a measure in chemistry that quantifies the ability of a solvent to act as a Lewis base. In electrolytes, the donor number (DN) measures a solvent's ability to donate electrons. A higher DN indicates that the solvent can more readily donate electrons, leading to stronger interactions with Na^+. Generally, solvents with high dielectric constants have higher DN values and thus interact more readily with Na^+. As a result, the solvation shell around Na^+ is predominantly composed of high-dielectric-constant solvents, reducing Na^+ interactions with anionic counterparts. Conversely, the outer layers of the solvation shell and other regions are mainly occupied by solvents with lower dielectric constants. These solvents often have lower viscosities, which can enhance the ionic conductivity of the electrolyte.

Currently, the solvents commonly used in sodium-ion batteries are primarily carbonate esters and ethers. Reports on the use of carboxylic esters and sulfur-containing solvents are relatively scarce. Neat carbonate solvents generally have intrinsic solution conductivities of less than 10^{-7} S/cm dielectric constants of ≥ 3, and are capable of dissolving sodium salts to high concentrations. Table 2 summarizes the properties of several commonly used solvents. Additional properties of electrolyte solvents and strategies for formulating electrolytes are covered in several comprehensive review articles [52, 53].

Additives

Electrolyte additives are components present in the electrolyte at relatively low concentrations (typically less than 10%). These additives can be gases, liquids, or solids, and are used in small amounts to address specific issues or improve certain aspects of battery performance. They are designed to optimize battery performance without significantly increasing production costs or altering the manufacturing process. Sometimes, commonly used organic solvents or salts with unique physicochemical properties can also be employed as additives in small quantities. Conceptually, the distinction between additives, solvents, and salts lies in their concentration within the electrolyte. Additives are used at lower concentrations, while higher concentrations of solvents or salts are classified accordingly.

Electrolyte additives for sodium-ion batteries can be categorized into inorganic and organic types based on their composition. Inorganic additives mainly consist of solid sodium salts, such as $NaBF_4$, NaDFOB, and NaBOB. In contrast, organic additives include compounds like vinylene carbonate, fluoroethylene carbonate, ethylene sulfite, succinonitrile (SN), and 1,3,2-dioxathiolane-2,2-dioxide (DTD).

Due to the various roles that different additives play in the electrolyte, sodium-ion battery additives can be categorized based on their functional characteristics. Film-forming additives are primarily used to enhance the stability of the cathode and anode electrolyte interphases (CEI and SEI). By optimizing both the SEI and cathode electrolyte interphase (CEI), Dahn et al. [55] reported that the electrolyte additives prop-1-ene-1,3-sultone (PES) and DTD significantly extend the lifespan of sodium-ion batteries, even at elevated temperatures of 40 and 55 °C. Remarkably, the sodium ion batteries demonstrated a longer lifespan compared to $LiFePO_4$/graphite cells under similar conditions (Fig. 28).

Flame-retardant additives help to reduce the flammability of the electrolyte. Fluorine (F), phosphorus (P), chlorine (Cl), and bromine (Br) are all effective flame-retardant elements. Among them, organic phosphorus-based flame retardant additives have been the most extensively studied in recent years. Examples include triethyl phosphate (TEP), trimethyl phosphate (TMP), and ethyl difluorophosphate (EFPN). These compounds are mostly liquid at room temperature and have some solubility in organic solvents. When heated, these flame retardants release phosphorus-containing free radicals, such as $P^•$, which can capture the $H^•$ radicals involved in the chain combustion reactions of organic materials. This process effectively terminates the

Table 2 Properties of selected solvents. Reproduced from Ref. [54], Copyright 2011 McGraw-Hill Education

Characteristic	EC*	PC*	DMC*	EMC*	DEC*	FEC*	MA*	MP*
Structure	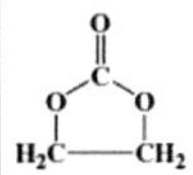	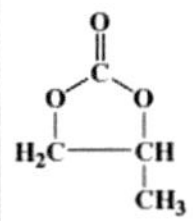	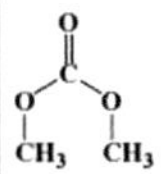	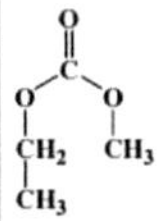	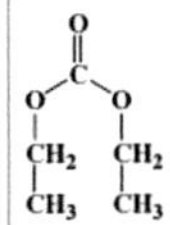	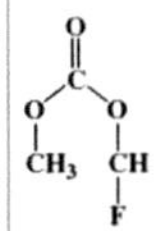	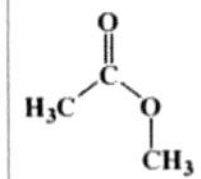	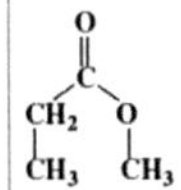
BP, °C	248 [53]	242 [53]	90	109	126	212	57	80
MP, °C	36.4 [53]	−48 [53]	4.6	−55	−74 [54]	20	−98	−88
Density, g/mL	1.32	1.20 [55]	1.07	1.0	0.97		0.932	0.915
Viscosity, cP	1.86 (40 °C)	2.5	0.59	0.65	0.75		0.38	0.43
Dielectric constant	89.6 (40 °C)	64.9 [55]	3.12	2.9	2.82		6.7	6.2
Donor number	16.4	15	8.7	6.5	8			
Mol. Wt.	88.1	102.1	90.1	104.1	118.1	106.5	74.08	88.11

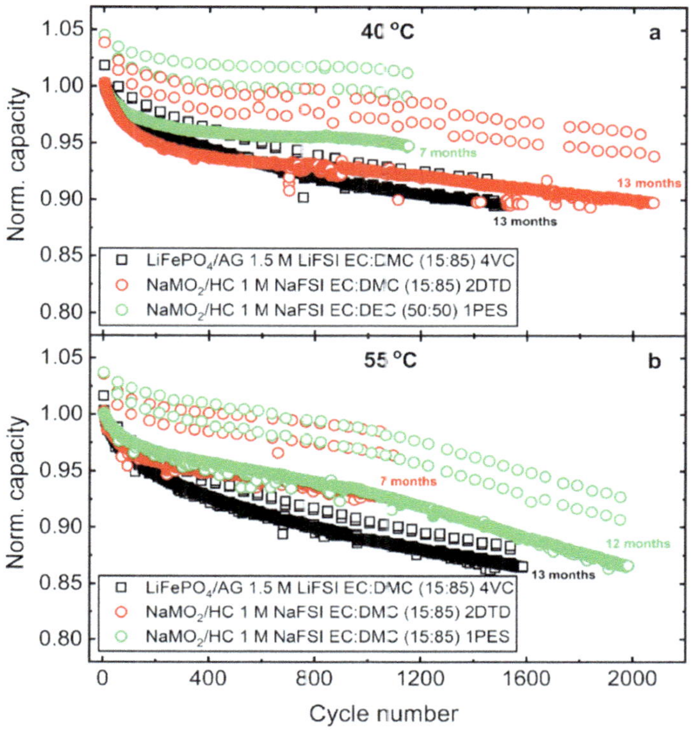

Fig. 28 Capacity retention of 250 mAh $NaMO_2$/hard carbon pouch cells at 40 °C (**a**) or 55 °C (**b**) between 2.0 and 3.8 V compared to $LiFePO_4$/graphite cell. All cells cycle with C/3 rate and C/20 check-up cycles. Reproduced from Ref. [55], Copyright 2024 IOP Science

chain reactions, making it difficult or impossible for the organic electrolyte to sustain combustion. Feng et al. [56] demonstrated that adding 5% EFPN to an electrolyte solution of $NaPF_6$ in EC + DEC significantly reduces its flammability (Fig. 29).

Overcharge protection additives are designed to prevent battery combustion or explosion in the event of overcharging. Overcharge protection additives can generally be classified into two types: redox shuttle agents and electrochemical polymerizers, each operating through a different mechanism. Redox shuttle agents are first oxidized at the positive electrode and then shuttled to the negative electrode, where they are reduced. This shuttle mechanism helps prevent overcharging by reducing excessive current flow. On the other hand, electrochemical polymerizers undergo a polymerization reaction when the battery voltage exceeds the polymerization voltage of the additive. This reaction forms a polymer film on the positive electrode surface and separator while releasing protons. The polymer film increases the internal resistance of the battery, slowing down and preventing further electrolyte decomposition and thermal runaway. The protons released are reduced to hydrogen gas at the anode, and if the pressure exceeds a certain threshold, it can trigger a pressure relief valve.

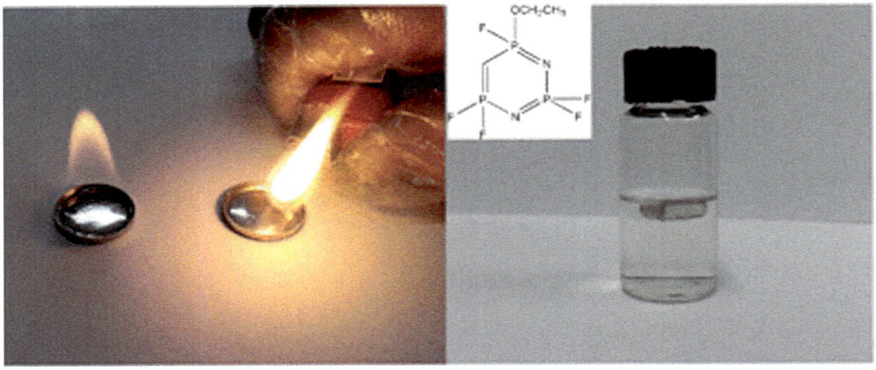

Fig. 29 Illustration of the chemical structure (middle), flammability test results (left), and sodium metal compatibility (right) for a solution of 5% EFPN in 1 M $NaPF_6$/EC–DEC (1:1, v/v). The left cell showcases the blank electrolyte, while the right cell displays the electrolyte with the addition of 5% EFPN. Reproduced from Ref. [56], Copyright 2015 Royal Society of Chemistry

Feng et al. [57] reported that biphenyl can serve as an overcharge protection additive in $Na_{0.44}MnO_2$/Na batteries. When the voltage reaches 4.3 V (vs. Na/Na^+), biphenyl undergoes electro-polymerization, effectively preventing voltage runaway by consuming the excess charge. This mechanism enables the battery to handle overcharge capacities up to 800%, demonstrating biphenyl's effectiveness in enhancing battery safety by mitigating the risks associated with overcharging (Fig. 30).

Each type of additive serves a specific purpose to improve the safety and performance of the battery.

Fig. 30 Cyclic voltammetry (CV) curves of 1 M $NaPF_6$ in EC/DEC (1:1 v/v) electrolyte with and without 3% biphenyl (BP) and the overcharge protection mechanism of BP (inner graph). Reproduced from Ref. [57], Copyright 2015 Royal Society of Chemistry

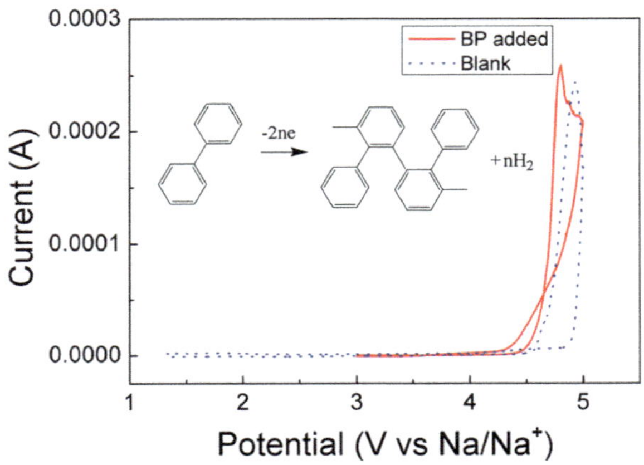

2.4 Separator Materials

Battery separators, which are thin (15–40 μm) microporous films, serve the crucial function of electrically isolating the positive and negative electrodes. As outlined by Dahn et al. [54], an effective separator should possess the following key characteristics:

- High tensile strength in the machine direction to enable automated winding,
- Resistance to yielding or shrinking in width,
- Puncture resistance against electrode materials,
- An effective pore size of less than 1 μm to prevent the migration of electrode materials and micro-short circuits,
- High wettability by the electrolyte,
- Compatibility and stability when in contact with both the electrolyte and electrode materials.

Given that sodium-ion battery technology closely follows the path of lithium-ion batteries, sodium-ion batteries currently use lithium-ion battery separators. At present, microporous polyolefin materials are the predominant choice for separators in commercially available liquid electrolyte cells due to their excellent mechanical properties, chemical stability, and cost-effectiveness. Although nonwoven materials have also been investigated, they have not been widely adopted, mainly due to difficulties in manufacturing thin, uniform materials with high strength. Therefore, this section will focus on discussing microporous polyolefin materials.

Microporous polyolefin materials currently in use are composed of polyethylene (PE), polypropylene (PP), or laminates of both as base films. These materials are produced through either a dry extrusion process or a wet solvent-based process. While the dry process is more environmentally friendly, the wet solvent process offers the advantage of producing thinner and more robust base films. These PP and PE base films exhibit excellent tensile strength (for example, the PP separator Celgard 2400 demonstrates 142 MPa along the machine direction (MD) and 14 MPa along the transverse direction (TD)). This superior tensile strength allows polyolefin separators to stay intact during battery assembly, significantly reducing the risk of battery breakage.

Additionally, classical tri-layer materials (PP/PE/PP) have been developed, where the polypropylene layers are designed to maintain the structural integrity of the film. The PE layer, with its lower melting point (135 °C), is intended to shut down the cell if it reaches an over-temperature condition, providing an essential safety feature.

Lithium-ion batteries and sodium-ion batteries have distinct chemistries, which necessitate different requirements for separators to ensure high-performance sodium-ion batteries:

- Electrolyte Compatibility: Propylene carbonate (PC), commonly used in SIBs but uncommon in lithium-ion batteries due to graphite exfoliation, cannot effectively wet PP or PE separators. To address this, it is worth considering the application of special surface coatings on the base films that allow the PC to wet more easily without undergoing oxidation during sodium ion battery operation.

- TM Dissolution: Pronounced dissolution of TMs can occur from sodium-ion metal oxide or Prussian blue analog positive electrodes. To prevent these dissolved TMs from depositing on the negative electrode and compromising cell performance, it is worth considering designing separators that can effectively capture and retain these dissolved TMs.
- Electrolyte Management in Cylindrical Cells: In high-energy-density cylindrical cells, electrolyte is densely packed and can be displaced during charging as electrode materials expand, only to be reabsorbed during discharge. This fluctuation can affect cell performance, especially in sodium-ion batteries with alloy-based negative electrodes due to greater volume changes. Improving the electrolyte retention capability of the separator is worth considering to address this issue.

3 Safety of Sodium-Ion Batteries

3.1 Temperature Dependence of the Reactions Between Charged Electrode Materials and Electrolyte

With the burgeoning commercial advancement of sodium-ion batteries, it is imperative that greater attention is devoted to their safety performance. Similar to lithium-ion batteries, all safety concerns with sodium-ion batteries ultimately arise from the interaction between the charged positive and negative electrode materials and the cell electrolyte at elevated temperatures. Figure 31 shows accelerating rate calorimetry data for the reactions of lithiated hard carbon (Li/HC) and sodiated hard carbon (Na/HC) negative electrode samples with carbonate-based electrolytes. The Na/HC samples exhibit worse thermal stability compared to the Li/HC samples in these carbonate-based electrolytes.

In addition to considering the effect of surface area, the degree of reactions between Na/HC and electrolytes heavily depends on electrolyte chemistries. Proper electrolyte additives like FEC can significantly decrease the self-heating rate (SHR) at certain temperatures (Fig. 32). To increase the volumetric energy density of sodium-ion batteries, sodium-alloying compounds such as phosphorus (P), germanium (Ge), tin (Sn), antimony (Sb), lead (Pb), and bismuth (Bi) can be used as negative electrode materials [24]. However, there is a lack of systematic studies on the thermal stability of these materials with electrolytes. It is recommended to study the reactions between these materials and electrolytes at elevated temperatures, considering electrolyte chemistries, the degree of sodiation, surface area, and their mixtures with hard carbon.

For the layered oxide positive electrode, the release of lattice oxygen at elevated temperatures is the primary cause of accelerating exothermic reactions. For instance,

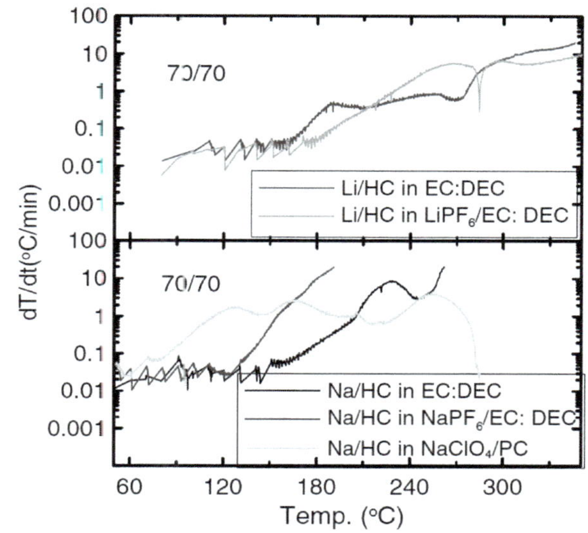

Fig. 31 Self-heating rate (SHR) versus temperature of 70 mg Li/HC or 70 mg Na/HC reacting with 70 mg of electrolyte or solvent. EC:DEC (1:2 v:v). Reproduced from Ref. [58], Copyright 2012 IOP Science

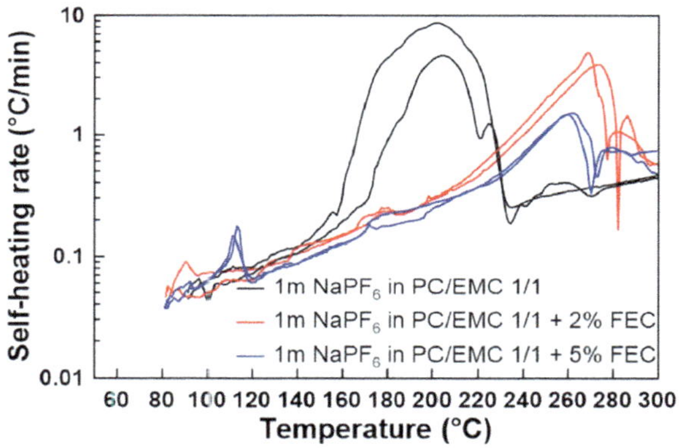

Fig. 32 SHR versus temperature for sodiated hard carbon reacting with 1 m $NaPF_6$ in PC/EMC (1/1 by vol.) with 2 wt% (red curve) and 5 wt% (blue curve) FEC compared with the control electrolyte (black curve). Reproduced from Ref. [59], Copyright 2012 IOP Science

similar to $LiCoO_2$, charged $NaCoO_2$ undergoes three successive exothermic reactions due to the release of lattice oxygen (Fig. 33), as detailed below [60].

First exotherm:

$$Na_{0.35}CoO_2 \rightarrow 1/2Na_{0.7}CoO_2 + 1/6Co_3O_4 + 1/6O_2(\text{combusts solvent})$$

Second exotherm:

$$1/2Na_{0.7}CoO_2 \rightarrow 1/6Co_3O_4 + 1/6O_2(\text{combusts solvent}) + 1/6O_2$$

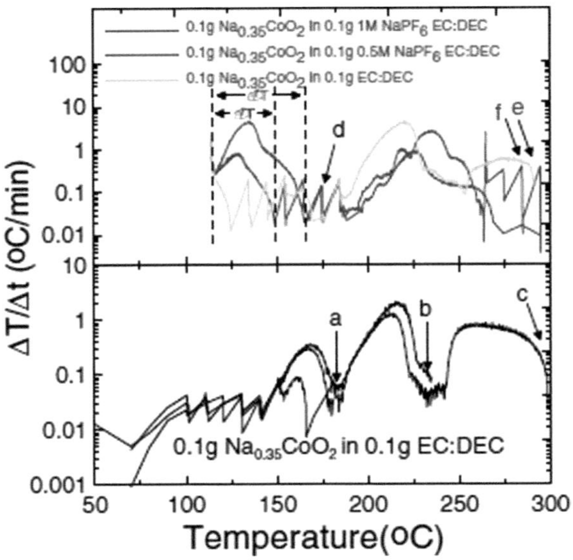

Fig. 33 Self-heating rate (SHR) versus temperature of 100 mg $Na_{0.35}CoO_2$ in the same mass of EC:DEC (1:2 v/v), 0.5 M $NaPF_6$ in EC:DEC (1:2 v/v) or 1 M $NaPF_6$ in EC:DEC (1:2 v/v). Reproduced from Ref. [60], Copyright 2012 IOP Science

Third exotherm:

$$1/3Co_3O_4 \rightarrow Co \text{ or } CoO$$

Surprisingly, the thermal stability of deintercalated $Na_{0.5}CrO_2$ is excellent, showing virtually no reactivity with non-aqueous solvents between room temperature and 350 °C (Fig. 34). Its reactivity is even lower than that of Li_0FePO_4 in the same solvent. $Na_{0.5}CrO_2$ is thought to transform into $NaCrO_2$ and P3-$CrO_{2-\delta}$ during its reaction with solvents or electrolytes, with minimal oxygen release contributing to its high thermal stability.

To understand the reactions between positive electrodes and electrolytes at elevated temperatures, it is essential to study the behavior of lattice oxygen in these electrodes. Future research should focus on the lattice oxygen activity in layered oxides with diverse transition metal chemistries and phosphate-based positive electrode materials.

As noted by Dahn et al. [54], calorimetry data, which is typically measured at relatively low temperatures and low heating rates, is best suited for assessing abuse tolerance under similar conditions (e.g., hot-box testing). However, in cases of extreme thermal abuse (such as nail penetration, crush tests, or internal shorts), other factors, including the total heat released during such events, must be considered when selecting materials for cell safety.

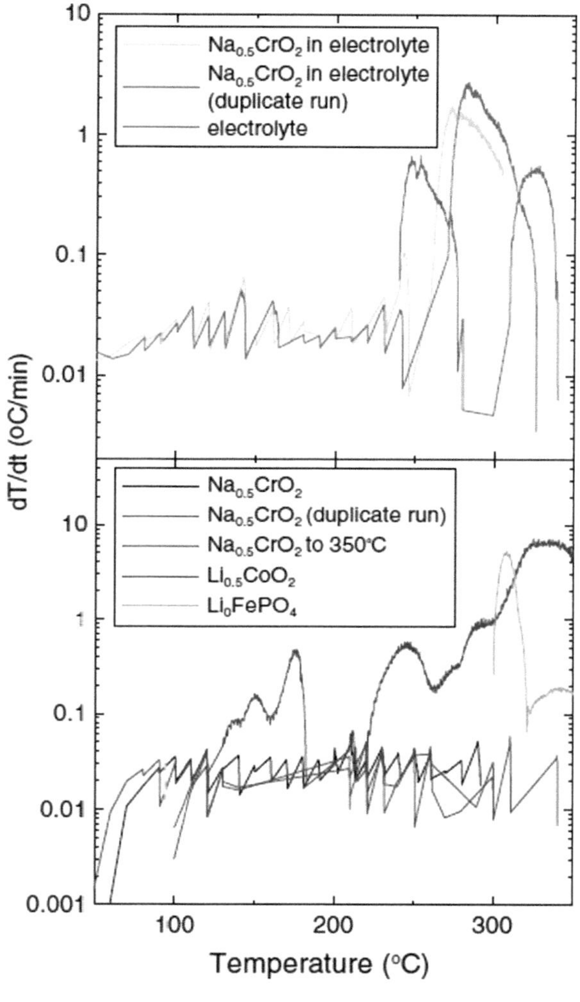

Fig. 34 Self-heating rate (SHR) versus temperature of 100 mg $Na_{0.5}CrO_2$, $Li_{0.5}CoO_2$, or Li_0FePO_4 in the same mass of EC: DEC (1:2 v/v) or 1 M $NaPF_6$ in EC: DEC (1:2 v/v), and the electrolyte itself. Reproduced from Ref. [61], Copyright 2012 IOP Science

3.2 *Regulatory Criteria for the Safety and Design of Sodium-Ion Batteries*

The safety of sodium-ion batteries is a top priority for everyone involved: consumers seek to avoid injuries, original equipment manufacturers aim to prevent costly product recalls, and battery producers strive for long-term business sustainability. Both cell manufacturers and original equipment manufacturers (OEMs) are committed to ensuring the highest standards of battery safety to meet these objectives. In the realm of lithium-ion batteries, there are well-established safety standards both domestically and internationally, which encompass a wide range of products from cells to modules and systems. These standards help to raise the market entry threshold for lithium-ion battery products, significantly enhancing consumer safety. However, in the field of sodium-ion batteries, standardization remains in its early stages.

Given the similarities in safety issues between lithium-ion batteries and sodium-ion batteries, it would be beneficial to refer to the established standardization system of lithium-ion batteries when developing standards and approving projects for sodium-ion batteries. For example, Underwriters Laboratories has incorporated sodium ion batteries into their testing services under the same standards as lithium-ion batteries for UN/DOT 38.3 (UN Manual of Tests and Criteria, Sect. 38.3, Recommendations on the Transport of Dangerous Goods) compliance [62].

According to Dahn et al. [54], the standardization of sodium ion battery safety testing can draw from established guidelines such as Underwriters Laboratories UL1642 "Standard for Li-ion Batteries," the International Electrotechnical Commission IEC 62133 "Secondary cells and batteries containing alkaline or other non-acid electrolytes," the Institute of Electrical and Electronics Engineers IEEE 1625 "Standard for Rechargeable Batteries for Multi-Cell Mobile Computing Devices," and IEEE 1725 "Standard for Rechargeable Batteries for Cellular Telephones." It is worth noting that Prussian blue analogue positive electrode materials could release hydrogen cyanide (HCN) at elevated temperatures. Therefore, testing for sodium ion battery toxicity, such as HCN release, should also be included in safety evaluations.

4 Conclusions and Future Trends

Sodium-ion batteries are increasingly being adopted in new and emerging markets, such as low-speed vehicles and grid energy storage, complementing lithium-ion batteries in the global push for electrification by utilizing earth-abundant elements. Their acceptance is driven by their ability to deliver strong performance in several areas, including long cycle life, extended storage life, and a broad operating temperature range, all within a safe and cost-effective framework. The choice of positive electrode materials allows for trade-offs in cost, power, energy, and thermal stability, depending on application needs.

Further advancements in sodium-ion battery performance are anticipated through the development of improved materials. Current research and development efforts are heavily focused on enhancing sodium-ion battery materials. Efforts to improve negative electrode materials are underway, with the aim of increasing capacity and enhancing volumetric energy density. Promising anode materials, such as those based on tin, lead, and phosphorus, are being investigated for their potential to significantly improve energy density. Additionally, innovations in traditional non-aqueous electrolytes and electrode coatings are aimed at extending both cycle and calendar life while minimizing temperature sensitivity. The goal is to maintain consistent battery life across a range of temperatures, reducing the typical degradation seen at elevated temperatures.

As sodium-ion batteries become more widely used in grid energy storage, safety is becoming a key focus area that requires deeper exploration. Research into aqueous and all-solid-state electrolyte systems represents a cutting-edge effort in this field, seeking to improve safety and performance. In addition to mitigating risks of thermal runaway, the potential toxicity of materials, such as Prussian blue cathodes, must also be considered, particularly due to the possible release of hydrogen cyanide (HCN) under certain conditions.

The price of lithium salts, such as lithium hydroxide (LiOH) and lithium carbonate (Li_2CO_3), is notoriously volatile and difficult to predict. When prices are low, the commercial incentive to pursue sodium-ion battery technology diminishes, as industries tend to favor established lithium-ion solutions like $LiFePO_4$/graphite cells, which offer a more attractive margin. This fluctuation presents a challenge for maintaining the momentum of sodium-ion battery commercialization during periods of low lithium prices. To address this, it's crucial to focus on the unique advantages of sodium-ion batteries beyond just cost, such as the abundance of sodium, supply chain stability, and suitability for specific applications like large-scale energy storage. Continued investments in improving sodium-ion battery performance, safety, and scalability will also help solidify its position in the market, regardless of lithium price fluctuations. Developing a robust and diversified sodium-ion ecosystem can reduce the industry's dependency on lithium pricing dynamics, ensuring long-term viability and competitiveness.

Exercise Questions

1. What is the primary mechanism by which sodium-ion (Na-ion) batteries operate?

Answer:

Sodium-ion batteries operate through the movement of sodium ions (Na^+) between the positive and negative electrodes. During charging, Na^+ ions are extracted from the positive electrode (often made of a layered metal oxide like $NaTMO_2$) and inserted into the negative electrode (typically hard carbon). During discharging, the ions move back to the positive electrode. This process is often referred to as the "rocking chair" mechanism because of the back-and-forth motion of sodium ions between the electrodes during cycling.

2. How do the open-circuit voltage (U_{ocv}) and operating voltage (U_w) differ in a sodium-ion battery?

Answer:

The open-circuit voltage (U_{ocv}) is the potential difference between the positive and negative electrodes when no current flows through the circuit. It represents the maximum possible voltage the battery can provide. The operating voltage (U_w), or discharge voltage, is the actual voltage observed when the battery is discharging under load. The operating voltage is always lower than the open circuit voltage due to the internal resistances, such as ohmic and polarization resistances, that the current must overcome within the battery during operation.

3. What role does internal resistance play in sodium-ion battery performance?

Answer:

Internal resistance in sodium-ion batteries affects their efficiency and operating voltage. It consists of two main components: ohmic resistance and polarization resistance.

- Ohmic resistance is the sum of resistances from the electrode materials, separator, electrolyte, and contact points within the battery.
- Polarization resistance arises from the electrochemical reactions at the electrodes and includes electrochemical polarization (due to the kinetics of the charge transfer reactions) and concentration polarization (caused by the diffusion of ions). Higher internal resistance reduces the battery's efficiency, lowers its operating voltage, and limits its ability to deliver high power.

4. What are the main structural types of oxides used as sodium-ion battery-positive electrode materials?

Answer:

The oxide-based positive electrode materials are primarily classified into layered oxides and tunnel-structured oxides.

5. How does the inductive effect of the anions of polyanionic compounds impact the performance of polyanionic positive electrode materials-based battery?

Answer:

The inductive effect of anions like PO_4^{3-} and SO_4^{2-} enhances the battery's operating voltage by increasing the redox potential of transition metals (TMs). This strong inductive effect can raise the energy density of the battery, making polyanionic compounds a promising choice for high-voltage sodium-ion batteries.

6. What advantages do Prussian blue analogs offer as positive electrode materials for sodium-ion batteries? List at least three.

- They have an open three-dimensional ion channel structure that enables fast sodium ion diffusion.
- Their Fe-CN-M framework is stable, maintaining the structural integrity during cycling, leading to a longer battery lifespan.
- Prussian blue analogs utilize earth-abundant elements including C, N, Fe, etc.

7. Compare and contrast the solid-state reaction and sol–gel methods used for synthesizing sodium-ion battery-positive electrode materials. What are the main benefits and drawbacks of each?

Answer:

- Solid-state reaction method: This is a simple and widely used method that involves mixing solid precursors and calcining them at high temperatures. Its main advantages include ease of operation and suitability for large-scale industrial production. However, it often leads to inhomogeneous materials with larger particle sizes and requires long reaction times.
- Sol–gel method: This method involves creating a gel from dissolved metal salts, followed by drying and calcination. It offers better control over particle size and morphology, resulting in more uniform compositions. However, it is more time-consuming and involves multiple synthesis steps.

8. Explain why graphite, which is widely used in lithium-ion batteries, is not suitable for sodium-ion batteries.

Answer:

Graphite is unsuitable for sodium-ion batteries because it exhibits very low sodium storage capacity. Sodium cannot intercalate into graphite in the same way lithium does. Studies, such as those by Asher et al. (1958) and Ge et al. (1988), revealed that sodium forms a compound, NaC_{64}, which corresponds to minimal sodium intercalation into graphite. Furthermore, density functional theory (DFT) calculations showed that sodium-graphite compounds are not thermodynamically stable compared to other alkali metals, which explains the poor performance of graphite as a negative electrode in sodium-ion systems.

9. Describe the "house of cards" model proposed by Dahn et al. for sodium storage in hard carbon. How does this model explain the electrochemical behavior observed in hard carbon?

Answer:

The "house of cards" model proposed by Dahn et al. in 2000 explains sodium storage in hard carbon by suggesting that small aromatic fragments (~40 Å in lateral extent) are randomly stacked, creating regions where layers are either parallel or randomly oriented. The sloping region in the electrochemical profile is due to sodium insertion between these parallel layers, while the plateau region near 0 V is attributed to sodium adsorption into the nanoporosity between the randomly stacked layers. This model successfully explains the electrochemical behavior of hard carbon, including the reversible sodium insertion process.

10. What are the key characteristics that an electrolyte must possess for optimal sodium-ion battery performance?

Answer:

The key characteristics include high ionic conductivity, wide liquid temperature range, good chemical stability, excellent electrochemical stability, good thermal stability, low cost, and environmental friendliness.

11. What are the roles of electrolyte additives in improving sodium-ion battery performance?

Answer:

Electrolyte additives improve the stability of the electrode–electrolyte interphase (SEI and CEI), reduce flammability, and provide overcharge protection.

12. Explain the function of separators in sodium-ion batteries and why microporous polyolefin materials are commonly used.

Answer:

Separators isolate the positive and negative electrodes while allowing ion transport. Microporous polyolefin separators are used for their mechanical strength, chemical stability, and cost-effectiveness.

13. What safety strategies can be employed to minimize the thermal risks associated with sodium-ion batteries, especially with sodiated hard carbon and alloy-based negative electrodes?

Answer:

Safety strategies include the use of additives such as FEC to reduce self-heating rates, the development of better thermal management systems, and the selection of safer electrode materials with enhanced thermal stability.

References

1. Yung-Fang Yu Y, Kummer JT (1967) Ion exchange properties of and rates of ionic diffusion in beta-alumina. J Inorg Nucl Chem 29(9):2453–2475
2. Coetzer J (1986) A new high energy density battery system. J Power Sources 18(4):377–380
3. Delmas C, Braconnier J-J, Fouassier C, Hagenmuller P (1981) Electrochemical intercalation of sodium in $NaxCoO_2$ bronzes. Solid State Ionics 3–4:165–169
4. Delmas C (2018) Sodium and sodium-ion batteries: 50 years of research. Adv Energy Mater 8(17):1703137
5. Stevens DA, Dahn JR (2000) High capacity anode materials for rechargeable sodium-ion batteries. J Electrochem Soc 147(4):1271
6. Ellis BL, Makahnouk WRM, Makimura Y, Toghill K, Nazar LF (2007) A multifunctional 3.5 V iron-based phosphate cathode for rechargeable batteries. Nat Mater 6(10):749–753

7. Komaba S, Murata W, Ishikawa T, Yabuuchi N, Ozeki T, Nakayama T, Ogata A, Gotoh K, Fujiwara K (2011) Electrochemical Na insertion and solid electrolyte interphase for hard-carbon electrodes and application to Na-ion batteries. Adv Funct Mater 21(20):3859–3867
8. Wang L, Lu Y, Liu J, Xu M, Cheng J, Zhang D, Goodenough JB (2013) A superior low-cost cathode for a Na-ion battery. Angew Chem Int Ed 52(7):1964–1967
9. (2014) Novel copper redox-based cathode materials for room-temperature sodium-ion batteries. Chin Phys B 23(11):118202
10. Mu L, Xu S, Li Y, Hu Y-S, Li H, Chen L, Huang X (2015) Prototype sodium-ion batteries using an air-stable and Co/Ni-free O3-layered metal oxide cathode. Adv Mater 27(43):6928–6933
11. Li Y, Hu Y-S, Qi X, Rong X, Li H, Huang X, Chen L (2016) Advanced sodium-ion batteries using superior low cost pyrolyzed anthracite anode: towards practical applications. Energy Storage Mater 5:191–197
12. Lu Y, Rong X, Hu Y-S, Chen L, Li H (2019) Research and development of advanced battery materials in China. Energy Storage Mater 23:144–153
13. Xing Z, Zhao W, Yu B, Wang Y, Zhou L, Xiong P, Chen M, Zhu J Electrolyte design strategies for aqueous sodium-ion batteries: progress and prospects. Small 2405442
14. Liu Q, Zhao X, Yang Q, Hou L, Mu D, Tan G, Li L, Chen R, Wu F (2023) The progress in the electrolytes for solid state sodium-ion battery. Adv Mater Technol 8(7):2200822
15. Delmas C, Fouassier C, Hagenmuller P (1980) Structural classification and properties of the layered oxides. Physica B+C 99(1):81–85
16. Pan H, Hu Y-S, Chen L (2013) Room-temperature stationary sodium-ion batteries for large-scale electric energy storage. Energy Environ Sci 6(8):2338–2360
17. Zheng Q, Yi H, Li X, Zhang H (2018) Progress and prospect for NASICON-type $Na_3V_2(PO_4)_3$ for electrochemical energy storage. J Energy Chem 27(6):1597–1617
18. Shishkin M, Sato H (2017) Ab initio study of stability of $Na_2Fe_2(SO_4)_3$, a high potential Na-ion battery cathode material. J Phys Chem C 121(37):20067–20074
19. Wu X, Wu C, Wei C, Hu L, Qian J, Cao Y, Ai X, Wang J, Yang H (2016) Highly crystallized $Na_2CoFe(CN)_6$ with suppressed lattice defects as superior cathode material for sodium-ion batteries. ACS Appl Mater Interfaces 8(8):5393–5399
20. Zhang H, Gao Y, Liu X-H, Yang Z, He X-X, Li L, Qiao Y, Chen W-H, Zeng R-H, Wang Y, Chou S-L (2022) Organic cathode materials for sodium-ion batteries: from fundamental research to potential commercial application. Adv Funct Mater 32(4):2107718
21. Wu C, Yang Y, Zhang Y, Xu H, He X, Wu X, Chou S (2024) Hard carbon for sodium-ion batteries: progress, strategies and future perspective. Chem Sci 15(17):6244–6268
22. Yin X, Sarkar S, Shi S, Huang Q-A, Zhao H, Yan L, Zhao Y, Zhang J (2020) Recent progress in advanced organic electrode materials for sodium-ion batteries: synthesis, mechanisms, challenges and perspectives. Adv Funct Mater 30(11):1908445
23. Guo S, Yi J, Sun Y, Zhou H (2016) Recent advances in titanium-based electrode materials for stationary sodium-ion batteries. Energy Environ Sci 9(10):2978–3006
24. Garayt MDL, Zhang L, Zhang Y, Obialor MC, Deshmukh J, Xing Y, Yang C, Metzger M, Dahn JR (2024) Practical alloy-based negative electrodes for Na-ion batteries. J Electrochem Soc 171(7):070523
25. Asher RC, Wilson SA (1958) Lamellar compound of sodium with graphite. Nature 181(4606):409–410
26. Ge P, Fouletier M (1988) Electrochemical intercalation of sodium in graphite. Solid State Ionics 28–30:1172–1175
27. Liu Y, Merinov BV, Goddard WA (2016) Origin of low sodium capacity in graphite and generally weak substrate binding of Na and Mg among alkali and alkaline earth metals. Proc Natl Acad Sci 113(14):3735–3739
28. Dahn JR, Zheng T, Liu Y, Xue JS (1995) Mechanisms for lithium insertion in carbonaceous materials. Science 270(5236):590–593
29. Zheng T, McKinnon WR, Dahn JR (1996) Hysteresis during lithium insertion in hydrogen-containing carbons. J Electrochem Soc 143(7):2137

30. Bommier C, Surta TW, Dolgos M, Ji X (2015) New mechanistic insights on Na-ion storage in nongraphitizable carbon. Nano Lett 15(9):5888–5892
31. Buiel E, Dahn JR (1999) Li-insertion in hard carbon anode materials for Li-ion batteries. Electrochim Acta 45(1):121–130
32. Senguttuvan P, Rousse G, Seznec V, Tarascon J-M, Palacín MR (2011) $Na_2Ti_3O_7$: lowest voltage ever reported oxide insertion electrode for sodium ion batteries. Chem Mater 23(18):4109–4111
33. Zhai H, Xia BY, Park HS (2019) Ti-based electrode materials for electrochemical sodium ion storage and removal. J Mater Chem A 7(39):22163–22188
34. Park SI, Gocheva I, Okada S, Yamaki J-I (2011) Electrochemical properties of $NaTi_2(PO_4)_3$ anode for rechargeable aqueous sodium-ion batteries. J Electrochem Soc 158(10):A1067
35. Sun Y, Zhao L, Pan H, Lu X, Gu L, Hu Y-S, Li H, Armand M, Ikuhara Y, Chen L, Huang X (2013) Direct atomic-scale confirmation of three-phase storage mechanism in $Li_4Ti_5O_{12}$ anodes for room-temperature sodium-ion batteries. Nat Commun 4(1):1870
36. Qiao S, Zhou Q, Ma M, Liu HK, Dou SX, Chong S (2023) Advanced anode materials for rechargeable sodium-ion batteries. ACS Nano 17(12):11220–11252
37. Zhao L, Zhao J, Hu Y-S, Li H, Zhou Z, Armand M, Chen L (2012) Disodium terephthalate ($Na_2C_8H_4O_4$) as high performance anode material for low-cost room-temperature sodium-ion battery. Adv Energy Mater 2(8):962–965
38. Wu X, Ma J, Ma Q, Xu S, Hu Y-S, Sun Y, Li H, Chen L, Huang X (2015) A spray drying approach for the synthesis of a $Na_2C_6H_2O_4$/CNT nanocomposite anode for sodium-ion batteries. J Mater Chem A 3(25):13193–13197
39. Abraham KM (2020) How comparable are sodium-ion batteries to lithium-ion counterparts? ACS Energy Lett 5(11):3544–3547
40. Rudola A, Sayers R, Wright CJ, Barker J (2023) Opportunities for moderate-range electric vehicles using sustainable sodium-ion batteries. Nat Energy 8(3):215–218
41. Yabuuchi N, Matsuura Y, Ishikawa T, Kuze S, Son J-Y, Cui Y-T, Oji H, Komaba S (2014) Phosphorus electrodes in sodium cells: small volume expansion by sodiation and the surface-stabilization mechanism in aprotic solvent. ChemElectroChem 1(3):580–589
42. https://www.usgs.gov/centers/national-minerals-information-center/recycling-statistics-and-information
43. Darwiche A, Dugas R, Fraisse B, Monconduit L (2016) Reinstating lead for high-loaded efficient negative electrode for rechargeable sodium-ion battery. J Power Sources 304:1–8
44. Crouch-Baker S, Deublein G, Tsai HC, Zhou LZ, Huggins RA (1990) Materials considerations related to sodium-based rechargeable cells for use above room temperature. Solid State Ionics 42(1):109–115
45. Ellis LD, Hatchard TD, Obrovac MN (2012) Reversible insertion of sodium in tin. J Electrochem Soc 159(11):A1801
46. Tan DHS (2024) Advanced sodium-ion batteries. In: 41st International battery seminar & exhibit
47. Obrovac MN, Chevrier VL (2014) Alloy negative electrodes for Li-ion batteries. Chem Rev 114(23):11444–11502
48. Yao Q, Zhu Y, Zheng C, Wang N, Wang D, Tian F, Bai Z, Yang J, Qian Y, Dou S (2023) Intermolecular cross-linking reinforces polymer binders for durable alloy-type anode materials of sodium-ion batteries. Adv Energy Mater 13(9):2202939
49. Kim D, Hwang C, Jeong J, Song W-J, Park S, Song H-K (2019) Bipolymer-cross-linked binder to improve the reversibility and kinetics of sodiation and desodiation of antimony for sodium-ion batteries. ACS Appl Mater Interfaces 11(46):43039–43045
50. Sarkar S, Mukherjee PP (2021) Synergistic voltage and electrolyte mediation improves sodiation kinetics in μ-Sn alloy-anodes. Energy Storage Mater 43:305–316
51. Lu H, Wu L, Xiao L, Ai X, Yang H, Cao Y (2016) Investigation of the effect of fluoroethylene carbonate additive on electrochemical performance of Sb-based anode for sodium-ion batteries. Electrochim Acta 190:402–408

52. Xu K (2014) Electrolytes and Interphases in Li-Ion batteries and beyond. Chem Rev 114(23):11503–11618
53. Xu K (2004) Nonaqueous liquid electrolytes for lithium-based rechargeable batteries. Chem Rev 104(10):4303–4418
54. Thomas BR (2011) Linden's handbook of batteries, 4th edn. McGraw-Hill Education, New York
55. Hijazi H, Ye Z, Zsoldos E, Obialor M, Black W, Azam S, Dahn JR, Metzger M (2024) Can layered oxide/hard carbon sodium-ion pouch cells with simple electrolyte additives achieve better cycle life than LFP/graphite cells? J Electrochem Soc 171(5):050521
56. Feng J, An Y, Ci L, Xiong S (2015) Nonflammable electrolyte for safer non-aqueous sodium batteries. J Mater Chem A 3(28):14539–14544
57. Feng J, Ci L, Xiong S (2015) Biphenyl as overcharge protection additive for nonaqueous sodium batteries. RSC Adv 5(117):96649–96652
58. Xia X, Dahn JR (2012) Study of the reactivity of Na/hard carbon with different solvents and electrolytes. J Electrochem Soc 159(5):A515
59. Shipitsyn V, Jayakumar R, Zuo W, Yin W, Huber E, Ma L (2023) The impact of fluoroethylene carbonate additive on charged sodium ion electrodes/electrolyte reactivity studied using accelerating rate calorimetry. J Electrochem Soc 170(11):110501
60. Xia X, Dahn JR (2012) A study of the reactivity of de-intercalated P2-$NaxCoO_2$ with non-aqueous solvent and electrolyte by accelerating rate calorimetry. J Electrochem Soc 159(5):A647
61. Xia X, Dahn JR (2011) $NaCrO_2$ is a fundamentally safe positive electrode material for sodium-ion batteries with liquid electrolytes. Electrochem Solid-State Lett 15(1):A1
62. https://www.ul.com/services/ev-battery-testing-compliance-regulatory-requirements-and-standards

Advances in Electrocatalyst Design for Hydrogen Production by Water Electrolysis

Sachin Kumar and Yang Yang

Abstract Hydrogen has been widely used in industrial processes. However, most of the hydrogen produced today comes in the form of grey hydrogen, which is obtained from the cracking of hydrocarbons. An alternative method for generating hydrogen is through water electrolysis, which produces both hydrogen and oxygen. However, the efficiency of hydrogen production is limited by the availability of effective electrocatalysts for the hydrogen evolution reaction (HER). In this chapter, we provide a comprehensive list of electrocatalysts used for HER and discuss how their intrinsic properties can be modified through various strategies. These modifications can tune the electronic structure, such as the d-band center, and ultimately influence their activity in hydrogen production.

1 Introduction

Hydrogen is widely used in industrial processes such as petroleum refining, metal treatment, fertilizer production, and the synthesis of various chemicals. Additionally, hydrogen plays a crucial role in rocket fuels and fuel cells for electricity generation and powering vehicles [1]. Currently, most hydrogen is produced through steam reforming [2], a process reliant on fossil fuels, which contributes to carbon emissions and exacerbates global warming. To reduce the environmental impact of fossil fuel reliance, there is an increasing focus on sustainable alternatives that provide beneficial ecological outcomes [3]. From this perspective, the electrochemical hydrogen evolution reaction (HER) is a promising route for producing green hydrogen (H_2). In electrochemical water splitting, electricity is applied to split water, producing oxygen at the anode and hydrogen at the cathode. One of the most eco-friendly methods for generating H_2 utilizes renewable energy sources.

S. Kumar · Y. Yang (✉)
University of Central Florida, Orlando, USA
e-mail: yang.yang@ucf.edu

S. Kumar
e-mail: sachin.kumar@ucf.edu

D. Shi (ed.), *Energy Materials Science and Engineering: From New Devices to AI Power Systems*, Green Energy and Technology, https://doi.org/10.1007/978-3-031-88550-1_7

HER typically proceeds via two pathways: water reduction, which occurs in an alkaline medium [4], and proton reduction, which takes place in an acidic medium [5], both leading to the evolution of H_2 gas. The specific reaction mechanism depends on whether the medium is alkaline or acidic. To ensure that HER is both highly efficient and energy-conserving, the development of an effective catalyst is very essential. Most systems used for HER rely on platinum group metals (PGMs), which significantly raise the overall cost. Consequently, there is a need to develop systems with ultra-low Pt loading or alternative catalysts to reduce costs. The current focus is on designing electrolyzers with low capital expenditure (CAPEX) with high efficiency. Electrolyzers are categorized into alkaline exchange membrane water electrolyzers (AEMWE), proton exchange membrane water electrolyzers (PEMWE), and alkaline water electrolyzers (AWE). PEMWE is suitable for on-site H_2 generation, while AWE is a well-developed and commercially used technology for large-scale H_2 production [6]. AEMWE, still in the developmental phase, aims to incorporate compact designs and use non-platinum group metals for hydrogen production. Both theoretical and experimental studies have shown substantial progress in HER electrocatalyst development.

This chapter presents a comprehensive overview of the hydrogen evolution reaction (HER) mechanism, focusing on fundamental concepts and advancements in electrocatalysts, ranging from Platinum Group Metals (PGMs) to Non-Platinum Group Metals (Non-PGMs). A key emphasis is placed on the role of electronic structure modifications in influencing HER performance. The discussion includes a detailed analysis of how Gibbs free energy (ΔG_{free}) and the variation in the d-band center affect catalytic activity. Furthermore, the chapter elaborates on the strategies employed to alter these parameters through doping with non-metal and metal elements, shedding light on the underlying electronic and structural changes.

The HER electrocatalysts are systematically categorized into PGM and Non-PGM groups, with each section providing significant insights into their unique properties and performance enhancements. For PGM catalysts, particular attention is given to their inherent high activity and strategies to further optimize their efficiency while minimizing their use. In the case of Non-PGM catalysts, the focus shifts to the development of cost-effective, earth-abundant alternatives that achieve comparable or superior HER activity through innovative structural and compositional modifications. Key achievements in catalyst development are highlighted, including examples of how specific modifications, such as heteroatom doping or alloying, have led to remarkable improvements in HER performance. The chapter also discusses the theoretical and experimental approaches used to tailor the electronic properties of catalysts, providing readers with a deeper understanding of the principles driving these advancements. Overall, this chapter offers a well-rounded perspective on HER catalysis, combining theoretical insights with practical considerations, making it a valuable resource for researchers seeking to enhance catalyst design and performance in hydrogen production technologies.

2 Fundamentals of the Hydrogen Evolution Reaction (HER)

In an electrolyser, the HER occurs at the cathodic site while the oxygen evolution reaction (OER) occurs at the anodic site.

2.1 Reaction Mechanism of HER

The reaction mechanism for the HER is illustrated in Fig. 1, with detailed explanations provided below.

Step 1: Volmer Reaction

In the Volmer reaction, the process differs based on the medium:

a. **Acidic Medium**: Protons (H^+) combine with electrons on the active site to form adsorbed hydrogen (H^*), as represented by the following reaction:

$$\text{acidic } H^+ + e^- + {}^* = H^* \left({}^* = \text{ active site}\right)$$

b. **Alkaline Medium**: Water (H_2O) acts as the reactant, leading to the formation of H^* on the active site and releasing hydroxide ions (OH^-):

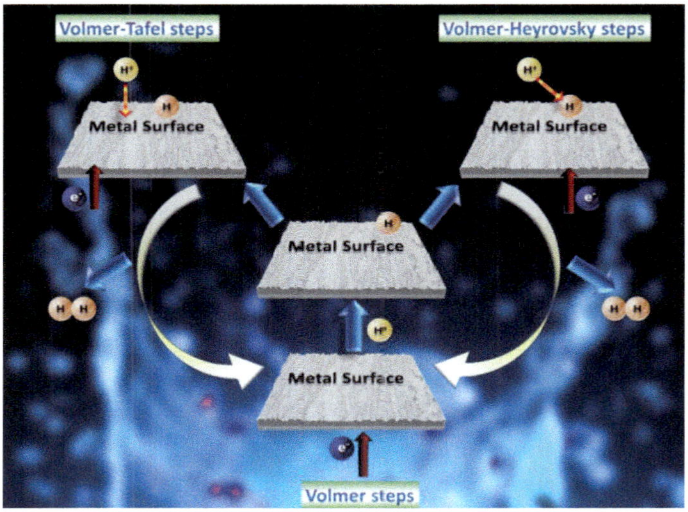

Fig. 1 The HER mechanism. Reproduced with permission [7]. Copyright 2024, American Chemical Society

$$H_2O + e^- + {}^* = H^* + OH^-; \left({}^* = \text{active site}\right)$$

This mechanism represents the formation of H* on the active site in both acidic and alkaline environments.

Step 2: Tafel or Heyrovsky Mechanism

Following the Volmer step, the generation of H_2 proceeds via one of two mechanisms:

(a) **Tafel Mechanism**: When the concentration of H* on the active site becomes sufficiently high, two H* species combine to form molecular hydrogen (H_2):

$$2H^* = H_2 + 2^*$$

This occurs in both acidic and alkaline media.

(b) **Heyrovsky Mechanism**: In this pathway, the adsorbed hydrogen (H*) reacts with another H^+ (in acidic medium) or $H_2O + e^-$ (in alkaline medium) to produce H_2:

Acidic medium:

$$H^+ + e^- + H^* = H_2 + {}^*$$

Alkaline medium:

$$H_2O + e^- + H^* = H_2 + {}^* + {}^- OH$$

Next, we will explore the predictive factors for determining whether the HER follows the Volmer-Tafel or Volmer-Heyrovsky pathway. Typically, the Tafel slope is experimentally measured to distinguish between these pathways.

A Tafel slope of less than 40 mV dec^{-1} generally indicates the Volmer–Tafel pathway, whereas a slope greater than 40 mV dec^{-1} suggests the Volmer–Heyrovsky pathway [7]. In the HER, the rate-determining step (RDS) can be Volmer, Heyrovsky, or Tafel. A Tafel slope of 120 mV dec^{-1} indicates that the Volmer step (adsorption) is the RDS. If the Tafel slope is 40 mV dec^{-1}, the RDS is likely the Heyrovsky step (desorption), and when the Tafel slope is approximately 30 mV dec^{-1}, the Tafel step (desorption) is the RDS [8].

2.2 *How to Determine the Pathway Mechanism and Factors That Influence It*

Considering MoS_2 as a catalyst for the HER, three variations—Bulk MoS_2, MoS_2/RGO, and MoS_2/HG—exhibit Tafel slopes of 93, 50, and 41 mV dec^{-1}, respectively. Since these values exceed 40 mV dec^{-1}, the reaction mechanism is predominantly

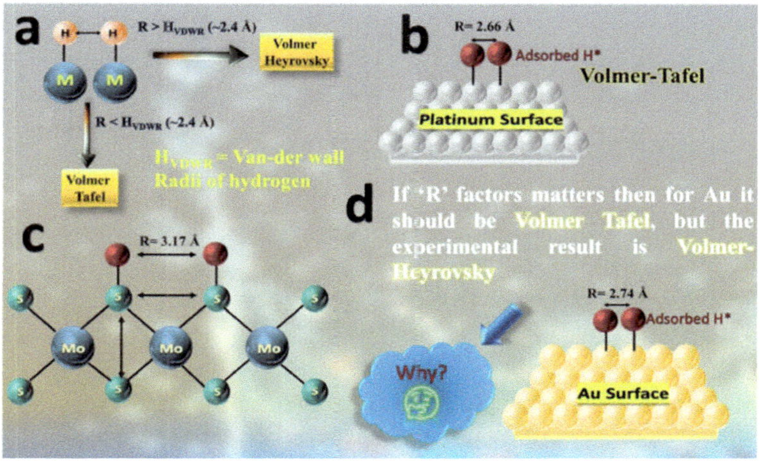

Fig. 2 Schematic representation of different atomic factors influencing the mechanistic outcomes of the Hydrogen Evolution Reaction (HER) in acidic conditions. Reproduced with permission [7]. Copyright 2024, American Chemical Society

Volmer-Heyrovsky [9]. In MoS_2, the distance between two active sites is larger than the van der Waals radius of adjacent bonds. Specifically, the distance between sulfur atoms, which serve as the active sites in MoS_2, is 3.17 Å. This larger spacing favors H_2 desorption through the Heyrovsky mechanism [7]. In contrast, for Pt/C, which is highly active for HER, the Volmer-Tafel mechanism is generally favored due to high hydrogen coverage on Pt surfaces [10]. The high activity of Pt/C can be attributed to the closely packed active sites, where adsorbed hydrogen atoms (H*) are in close proximity, facilitating H_2 desorption via the Tafel mechanism (Fig. 2). The Pt–Pt distance is approximately 2.66 Å, which is near the van der Waals radius of two hydrogen atoms [7].

2.3 *Parameter for Evaluation of the Material to Be an Active Electrocatalyst for the HER*

2.3.1 Overpotential

The Nernst equation can be written as

$$E_{\mathrm{HER}} = E^{0}_{\mathrm{H_2/H^+}} - \frac{RT}{F} In\left(\frac{a^{+}_{\mathrm{H}}}{P^{1/2}_{\mathrm{H_2}}}\right)$$

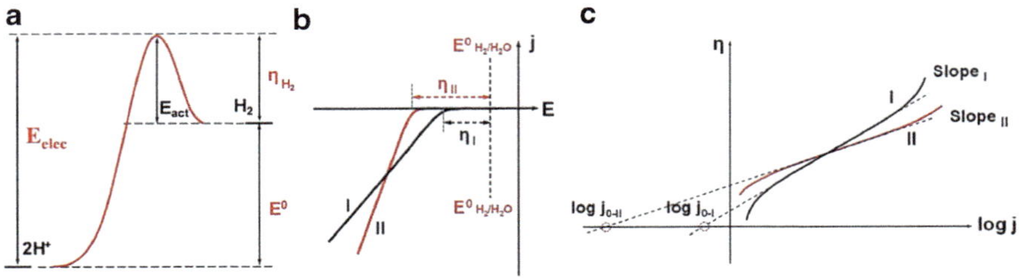

Fig. 3 **a** HER shows the activation barrier, **b** linear sweep voltammetry curves of two different catalysts with their respective onset potentials indicated, and **c** exchange current density and Tafel slope of the two different catalysts. Reproduced with permission [11]. Copyright 2012, RSC Publishing

When calculating the potential relative to the Standard Hydrogen Electrode (SHE), it is expressed as $E_{\text{HER}} = -0.059$ pH V versus SHE. Since the SHE remains constant, the potential for the HER varies with the pH. However, if the SHE is replaced with a Reversible Hydrogen Electrode (RHE), the potential for HER becomes $E_{\text{HER}} = 0$ V versus RHE, as the -0.059 pH V term is effectively cancelled out.

The Nernst equation describes the thermodynamic potential at which a reaction begins. For HER, this thermodynamic potential is 0 V versus RHE. Figure 3a–c illustrates the ideal HER scenario alongside the overpotential associated with the HER reaction of two different catalysts. For HER to occur, it requires a certain activation energy to overcome the reaction barrier, as shown in Fig. 3a. In practice, overpotential is always present, meaning the reaction does not occur at 0 V versus RHE. In Fig. 3b, the ideal case is represented by the dashed line, while the onset potentials of two different catalysts are indicated by η_1 and η_2. Therefore, overpotential (η) will be calculated as $\eta = E^0{}_{\text{H2/H2O}} - E_{\eta 1}$. Figure 3c represents the Tafel slopes of two different catalysts along with their exchange current densities, which can be measured at the intersection with the X-axis where the overpotential is zero.

Overpotential can be divided into three components: (a) kinetic, (b) concentration, and (c) ohmic. Kinetic overpotential depends primarily on the intrinsic properties of the catalysts. Platinum (Pt) is the only metal known to exhibit the highest electrochemical activity for HER [12]. Concentration overpotential, also referred to as mass transport limitation, arises when difference occurs between the bulk solution and the electrode surface [13]. It involves the following three steps. Migration of H^+ ions from the electrolyte to the boundary of the electrocatalyst. Electrocatalytic activity at the electrode surface, encompassing reactant attachment, electron exchange at the interface, and the release of intermediates/products. Product displacement from the electrode boundary to the electrolyte or the atmosphere, including the release of gas bubbles and the diffusion of dissolved hydrogen [14]. Ohmic overpotential is caused by the resistance due to ion diffusion between the reference and the working electrode. It can be reduced by decreasing the distance between the reference and working electrodes. It can be further reduced by increasing the ionic conductivity of the electrolyte. Another overpotential is the bubble overpotential in which the bubble blocks the pathway for ion conduction, reduces concentration at electrode–electrolyte interface and lowers the electrochemical area available for the reaction [15].

2.3.2 Tafel, Exchange Current Density and Energy of Hydrogen Adsorption

Rate determining step can be easily diagnosed with the Tafel and possible mechanism operating in the electrocatalyst. Tafel slope can be derived by Tafel equation, i.e., $\eta = b \log(j) + a$, where η is the overpotential, j is the current density, b is the Tafel slope and a is the Tafel constant [13]. Tafel plot is obtained by plotting the logarithm of current density versus overpotential. A lower Tafel slope indicates that a catalyst is capable of generating current at a reduced overpotential [16]. The overpotential should exceed *RT/F*; otherwise, the relationship becomes ohmic. In other words, to maintain the linearity of the equation, the overpotential must be greater than *RT/F* [17].

Exchange current density is measured when the overpotential is zero. It can be measured experimentally by extrapolating the linear part of the Tafel plot to X-axis = 0, where the overpotential is zero [18]. A higher exchange current density indicates a more effective catalyst. It can be understood as a descriptor of how an electrocatalyst transfers electrons for the electrochemical reaction and is related to a higher rate of the electrochemical reaction [19].

The log (i_0/exchange current density) has been plotted against the hydrogen chemisorption energy and free energy for the hydrogen adsorption as shown in Fig. 4 [20]. Figure 4 clearly shows the volcano plot relationship between log (i_0/exchange current density) and chemisorption energies. Most of the electrochemically active elements are on the left side of the volcano plot, while inactive elements are on the right side. For a catalyst to be active, it must neither adsorb hydrogen too strongly nor too weakly; it must achieve an optimal balance. This can be further explained in terms of the free energy and the reaction coordinate, as shown in Fig. 5, which clearly indicates that the free energy of the intermediates is thermoneutral compared to other metals. This further indicates that H-adsorption on Pt is neither too weak nor too strong, whereas in the case of Ni and Mo, H-adsorption occurs too strongly while in case of Au, it is too weak.

2.3.3 Specific and Mass Activities

Specific and mass activities are key parameters for evaluating the electrochemical performance of a catalyst. The catalytic performance is typically assessed by plotting the I-V curve. To determine the mass activity of the catalyst, the current was normalized by the total mass loading of the catalyst ($A/g_{\text{total-catalyst-loading}}$) [21]. To measure the specific activity of the electrocatalyst, the current was normalized by the electrochemical active surface area (ECSA) ($A/cm^2{}_{\text{ECSA-Catalyst}}$) [22]. Mass loading and

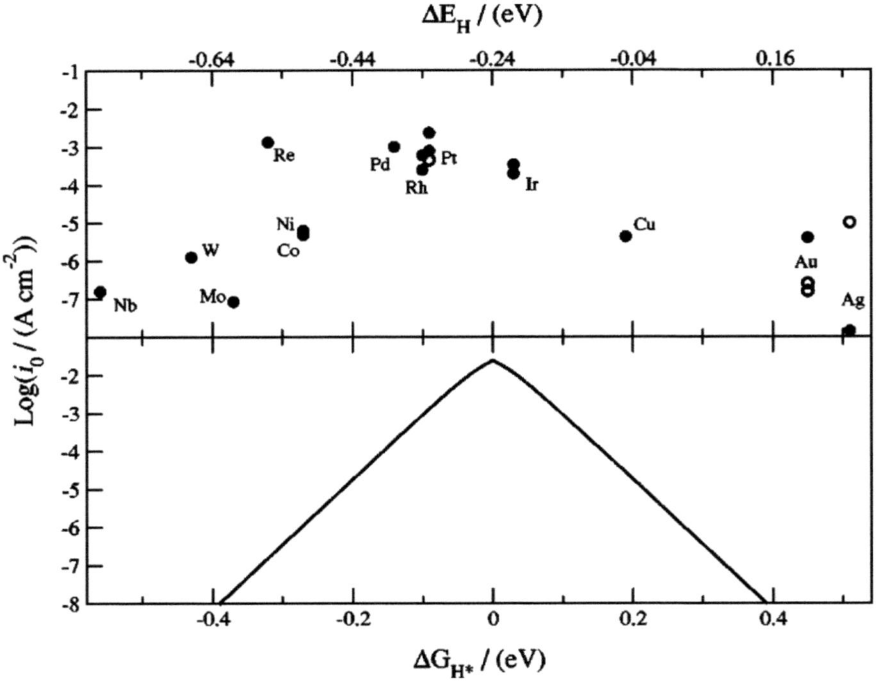

Fig. 4 (Top): The enthalpy of hydrogen chemisorption is plotted as a function of the experimentally derived log (i_0/exchange current density). Single-crystal data is indicated by open symbols for Au(111), Ag(111), and Pt(111). (Bottom): The free energy of hydrogen chemisorption is plotted as a function of log (i_0/exchange current density). Reproduced with permission [20]. Copyright 2005, IOP Publishing, Ltd.

specific activity are well-known parameters for evaluating the activity of electrocatalysts. Mass and specific activities (in terms of ECSA) are much more reliable than the specific activity calculated by normalizing the geometrical area of the electrodes. Geometrical areas are often considered to be planar and smooth surfaces, but they can also be rough and porous. The specific activity of the electrocatalyst also reflects the intrinsic activity of the electrocatalyst. Specific activity in terms of ECSA of the electrocatalyst and mass activity is much more reliable than that of the geometrically normalized current densities.

2.3.4 Turnover Frequency (TOF)

Turnover Frequency (TOF) is another parameter used to distinguish the electrochemical activity of a catalyst. It is defined as the number of reactant molecules converted or hydrogen molecules generated per active site per second [23]. Higher TOF values

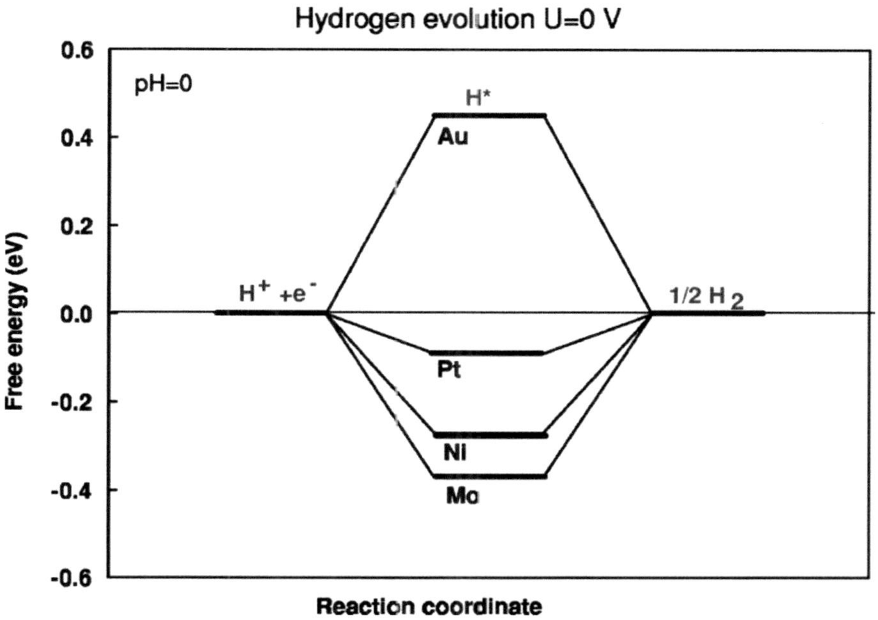

Fig. 5 The Gibbs free energy of the intermediates at a potential of 0 V, corrected for entropy and zero-point energy. DFT calculations were used to determine the energy of the adsorbed intermediate states. Reproduced with permission [20]. Copyright 2005, IOP Publishing, Ltd.

indicate greater intrinsic activity of the catalyst. The catalytic surface area of the catalyst differs from its geometrical surface area due to surface roughness and porosity [24]. Consider the case of amorphous MoS_2 deposited by Atomic Layer Deposition (ALD) on Au. ALD-MoS_2 shows a very low exchange current density J_0 compared to other forms of MoS_2 due to its low porosity and surface area. The low exchange current density can be mitigated by using a high surface area substrate. However, the TOF for ALD-MoS_2 is 3 H_2/s at an overpotential of 0.215 V per active site, which is the highest compared to other MoS_2 reported at that time [24]. This indicates that the intrinsic activity of the amorphous MoS_2 is much higher than other MoS_2.

TOF can be calculated using the following equation:

TOF $= \frac{J(\mathrm{A\,cm^{-2}})}{nFN}$; where N is the number of active sites, J is the current density, n is the number of electrons involved in the electrode reaction which is 2 in the case of HER and F is the Faradic constant.

3 Design Strategies of Electrocatalysts for High-Current–Density Water Electrolysis: d-Band Centre and Gibbs Free Energy

In catalysts, the d-band center is crucial for regulating adsorption and desorption energies, as well as the molecular activation barrier. Interactions between the d-orbitals of the catalyst and the orbitals of adsorbed species significantly influence catalytic activity at the surface. Modifying d-orbital characteristics through methods such as alloying or non-metal doping can shift the d-band center relative to the Fermi level, resulting in either stronger or weaker hydrogen binding. Here, we present a detailed overview of both Non-Platinum Group Metal (N-PGM) and Platinum Group Metal (PGM) electrocatalysts that are active for HER, highlighting how these modifications alter the electronic structure and their impact on HER activity.

3.1 Non-platinum (Pt) Group Metals (Non-PGM)

Platinum (Pt) is widely recognized as one of the most effective and efficient electrocatalysts for the hydrogen evolution reaction (HER). However, its high cost and limited availability pose significant challenges, driving the need for alternative electrocatalysts to reduce the overall system cost and improve scalability. In this context, extensive efforts have been directed toward developing non-platinum group metal (non-PGM) catalysts. In this chapter, these catalysts are broadly categorized into three main types: metal and metal oxides/sulphides, non-metal-doped metals and metal oxides, and single-atom non-platinum group metal catalysts.

3.1.1 Metal/Metal Oxide/Metal Sulphides

Luo et al. [25] synthesized a nickel-based electrocatalyst, (hydroxide = h) h-NiMoFe, which demonstrates high activity for the HER. At a current density of 10 mA cm^{-2}, the observed overpotential was approximately 14 mV, which is significantly lower than that of Pt/C (59 mV) and Ni (309 mV). Various characterization techniques such as X-ray absorption spectroscopy (XAS) and X-ray photoelectron spectroscopy (XPS) have established that in the case of h-NiMoFe, the generation of $Ni(OH)_2$ is significantly higher, even at elevated current densities, which enhances water dissociation on the nickel surface. DFT calculations revealed the formation of $Fe–(OH)_4–Ni_4$ (Fig. 6a, b). The bond between Fe–OH–Ni can extract electrons from the Ni atoms, leading to unoccupied states. The d-band center of Ni in h-NiMoFe was found to be more negative than that of NiMo without Fe doping (Fig. 6c) and the adsorption free energy ($\triangle G_{free}$) of h-NiMoFe was close to 0 eV (Fig. 6d). The stronger hydrogen bonding is one of the reasons for the higher intrinsic activity compared to other catalysts.

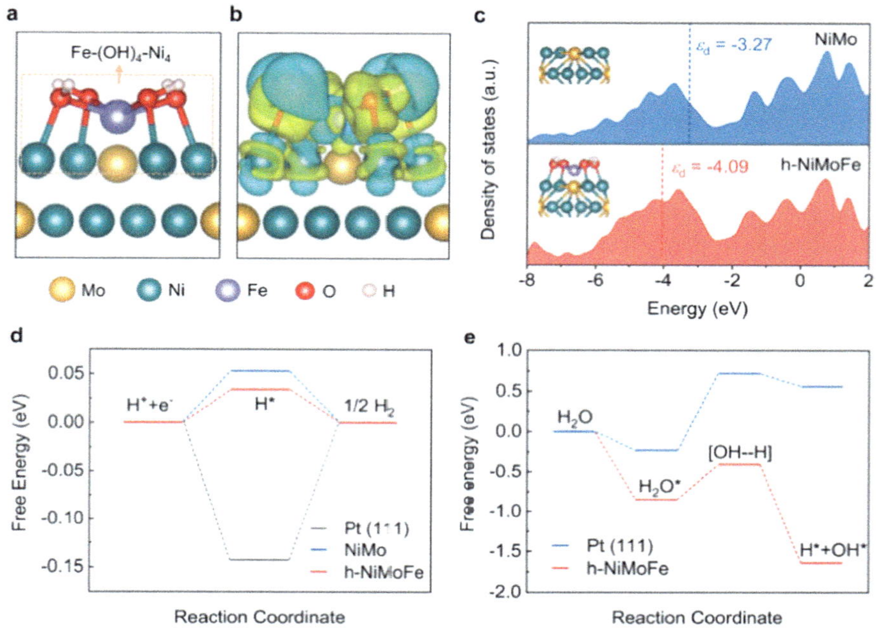

Fig. 6 DFT calculations: **a** the optimized structure of a Fe–$(OH)_4$–Ni_4 unit on a Ni_4Mo (002) surface, and **b** the associated charge density difference, with yellow representing electron accumulation and cyan indicating electron depletion. **c** The computed partial density of states (PDOS) for Ni in the NiMo (Ni_4Mo(002)) slab (top) and the h-NiMoFe slab (bottom), with the Ni d-band centers marked by dashed lines. Adsorption energies for **d** hydrogen on h-NiMoFe and control samples, and **e** dissociated water (H_2O) on h-NiMoFe and Pt (111) are also presented. Reproduced with permission [25]. Copyright 2021, RSC Publishing

HER was conducted in an alkaline medium, where water dissociation is the most crucial step. DFT calculations reveal that, compared to Pt (111), h-NiMoFe exhibits faster kinetics and a lower energy barrier for water dissociation (Fig. 6e).

Similarly, metal oxides have been synthesized that show enhanced HER activity. Chen et al. [26] synthesized Co_3O_4-VO_x layered nanosheets and investigated the promotional role of vanadium species on Co_3O_4. During electrolysis, Co_3O_4 is converted into $Co(OH)_2$, while VO_x transforms into VO_4^{3-} and re-adsorbs onto the surface of $Co(OH)_2$. Volmer–Heyrovsky mechanism is followed as indicated by the Tafel slope (Fig. 7a). DFT calculations revealed that due to the semiconductor nature of $Co(OH)_2$, its electronic density of states (DOS) is discontinuous at the Fermi level. However, when VO_4^{3-} is adsorbed onto the $Co(OH)_2$ surface, the Fermi level improves, leading to increased electron mobility and enhanced charge transfer kinetics (Fig. 7b, c). Additionally, the d-band center of $Co(OH)_2$ is located at −2.72 eV, compared to Co_3O_4-VO_x, which is positioned at −2.98 eV (Fig. 7d). The adsorption energy of H_2O* and intermediates such as H*–OH* is significantly lower for $Co(OH)_2$-VO_4^{3-}, resulting in stronger adsorption, whereas in the case of $Co(OH)_2$, weaker adsorption hinders the HER in an alkaline medium. The free energy for H-OH dissociation over $Co(OH)_2$ is 1.66 eV, which is considerably higher than the 1.36 eV observed for $Co(OH)_2$-VO_4^{3-}. For comparison, the dissociation energy of H-OH over Pt(111) ranges from 0.8 to 1.1 eV (Fig. 7e).

MXene is 2-D material of early transition metal carbides and carbonitrides. Composites of MXene and Metal Oxides have shown enhanced HER activity. Cheng et al. [27], synthesized the Mo-CoOx@MXene-Al catalyst, in which single-atom Mo is anchored to clusters of CoOx on an MXene substrate. The catalytic activity of this system is comparable to that of commercially available PtRu/C. The three components—Mo, CoOx, and MXene-Al—are interconnected by bridging oxygen atoms. The Gibbs free energy (ΔG_{H*}) of hydrogen adsorption for various substrates, including Mo-CoOx@MXene-Al, CoOx@MXene-Al, and Mo@MXene-Al, was calculated as −0.004 eV, −0.53 eV, and −2.7 eV, respectively. This indicates that

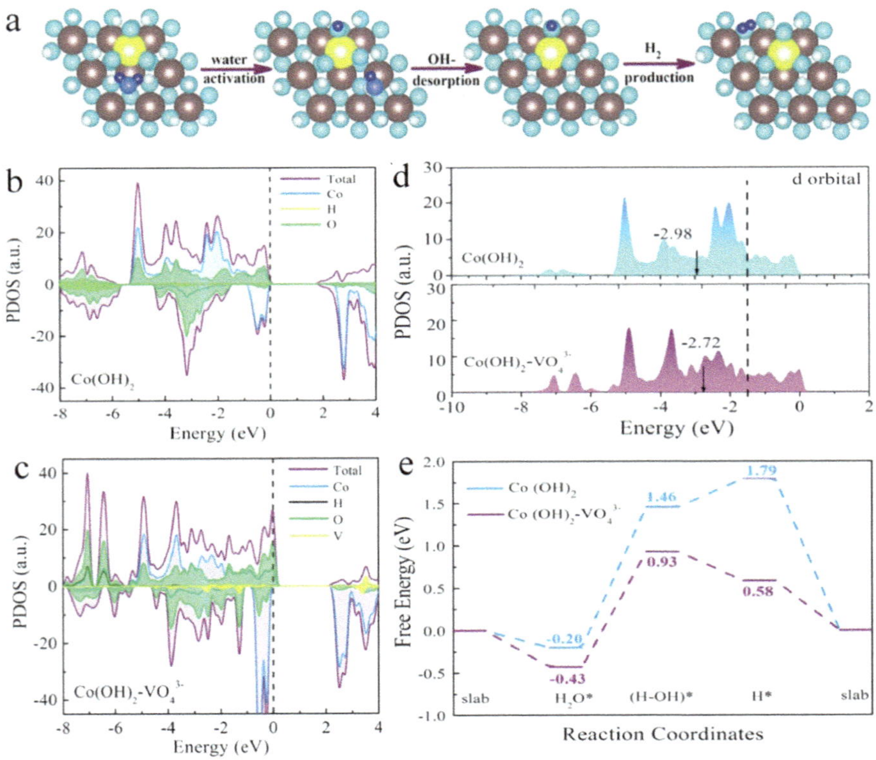

Fig. 7 **a** Mechanism of hydrogen generation on $Co(OH)_2$-VO_4^{3-}. The computed density of states (DOS) for **b** $Co(OH)_2$ and **c** $Co(OH)_2$-VO_4^{3-}. **d** A comparative analysis of the projected density of states (PDOS) for the Co d-orbital in $Co(OH)_2$ and $Co(OH)_2$-VO_4^{3-}. **e** The respective free energy profiles for the HER on $Co(OH)_2$ and $Co(OH)_2$-VO_4^{3-}. Reproduced with permission [26]. Copyright 2023, American Chemical Society

Mo-CoOx@MXene-Al has an optimal ΔG_{H*} value close to 0 eV. Furthermore, the d-band center suggests that Mo-CoOx@MXene-Al exhibits the weakest hydrogen adsorption energy, significantly enhancing the HER. High-entropy alloys (HEAs), an emerging class of materials composed of five or more metallic elements in near-equiatomic or different ratios, have demonstrated improved catalytic properties when synthesized as nanoparticles. Liu et al. [28] synthesized a non-noble metal HEA composed of CuNiMoWCo, grown on Cu foam. The d-band center of Ni (111) was measured at −1.27 eV in H* sites, indicating stronger hydrogen binding compared to Ni-HEA, which has a d-band center of −2.0 eV and exhibits weaker hydrogen binding. Conversely, for OH* sites, the d-band center of Ni in Ni-HEA shifts upward relative to Ni (111), resulting in stronger OH* adsorption. This adjustment significantly enhances the HER performance in case of HEAs. Heterostructures, due to synergistic effects arising from interactions between distinct domains, exhibit improved catalytic properties. Hu et al. [29] developed heterostructures combining MoS_2 with metal oxides such as Ni_2O_3H, Co_3O_4, and Fe_2O_3. Among these, MoS_2/Ni_2O_3H demonstrated superior HER efficiency compared to the other heterostructures. This improved catalytic activity is linked to a narrower t_2g–e_g energy gap in MoS_2/Ni_2O_3H, which is smaller than the corresponding gap in Co_3O_4 and Fe_2O_3, thereby promoting an efficient HER process. The reduced t_2g–e_g gap enables faster electron transfer, overcoming the larger transfer barriers observed in Co_3O_4 and Fe_2O_3. Doping is another effective strategy to modulate the intrinsic properties of a material. Ling et al. [30] synthesized a dual-doped CoO material incorporating both Ni and Zn. This dual doping introduced electronic modifications that transformed CoO into a highly active catalyst for the HER. Zn doping improved the material's electrical conductivity, while Ni doping optimized the hydrogen-binding energy, facilitating efficient hydrogen adsorption and desorption. Chen et al. [31] synthesized Ni_4Mo/$NiMoO_4$ ($NiMoO_x$) nanoparticles supported on CMK-3, a mesoporous carbon material. In this structure, Mo serves as the preferred site for hydrogen adsorption. Density functional theory (DFT) calculations revealed that the ΔG_{H*} for $NiMoO_x$ is 0.038 eV, significantly lower than the values for Ni_4Mo (−0.261 eV) and $NiMoO_x$ (0.305 eV). This value indicates that $NiMoO_x$ achieves an optimal hydrogen binding strength—neither too strong nor too weak—whereas Ni_4Mo binds H too strongly and $NiMoO_4$ binds H too weakly.

3.1.2 Non-metal Doped Metal/Metal Oxide

Feng et al. [32] synthesized P-doped cobalt iron double hydroxide (Cu NDs/P-FeCoLDH) over copper dots, which exhibits significantly higher HER activity compared to FeCoLDH. Theoretical calculations were carried out considering the dissociation of H_2O into H_2. The results show that the adsorption energy of H* is more favorable, contributing to the high HER activity of Cu NDs/P-FeCoLDH.

From Fig. 8, it is clear that the adsorption of H_2O is much stronger on Cu NDs/P-FeCoLDH compared to FeCoLDH, while the free energy for hydrogen desorption is less negative, resulting in the rapid release of H_2. This variation in free energy can be attributed to phosphorus (P) incorporation, which leads to the formation of P–O bonds. These bonds weaken the interaction with adsorbed hydrogen atoms (H_{ads}), while copper (Cu) nanodots accelerate water decomposition.

Similarly, Stefan et al. [33] synthesized a $(Fe_xNi_{1-x})_9S_8$ electrocatalyst for the HER and studied the effect of the Ni ratio on HER activity. It was found that increasing the iron content in the nickel structure reduced HER activity. The highest activity was observed at a Ni and Fe ratio of 1:1, attributed to the optimal number of active sites (Fig. 9a–d). DFT calculations were performed to investigate the factors responsible for HER activity. The surface energies of $Fe_{4.5}Ni_{4.5}S_8$ (111), $Fe_6Ni_3S_8$ (111), and $Fe_3Ni_6S_8$ (111) were found to be 0.054, 0.048, and 0.042 eV/Å^2, respectively. The surface energy of $Fe_{4.5}Ni_{4.5}S_8$, with a 1:1 Ni: Fe ratio, was significantly higher than that of the other electrocatalysts. This higher surface energy is directly linked to its enhanced HER activity.

Hao et al. [34] synthesized cobalt phosphide (CoP) over graphene nanoscrolls supported on carbon nanotubes (CNTs) using chemical vapor deposition. These graphene nanoscrolls encapsulate NiM (M = Zn, Mo, Ni, Co) bimetallic nanoalloys. Their findings revealed that a lower d-band center, relative to the Fermi level, results in a reduction of adsorption surface energy. The d-band center values follow this decreasing order: -1.07 eV (CoP) > -1.16 eV (NiCo@C-CoP) > -1.21 eV (Ni@C-CoP) > -1.23 eV (NiMo@C-CoP) > -1.27 eV (NiZn@C-CoP). This trend aligns with the observed HER activity, with NiZn@C-CoP demonstrating the highest performance. Apart from P and S doping, dual-doping is another effective strategy for enhancing electrochemical activity. Chang et al. [35] synthesized dual-doped Fe, P nanoporous films (Fe, P-$NiSe_2$ NFs). The phosphorus (P) doping enhances electron conductivity and protects the selenium from surface dissolution. Nickel (Ni) serves

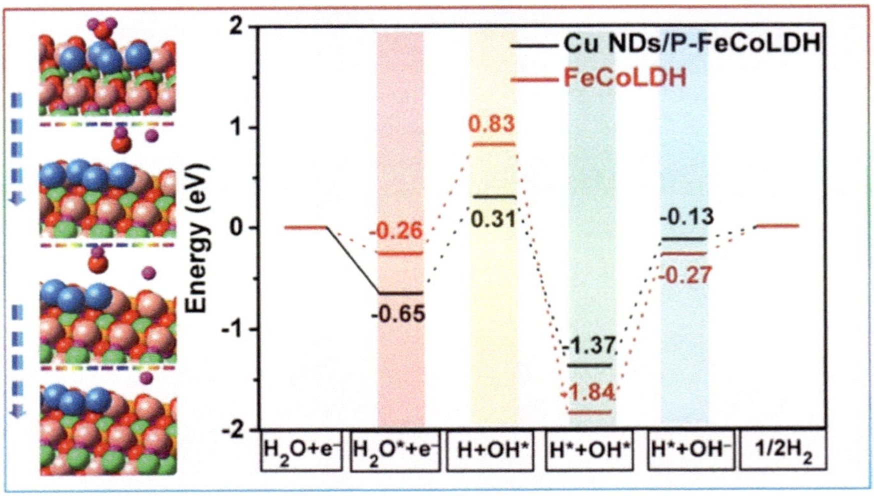

Fig. 8 Theoretical free energy calculation on the FeCo-LDH and Cu NDs/P-FeCo-LDH. Reproduced with permission [32]. Copyright 2021, Elsiever

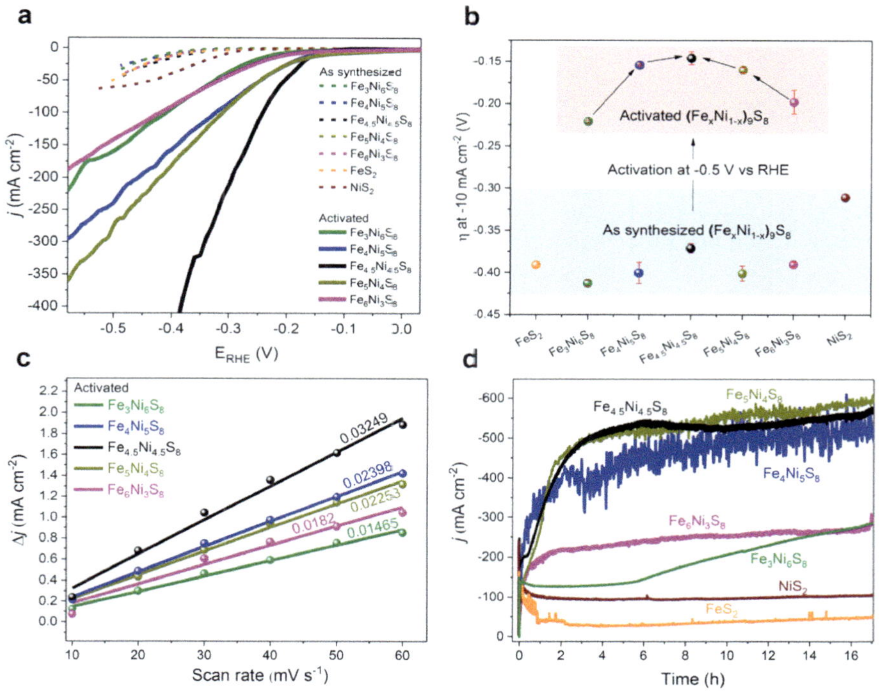

Fig. 9 **a** Voltammograms for FeS_2, NiS_2, and synthesized and activated pentlandite materials ($(FexNi_{1-x})_9S_8$) were obtained at a sweep speed of 5 mV s^{-1}. **b** A graph of catalytic performance showing the overpotential (η, V) necessary to reach a current density of − 10 mA cm^{-2}, plotted against the Fe ratio for both synthesized (with error margins from three repeated tests indicated in red) and activated ($(Fe_xNi_{1-}x)_9S_8$ samples. **c** Slopes of electrochemical surface area (ECSA) measured across scan rates from 10 to 60 mV s^{-1}. **d** A stability test was performed over 17 h at a fixed potential of −0.5 V versus RHE. The scan fidelity here is impacted by hydrogen bubble formation on the electrode surface during HER operation. Reproduced with permission [33]. Copyright, 2018, American Chemical Society

as the active site for the oxygen evolution reaction (OER), while iron (Fe) acts as the active site for the HER. A current density of 0.8 A cm^{-2} was achieved at 1.8 V, demonstrating high selectivity and long-term stability over 200 h. Cao et al. [36] synthesized $Co_{0.6}Mo_{1.4}N_2$, which possesses a mixed-layer close-packed structure. This structure comprises octahedral sites occupied by $Co2^+$ and $Mo3^+$, as well as trigonal prismatic sites where Mo displays oxidation states varying from $>3^+$ to $\leq 4^+$. The enhanced HER activity of this catalyst can be attributed to the alternating substitution of Mo by Co in the octahedral sites and the layered structure, which allows for fine-tuning of Mo's catalytic properties by the 3d transition metal. Similarly, heterostructures based on non-metal catalysts have exhibited promising results. Fu et al. [37] synthesized a Ni_3S_2/Cr_2S_3 hybrid electrocatalyst over Ni foam, which demonstrates remarkable performance with a current density of 3.5 A cm^{-2} at an overpotential of 251 ± 3 mV in an alkaline medium. Density functional theory (DFT) calculations were conducted on the experimentally identified $Cr_2S_3(11\overline{3})$ and $Ni_3S_2(110)$ surfaces to understand their catalytic behavior. The DFT results reveal a water dissociation energy barrier of 1.47 eV for $Ni_3S_2(110)$ compared to a significantly lower barrier of 0.68 eV for $Cr_2S_3(11\overline{3})$. This finding indicates that water dissociation occurs more readily at Cr_2S_3 sites. The Cr_2S_3 sites facilitate the catalytic acceleration of water dissociation through the reaction (Cr_{5c}-H_2O^* + S3c(#) → Cr_{5c}-OH^* + S_{3c}-$H^{\#}$). S_{3c}-$H^{\#}$ species formed during this process is rapidly transferred to the Ni_3S_2 sites via a hydrogen spillover effect. This interaction enables an efficient coupling of the catalytic properties of both components, with Cr_2S_3 enhancing water dissociation and Ni_3S_2 facilitating subsequent hydrogen evolution. The overall process is consistent with the Tafel mechanism, highlighting the synergy between the two materials in improving HER performance.

3.1.3 Single-Atom Non-platinum Group Metal

Zhou et al. [38] synthesized a single-atom Mo-tip structure on 1T-phase transition metal dichalcogenides (TMDs), specifically 1T-CrS_2. This structure generated a strong electric field around the Mo sites, allowing electrons to be efficiently transferred from 1T-CrS_2, which significantly enhanced the HER performance. The hydrogen adsorption energy (ΔG_H*) was observed to be lowest for the Mo single-atom/1T-CrS_2 system at −0.06 eV, close to the optimal value of zero, compared to 0.24 eV for 1T-CrS_2 alone. Kundu et al. [39] synthesized a mononuclear bismuth metal complex (BMC) on graphene oxide (GO) sheets. Among various loadings of BMC—3, 5, and 7 wt%—the 5 wt% composition demonstrated superior activity. This catalyst exhibited enhanced performance not only in 1 M KOH but also in river water. Ji et al. [40] synthesized Cu atoms on both 2H-MoS_2 and 1T-MoS_2 surfaces. Cu atoms on the metallic phase (1T-MoS_2) exhibited superior performance compared to those on 2H-MoS_2. The Cu atoms on 1T-MoS_2 showed better charge transfer, ranging from −0.27 to −0.38 e, compared to the 2H-MoS_2 surface, which displayed values of −0.19 and −0.18 e. This higher negative surface charge on the Cu/1T-MoS_2 surface facilitates the release of adsorbed H-atoms by weakening the S–H bond on the MoS_2 surface.

3.2 Platinum Group Metals (PGM)

3.2.1 Single Atom PGM

Platinum is considered the most efficient metal for HER electrocatalysis. Therefore, significant efforts have been directed toward reducing the Pt loading while maintaining high catalytic performance. Ma et al. [41] have synthesized Pt atoms substituted at W positions in Tungsten carbide (WC) (Pt_{doped}@WC) and compared it with the Pt atoms adsorbed over the surface of the WC ($Pt_{adsorbed}$@WC). Density Functional Theory had been carried out to measure the electronic properties change by substituting the Pt with W (Fig. 10a–e). It has been revealed that the energy levels of Pt 5d in both Pt-doped@WC and Pt (111) are almost similar but are significantly higher in Pt-adsorbed@WC (Fig. 10b). The energy level of Pt-doped@WC (i.e., $\triangle G_{H*}$) from Fig. 10c closely matches that of the Pt(111) surface, whereas for Pt-adsorbed@WC, it is significantly higher.

In the case of the alkaline medium, the dissociation barrier for H_2O over Pt-doped@WC (Fig. 10d) is much lower than that of Pt(111), WC(111), and $Pt_{-adsorbed}$@WC_x. Additionally, H* is directly adsorbed onto Pt in Pt_{-doped}@WC, whereas an extra energy of approximately 0.47 eV is required to transfer H* from W in $Pt_{-adsorbed}$@WC. Figure 11a–e clearly illustrates the activity of various catalysts.

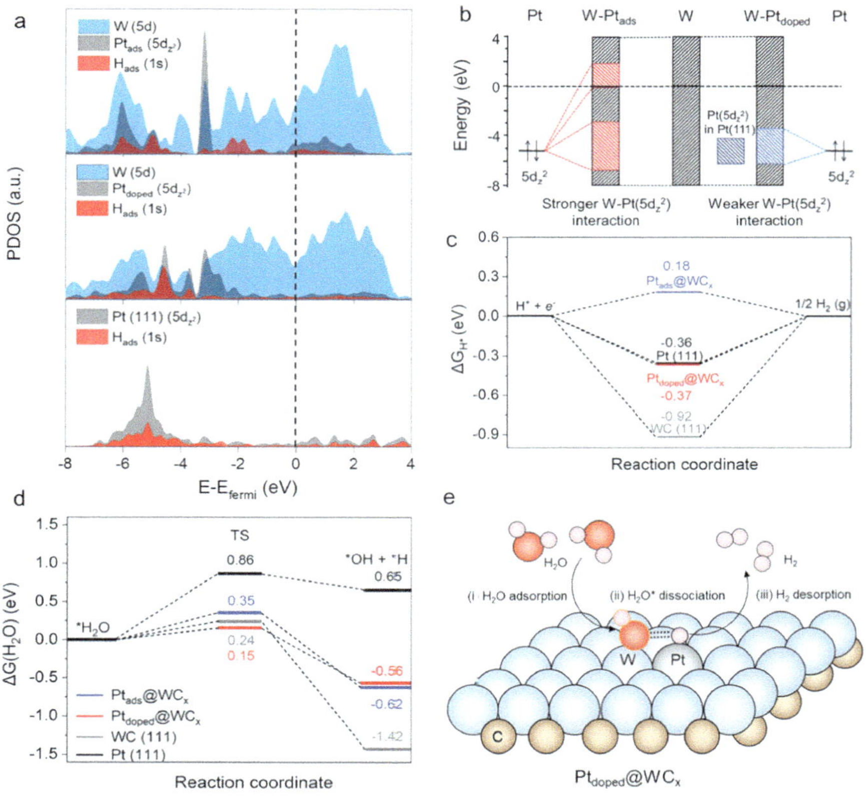

Fig. 10 HER mechanisms and free energy predictions for various catalysts: **a** PDOS analysis showing the interaction between W 5d, Pt 5dz^2, and H 1s orbitals for hydrogen adsorption on Pt sites, with the catalysts listed as Pt_{ads}@WC. Reproduced with permission [41]. Copyright 2022, John Wiley and Sons

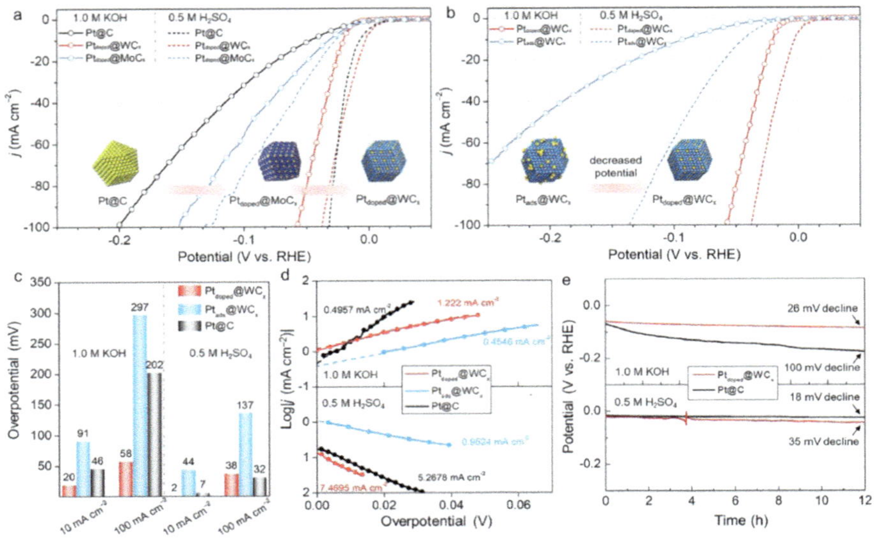

Fig. 11 HER catalytic performance: **a** LSV curves for Pt-doped@WCx, Pt-doped@MoCx, and commercial 20% Pt@C in 1.0 M KOH and 0.5 M H2SO4 with IR compensation (scan rate: 10 mV s $^{-1}$ in a three-electrode setup). **b** LSV curves for Pt-doped@WCx and Ptads@WCx in 1.0 M KOH. Reproduced with permission [41]. Copyright 2022, John Wiley and Sons

Pt_{-doped}@WC exhibits the highest activity for the HER in both acidic and alkaline media, with the HER kinetics being significantly faster in the acidic medium compared to the alkaline. Pt_{-doped}@WC shows a mass activity of 49.5 A mg^{-Pt} which is 40 times much higher than that of the Pt@C in an alkaline medium (Fig. 11a–c).

Zhou et al. [42] synthesized single-atom Pt anchored to a metal/metal oxide interface, demonstrating its efficiency as a catalyst for the HER. Theoretical calculations indicate that Pt anchored to the Ni/NiO interface possesses more free electrons compared to Pt anchored separately on Ni or NiO. These additional free electrons enhance hydrogen adsorption (H* ads) and H_2 evolution. Furthermore, the study reveals that d-electrons are highly concentrated near the Fermi level, which contributes to H_2O activation and significantly boosts catalytic activity. Similar to single-atom Pt catalysts (Pt-SAC), single-atom Ru catalysts (Ru-SAC) have also been synthesized. Huang [43] synthesized single-atomic Ru on microporous N-doped carbon (NC), Ru nanoclusters (Ru NC) on NC, and Ru nanoparticles (Ru NP) on NC. Density Functional Theory (DFT) calculations reveal that Ru-SAs/NC has a free energy of 0.97 eV, signifying strong hydrogen adsorption. In contrast, Ru NCs/NC exhibits a much lower free energy of 0.33 eV, which is close to 0 eV, indicating weaker hydrogen binding to the surface. This weaker binding is advantageous, as it facilitates the HER.

Wang et al. [44] synthesized single-atom Ru doped into the basal plane of 2H-MoS_2. The $Ru_{0.10}$@2H-MoS_2 catalyst, containing approximately 0.1 mmol of Ru, exhibits enhanced HER performance. This improved activity is attributed to the presence of sulfur vacancies and Ru doping, which together provide an optimal ΔG_H* for hydrogen adsorption and desorption (Fig. 12a–d). Yu et al. [45] synthesize Ir-g-CN and Ru-g-CN where single atoms of Ir and Ru are anchored onto graphitic carbon nitride (g-CN). Both Ir-g-CN and Ru-g-CN exhibit improved electrocatalytic HER performance. Density functional theory (DFT) calculations indicate that the inclusion of Ir and Ru into the g-CN matrix significantly narrows the bandgap. This modification enhances charge transfer at the surface where the reaction occurs, thereby boosting HER activity.

3.2.2 PGM Nanoclusters/Nanoparticles

Da et al. [46] synthesized p-Pt_3V over carbon cloth using liquid nitrogen, thermal shock treatment, and a direct current technique. The as-synthesized p-Pt_3V was further anodized to generate a-Pt_3V. The a-Pt_3V exhibited significantly higher mass activity compared to both p-Pt_3V and the state-of-the-art commercial Pt/C. Specifically, a-Pt_3V delivered current densities of 10 and 100 mA cm^{-2} at overpotentials of 20 and 79 mV, respectively, which are much lower than those for 20% Pt/C (30 and 235 mV). The measured Tafel slope for a-Pt_3V was 36 mV dec^{-1}, considerably lower than p-Pt_3V (94 mV dec^{-1}) and Pt/C (43 mV dec^{-1}). The d-band centre was estimated using ultraviolet photoelectron spectroscopy (UPS), revealing that the work function of a-Pt_3V was much lower than that of p-Pt_3V and Pt/C (Fig. 13a), indicating faster electron transfer. To further explain the enhanced electrocatalytic activity, DFT calculations were performed and the corresponding model is shown in Fig. 13b, c. The d-band centre of a-Pt_3V shifted negatively compared to Pt, which is critical for evaluating adsorption strength (Fig. 13e). A close-to-zero value for $\triangle G_H$* is optimal for balancing adsorption and desorption during catalysis. The $\triangle G_H$* for a-Pt_3V was calculated to be 0.26 eV, lower than that of Pt (−0.3 eV), indicating enhanced hydrogen adsorption (Fig. 13f). Additionally, the subsurface V in a-Pt_3V donates electrons to Pt, further lowering the d-band centre and improving catalytic performance. Figure 13g clearly illustrates the relationship between the d-band center and ΔG_{H*}. V-doping lowers the 4d energy level of Pt, enhancing HER by weakening the binding strength of H* on Pt sites. Figure 13h illustrates the vacancy formation energy (E_{vac}), a parameter used to assess the resistance to Pt dissolution. A higher E_{vac} indicates a greater likelihood of dissolution resistance. The elevated E_{vac} for a-Pt_3V suggests improved durability compared to Pt (111).

Jialun et al. [47] synthesized a PtNiNb alloy, which exhibited enhanced mass activity and stability compared to the state-of-the-art Pt/C (20%). The HER activity is described by the hydrogen adsorption Gibbs free energy ($\triangle G_H$*). The $\triangle G_H$* for the PtNi slab at the Ni–Ni bridge site is 0.046 eV, significantly lower than that of the Pt (111) slab (−0.266 eV), highlighting the positive impact of Ni incorporation. Furthermore, Nb incorporation in PtNiNb optimizes the binding energy to −0.010 eV, which is even lower than that achieved with Ni, indicating the beneficial role of Nb in the system. Xu et al. [47] synthesized ultrasmall ruthenium (Ru) nanoparticles

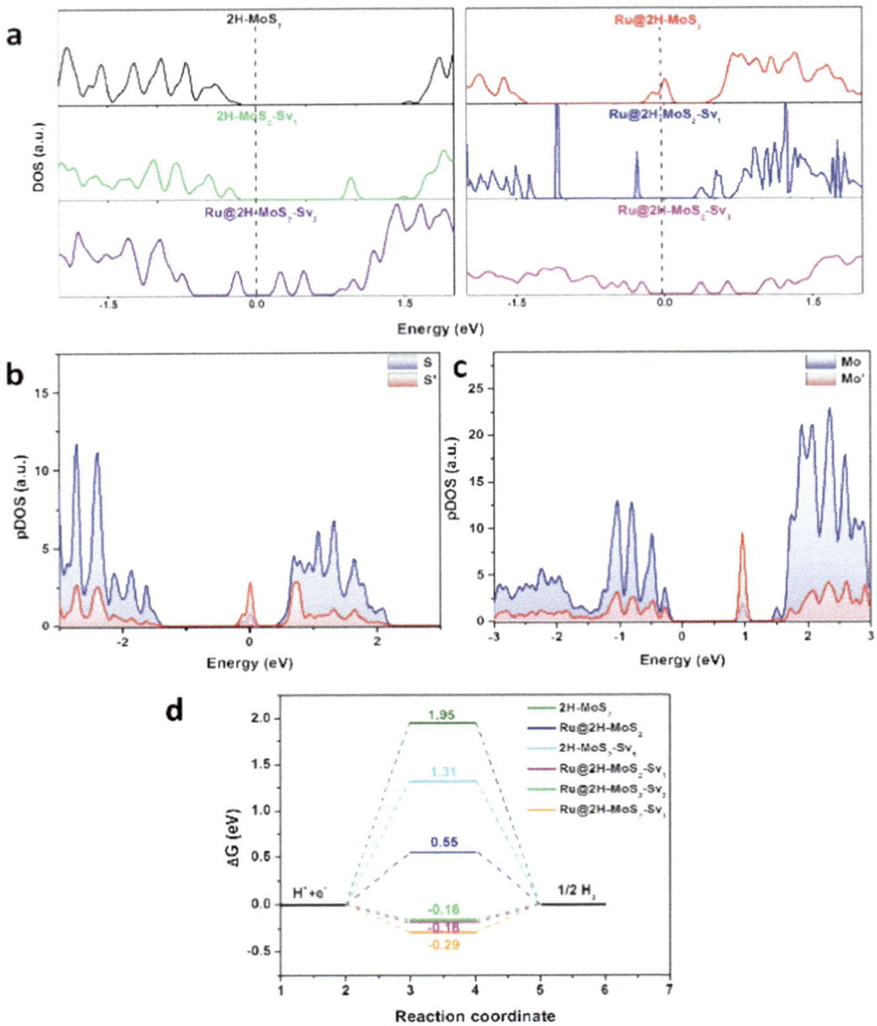

Fig. 12 **a** The overall density of states (DOS) for 2H-MoS_2, Ru@2H-MoS_2, 2H-MoS_2-Sv1, and Ru@2H-MoS_2 with different sulfur vacancy configurations. The partial density of states (PDOS) for the **b** 3p orbital of the sulfur atom (adjacent to Ru) and the sulfur atom (distant from Ru) in Ru@2H-MoS_2, and **c** 4d orbital of molybdenum (saturated site) and Mo′ (unsaturated site) in 2H-MoS_2-Sv1. **d** Diagrams of free energy profiles for hydrogen evolution across pure 2H-MoS_2, Ru@2H-MoS_2, 2H-MoS_2-Sv1, Ru@2H-MoS_2-Sv1, Ru@2H-MoS_2-Sv2, and Ru@2H-MoS_2-Sv3. Reproduced with permission [44]. Copyright 2021, Elsevier

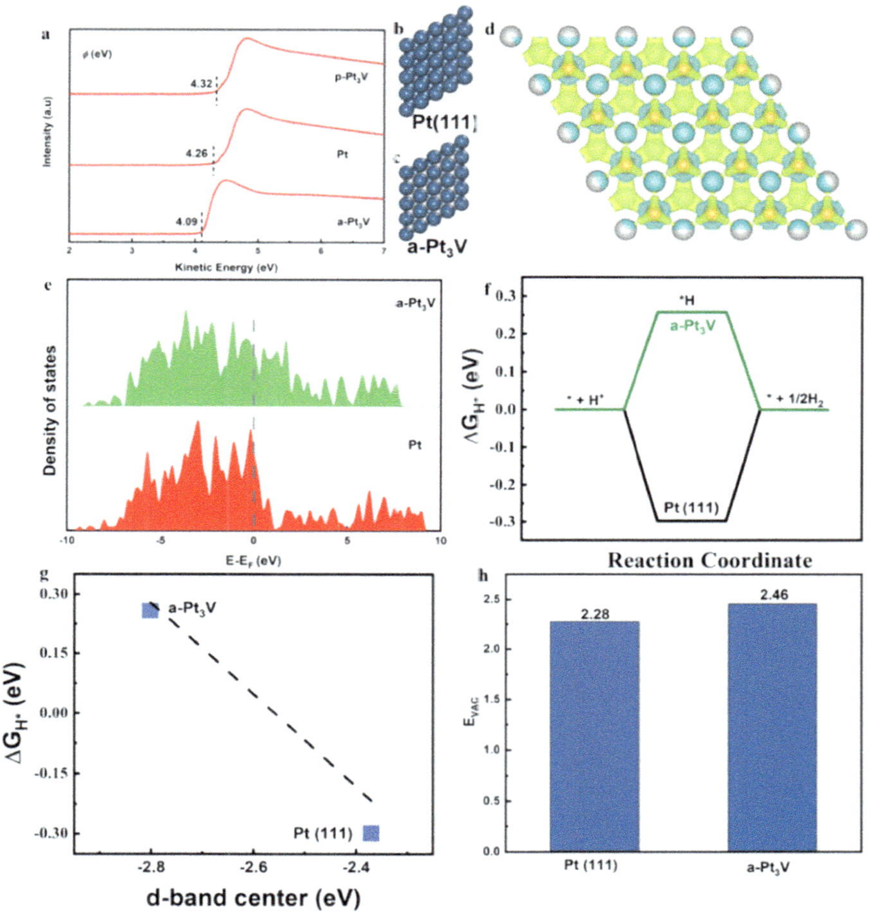

Fig. 13 **a** UPS spectra of a-Pt_3V, p-Pt_3V, and Pt. **b** Top view of the Pt (111) surface and **c** a-Pt_3V, where Pt atoms are shown in blue and V atoms in grey. **d** Differential charge density map for a-Pt_3V, with cyan regions representing charge accumulation and yellow regions showing charge depletion. **e** PDOS for surface Pt atoms in Pt (111) and Pt_3V. **f** Gibbs free energy profile for Pt sites. **g** Relationship between the d-band center of active Pt atoms and *H adsorption energy. **h** Dissolution potential of Pt atoms from Pt (111) and Pt_3V. Reproduced with permission [46]. Copyright 2023, John Wiley and Sons

on a boron and nitrogen co-doped polar carbon (B-N-PC) surface. The Ru/(B-N)-PC catalyst exhibited significantly better performance than Pt/C, with an overpotential of 15 mV at 10 mA cm^{-2} and a Tafel slope of 22.6 mV dec^{-1}. Both experimental results and theoretical calculations indicate electron transfer from nitrogen to ruthenium, followed by back-transfer from ruthenium to boron, which facilitates optimal hydrogen adsorption energy. The Gibbs free energy for hydrogen adsorption ($\triangle G_H$*) was found to be −0.383 eV, considerably lower than that of Ru/PC (porous carbon).

Wan et al. [48] synthesized amorphous nickel hydroxide ($Ni(OH)_2$), which forms a shell around Pt tetrapods. This shell not only facilitates water dissociation but also promotes proton conduction to the Pt active sites. Density functional theory (DFT) calculations indicate that the energy barrier for water dissociation on Pt (111) is 0.92 eV, with a Gibbs free energy (ΔG) of 0.70 eV, both significantly higher than those for $Ni(OH)_2$, where the dissociation energy is 0.49 eV and ΔG is 0.33 eV. At $Ni(OH)_2$, water dissociates at a nickel site coordinated with four oxygen atoms, where H^+ is transferred to a neighboring oxygen, forming a Ni-$(OH)^+$-Ni bridge. The proton then easily migrates to other oxygen atoms within the Ni-OxHy framework. The hydroxyl ions generated are readily replaced by new water molecules, allowing the water dissociation cycle to continue and maintain charge balance. Moreover, the amorphous shell enhances stability by preventing Pt particles from being easily poisoned by contaminants. Geng et al. [49] developed a Ru-modified MoS_2 material. The Ru-modified MoS_2 exhibits a notable enhancement in catalytic activity compared to pure MoS_2. The Ru-MoS_2 system shows a higher electron density near the Fermi level, which improves its electrical conductivity relative to MoS_2. The electron transfer from Ru to MoS_2 results in a reduction of the d-band center. Specifically, the d-band center shifts from -1.67 eV in MoS_2 to -2.23 eV in Ru-MoS_2, indicating increased filling of the anti-bonding states and improved performance for the HER. Zhou [50] developed a nanorod array of Ru@Cu-TiO_2/Cu to drive the HER in alkaline conditions. The synthesis process entailed electrodepositing Ru particles onto vertically structured Cu-TiO_2/Cu nanorods. Ru plays a pivotal role in breaking water molecules, a function that is further amplified by adjacent Cu clusters. Moreover, Cu is instrumental in the desorption of OH* and H_2, which is critical for averting surface poisoning. Ultimately, the synergy at the Ru-Cu interface optimizes the reaction energy landscape, dramatically boosting the electrocatalytic efficiency. Zhao et al. [51] synthesized ruthenium nanoparticles (Ru NPs) supported on titanium nitride (TiN). TiN, as a support material for Ru, weakens hydrogen adsorption, thereby enhancing the HER activity. The researchers also plotted a thermodynamic Sabatier-type activity plot, where activity is defined as the inverse of the HER overpotential. For Ru alone, the binding energy lies on the left side of the volcano plot, indicating strong H adsorption with η_{HER} values between 0.32 and 0.17 V. In contrast, Ru/TiN shows binding energies distributed on both the left and right sides of the volcano plot, ranging from 0.1 to 1.05 eV. The volcano plot also shows that TiN binds hydrogen weakly, with $\eta_{HER} = -0.73$ V. Additionally, Pt (111) binds hydrogen more strongly, with a binding energy of approximately 0.1 eV, compared to the optimal value of around 0.83 eV. Mu et al. [52] synthesized RuRh bimetallene nanoring as an efficient catalyst for the HER in both acidic and alkaline mediums. ΔG_H* was found to be optimum for $RuRh_2$ bimetallene/Ru (Rh). Electronic structure modulation due to alloy formation, strain effect provides the optimum ΔG_H* which enhances the HER significantly. Liu et al. [53] synthesized PtCu nanospheres anchored on a nano-array of WO_3 on Cu-foam. The PtCu/WO_3@CF catalyst demonstrated ultra-high activity compared to commercial Pt/C. The ΔG_{H*} for hydrogen adsorption on WO_3(001) was calculated to be 0.42 eV, significantly larger than that for hydrogen adsorption on the WO_3(001) heterojunction at Pt(001)/WO_3(001). This clearly indicates that hydrogen adsorption is enhanced at the interface between Pt and WO_3. Additionally, hydrogen adsorption over Pt was more optimized at the heterojunction compared to Pt (001) sites alone, indicating improved HER activity at the heterojunction under acidic conditions. Shi et al. [54] synthesized single-atom Pt on various substrates, including WSe_2, $MoSe_2$, MoS_2, and WS_2, to study the modulation of its electronic structure. Their findings showed that as the oxidation state of Pt increases, hydrogen adsorption ability improves while HER activity decreases. However, in an alkaline medium, HER activity increases with the oxidation state of Pt. Wu et al. [55] synthesized ultrathin Ir nanosheets containing both amorphous and crystalline phases. The boundaries between these phases exhibit a tensile strain of 4%, resulting in significantly improved activity compared to either the crystalline or amorphous phases of Ir and Pt/C. This strain modulates the d-band center to achieve optimal H-adsorption energy, thereby enhancing HER activity. Lin et al. [56] synthesized an intermetallic alloy of $Ir_{1-x}Rh_xSb$, where the active sites are Ir and Rh atoms, negatively charged due to electron transfer from Sb. This electron transfer from Sb enhances hydrogen adsorption ability, thereby improving HER activity.

Kim et al. [3] synthesized F-doped SnO_2 (IV) with dispersed Pt nanoclusters. Pt on F-SnO_2 exhibits much higher stability compared to state-of-the-art Pt/C, attributed to a strong metal-support interaction. F-SnO_2@Pt has an optimized binding energy (ΔG_H*) for hydrogen adsorption compared to SnO@Pt and F-SnO_2 alone, which enhances HER activity. Zhao et al. [57] synthesized Pt-O-Pt atomic clusters on Co/N co-doped carbon (CoNC) (Pt ACs/CoNC). Pt ACs/CoNC demonstrates a mass activity six times greater than that of Pt/C. The presence of the Pt-O-Pt bond results in electron density that is more localized on the Pt atom compared to the Pt-O-Co bond, where the electron density is less localized on the Pt atom. A higher electron density on the Pt is considered beneficial for HER activity. Also, Pt ACs/CoNC shows much better activity than Pt single atom (Pt SAs/CoNC) due to above said reason. Tan et al. [58] synthesized Pt nanoparticles supported on MgO nanosheets. These MgO nanosheets exhibit oxygen vacancies, which facilitate water dissociation to form H_3O^+ and enable charge transfer to Pt, resulting in negatively charged Pt ($Pt\delta^-$). The electrostatic interaction between $Pt\delta^-$ and H_3O^+ creates a localized acidic environment around Pt, as confirmed by Raman spectroscopy, synchrotron radiation infrared spectroscopy, and X-ray absorption spectroscopy. Wang et al. [59] synthesized noble metal clusters on $Ti_3C_2T_x$ MXenes. They found that MXenes with Ti vacancy sites supporting Ru clusters exhibit superior catalytic activity compared to other catalysts, attributed to a strong support interaction. This interaction is particularly strong due to the interaction between Ru and C, which acts as a ligand within the MXene lattice. This interaction provides optimized binding energy for various adsorption and dissociation steps and enhances the HER significantly.

4 Conclusion

In summary, this chapter provides a comprehensive exploration of the hydrogen evolution reaction (HER), focusing on its fundamental mechanisms and recent advancements in electrocatalysis, which play a critical role in enhancing the efficiency of electrolysis systems. The HER mechanism is discussed through its two primary pathways—the Volmer-Tafel and Volmer-Heyrovsky routes—emphasizing the stepwise electron-proton transfer dynamics that are essential for understanding catalytic behavior. A key aspect of this chapter is the detailed examination of electrocatalyst development, encompassing a wide range of materials from noble to non-noble metals. The discussion highlights their catalytic properties, emphasizing strategies to improve activity, durability, and cost-effectiveness. Advanced electrocatalysts with superior performance compared to state-of-the-art materials are showcased, underscoring the significant progress made in the field. Central to these advancements is the role of electronic modifications in enhancing catalytic efficiency. The doping of metals or non-metals is shown to effectively shift the d-band center relative to the Fermi level, which directly influences the adsorption and desorption strengths of hydrogen intermediates. This electronic tailoring enables improved reaction kinetics and enhanced HER performance. Catalysts exhibiting a Gibbs free energy of hydrogen adsorption (ΔG_{H^*}) close to zero are identified as the most efficient, as they achieve an optimal balance where hydrogen binding is neither too strong nor too weak, thus maximizing activity. These findings highlight the critical importance of electronic structure engineering in the design of next-generation HER electrocatalysts. By leveraging these strategies, researchers can develop materials with exceptional catalytic performance, contributing to more efficient and sustainable hydrogen production technologies. This chapter provides a foundation for further innovation in HER electrocatalysis, paving the way for advancements in renewable energy systems.

Questions and Answers

Question 1: $\mathbf{E_{HER} = E_{(H_2/H^+)^0} - (RT/F)\ ln((a_{H^+})/(P_{H_2}{}^{1/2})}$ Calculate E_{HER} at 298 K for a hydrogen gas partial pressure of $P_{H_2} = 0.5$ atm and hydrogen ion activity a_ H⁺ = 0.001. Assume $E_{H_2/H^{+0}} = 0$ V, $R = 8.314$ J/(mol K), and $F = 96{,}485$ C/mol. Ans: $E_{HER} = 0.171$ V

Question 2: $\mathbf{b = RT/(\alpha * F)}$ The Tafel slope b for a catalyst is given by: where α is the transfer coefficient, $R = 8.314$ J/(mol K), $T = 298$ K, and $F = 96{,}485$ C/mol. Calculate the value of α if the Tafel slope b is 0.120 V/decade. Ans: $\alpha = 0.5$

Question 3: Using the Turnover Frequency (TOF) equation, where $J = 0.05$ A/cm^2, $n = 2$, $F = 96{,}485$ C/mol, and $N = 5 \times 1015$ sites/cm^2, calculate the TOF in s^{-1} per active site and interpret the result in terms of catalyst activity. Ans: TOF = 1.04 s^{-1}

Question 4: Given that the Tafel equation is $\eta = b\log(j) + a$, where the Tafel slope b changes with temperature. If b is proportional to $RT/(\alpha * F)$, calculate the change in b when the temperature increases from 298 to 350 K, assuming the transfer coefficient α remains constant. Ans: $b_{350\ K} = 0.1409$ V/decade

Question 5: A catalyst has an exchange current density j_0 of 0.001 A/cm^2 at a pH of 7. Calculate the pH effect on E_{HER} if the pH is increased to 10, using the relationship $E_{HER} = -0.059 * \text{pH}$ V versus SHE. Ans: $\Delta E^{HER} = -0.177$ V

Question 6: What is the significance of the d-band center in determining the catalytic activity of transition metal-based catalysts, and how does it influence adsorption energies during HER?

Answer: The d-band center is a key electronic property that influences the interaction between the catalyst surface and adsorbates. A lower d-band center, relative to the Fermi level, typically weakens the adsorption energy of intermediates, allowing for optimal adsorption and desorption steps critical for HER. This balance improves reaction kinetics, making the catalyst more efficient.

Question 7: How does doping with heteroatoms improve the performance and stability of electrocatalysts for HER and OER?

Answer: Doping with heteroatoms can modify the electronic structure of the catalyst, enhancing charge transfer and optimizing active site properties. It also contributes to structural stability by preventing degradation during prolonged operation, as seen in elements like phosphorus protecting selenium from surface dissolution.

Question 8: What does a low overpotential indicate about an electrocatalyst's efficiency, and why is this parameter significant for HER?

Answer: A low overpotential reflects the energy efficiency of an electrocatalyst, as it indicates that minimal extra voltage is required to drive the reaction. This efficiency is crucial for sustainable hydrogen production, making low-overpotential materials highly desirable for HER.

Question 9: How do surface modifications, such as the formation of hydroxide species, enhance water dissociation and HER activity, particularly in alkaline media?

Answer: Surface modifications, such as hydroxide formation, create active sites that lower the energy barrier for water dissociation. These sites facilitate the breakdown of water molecules into hydrogen, enhancing HER efficiency in environments with low proton availability, such as alkaline media.

Question 10: How do atomic or molecular arrangements, such as specific bonding structures, impact the electronic properties of electrocatalysts and contribute to their HER efficiency?

Answer: Unique bonding arrangements, such as Fe–OH–Ni, modify the electronic environment by extracting or redistributing electrons, creating unoccupied states that enhance reactivity. These electronic adjustments make the catalyst more efficient in facilitating adsorption, reaction, and desorption steps.

Question 11: Why is an adsorption free energy (ΔG_{free}) near zero considered optimal for HER activity, and how can catalyst composition and structure be tailored to achieve this?

Answer: A ΔG_{free} near zero ensures that hydrogen intermediates are neither too strongly nor too weakly bound, balancing adsorption and desorption. Catalyst composition, such as doping or alloying, can adjust electronic properties to achieve this optimal balance, improving HER kinetics.

Question 12: Why is water dissociation a critical step for HER in alkaline media, and what strategies can be employed to enhance its kinetics?

Answer: In alkaline media, where proton availability is low, water dissociation provides the necessary hydrogen for HER. Enhancing water dissociation can be achieved by modifying surface properties or introducing catalytic sites, such as $Ni(OH)_2$, which lower the activation energy for this step.

Question13: How does phosphorus (P) doping contribute to the performance of Fe, P-$NiSe_2$ nanoporous films in catalyzing HER and OER?

Answer: Phosphorus doping in Fe, P-$NiSe_2$ nanoporous films enhances the material's electron conductivity, which is essential for efficient charge transfer during catalytic reactions. Additionally, it protects selenium (Se) from surface dissolution, which is crucial for maintaining the catalyst's structural integrity and performance over time. This structural stability enables long-term operation, as demonstrated by the current density of 0.8 A cm^{-2} at 1.8 V and the stability observed over 200 h.

Question 14: Why is the $Co_{0.6}Mo_{1.4}N_2$ catalyst with a mixed-layer close-packed structure effective for HER, and how does the structure influence its catalytic properties?

Answer: The $Co_{0.6}Mo_{1.4}N_2$ catalyst's effectiveness for HER arises from its unique mixed-layer close-packed structure, which includes octahedral and trigonal prismatic sites occupied by $Co2^+$ and Mo with varying oxidation states. This layered structure permits fine-tuning of Mo's catalytic properties through controlled Co substitution in the octahedral sites. The distribution of oxidation states and the presence of 3d transition metal (Co) in this configuration enable a more flexible adjustment of electronic and structural characteristics, promoting better HER performance.

Question 15: How does the encapsulation of NiM (where M = Zn, Mo, Ni, Co) nanoalloys within graphene nanoscrolls contribute to the performance of CoP catalysts?

Answer: The encapsulation of NiM nanoalloys within graphene nanoscrolls supported on carbon nanotubes (CNTs) provides several advantages for the CoP catalyst's performance. The graphene nanoscrolls offer structural stability and increase the surface area, enhancing active site accessibility for HER. Additionally, the specific choice of bimetallic combinations (e.g., Zn, Mo, Ni, Co) influences the electronic structure and, consequently, the d-band center, allowing for a tunable approach to optimize adsorption energies and catalytic performance. This combination improves HER activity, as seen in NiZn@C-CoP, which benefits from the synergistic effects of Ni and Zn in its encapsulated form.

Question 16. What does the low overpotential of h-NiMoFe at 10 mA cm^{-2} indicate about its efficiency as an electrocatalyst for HER?

Answer: The low overpotential (14 mV) of h-NiMoFe compared to Pt/C (59 mV) and Ni (309 mV) suggests that h-NiMoFe is highly efficient in catalyzing the hydrogen evolution reaction. Overpotential is the extra voltage required beyond the theoretical voltage to drive the reaction. A lower overpotential indicates that the catalyst requires less energy to achieve the desired reaction rate, making it more energy-efficient and suitable for practical applications in hydrogen production.

Question 17. Why is the formation of $Ni(OH)_2$ significant for HER activity in h-NiMoFe, and how does it influence water dissociation?

Answer: The formation of $Ni(OH)_2$ on the catalyst surface is essential because it enhances water dissociation, a key step in HER, especially in alkaline media. $Ni(OH)_2$ serves as an active site that facilitates the breakdown of water molecules, providing hydrogen atoms necessary for the reaction. In h-NiMoFe, the generation of $Ni(OH)_2$ remains stable even at elevated current densities, ensuring consistent water dissociation and contributing to the overall catalytic efficiency.

Question 18. What role does the $Fe–(OH)_4–Ni_4$ structure play in the electronic properties of h-NiMoFe, and why is it beneficial for HER?

Answer: The $Fe–(OH)_4–Ni_4$ structure alters the electronic properties by creating bonds (Fe–OH–Ni) that can extract electrons from Ni atoms. This electron extraction leads to unoccupied states in Ni, making the catalyst more reactive. The presence of unoccupied states allows Ni to participate more effectively in electron transfer during the HER, enhancing the overall catalytic activity. This structure stabilizes the electronic environment required for optimal hydrogen adsorption and reaction kinetics.

Question 19. How do the unoccupied Ni states created by the Fe–OH–Ni bonds in h-NiMoFe influence HER activity?

Answer: The unoccupied Ni states created by the electron-withdrawing effect of Fe–OH–Ni bonds increase the reactivity of the Ni atoms, making them more capable of engaging in electron transfer during the HER. These unoccupied states allow Ni to interact more actively with hydrogen atoms and facilitate the adsorption process, which is vital for efficient hydrogen evolution. This electronic configuration enhances the catalyst's intrinsic activity, contributing to its superior performance compared to other materials.

Question 20: In what ways do the synthesis methods (e.g., chemical vapor deposition for CoP, dual-doping for Fe, P-$NiSe_2$ NFs) impact the structural and functional properties of these catalysts?

Answer: The synthesis methods play a crucial role in defining the structural and functional properties of these catalysts. For instance, chemical vapor deposition allows precise control over the growth and integration of CoP on CNT-supported graphene nanoscrolls, ensuring uniformity and stability essential for enhanced HER performance. Similarly, dual-doping (Fe, P) for $NiSe_2$ NFs introduces specific elements (Fe and P) that are strategically placed to improve electron conductivity, stability, and selectivity, targeting both HER and OER. These tailored synthesis techniques allow fine-tuning of electronic, structural, and catalytic properties, directly impacting the overall efficiency and durability of the catalysts.

References

1. EIA (2020) Use-of-hydrogen @ www.Eia.Gov
2. Kaiwen L, Bin Y, Tao Z (2018) Economic analysis of hydrogen production from steam reforming process: a literature review. Energy Sources Part B Econ Plan Policy 13(2):109–115. https://doi.org/10.1080/15567249.2017.1387619
3. Kim T, Roy SB, Moon S, Yoo SH, Choi H, Parale VG, Kim Y, Lee J, Jun SC, Kang K et al (2022) Highly dispersed Pt clusters on F-doped Tin(IV) oxide aerogel matrix: an ultra-robust hybrid catalyst for enhanced hydrogen evolution. ACS Nano 16(1):1625–1638. https://doi.org/10.1021/acsnano.1c10504
4. Mahmood N, Yao Y, Zhang JW, Pan L, Zhang X, Zou JJ (2018) Electrocatalysts for hydrogen evolution in alkaline electrolytes: mechanisms, challenges, and prospective solutions. Adv Sci 5(2). https://doi.org/10.1002/advs.201700464
5. Murthy AP, Madhavan J, Murugan K (2018) Recent advances in hydrogen evolution reaction catalysts on carbon/carbon-based supports in acid media. J Power Sources 398(May):9–26. https://doi.org/10.1016/j.jpowsour.2018.07.040
6. Baek DS, Lee J, Lim JS, Joo SH (2021) Nanoscale electrocatalyst design for alkaline hydrogen evolution reaction through activity descriptor identification. Mater Chem Front 5(11):4042–4058. https://doi.org/10.1039/d1qm00183c
7. Karmakar A, Durairaj M, Madhu R, De A, Dhandapani HN, Spencer MJS, Kundu S (2024) From proximity to energetics: unveiling the hidden compass of hydrogen evolution reaction. ACS Mater Lett 6(7):3050–3062. https://doi.org/10.1021/acsmaterialslett.4c00761
8. Chia X, Adriano A, Lazar P, Sofer Z, Luxa J, Pumera M (2016) Layered platinum dichalcogenides (PtS2, PtSe2, and PtTe2) electrocatalysis: monotonic dependence on the chalcogen size. Adv Funct Mater 26(24):4306–4318. https://doi.org/10.1002/adfm.201505402
9. Han X, Tong X, Liu X, Chen A, Wen X, Yang N, Guo XY (2018) Hydrogen evolution reaction on hybrid catalysts of vertical MoS_2 nanosheets and hydrogenated graphene. ACS Catal 8(3):1828–1836. https://doi.org/10.1021/acscatal.7b03316
10. Cheng N, Stambula S, Wang D, Banis MN, Liu J, Riese A, Xiao B, Li R, Sham TK, Liu LM et al (2016) Platinum single-atom and cluster catalysis of the hydrogen evolution reaction. Nat Commun 7:1–9. https://doi.org/10.1038/ncomms13638
11. Zeng M, Li Y (2015) Recent advances in heterogeneous electrocatalysts for the hydrogen evolution reaction. J Mater Chem A 3(29):14942–14962. https://doi.org/10.1039/c5ta02974k
12. Hansen JN, Prats H, Toudahl KK, Mørch Secher N, Chan K, Kibsgaard J, Chorkendorff I (2021) Is there anything better than Pt for HER? ACS Energy Lett 6(4):1175–1180. https://doi.org/10.1021/acsenergylett.1c00246

13. Zhou F, Zhou Y, Liu GG, Wang CT, Wang J (2021) Recent advances in nanostructured electrocatalysts for hydrogen evolution reaction. Rare Met 40(12):3375–3405. https://doi.org/10.1007/s12598-021-01735-y
14. Zhang C, Xu Z, Han N, Tian Y, Kallio T, Yu C, Jiang L (2023) Superaerophilic/superaerophobic cooperative electrode for efficient hydrogen evolution reaction via enhanced mass transfer. Sci Adv 9(3). https://doi.org/10.1126/sciadv.add6978
15. Angulo AE, Frey D, Modestino MA (2022) Understanding bubble-induced overpotential losses in multiphase flow electrochemical reactors. Energy Fuels 36(14):7908–7914. https://doi.org/10.1021/acs.energyfuels.2c01543
16. Luo Y, Zhang Y, Zhu J, Tian X, Liu G, Feng Z, Pan L, Liu X, Han N, Tan R (2024) Material engineering strategies for efficient hydrogen evolution reaction catalysts. Small Methods 2400158:1–18. https://doi.org/10.1002/smtd.202400158
17. Murthy AP, Theerthagiri J, Madhavan J (2018) Insights on Tafel constant in the analysis of hydrogen evolution reaction. J Phys Chem C 122(42):23943–23949. https://doi.org/10.1021/acs.jpcc.8b07763
18. Zhao G, Rui K, Dou SX, Sun W (2018) Heterostructures for electrochemical hydrogen evolution reaction: a review. Adv Funct Mater 28(43):1–26. https://doi.org/10.1002/adfm.201803291
19. Liu F, Shi C, Guo X, He Z, Pan L, Huang ZF, Zhang X, Zou JJ (2022) Rational design of better hydrogen evolution electrocatalysts for water splitting: a review. Adv Sci 9(18). https://doi.org/10.1002/advs.202200307
20. Nørskov JK, Bligaard T, Logadottir A, Kitchin JR, Chen JG, Pandelov S, Stimming U (2005) Trends in the exchange current for hydrogen evolution. J Electrochem Soc 152(3):J23. https://doi.org/10.1149/1.1856988
21. Anantharaj S, Ede SR, Karthick K, Sam Sankar S, Sangeetha K, Karthik PE, Kundu S (2018) Precision and correctness in the evaluation of electrocatalytic water splitting: revisiting activity parameters with a critical assessment. Energy Environ Sci 11(4):744–771. https://doi.org/10.1039/c7ee03457a
22. Zhou Z, Wei L, Wang Y, Karahan HE, Chen Z, Lei Y, Chen X, Zhai S, Liao X, Chen Y (2017) Hydrogen evolution reaction activity of nickel phosphide is highly sensitive to electrolyte PH. J Mater Chem A 5(38):20390–20397. https://doi.org/10.1039/c7ta06000a
23. Kozuch S, Martin JML (2012) "Turning over" definitions in catalytic cycles. ACS Catal 2(12):2787–2794. https://doi.org/10.1021/cs3005264
24. Shin S, Jin Z, Kwon DH, Bose R, Min YS (2015) High turnover frequency of hydrogen evolution reaction on amorphous MoS_2 thin film directly grown by atomic layer deposition. Langmuir 31(3):1196–1202. https://doi.org/10.1021/la504162u
25. Luo Y, Zhang Z, Yang F, Li J, Liu Z, Ren W, Zhang S, Liu B (2021) Stabilized hydroxide-mediated nickel-based electrocatalysts for high-current-density hydrogen evolution in alkaline media. Energy Environ Sci 14(8):4610–4619. https://doi.org/10.1039/d1ee01487k
26. Chen XH, Li T, Li X, Lei J, Li NB, Luo HQ (2023) Oxyanion-coordinated Co-based catalysts for optimized hydrogen evolution: the feedback of adsorbed anions to the catalytic activity and mechanism. ACS Catal 13(10):6721–6729. https://doi.org/10.1021/acscatal.3c00598
27. Jiang HJ, Underwood TC, Bell JG, Ranjan S, Sasselov D, Whitesides GM (2017) Mimicking lighting-induced electrochemistry on the early earth. Proc Natl Acad Sci 2017:120. https://doi.org/10.1073/pnas
28. Liu S, Wang Y, Jiang T, Jin S, Sajid M, Zhang Z, Xu J, Fan Y, Wang X, Chen J et al (2024) Non-noble metal high-entropy alloy-based catalytic electrode for long-life hydrogen gas batteries. ACS Nano 18(5):4229–4240. https://doi.org/10.1021/acsnano.3c09482
29. Hu J, Zhang C, Zhang Y, Yang B, Qi Q, Sun M, Zi F, Leung MKH, Huang B (2020) Interface modulation of MoS_2/metal oxide heterostructures for efficient hydrogen evolution electrocatalysis. Small 16(28):1–9. https://doi.org/10.1002/smll.202002212
30. Ling T, Zhang T, Ge B, Han L, Zheng L, Lin F, Xu Z, Hu WB, Du XW, Davey K et al (2019) Well-dispersed nickel- and zinc-tailored electronic structure of a transition metal oxide for highly active alkaline hydrogen evolution reaction. Adv Mater 31(16):1–7. https://doi.org/10.1002/adma.201807771

31. Chen Y, Yue K, Zhao JW, Cai Z, Wang X, Yan Y (2023) Effective Modulating of the Mo dissolution and polymerization in $Ni_4Mo/NiMoO_4$ heterostructure via metal-metal oxide-support interaction for boosting H_2 production. Chem Eng J 466(April):143097. https://doi.org/10.1016/j.cej.2023.143097
32. Feng H, Yu J, Tang L, Wang J, Dong H, Ni T, Tang J, Tang W, Zhu X, Liang C (2021) Improved hydrogen evolution activity of layered double hydroxide by optimizing the electronic structure. Appl Catal B Environ 297(June):120478. https://doi.org/10.1016/j.apcatb.2021.120478
33. Piontek S, Andronescu C, Zaichenko A, Konkena B, Junge Puring K, Marler B, Antoni H, Sinev I, Muhler M, Mollenhauer D et al (2018) Influence of the Fe: Ni ratio and reaction temperature on the efficiency of $(Fe_xNi_{1-x})_9S_8$ electrocatalysts applied in the hydrogen evolution reaction. ACS Catal 8(2):987–996. https://doi.org/10.1021/acscatal.7b02617
34. Luo H, Zhang X, Zhu H, Zhang K, Yang F, Xu K, Yu S, Guo D (2023) Tailoring D-band center over electron traversing effect of NiM@C-CoP (M=Zn, Mo, Ni, Co) for high-performance electrocatalysis hydrogen evolution. J Mater Sci Technol 166:164–172. https://doi.org/10.1016/j.jmst.2023.05.028
35. Chang J, Wang G, Yang Z, Li B, Wang Q, Kuliiev R, Orlovskaya N, Gu M, Du Y, Wang G et al (2021) Dual-doping and synergism toward high-performance seawater electrolysis. Adv Mater 33(33). https://doi.org/10.1002/adma.202101425
36. Cao B, Veith GM, Neuefeind JC, Adzic RR, Khalifah PG (2013) Mixed close-packed cobalt molybdenum nitrides as non-noble metal electrocatalysts for the hydrogen evolution reaction. J Am Chem Soc 135(51):19186–19192. https://doi.org/10.1021/ja4081056
37. Fu HQ, Zhou M, Liu PF, Liu P, Yin H, Sun KZ, Yang HG, Al-Mamun M, Hu P, Wang HF et al (2022) Hydrogen spillover-bridged Volmer/Tafel processes enabling ampere-level current density alkaline hydrogen evolution reaction under low overpotential. J Am Chem Soc 144(13):6028–6039. https://doi.org/10.1021/jacs.2c01094
38. Zhou W, Jiang Z, Chen M, Li Z, Luo X, Guo M, Yang Y, Yu T, Yuan C, Wang S (2021) Directly anchoring non-noble metal single atoms on 1T-TMDs with tip structure for efficient hydrogen evolution. Chem Eng J 2022(428):131210. https://doi.org/10.1016/j.cej.2021.131210
39. Kundu D, Hazra A, Bhattacharjee S, Dutta J, Murmu NC, Bhaumik A, Banerjee P (2024) Integration of a bismuth-based tris-mononuclear complex with 2D functional materials for highly efficient and durable aqueous electrocatalytic hydrogen evolution. ACS Appl Mater Interfaces 16(22):28423–28434. https://doi.org/10.1021/acsami.4c02234
40. Ji L, Yan P, Zhu C, Ma C, Wu W, Wei C, Shen Y, Chu S, Wang J, Du Y et al (2019) One-pot synthesis of porous 1T-phase MoS_2 integrated with single-atom Cu doping for enhancing electrocatalytic hydrogen evolution reaction. Appl Catal B Environ 251:87–93. https://doi.org/10.1016/j.apcatb.2019.03.053
41. Ma T, Cao H, Li S, Cao S, Zhao Z, Wu Z, Yan R, Yang C, Wang Y, van Aken PA et al (2022) Crystalline lattice-confined atomic Pt in metal carbides to match electronic structures and hydrogen evolution behaviors of platinum. Adv Mater 34(41):1–11. https://doi.org/10.1002/adma.202206368
42. Zhou KL, Wang Z, Han CB, Ke X, Wang C, Jin Y, Zhang Q, Liu J, Wang H, Yan H (2021) Platinum single-atom catalyst coupled with transition metal/metal oxide heterostructure for accelerating alkaline hydrogen evolution reaction. Nat Commun 12(1):1–10. https://doi.org/10.1038/s41467-021-24079-8
43. Huang X, Lu R, Cen Y, Wang D, Jin S, Chen W, Geoffrey I, Waterhouse N, Wang Z, Tian S et al (2023) Micropore-confined Ru nanoclusters catalyst for efficient PH-universal hydrogen evolution reaction. Nano Res 16(7):9073–9080. https://doi.org/10.1007/s12274-023-5711-1
44. Wang J, Fang W, Hu Y, Zhang Y, Dang J, Wu Y, Chen B, Zhao H, Li Z (2021) Single atom Ru doping 2H-MoS_2 as highly efficient hydrogen evolution reaction electrocatalyst in a wide PH range. Appl Catal B Environ 298:120490. https://doi.org/10.1016/J.APCATB.2021.120490
45. Yu Z, Li Y, Torres-Pinto A, LaGrow AP, Diaconescu VM, Simonelli L, Sampaio MJ, Bondarchuk O, Amorim I, Araujo A et al (2022) Single-atom Ir and Ru anchored on graphitic carbon nitride for efficient and stable electrocatalytic/photocatalytic hydrogen evolution. Appl Catal B Environ 310(February). https://doi.org/10.1016/j.apcatb.2022.121318

46. Da Y, Jiang R, Tian Z, Chen G, Xiao Y, Zhang J, Xi S, Deng Y, Chen W, Han X et al (2023) Development of a novel Pt_3V alloy electrocatalyst for highly efficient and durable industrial hydrogen evolution reaction in acid environment. Adv Energy Mater 13(16):1–8. https://doi.org/10.1002/aenm.202300127
47. Gu J, Li L, Xie Y, Chen B, Tian F, Wang Y, Zhong J, Shen J, Lu J (2023) Turing structuring with multiple nanotwins to engineer efficient and stable catalysts for hydrogen evolution reaction. Nat Commun 14(1):1–13. https://doi.org/10.1038/s41467-023-40972-w
48. Wan C, Zhang Z, Dong J, Xu M, Pu H, Baumann D, Lin Z, Wang S, Huang J, Shah AH et al (2023) Amorphous nickel hydroxide shell tailors local chemical environment on platinum surface for alkaline hydrogen evolution reaction. Nat Mater 22(8):1022–1029. https://doi.org/10.1038/s41563-023-01584-3
49. Geng S, Tian F, Li M, Liu Y, Sheng J, Yang W, Yu Y, Hou Y (2022) Activating interfacial S sites of MoS_2 boosts hydrogen evolution electrocatalysis. Nano Res 15(3):1809–1816. https://doi.org/10.1007/s12274-021-3755-7
50. Zuo Y, Bellani S, Saleh G, Ferri M, Shinde DV, Zappia MI, Buha J, Brescia R, Prato M, Pascazio R et al (2023) Ru-Cu nanoheterostructures for efficient hydrogen evolution reaction in alkaline water electrolyzers. J Am Chem Soc 145(39):21419–21431. https://doi.org/10.1021/jacs.3c06726
51. Zhao J, Urrego-Ortiz R, Liao N, Calle-Vallejo F, Luo J (2024) Rationally designed Ru catalysts supported on TiN for highly efficient and stable hydrogen evolution in alkaline conditions. Nat Commun 15(1):1–14. https://doi.org/10.1038/s41467-024-50691-5
52. Mu X, Gu J, Feng F, Xiao Z, Chen C, Liu S, Mu S (2021) RuRh bimetallene nanoring as high-efficiency PH-universal catalyst for hydrogen evolution reaction. Adv Sci 8(2):1–8. https://doi.org/10.1002/advs.202002341
53. Liu L, Wang Y, Zhao Y, Wang Y, Zhang Z, Wu T, Qin W, Liu S, Jia B, Wu H et al (2022) Ultrahigh Pt-mass-activity hydrogen evolution catalyst electrodeposited from bulk Pt. Adv Funct Mater 32(20):26–28. https://doi.org/10.1002/adfm.202112207
54. Shi Y, Ma ZR, Xiao YY, Yin YC, Huang WM, Huang ZC, Zheng YZ, Mu FY, Huang R, Shi GY et al (2021) Electronic metal-support interaction modulates single-atom platinum catalysis for hydrogen evolution reaction. Nat Commun 12(1):1–11. https://doi.org/10.1038/s41467-021-23306-6
55. Wu G, Han X, Cai J, Yin P, Cui P, Zheng X, Li H, Chen C, Wang G, Hong X (2022) In-plane strain engineering in ultrathin noble metal nanosheets boosts the intrinsic electrocatalytic hydrogen evolution activity. Nat Commun 13(1). https://doi.org/10.1038/s41467-022-31971-4
56. Lin Z, Xiao BB, Huang M, Yan L, Wang Z, Huang Y, Shen S, Zhang Q, Gu L, Zhong W (2022) Realizing negatively charged metal atoms through controllable D-electron transfer in ternary $Ir_{1-x}Rh_xSb$ intermetallic alloy for hydrogen evolution reaction. Adv Energy Mater 12(25):1–7. https://doi.org/10.1002/aenm.202200855
57. Zhao Y, Kumar PV, Tan X, Lu X, Zhu X, Jiang J, Pan J, Xi S, Yang HY, Ma Z et al (2022) Modulating Pt-O-Pt atomic clusters with isolated cobalt atoms for enhanced hydrogen evolution catalysis. Nat Commun 13(1):1–11. https://doi.org/10.1038/s41467-022-30155-4
58. Tan H, Tang B, Lu Y, Ji Q, Lv L, Duan H, Li N, Wang Y, Feng S, Li Z et al (2022) Engineering a local acid-like environment in alkaline medium for efficient hydrogen evolution reaction. Nat Commun 13(1):1–9. https://doi.org/10.1038/s41467-022-29710-w
59. Wang X, Ding J, Song W, Yang X, Zhang T, Huang Z, Wang H, Han X, Hu W (2023) Cation vacancy clusters in $Ti_3C_2T_x$ MXene induce ultra-strong interaction with noble metal clusters for efficient electrocatalytic hydrogen evolution. Adv Energy Mater 13(23):1–11. https://doi.org/10.1002/aenm.202300148

Thermoelectric Materials and Devices for Applications Below 250 °C

Zhongxin Liang and Zhifeng Ren

Abstract In this chapter, we first provide a comprehensive overview of thermoelectric effects, followed by strategies for optimizing transport properties and elucidating their underlying mechanisms. We then present an in-depth discussion of recent advances in near-room-temperature thermoelectric materials and devices, aiming to inspire further research toward the development of next-generation commercial modules with enhanced performance across a wide range of applications.

1 Introduction

Thermoelectric (TE) materials have the remarkable ability to directly convert heat into electricity and vice versa, making them highly promising for waste heat recovery and solid-state cooling applications. Three fundamental TE effects, along with Joule heating, form the fundamental principles of TE technology [1–3]. The first is the Seebeck effect, discovered in 1821 by Thomas Johann Seebeck. He observed that a voltage (ΔV) is generated between the junctions of two dissimilar conductors (a and b) subjected to a temperature difference (ΔT), as illustrated in Fig. 1a. The differential Seebeck coefficient of two conductors (S_{ab}) quantifies this effect and is given by: $S_{ab} = \frac{\Delta V}{\Delta T}$. Shortly thereafter, Jean Charles Athanase Peltier identified the second of the three TE effects, known as the Peltier effect. It describes the absorption or release of heat at the junction of two materials when an electric current (I) passes through it, as shown in Fig. 1b. The differential Peltier coefficient (π_{ab}) is defined

Z. Liang
Department of Physics, Texas Center for Superconductivity at the University of Houston (TcSUH), University of Houston, Houston, TX, USA
e-mail: zliang20@central.uh.edu

Z. Ren (✉)
Paul C. W. Chu and May P. Chern Endowed Chair in Condensed Matter Physics, Department of Physics, Texas Center for Superconductivity at the University of Houston (TcSUH), University of Houston, Houston, TX, USA
e-mail: zren@uh.edu
URL: https://www.uh.edu/nsm/physics/people/profiles/zhifeng-ren/

D. Shi (ed.), *Energy Materials Science and Engineering: From New Devices to AI Power Systems*, Green Energy and Technology, https://doi.org/10.1007/978-3-031-88550-1_8

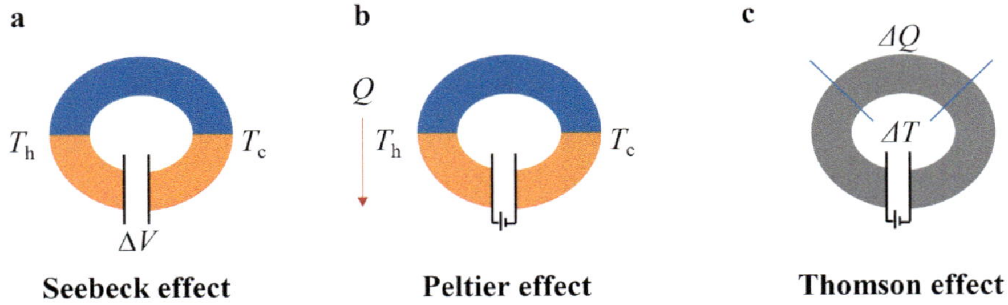

Fig. 1 Schematic illustrations of **a** the Seebeck effect, **b** the Peltier effect, and **c** the Thomson effect

as: $\pi_{ab} = \frac{dQ/dt}{I}$, where Q is heat energy and t is time. Finally, Lord Kelvin proposed the third effect, the Thomson effect, which accounts for the reversible absorption or release of heat along a conductor when an electric current flows through it in the presence of a temperature gradient (Fig. 1c). The Thomson coefficient (β) is expressed as: $\beta = \frac{dQ/dt}{I\Delta T}$. These three effects are interconnected, and their relationships are described as follows: $\pi_{ab} = S_{ab}T$ and $\beta = \frac{dS}{dT}T$.

The maximum heat-to-electricity conversion efficiency of a TE power generator is jointly determined by the Carnot efficiency and the constituent material's properties: $\eta_{max} = \frac{T_h - T_c}{T_h}\frac{\sqrt{1+\overline{zT}}-1}{\sqrt{1+\overline{zT}}+\frac{T_c}{T_h}}$, where T_h and T_c are the hot- and cold-side temperatures, respectively, of the device. The figure of merit zT of a TE material is a dimensionless parameter that combines the material's key properties: $zT = \frac{S^2\sigma}{\kappa_e+\kappa_l}T$, where S, σ, κ_e, and κ_l are the Seebeck coefficient, electrical conductivity, electronic thermal conductivity, and lattice thermal conductivity, respectively. The electrical properties-related term in the numerator is defined as the power factor (PF $= S^2\sigma$) as it directly determines the output power of a corresponding device. Clearly, a high zT value requires good electrical conductivity, a large Seebeck coefficient, and low thermal conductivity. Unfortunately, these parameters are highly interrelated through the Fermi level, posing an intrinsic challenge to boost the zT value, as illustrated in Fig. 2. Therefore, any performance improvements in TE materials always require a careful balancing of these parameters, as well as the leveraging of their limited synergies.

In this chapter, we first introduce the key transport properties in TE materials. A comprehensive summary of various strategies for optimizing individual parameters and the underlying mechanisms driving these improvements is provided. We also discuss the latest advancements in near-room-temperature TE power generators and cooling devices, emphasizing the importance of tailoring materials for specific application scenarios. We hope this chapter can serve as a resource for researchers and students interested in TE materials and devices by offering a clear understanding of the fundamental principles, optimization strategies, and cutting-edge advances in the field.

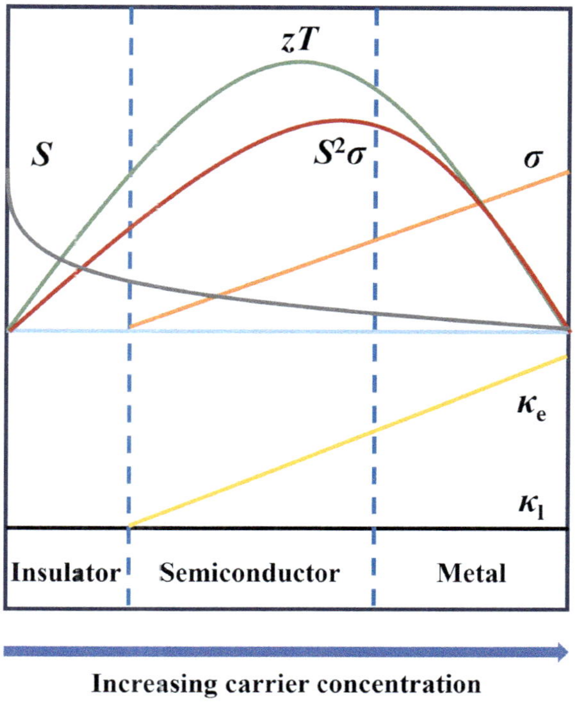

Fig. 2 Schematic illustration of the relationship between TE transport parameters and their dependence on carrier concentration

2 Material Properties

In this section, we present commonly employed strategies for optimizing the transport properties of TE materials, aiming to provide a systematic overview and insights into the strategy-structure–property relations. The discussion is divided into three parts: introducing a molecular orbital scheme for TE compounds, optimizing electronic transport properties to achieve a high power factor, and suppressing thermal transport.

2.1 A Molecular Orbital Scheme

Before introducing specific optimization strategies for each parameter, here we first provide a molecular orbital scheme based on band theory with the aim that the underlying mechanisms can be intuitively understood. Taking a polar-covalent solid MX (where M is an atom of an element with lower electronegativity and X is one with higher electronegativity) as an example, Fig. 3 illustrates a simplified model of its band formation [4, 5]. Element X, with a larger ionization energy and higher electronegativity, has a lower atomic orbital energy. The energy difference between the atomic orbitals of M and X is set as $2A$. As their atomic orbitals interact, a hybrid

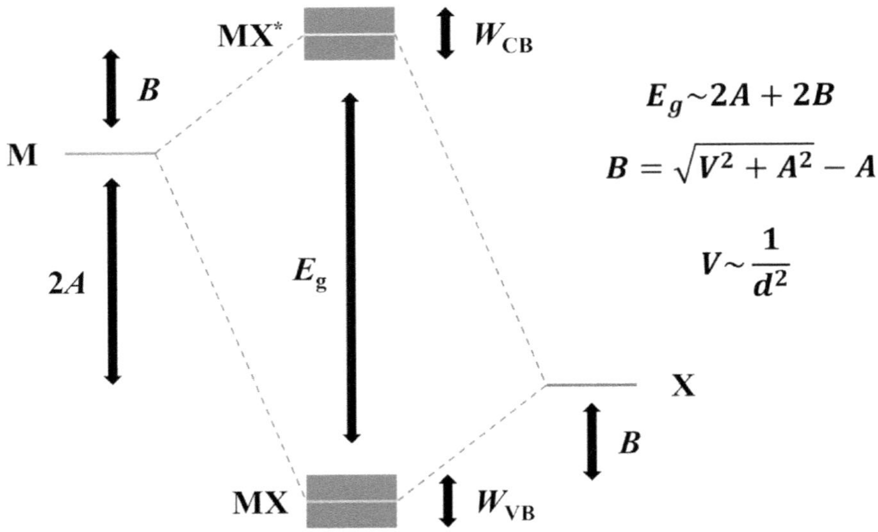

Fig. 3 Molecular orbital scheme for the band states in a polar-covalent solid MX

bonding orbital (MX) and an antibonding orbital (MX^*) form. The bonding energy (B) is determined by the stabilization of MX or the destabilization of MX^* relative to the atomic orbitals. For simplicity, MX represents the valence band and MX^* corresponds to the conduction band. The energy difference between MX and MX^* defines the band gap E_g of the compound. Clearly, as the difference in electronegativity between M and X increases, E_g becomes larger.

To describe the overlap interaction of the orbitals, we define V as the strength of the nearest-neighbor coupling, which increases as the atomic separation d decreases: $V \sim 1/d^2$. A larger orbital overlap leads to greater band dispersion, or a larger bandwidth W, which corresponds to a lower effective mass. It can generally be concluded that, as A decreases (indicating a more covalent character), the orbital overlap interaction V becomes more pronounced since neighboring orbitals are closer in energy, resulting in broader band dispersion. This explains why most TE materials are covalent compounds, as their low effective mass and resulting high mobility arise from this enhanced orbital overlap. In addition, increased bandwidth can narrow the band gap; in an extreme case, when W exceeds the sum of A and B, the material exhibits metallic or semimetallic behaviors.

We now use $Mg_3(Sb, Bi)_2$ compounds as examples to demonstrate the model introduced above. The smaller electronegativity of Bi implies its higher atomic orbital energy in comparison to Sb. Consequently, the A value for Mg–Bi is smaller than that for Mg–Sb, resulting in a larger E_g in Mg_3Sb_2. Experimental evidence confirms that intrinsic Mg_3Bi_2 behaves as a semimetal, while Mg_3Sb_2 is a semiconductor [6]. Furthermore, due to the reduced polarity of the Mg–Bi bonds and the larger atomic size of Bi, the orbital overlap V increases, leading to broader band dispersion W in

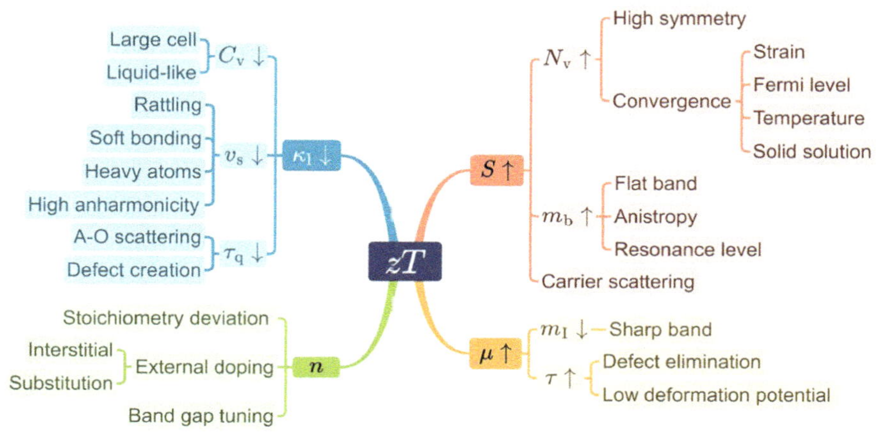

Fig. 4 Summary illustration of TE transport property optimization strategies. See the discussion for definitions of variables

the energy space, corresponding to a steeper parabolic shape in *k*-space and thus a smaller band effective mass. Experimental studies confirm a higher density-of-states effective mass of ~1.2 m_e in n-type Mg_3Sb_2 compared to ~0.7 m_e in Mg_3Bi_2 [7].

It is important to note that, in real compounds, the correlations between the parameters discussed above are often complex due to potential intricate hybridization effects, making a simple bonding scheme insufficient for completely accurate predictions. Nevertheless, this analysis does offer valuable insights and can serve as a foundational guide for understanding the optimization strategies outlined below, as well as for uncovering structure–property relationships in TE materials.

Before embarking on a systematic discussion of various strategies to optimize TE transport properties, Fig. 4 provides a summary illustration to help readers better comprehend the logic underlying these strategies.

2.2 *Enhancing the Seebeck Coefficient*

Achieving a maximum zT value requires an optimal positioning of the Fermi level, leading to an optimal Seebeck coefficient. For typical TE materials governed by the acoustic phonon scattering mechanism, it can be derived that the ideal absolute Seebeck coefficient ranges from 170 to 220 μV/K. This range reflects a partly degenerate electronic structure, where the Fermi level is close to the band edge. However, in superior TE materials with exceptionally high zT, the optimal Seebeck coefficient tends to be significantly larger [8]. Such materials behave more like non-degenerate semiconductors with much lower carrier concentrations and greater carrier mobility.

According to Bethe-Sommerfeld's equation, enhancing the Seebeck coefficient depends on increasing the energy dependence of either carrier concentration (n) or carrier mobility (μ): $S = \frac{\pi^2}{3}\frac{k_B}{e}(k_B T)\left[\frac{1}{n(E)}\frac{dn(E)}{dE} + \frac{1}{\mu(E)}\frac{d\mu(E)}{dE}\right]_{E=E_F}$, where k_B is the Boltzmann constant, e is the electron charge, and E_F is the Fermi energy [9]. The first term within the square brackets, representing the potential energy component, can be enhanced by locally increasing the density-of-states effective mass $m_d{}^*$, which is determined by band degeneracy (N_v) and band effective mass m_b as: $m_d^* = N_v^{2/3} m_b$. Accordingly, increasing the band effective mass by locally enhancing the density of states is advantageous and typically achieved through mechanisms such as introducing a resonance level or flattening the band. In addition, identifying materials with inherently high band degeneracy (e.g., high-symmetry compounds) or promoting band convergence can result in a significantly enhanced Seebeck coefficient. The second term within the square brackets, the kinetic energy component, can be optimized by tuning carrier scattering mechanisms or creating energy barriers to preferentially scatter low-energy carriers.

Achieving band convergence (i.e., increasing N_v), can be realized through various approaches, including temperature modulation, deepening of the Fermi level, and solid solution formation. For solid solution-induced band convergence, it is crucial to first determine the dominant contributions of the up-lying and down-lying bands. Once these band characteristics are identified, their positions can be tailored by introducing specific perturbations to the band structure through compositional modifications. The choice of substitution atoms can be guided by factors such as electronegativity, electronic configuration, and atomic size, as previously described. However, it is crucial to recognize that the positive effects of band convergence may be diminished if carriers experience strong intervalley scattering, which becomes particularly pronounced when the bands are aligned at the same k point.

2.3 *Optimizing Carrier Concentration*

Electrical conductivity is primarily influenced by carrier concentration and carrier mobility based on the equation: $\sigma = ne\mu$. Mixed conduction behavior is commonly observed in TE materials at high temperatures due to their small band gap, and such bipolar conduction adversely impacts the Seebeck coefficient and increases the total thermal conductivity of these materials, thereby decreasing their overall zT. Consequently, here we only focus on materials with single-type carriers, either holes (p-type) or electrons (n-type). As shown in Fig. 2, efficient TE materials are highly doped semiconductors or semimetals, with carrier concentrations in the range of 10^{18}–10^{21} cm^{-3}. As such, optimizing the carrier concentration is critical for achieving a high zT. Such optimization is commonly accomplished through stoichiometry adjustment, external doping, and band gap tuning.

Intrinsic vacancies or interstitial atoms in an as-prepared compound can influence its carrier concentration. For a chemical compound MX as described in Sect. 2.1, a vacancy at the M site (V_M), an interstitial X atom (X_i), and substitution of an X atom by an M antisite defect (M_X) can each introduce holes, leading to p-type conduction, while a vacancy at the X site (V_X), an interstitial M atom (M_i), and substitution of a M atom by a X antisite defect (X_M) can each introduce electrons, leading to n-type conduction. Due to the presence of these intrinsic defects, a prepared TE compound with stoichiometric composition typically exhibits a specific conduction type, and the dominant carriers may shift over a range of temperatures due to thermal excitation or defect evolution. For instance, a prepared Mg_3Sb_2 sample tends to be p-type conductive at room temperature due to the presence of V_{Mg} resulting from Mg loss throughout the synthesis process. Similar behavior is observed in other materials, such as Zn_4Sb_3 (V_{Zn}), GeTe (V_{Ge}), and $AgSbTe_2$ (V_{Ag}). In practice, adjusting the stoichiometry may allow for controlling the type of conduction and the carrier concentration. In the case of Mg_3Sb_2, by adding excess Mg or annealing the sample in Mg vapor, V_{Mg} can be effectively filled, switching the sample to be n-type conductive, which typically offers better TE performance due to the higher band degeneracy in the conduction band. However, it is not always feasible to switch conduction types by simply adding more of a specific element, which might be due to the fact that certain vacancies may have much lower formation energies that would pin the Fermi level close to one end, disabling the compensation and doping. For example, the synthesis of n-type $CaZn_2Sb_2$ has not yet been reported, despite the high *PF* predicted. In such cases, rather than filling the vacancy, the stoichiometry deviation may tend to induce a secondary phase. Therefore, exploring more strategies beyond simple molar ratio adjustment or introducing ultrahigh alloying content to achieve controllable conduction type remains imperative. Before employing appropriate interventions, the mechanisms that hinder favorable doping must first be identified. Various post-treatment methods, nonequilibrium synthesis, or aggressive doping techniques are then expected to break these limits to achieve an optimal doping level.

External doping is a widely adopted method for modulating carrier concentration to achieve an optimal level within a specific temperature range. Based on the doping sites, it can be classified as substitutional or interstitial doping (modulation doping is not discussed here). Although the doping ratio in most TE materials for near-room-temperature applications is typically very low, usually less than 0.5 at%, its impact on carrier transport can be significant beyond merely affecting carrier concentration. This influence arises because the introduction of dopants may substantially disrupt carrier transport, particularly when the dopant is introduced at sites with orbitals that dominate carrier transport. For a TE compound MX, the conduction band and valence band are predominantly contributed by the M and X elements, respectively. Consequently, for p-type (n-type) conduction, doping at the X (M) site tends to cause greater disruption to carrier mobility compared to doping at the M (X) site.

Even for an identical doping site, the resulting TE properties, including carrier density and carrier mobility, as well as the dopant's solubility, may vary significantly depending on the choice of doping element, even within the same group. This can be rationalized by three main aspects: differences in the substitutional defect formation energy, the formation energy of the secondary phases consisting of dopant and host elements, and the chemical bonding strength or charge transfer between the dopant and the host. The concepts of electronegativity, ionic radius, valence state, and coefficient of diffusion may provide some guidance in the dopant choice. For example, Zhang et al. investigated the effect of chalcogen dopants Q (Q = S, Se, and Te) in Mg_3Sb_2, and found that a smaller electronegativity difference between Mg and Q led to higher carrier density and mobility [10]. This improvement was attributed to the lower formation energy of Q_{Sb} defects and the enhanced orbital overlap. In another study focusing on ZrNiSn and NbFeSb half-Heusler compounds, it was observed that a smaller ionic radius difference between the dopant and host atoms is crucial for achieving high carrier mobility in the doped samples [11].

Due to the limited space within the unit cell, the amounts and types of elements available for interstitial doping are highly constrained, yet their effects can be substantial. A notable example is the Cu-doped n-type Bi_2Te_3, in which Cu occupies the interstitial site and significantly enhances carrier mobility from 173 $cm^2\ V^{-1}\ s^{-1}$ in the pristine sample to 245 $cm^2\ V^{-1}\ s^{-1}$ despite a threefold increase in carrier density in the doped sample [12]. Xu et al. categorized interstitial dopants in TE materials based on their effects on carrier transport and phonon scattering [13]. For carrier transport, interstitial elements are categorized as interlayer interstitials (Cu_i or Ag_i in Bi_2Te_3 and Cu_i in $SnSe_2$), dynamic interstitials (Cu_i or Ag_i in PbQ), and liquid interstitials (Zn_i in β-Zn_4Sb_3). For phonon transport, they can act as rattling interstitials (Ti_i, In_i, In_i-Ce_i, or Sr_i-Ba_i-Yb_i-In_i in $CoSb_3$) or decoupling interstitials (Cu_i, Ag_i, Zn_i, Ni_i, or Al_i in PbQ and Cu_i in GeTe). Despite these promising effects, interstitial doping presents two primary challenges: (1) it is relatively difficult to identify the exact presence and quantify the amount of interstitial dopant and (2) many interstitial atoms are small and can dynamically diffuse out of the interstitial site at elevated temperatures. Experimentally, the latter often manifests as an increase in electrical conductivity after a thermal cycle, accompanied by gradual performance degradation upon cooling. Consequently, it is important to gain deeper understanding of the temperature-dependent structural evolution in, and improve the stability of, interstitial-doped materials.

It should be noted that the above discussion assumes an ideal scenario of complete doping into the matrix without the formation of a secondary phase. However, dopant-induced elemental aggregation near the grain boundary and the formation of nano-precipitates can also significantly impact carrier transport behaviors for decoupling electron and phonon transport in some cases.

2.4 Improving Carrier Mobility

A material's carrier mobility is governed by its effective mass and the scattering process according to $\mu = \frac{e<\tau>}{m_I}$, where $<\tau>$ is the average relaxation time and m_I is the inertial effective mass. We first look into the effect of band structure on carrier mobility. Clearly, a small effective mass is favorable for high mobility. Therefore, sharpening the band is a natural approach to improve mobility, despite its negative effect on the Seebeck coefficient. It has been demonstrated that a lighter band is generally favorable for the overall improvement in zT given an optimized carrier concentration and the absence of other effects such as strong carrier scattering. As stated previously, an enhanced orbital overlap stemming from shorter bonds and a low degree of polarity could lead to a smaller band effective mass and thus high carrier mobility. In some cases, higher crystal symmetry can significantly enhance carrier mobility by increasing the delocalization of electronic states. For example, in SnSe, a phase transition from the low-symmetry *Pnma* phase to the high-symmetry *Cmcm* phase results in improved mobility [14]. Another approach leverages band anisotropy, where a high degree of band anisotropy can result in a large ratio between the band effective mass and the inertial effective mass, simultaneously enabling a high Seebeck coefficient and elevated carrier mobility. Examples of materials benefiting from this complex Fermi surface effect include Bi_2Te_3, PbTe, and Mg_3Sb_2.

Important scattering mechanisms for carrier transport include acoustic phonon scattering, ionized impurity scattering, alloy scattering, grain boundary scattering, and polar optical phonon scattering. Under different scattering mechanisms, the temperature dependence of a material's mobility varies significantly. In practice, mixed scattering behaviors are frequently observed in TE materials, making quantitative analysis of a specific scattering mechanism complicated. When acoustic phonon scattering dominates, reducing the strength of phonon-electron coupling in a material (characterized by a lower deformation potential) can effectively minimize carrier scattering caused by lattice vibrations. A high carrier mobility can be achieved by minimizing multi-scale defects, such as point defects, dislocations, grain boundaries, and nanoprecipitates, through precise control of composition, processing, and post-treatment. For instance, transition-metal doping effectively reduces Mg vacancies in n-type Mg_3Sb_2, boosting its carrier mobility near room temperature [7]. Additionally, growing single-crystalline samples and long-term annealing can reduce compositional inhomogeneity and other defects, thereby eliminating carrier scattering that would result from these imperfections.

2.5 Reducing Lattice Thermal Conductivity

In experiments, lattice thermal conductivity is often obtained by subtracting electronic lattice thermal conductivity from total thermal conductivity (κ): $\kappa_l = \kappa - \kappa_e$, assuming the absence of bipolar conduction. The electronic thermal conductivity is typically given by the Wiedemann–Franz law: $\kappa_e = L_{Lorenz}\sigma T$, where L_{Lorenz} is the Lorenz number.

As illustrated in Fig. 2, lattice thermal conductivity κ_l stands out as the only relatively independent parameter in TE materials, meaning that reducing κ_L minimally affects other properties, thereby leading to an improvement in zT. However, it is misleading to assume that an ultra-low κ_l would automatically yield a maximum zT. First, significant reductions in κ_l typically involves introducing various phonon scattering centers, which can also scatter charge carriers. Second, theoretical derivations and experimental data consistently show that the κ_l/κ ratio for most current TE materials remains around 0.5 [15]. Nonetheless, reducing κ_l offers a straightforward and effective strategy for enhancing zT.

In lattice conduction, heat transport occurs through the vibrations of the crystal lattice, referred to as phonons. A simplified kinetic expression for lattice thermal conductivity is $\kappa_l = \frac{1}{3}C_V v_s l = \frac{1}{3}C_V {v_s}^2 \tau_q$, which involves several parameters: the specific heat capacity (C_V); sound velocity (v_s); the phonon mean free path (l); phonon relaxation time (τ_q).

Specific heat represents the thermal energy stored within a solid and is generated by both acoustic and optical phonons. In a crystalline solid at very low temperatures (significantly below the Debye temperature), the specific heat capacity scales with temperature cubed, $C_V \sim T^3$. At high temperatures, it approaches a constant value following the Dulong-Petit law. However, in certain TE materials with soft bonding, the specific heat data at low temperatures show significant deviation from the Debye model and require additional Einstein modes for fitting, suggesting an unneglectable contribution from low-frequency optical phonons. In general, materials with large unit cells containing many atoms exhibit lower heat capacity since fewer acoustic phonons contribute to heat conduction. Furthermore, materials with liquid-like conduction behavior, such as Cu_2Te and Cu_2Se, display an unexpectedly low specific heat due to the suppression of their transverse vibrational modes [16].

Under the assumption of acoustic phonon scattering, the lattice thermal conductivity can be expressed as $\kappa_l \sim \frac{M v_s^3}{T \Omega^{2/3} \Upsilon^2}\left(\frac{1}{N^{1/3}}\right)$, where M, v_s, Ω, and γ are the average mass, mean speed of sound, average volume per atom, and Grüneisen parameter, respectively [4]. Clearly, a decreased sound velocity (v_s) contributes to a lower κ_l , where v_s is related to bonding strength and atomic mass following $v_s \sim \sqrt{k/M}$, where k represents the bonding strength. Thus, heavier atoms and weak bonding are beneficial for obtaining a low κ_l. For example, skutterudites show a low sound velocity due to flattened phonon dispersion relations and resonant scattering induced by their large and heavy filler atoms. It should be noted that a higher coordination number often results in looser atomic packing and increased bond lengths because of greater electrostatic repulsion. In addition to bonding strength, high anharmonicity in lattice vibrations stemming from asymmetric bonding or local lattice distortions, represented by a large Grüneisen parameter, also leads to lower κ_l. Experiments show that compounds with lone-pair electrons, such as ABX_2 (groups I–V–VI_2) compounds, often exhibit large Grüneisen parameters [16]. The much higher Grüneisen parameter of 1.83 in Mg_3Sb_2 compared to 1.44 in its ternary analog $CaMg_2Sb_2$, despite Mg having a lighter atomic mass, is ascribed to the unstable Mg-Sb bonding in Mg_3Sb_2 that significantly reduces sound velocity and expands the phonon scattering phase space [17].

The phonon mean free path in TE materials can be significantly reduced by various phonon scattering mechanisms, including phonon–phonon Umklapp scattering (U), point defect scattering (PD), grain boundary scattering (GB), dislocation scattering (DIS), and electron–phonon scattering (E-P). The phonon relaxation time is collectively influenced by these scattering processes: ${\tau_q}^{-1} = {\tau_U}^{-1} + {\tau_{PD}}^{-1} + {\tau_{GB}}^{-1} +$

$\tau_{DIS}{}^{-1} + \tau_{E\text{-}P}{}^{-1} + \ldots$. Phonon–phonon Umklapp scattering, based on the anharmonicity in the lattice potential, has a scattering probability that increases with the square of the phonon frequency (ω^2) and linearly with temperature at high temperatures, resulting in $\kappa_l \sim T^{-1}$. However, this linear trend is not always observed in complex TE materials like Mg_3Sb_2 due to their unique temperature-dependent structure and phonon dispersion relations [18]. With decreasing temperature, the influence of static defects on phonon scattering becomes more prominent. Point defect scattering, often achieved through the introduction of mass and strain fluctuations, is a time-honored strategy for enhancing phonon scattering in many TE materials. It is particularly effective at scattering high-frequency phonons due to its ω^4 dependence. In isovalent alloy systems, lattice thermal conductivity can be significantly lowered because of the high solubility limit, as shown in Si-Ge, Bi-Sb, and Mg_3Sb_2-Mg_3Bi_2. Although aliovalent doping typically involves a much lower dopant amount to ensure optimized carrier concentration, it can lead to an unexpectedly low lattice thermal conductivity when the dopant significantly alters the phonon dispersion and reduces sound velocity, which is simply lattice softening. Low-frequency phonons are especially sensitive to the rattling effects seen in materials with loosely bound atoms, such as skutterudites and clathrates. Ideal grain boundary scattering is independent of phonon frequency but largely depends on grain size. However, when the grain-boundary misfit angle is less than 15°, grain boundaries can strongly interact with low-frequency phonons over a wide temperature range, making them particularly effective for reducing thermal conductivity at low temperatures. Experiments on various materials have shown that nanostructuring can enable deliberate phonon scattering while preserving carrier mobility due to the longer mean free paths of phonons compared to charge carriers [19]. In general, a full-spectrum phonon scattering is desirable, and systematically quantifying the contributions of different scattering sources in a specific composition would provide deeper insight into these scattering processes.

3 Power Generators and Coolers for Near-Room-Temperature Applications

The discovery of Bi_2Te_3 in the 1950s catalyzed a surge in research to develop TE cooling devices. However, it soon became evident that its low efficiency and high cost limited its large-scale application. In recent decades, concerns over the environmental impact of massive fossil fuel consumption have intensified interest in improving energy efficiency, with TE power generation using waste heat emerging as a promising approach. Moreover, TE generators have no moving parts and do not require maintenance. Despite continuous progress toward improving zT values, the market for heat recovery using TE technology at high temperatures remains in its infancy. Consequently, recent research efforts have been focused on enhancing the near-room-temperature performance of TE materials to improve cooling systems or to recover low-grade heat (<523 K).

In this section, we first introduce the key parameters that characterize the performance of TE power generators and coolers, followed by a detailed discussion of the material factors influencing these parameters. We subsequently present a summary and comparison of state-of-the-art TE compounds, highlighting those with potential for next-generation devices. Rather than detailing recent optimization efforts for each compound, we focus on their suitability for different devices and outline the remaining challenges.

For TE power generators, two crucial parameters are conversion efficiency and output power density (ω). As previously discussed, conversion efficiency depends on the zT values of constituent materials, while output power density is expressed as a function of PF: $\omega_{\max} = \frac{f(T_h - T_c)^2}{4L}$PF, with high PF ideally ensuring a high output power density at a given temperature difference.

Figure 5a, b highlights TE materials with zT values exceeding 1.0 within the temperature range of 300–523 K, such as p-Bi_2Te_3, p-MgAgSb, p-$AgSbTe_2$, p-AgCuTe, p-SnSe, n-Bi_2Te_3, n-$Mg_3(Bi, Sb)_2$, n-Ag_2Se, and n-PbSe, as promising candidates in this range. Considering material cost and toxicity, p-Bi_2Te_3, p-SnSe, n-Bi_2Te_3, and n-$Mg_3(Bi, Sb)_2$ emerge as particularly viable for large-scale applications, suggesting that further studies on these materials are essential for developing high-η TE generators. Nevertheless, exploring alternative materials remains valuable. Breaking performance records in these materials not only promotes their use in cost-insensitive applications, such as aerospace exploration or micro-scale cooling devices where manufacturing costs are predominant, but also provides unique insights for designing new materials.

Alternatively, if prioritizing output power over conversion efficiency as the key criterion, high-PF materials should be the focus. For a TE generator with dimensions of 4×4 cm^2 and a fill factor of 35%, achieving a target output power of 5 W (ω = 0.3 W cm^{-2}) under a cold-side temperature of 323 K and a hot-side temperature of 473 K requires an average PF of approximately 35 μW cm^{-1} K^{-2} for the TE materials (accounting for both p- and n-type materials and temperature dependence). This estimate assumes ideal conditions without interfacial resistance or side heat loss, indicating that an even higher PF may be necessary.

NbFeSb, SnSe, and Bi_2Te_3 are p-type materials that meet the above PF criterion, with NbFeSb showing particular promise (Fig. 5c). For n-type conduction, materials like $YbAl_3$, CuNi, Mg_2Sn, $CoSb_3$, TiNiSn, and PbSe qualify (Fig. 5d). For maintaining the structural integrity of a device during numerous thermal cycles, its constituent p- and n-type materials should exhibit similar coefficients of thermal expansion (CTE). Thus, when using NbFeSb as p-type material, it can be paired with a n-type material such as $CoSb_3$, Mg_2Sn, TiNiSn, or $YbAl_3$ for a robust, high-ω, and decent-η TE generator. Recently Zhu et al. demonstrated an impressive power density of 1.8 W cm^{-2} and a decent conversion efficiency of 3% with a unicouple comprising p-type $Nb_{0.55}Ta_{0.40}Ti_{0.05}FeSb$ and n-type $Zr_{0.44}Hf_{0.44}Ti_{0.12}NiSn_{0.99}Sb_{0.01}$ at a temperature difference of approximately 180 K [20]. A high output power of 10

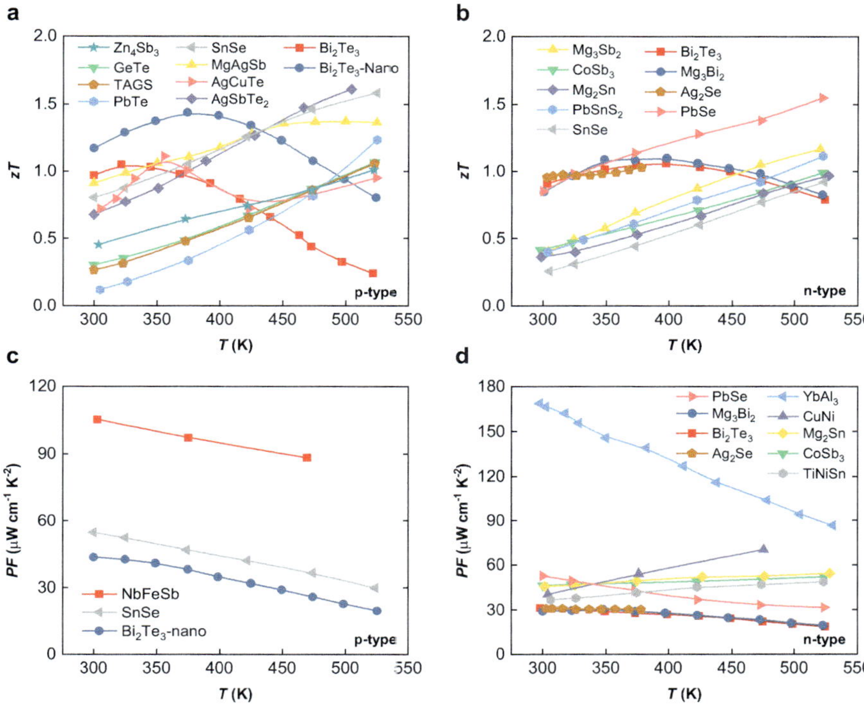

Fig. 5 Temperature-dependent zT of state-of-the-art **a** p-type [14, 19, 21–27] and **b** n-type [12, 28–35] TE materials for near-room-temperature applications. Temperature-dependent PF of **c** p-type [14, 19, 36] and **d** n-type [12, 29–31, 33, 35, 37–39] TE materials for developing next-generation high-power generators

W is predicted from this unicouple under the same size, fill factor, and temperature conditions specified above. Overall, there is significant potential in studying these materials to develop high-output TE generators that can replace commercial Bi_2Te_3 devices.

Beyond optimizing TE properties, developing a high-performance TE generator requires attention to several additional critical factors, such as the contact layer between the TE materials and the electrodes, as well as the stability and mechanical robustness of the constituent materials. Additionally, in device design, performance estimations often assume a constant temperature difference, but a constant heat flow is more realistic for practical applications. Exploring device performance under this condition is also an active research area.

We now shift our focus to TE coolers, the performance of which is characterized by the maximum temperature difference ($\Delta T_{\max}$), the coefficient of performance (COP), and cooling density (q_c) [6, 40]. The $\Delta T_{\max}$ of material is primarily determined by its zT and the operation temperature: $\Delta T_{\max} = T_h - \frac{\sqrt{1+2zT_h}-1}{z}$. As shown in Fig. 6a, p- and n-type Bi_2Te_3 alloys can currently deliver $\Delta T_{\max}$s of 89 K and 76 K, respectively, at a hot-side temperature of 323 K, which explains why most commercial TE devices reach a maximum temperature difference of 70–80 K. With a given 50% enhancement in the *zT* of p- and n-type Bi_2Te_3 compounds, the maximum temperature differences could reach 110 K and 96 K, respectively, potentially enabling a ΔT_{max} of 100 K at the device level. Given that p-type Bi_2Te_3 surpasses other materials around 323 K (Fig. 5a), ongoing research to further improve its performance remains highly important, despite intensive efforts to optimize this alloy over the past six decades. Among n-type materials, options such as PbSe, Mg_3Bi_2, and Ag_2Se demonstrate performance comparable to n-type Bi_2Te_3, offering a range of promising candidates for future breakthroughs in efficient cooling.

For a TE cooler, the COP represents the ratio of heat pumped to the electrical energy applied, a measure of its efficiency. The maximum COP, which depends directly on the *zT* of the constituent materials according to COP_{max} =

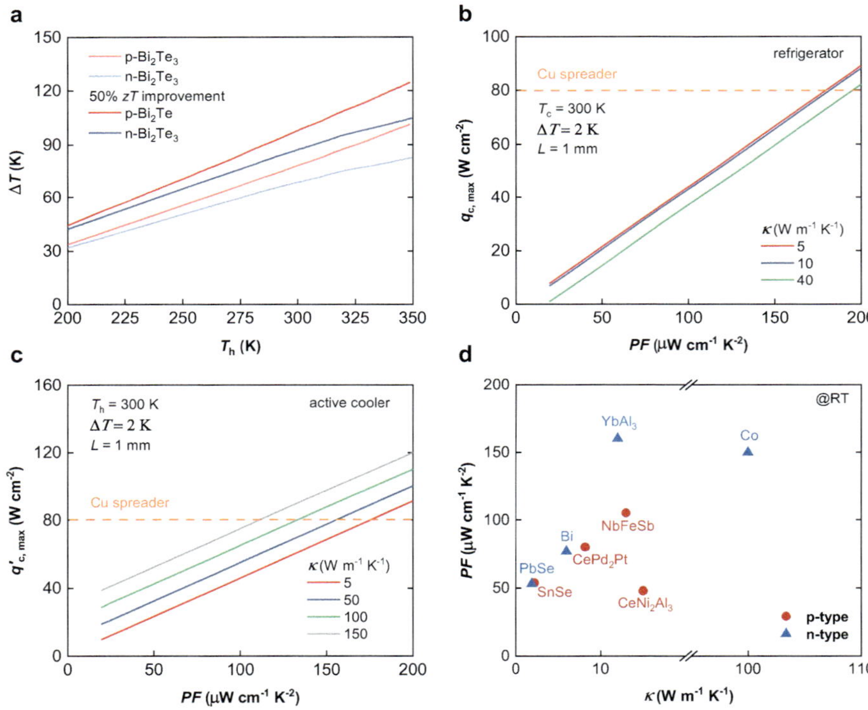

Fig. 6 **a** Maximum temperature difference as a function of hot-side temperature for commercial Bi_2Te_3 alloys and hypothetical Bi_2Te_3 alloys with a 50% increase in *zT*. Calculated maximum cooling density as a function of power factor in a TE **b** refrigerator and **c** active cooler for different thermal conductivity values. Dashed line: maximum cooling density of a copper plate for comparison. **d** Benchmark high-power factor TE materials and the total thermal conductivity of each at room temperature [34–37, 42–45]

$\frac{T_c}{T_h-T_c}\frac{\sqrt{1+\overline{zT}}-T_h/T_c}{\sqrt{1+\overline{zT}}+1}$, is limited by the relatively low zT values achieved in existing TE materials. As a result, TE refrigerators generally have a COP of around 1, significantly lower than vapor compression systems, which can easily reach a COP of 4. This disparity makes TE refrigeration less economically viable for large-scale cooling applications. However, increasing COP is still meaningful for using TE cooling in small-scale temperature control, and a 100% improvement in the average zT of current TE materials could lead to a COP of 2.

Another critical parameter for TE refrigeration is q_c, which quantifies the amount of heat pumped per unit area. The maximum heat pumping rate is given by $q_{c,\,max} = \frac{\mathrm{PF}T_c^2}{2L} - \frac{\kappa \Delta T}{L}$, where L is the TE leg length. Clearly, high $q_{c,\,max}$ can only be achieved under a small ΔT, making it particularly relevant for hot-spot cooling. In such a scenario, the cooling power depends heavily on the material's PF and decreases only slightly with high κ. Given a TE refrigerator with a leg length of 1 mm, the relationship between cooling density, power factor, and thermal conductivity is illustrated in Fig. 6b for a small ΔT of 2 K and a T_c of 300 K. Notably, cooling density increases sharply with increasing power factor, while the impact of thermal conductivity is modest. However, comparing the cooling density of a TE refrigerator to that of a copper heat spreader ($\kappa = 400$ W m^{-1} K^{-1}) with the same dimensions reveals that achieving similar performance would demand an ultrahigh room-temperature PF of around 200 μW cm^{-1} K^{-2}, while larger temperature differences would exacerbate this gap. Assuming a fill factor of 50% for the TE refrigerator, the PF value should be doubled. Unfortunately, no current materials meet this criterion. Also, using a TE cooler for hot-spot cooling would introduce additional Joule heating, which requires external heat management.

To address this, Zebarjadi proposed a TE active cooler, wherein active cooling via the Peltier effect and passive cooling via natural heat conduction jointly work to transport heat from the hot side to the cold side [41]. In this case, the cooling density is enhanced with increases in both the power factor and thermal conductivity of the constituent materials according to $q'_{c,\,max} = \frac{\mathrm{PF}T_h^2}{2L} + \frac{\kappa \Delta T}{L}$. We can then simply define a term denoted as effective thermal conductivity: $\kappa_{\mathrm{eff}} = \frac{\mathrm{PF}T_h^2}{2\Delta T} + \kappa$ to characterize the active heat pumping ability of a material, which contrasts with the conventional requirement of decoupling electron and phonon transport. This shift seems to allow for greater freedom in searching for efficient TE materials. Nevertheless, even at a small temperature difference of 2 K, a TE material with a room-temperature power factor of 150 μW cm^{-1} K^{-2} would require a thermal conductivity of at least 80 W m^{-1} K^{-1} to surpass the cooling density provided by a copper plate (Fig. 6c). While Co readily meets this standard (Fig. 6d), its cooling density would still not be comparable when taking the fill factor of a TE cooler into account. Optimizing TE leg geometry, such as by incorporating heat convection channels or inserting high-κ materials into the active cooler's empty spaces, could offer incremental improvements, but material innovations remain necessary. In the future, exploring other materials including intermetallic compounds, strongly correlated systems, one-dimensional conductors, and metallic carbon-based compounds may hold promises for advancing this field. However, from a commercialization standpoint, the competitiveness of TE active coolers remains uncertain when compared to thermal management solutions utilizing graphite foil, copper foil, and vapor chambers, which are commonly used in 5G mobile phones.

It is worth mentioning that additional effects, such as the electron expansion effect and the Thomson effect, can combine with the Peltier effect to enhance cooling performance in certain materials by taking advantage of their unique external field response or structural evolution [46, 47]. However, further research is required to develop efficient materials leveraging these coupling effects that can rival the current state-of-the-art Peltier materials.

4 Summary

Over the past few decades, significant advancements have been made in tuning the transport properties of TE materials and in developing robust TE devices. We have provided a comprehensive summary of carrier and phonon transport in TE materials and the strategies used to tune these properties. Performance improvements are generally driven by two factors: first, a deep understanding of the bonding-structure–property relationships in materials, enabling the employment of rational optimization strategies; and second, the design of appropriate microstructures, effectively decoupling electron–phonon transport. Recent advancements in developing competitive room-temperature performance in materials that can replace commercial Bi_2Te_3 alloys suggest the near-term potential of next-generation TE generators and coolers with improved conversion efficiency, output power density, temperature difference, and cooling density. Several promising materials central to these advances have been discussed. Continuous efforts to discover new materials and push the boundaries of their properties are essential for advancing their application. With an ever-expanding toolkit of property optimization strategies and insights drawn from diverse research fields, we are optimistic about the future impact of TE technology in fostering a sustainable society for all.

Acknowledgements The authors would like to thank Mr. Troy J. Christensen at TcSUH for constructive and thorough revisions. Zhongxin Liang thanks Rui Shu, Yu Wang, and Congcong Xu for the discussion.

Problems

1. For a degenerated TE material with an electrical resistivity of $5.2 \times 10^{-5}\ \Omega$ m and a total thermal conductivity of 0.75 W m^{-1} K^{-1} at 773 K, calculate the ratio of lattice thermal conductivity to total thermal conductivity. Assume a Lorenz number of 1.5×10^{-8} W Ω K^{-2} and no bipolar conduction.

2. The lattice thermal conductivity of a TE material is 1.8 W m^{-1} K^{-1} at 423 K and 1.4 W m^{-1} K^{-1} at 523 K. Estimate the lattice thermal conductivity at 673 K if phonon-phonon scattering dominates thermal transport.
3. Can you explain why nanostructuring might help to improve the zT value?

Solutions

1. $\kappa_e = \frac{LT}{\rho} = \frac{1.5\times10^{-8}\times773}{5.2\times10^{-5}} = 0.22\ (\mathrm{W\,m^{-1}\,K^{-1}})$,
   $\kappa_l = \kappa - \kappa_e = 0.75 - 0.22 = 0.53\ (\mathrm{W\,m^{-1}\,K^{-1}})$,
   $\frac{\kappa_l}{\kappa} = \frac{0.53}{0.75} = 0.71 = 71\%$.
2. When phonon–phonon scattering dominates thermal transport, the lattice thermal conductivity decreases linearly with temperature, suggesting $\kappa_l \times T = \text{constant}$. At 423 K, the constant is $1.8\times423 = 761.4$. At 523 K, the constant is $1.4\times523 = 732.2$. Take the average: $\frac{761.4+732.2}{2} = 746.8$. Therefore, when T is 673 K, we can get the lattice thermal conductivity: $\kappa_l = 746.8/673 = 1.1\ (\mathrm{W\,m^{-1}\,K^{-1}})$.
3. Nanostructuring can be advantageous for suppressing thermal transport while enhancing carrier transport, thereby improving the zT. First, nanostructuring introduces a large number of grain boundaries, which effectively scatter low- and mid-frequency phonons, thus reducing lattice thermal conductivity. Second, interfacial energy barriers or quantum confinement effects can enhance the Seebeck coefficient by modifying the energy distribution of carriers. Finally, since the mean free path of electrons is typically much smaller, electron transport is less affected by grain boundaries, preserving high electrical conductivity.

References

1. Goldsmid HJ (2010) Introduction to thermoelectricity. Springer, Berlin
2. Anatychuk L, Burkov A, Goldsmid J, Grin Y, Koumoto K, Narducci D, Nolas GS (2024) 200 years of thermoelectricity. Springer, Cham
3. Xu C, Liang Z, Song S, Ren Z (2022) In digital encyclopedia of applied physics. Wiley-VCH Verlag GmbH & Co. KGaA
4. Zeier WG, Zevalkink A, Gibbs ZM, Hautier G, Kanatzidis MG, Snyder GJ (2016) Angew Chem Int Ed 55:6826
5. Tang G, Ghosez P, Hong J (2021) J Phys Chem Lett 12:4227
6. Mao J, Zhu H, Ding Z, Liu Z, Gamage GA, Chen G, Ren Z (2019) Science 365:495
7. Shang H, Liang Z, Xu C, Mao J, Gu H, Ding F, Ren Z (2020) Research 2020:1219461
8. Zhi S, Jia J, Zhang Q, Cao F, Liu X, Mao J (2022) Mater Today Phys 22:100618
9. Mao J, Liu Z, Zhou J, Zhu H, Zhang Q, Chen G, Ren Z (2018) Adv Phys 67:69
10. Zhang J, Song L, Borup KA, Jørgensen MRV, Iversen BB (2018) Adv Energy Mater 8:1702776
11. Zhou J, Zhu H, Song Q, Ding Z, Mao J, Ren Z, Chen G (2022) Nat Commun 13:2482
12. Liu WS et al (2011) Adv Energy Mater 1:577
13. Xu L, Yin Z, Xiao Y, Zhao LD (2024) Adv Mater 36:e2406009
14. Qin B, Wang D, He W, Zhang Y, Wu H, Pennycook SJ, Zhao LD (2019) J Am Chem Soc 141:1141

15. Song S, Xu C, Liang Z, Ren Z (2021) Appl Phys Lett 119:180501
16. Kim H, Park G, Park S, Kim W (2021) ACS Nano 15:2182
17. Ding J, Lanigan-Atkins T, Calderon-Cueva M, Banerjee A, Abernathy DL, Said A, Zevalkink A, Delaire O (2021) Sci Adv 7:eabg1449
18. Zhu Y et al (2020) Research 2020:4589786
19. Poudel B et al (2008) Science 320:634
20. Zhu H, Li W, Nozariasbmarz A, Liu N, Zhang Y, Priya S, Poudel B (2023) Nat Commun 14:3300
21. Zhao H et al (2014) Nano Energy 7:97
22. Liu Z et al (2018) Proc Natl Acad Sci U S A 115:5332
23. Zhang Y et al (2023) Adv Mater 35:e2208994
24. Jiang J, Zhu H, Niu Y, Zhu Q, Song S, Zhou T, Wang C, Ren Z (2020) J Mater Chem A 8:4790
25. Zhu Y, Hu L, Zhan S, Ina T, Gao X, Hong T, Zhao L-D (2022) Energy Environ Sci 15:3958
26. Levin EM, Bud'ko SL, Schmidt-Rohr K (2012) Adv Funct Mater 22:2766
27. Zhu G, Liu W, Lan Y, Joshi G, Wang H, Chen G, Ren Z (2013) Nano Energy 2:1172
28. Shuai J et al (2017) Energy Environ Sci 10:799
29. Meng X et al (2017) Adv Energy Mater 7:1602582
30. Mao J, Kim HS, Shuai J, Liu Z, He R, Saparamadu U, Tian F, Liu W, Ren Z (2016) Acta Mater 103:633
31. Liu Z, Gao W, Oshima H, Nagase K, Lee CH, Mori T (2022) Nat Commun 13:1120
32. Zhan S et al (2024) Adv Energy Mater 2403194
33. Jiang F et al (2024) Adv Funct Mater 2415000
34. Chang C et al (2018) Science 360:778
35. Qin Y et al (2024) Science 383:1204
36. He R et al (2016) Proc Natl Acad Sci U S A 113:13576
37. Rowe DM, Kuznetsov VL, Kuznetsova LA, Gao M (2002) J Phys D: Appl Phys 35:2183
38. Mao J et al (2015) Nano Energy 17:279
39. Ren W et al (2019) Adv Electron Mater 5:1900166
40. Mao J, Chen G, Ren Z (2021) Nat Mater 20:454
41. Zebarjadi M (2015) Appl Phys Lett 106:203506
42. Watzman SJ, Duine RA, Tserkovnyak Y, Boona SR, Jin H, Prakash A, Zheng Y, Heremans JP (2016) Phys Rev B 94:144407
43. Gallo CF, Chandrasekhar BS, Sutter PH (1963) J Appl Phys 34:144
44. Boona SR, Morelli DT (2012) Appl Phys Lett 101:101909
45. Szajek A, Kowalczyk A (2019) Appl Phys A 125:757
46. Zebarjadi M, Akbari O (2023) Entropy 25:1540
47. Zebarjadi M, Tonni FF, Yazawa K, Shakouri A (2024) Mater Today Phys 48:101561

Energy and Power Systems

Modeling and Stability Analysis of Microgrids Integrated with Power-Electronics-Interfaced Distributed Energy Resources

Yan Li

Abstract The increasing integration of power-electronics-interfaced distributed energy resources (DERs) is transforming microgrids, offering flexibility while introducing challenges in modeling, control, and stability. This chapter provides a comprehensive study of these issues, focusing on inverter-based microgrids. Microgrids are introduced with an emphasis on their key features, operational flexibility, and challenges arising from power-electronics-based generation. The mathematical modeling of inverters in the d-q frame is discussed, along with the widely adopted double-loop control strategy for regulating voltage and current dynamics. Microgrid modeling using differential–algebraic equations (DAEs) is explored, and droop control is presented as a fundamental decentralized method for power-sharing and voltage-frequency regulation. Finally, small-signal stability analysis is examined, identifying key factors affecting microgrid stability and strategies for mitigating instabilities. By integrating power electronics, control theory, and stability analysis, this chapter provides a practical framework for understanding and improving microgrid operation, offering valuable insights for researchers and engineers working on next-generation power systems.

1 Introduction of Microgrids

The stable and resilient operation of electric power grids has become a critical concern in modern power engineering, particularly as the traditional centralized, top-down energy infrastructure, dominated by large-scale fossil fuel power plants, is rapidly becoming outdated and environmentally unsustainable. This centralized architecture is inherently vulnerable to extreme events, including severe weather conditions (e.g., hurricanes, storms, wildfires, and earthquakes), equipment failures, and cyberattacks. Despite substantial investments in grid modernization, the increasing complexity

---

Y. Li (✉)
The Pennsylvania State University, University Park, PA, USA
e-mail: yql5925@psu.edu

D. Shi (ed.), *Energy Materials Science and Engineering: From New Devices to AI Power Systems*, Green Energy and Technology, https://doi.org/10.1007/978-3-031-88550-1_9

and interconnectedness of energy systems demand further advancements to ensure reliability and resilience.

Recent grid failures underscore the urgent need for improvements. In February 2021, Texas experienced one of the most severe power crises in its history due to three back-to-back winter storms [1]. The unprecedented cold weather caused widespread equipment failures, leading to the worst energy infrastructure collapse in Texas history. More than 4.5 million homes and businesses lost power for several days [2], resulting in an economic disaster and a humanitarian crisis [2]. Similarly, on July 13, 2019, a major blackout in New York City's Manhattan plunged a significant portion of the borough into darkness for up to 5 h [3]. The failure was traced to a 13 kV cable burnout at West 64th Street and West End Avenue, affecting over 73,000 customers and causing tens of millions of dollars in economic losses.

These incidents illustrate the growing vulnerabilities of conventional power grids and the urgent need for transformation. To achieve a sustainable and resilient energy future, modern grid architectures must evolve. Decarbonization has become a central objective of national policies, development strategies, and regulatory frameworks, accelerating the shift toward renewable energy-based solutions. In this context, microgrids have emerged as a promising alternative, offering localized, flexible, and resilient energy solutions (Fig. 1).

Microgrids: A Pathway to Sustainable and Resilient Energy Systems

A microgrid is a localized, small-scale power system consisting of interconnected loads, distributed energy resources (DERs), energy storage systems, and control mechanisms within a well-defined electrical boundary [4]. Microgrids can seamlessly switch between grid-connected and islanded modes, enabling independent operation when external grid failures occur [5]. Typically deployed at the distribution level,

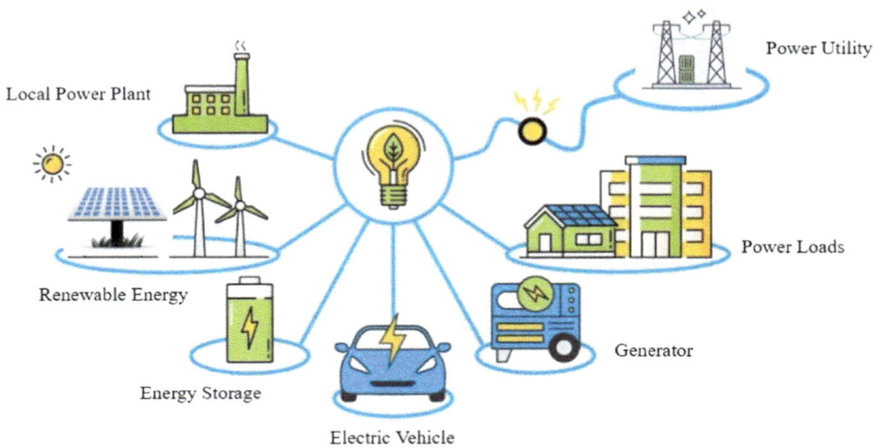

Fig. 1 Microgrid system with renewable energy

microgrids enhance grid resilience by integrating renewable energy sources such as solar and wind while maintaining reliability through advanced control strategies and energy storage.

As the energy sector transitions toward decentralized and intelligent power systems, microgrids are playing an increasingly pivotal role in improving grid flexibility, accommodating renewable integration, and advancing sustainability goals. Their ability to operate autonomously, support critical infrastructure, and provide reliable power during grid failures makes them an essential building block of the future energy landscape. By leveraging modern technologies, such as machine learning, optimization, and quantum computing, microgrids can be further enhanced to maximize their efficiency, stability, and economic viability.

Advantages of Adopting Microgrids

The adoption of microgrid technology is driven by several key factors that address the limitations of conventional power systems while enhancing reliability, sustainability, and economic efficiency. As energy systems evolve toward decentralization and resilience, microgrids provide a transformative solution capable of addressing modern energy challenges.

One of the primary motivations for microgrid deployment is grid resilience. Traditional power grids are vulnerable to natural disasters, cyberattacks, and equipment failures, often resulting in widespread blackouts. Microgrids, with their ability to operate autonomously in islanded mode, provide localized power generation that ensures continuous electricity supply even when the main grid experiences disruptions. By incorporating distributed energy resources (DERs) and energy storage, microgrids can effectively mitigate cascading failures and improve overall system stability.

Microgrids also play a crucial role in achieving sustainability goals. As the global push for decarbonization intensifies, integrating renewable energy sources such as solar, wind, and hydropower into microgrids significantly reduces greenhouse gas emissions. These systems support climate initiatives by lowering dependency on fossil fuels while promoting clean and sustainable energy use. Furthermore, advanced energy management systems within microgrids enhance efficiency by dynamically adjusting power generation and consumption based on real-time demand and weather conditions.

From an economic perspective, microgrids offer substantial cost-saving benefits. By enabling localized energy generation and storage, microgrids reduce reliance on centralized grid infrastructure, lowering transmission losses and energy costs. Additionally, demand-side management techniques, such as peak shaving and load shifting, allow consumers to optimize energy usage and minimize electricity expenses. In commercial and industrial settings, microgrids enhance energy security by preventing costly downtime due to grid failures.

Beyond urban applications, microgrids are critical for expanding energy access in remote, rural, and underserved regions. In areas where extending traditional grid infrastructure is economically infeasible, microgrids offer a reliable and cost-effective alternative for electrification. By providing stable and affordable electricity, microgrids improve healthcare, education, and economic development, fostering social and economic growth in communities that previously lacked access to modern energy services.

As energy demands continue to rise, microgrids stand out as a key enabler of next-generation power systems. Their ability to enhance resilience, integrate renewable energy, reduce costs, and expand access to electricity makes them an essential component of the future energy landscape. The ongoing advancement of smart control, optimization, and machine learning techniques will further enhance microgrid efficiency, making them a viable and scalable solution for global energy challenges.

1.1 Key Features of Microgrids

Microgrids are distinguished by several key features that enable their adaptability, resilience, and efficiency in modern power systems. Unlike conventional power grids, which rely on large-scale centralized generation, microgrids integrate DERs through power-electronic interfaces [6, 7]. These interfaces provide fast and flexible dispatch and control, allowing microgrids to dynamically regulate power flow and optimize energy use. However, these power-electronic interfaces present a double-edged challenge—while they enable high-speed control, they significantly reduce system inertia, making microgrids more susceptible to disturbances. The loss of inertia can compromise stability and resilience, particularly when microgrids operate in islanded mode, where they are disconnected from the bulk power grid [8]. As a result, stability analysis and enhancement have become critical aspects of microgrid development, especially as DERs play a growing role in the transition toward sustainable and decarbonized energy systems.

Renewable Energy Integration

A defining feature of microgrids is their ability to integrate renewable energy sources, including solar photovoltaic (PV) panels, wind turbines, and hydroelectric systems. These clean energy technologies play a vital role in reducing greenhouse gas emissions and supporting global decarbonization efforts. However, the inherent variability and intermittency of renewables present operational challenges, requiring advanced forecasting, real-time monitoring, and intelligent control mechanisms to ensure grid stability and reliability [9, 10]. The combination of power-electronic inverters, real-time optimization, and predictive analytics is essential for effectively managing renewable energy fluctuations within microgrid environments.

Decentralized Control

Microgrids can operate with decentralized control architectures [11], which reduce dependency on centralized grid operators and allow for real-time, localized decision-making. Unlike conventional power systems that rely on hierarchical control structures, microgrids use distributed control strategies, such as droop control, multi-agent systems, and consensus-based methods, to coordinate DERs, balance supply and demand, and respond dynamically to changes in load or generation. This decentralized approach enhances the flexibility and robustness of microgrid operations, making them more adaptable to evolving grid conditions.

Energy Storage Systems (ESS)

To compensate for the variability of renewable energy, battery energy storage systems (BESS) are widely deployed within microgrids [12]. These systems provide energy balancing, peak shaving, and backup power, ensuring that supply and demand remain stable. BESS also enhances frequency regulation by injecting or absorbing power as needed, thereby improving overall grid inertia in the absence of conventional rotating generators. Other forms of energy storage, such as flywheels, pumped hydro, and hydrogen storage, further contribute to improving microgrid reliability and enabling smoother renewable energy integration.

Operational Flexibility: Grid-Connected and Islanded Modes

Microgrids possess the unique capability to seamlessly transition between grid-connected and islanded modes [13], ensuring continuity of service during grid outages. In grid-connected mode, microgrids interact with the main power grid, exchanging power and participating in ancillary services, such as frequency regulation and demand response. However, during extreme events—such as hurricanes, earthquakes, or cyberattacks—microgrids can autonomously switch to islanded mode, operating independently to maintain power supply for critical loads. This operational flexibility enhances resilience and energy security, particularly in regions prone to natural disasters or grid infrastructure vulnerabilities.

With these features, microgrids represent a transformative solution for modern power systems, enabling resilient, sustainable, and decentralized energy management. The continued development of intelligent control mechanisms, advanced optimization techniques, and secure communication infrastructures will further enhance their effectiveness in shaping the future of energy distribution and consumption.

1.2 The Rise of Power-Electronics-Interfaced DERs

A significant technological advancement in microgrids is the widespread integration of power-electronics-interfaced DERs [14]. Unlike traditional synchronous generators, which rely on mechanical inertia to regulate voltage and frequency, power-electronic interfaces provide greater flexibility and precision in controlling electrical parameters. These interfaces enable seamless integration of renewable energy sources, such as PV, wind turbines, and battery storage systems, into modern power grids. By leveraging advanced control algorithms, inverter-based resources can dynamically adjust power output, respond to fluctuations in supply and demand, and support grid stability.

However, despite these advantages, power-electronics-interfaced DERs introduce several technical challenges that require careful consideration in microgrid design and operation. One of the primary challenges is their nonlinear dynamics. Unlike conventional generators, power-electronic converters operate through switching mechanisms, leading to complex and nonlinear behavior in power system interactions. These nonlinearities complicate stability analysis and control design, making it difficult to predict system responses to disturbances. Moreover, power electronics lack the inherent damping characteristics of synchronous machines, increasing the risk of instability and oscillatory behavior if not properly managed.

Another major concern is the reduction of system inertia. Traditional power systems rely on rotating inertia from large synchronous generators to naturally dampen frequency fluctuations. However, power-electronic converters do not possess mechanical inertia, meaning that sudden changes in load or generation can cause rapid and severe frequency deviations. Without sufficient inertia, microgrids become more susceptible to disturbances, requiring advanced control strategies, such as virtual inertia emulation and fast-response storage systems, to maintain grid stability.

Additionally, the interaction between power-electronics-based DERs and legacy grid components presents coordination challenges. Many power systems still depend on conventional grid infrastructure, including transformers, transmission lines, and electromechanical protection devices, which were not originally designed to accommodate high penetrations of power-electronic-based generation. Ensuring seamless integration between these legacy components and modern inverter-based resources requires careful system planning, adaptive protection schemes, and advanced communication protocols. Poor coordination can lead to voltage instability, harmonic distortions, and misoperation of protective relays, all of which threaten grid reliability.

Despite these challenges, the growing adoption of power-electronics-interfaced DERs represents an essential step toward modernizing power systems. Advances in grid-forming inverters, real-time optimization, and AI-driven control algorithms continue to improve the stability, efficiency, and resilience of microgrids. By addressing these technical hurdles, power electronics will play a pivotal role in enabling flexible, adaptive, and sustainable energy systems.

1.3 Challenges in Modeling and Stability Analysis

While microgrids offer numerous benefits, their dynamic, heterogeneous, and decentralized nature presents significant challenges in modeling, stability analysis, and control coordination [15]. Unlike traditional power systems, which are relatively centralized and well-structured, microgrids integrate diverse energy resources, including renewable generation, energy storage systems, and controllable loads, each with unique operational characteristics and control requirements. This diversity, combined with power-electronics interfaces and decentralized decision-making, makes accurate modeling and robust stability analysis critical for reliable operation.

One of the primary challenges is modeling complexity. The integration of heterogeneous DERs, energy storage, and varying load demands results in a system with nonlinear, time-varying dynamics that are difficult to capture using traditional power system models. While classical models rely on differential–algebraic equations (DAEs) to represent system behavior, microgrids require high-fidelity, scalable models that incorporate fast-switching inverter dynamics, stochastic renewable fluctuations, and adaptive control mechanisms. Additionally, scalability is a concern—microgrid models must balance computational efficiency with accuracy, ensuring that real-time simulations and control implementations remain feasible as the network grows in size and complexity.

Another fundamental challenge lies in stability analysis. Unlike conventional grids, where large synchronous machines inherently provide inertia and damping, microgrids often rely on power-electronics-based DERs, which lack physical inertia and are susceptible to rapid fluctuations in voltage and frequency. Stability concerns are exacerbated by variable load conditions, unpredictable renewable generation, and fault scenarios. Without proper stability assessment tools, undesirable oscillations, voltage collapses, and frequency deviations may lead to system-wide failures. Advanced small-signal stability analysis, Lyapunov-based methods, and real-time data-driven techniques are required to predict and mitigate instability risks in microgrids.

The decentralized control architecture of microgrids further complicates stability assurance, necessitating coordinated control strategies to manage interactions among distributed components. Unlike traditional grids, where a central operator regulates power flow and frequency, microgrids require hierarchical or distributed control frameworks that synchronize DERs, energy storage, and loads while responding autonomously to disturbances. Multi-agent control, consensus algorithms, and reinforcement learning techniques are emerging solutions to coordinate decision-making across distributed units, ensuring that local optimizations align with overall system stability objectives.

As microgrid deployments continue to expand, addressing these modeling, stability, and control challenges is crucial for enhancing grid resilience, ensuring reliable operation, and unlocking the full potential of decentralized energy systems. Future research efforts will need to focus on scalable physics-based models, machine learning-driven stability assessments, and intelligent control architectures to make microgrids more adaptive, efficient, and robust in the face of growing complexity.

2 Power Electronics Interfaces and Control

The increasing integration of DERs, such as solar photovoltaics (PV), wind turbines, and battery storage systems, has transformed the operation of microgrids. Unlike traditional synchronous generators, which directly interface with the power system, most DERs rely on power electronic interfaces, primarily inverters as shown in Fig. 2, to regulate power exchange between DERs and the grid. These power-electronic converters provide a high degree of controllability, enabling precise voltage and frequency regulation, dynamic power flow control, and grid support functionalities.

However, they also introduce new challenges in terms of stability, system inertia reduction, and control coordination.

Among the various types of power electronic interfaces, voltage source inverters (VSIs) play a crucial role in microgrid operation. Inverters act as the bridge between DC energy sources (such as PV panels or batteries) and the AC power grid, converting DC power into AC power while ensuring appropriate voltage and frequency characteristics and enabling bidirectional power flow, crucial for battery energy storage and vehicle-to-grid (V2G) applications. They can operate in two primary modes [16]:

- Grid-following mode: The inverter synchronizes with the existing grid and injects power based on the measured voltage and frequency.
- Grid-forming mode: The inverter operates autonomously in islanded conditions, regulating voltage and frequency independently, similar to a synchronous generator.

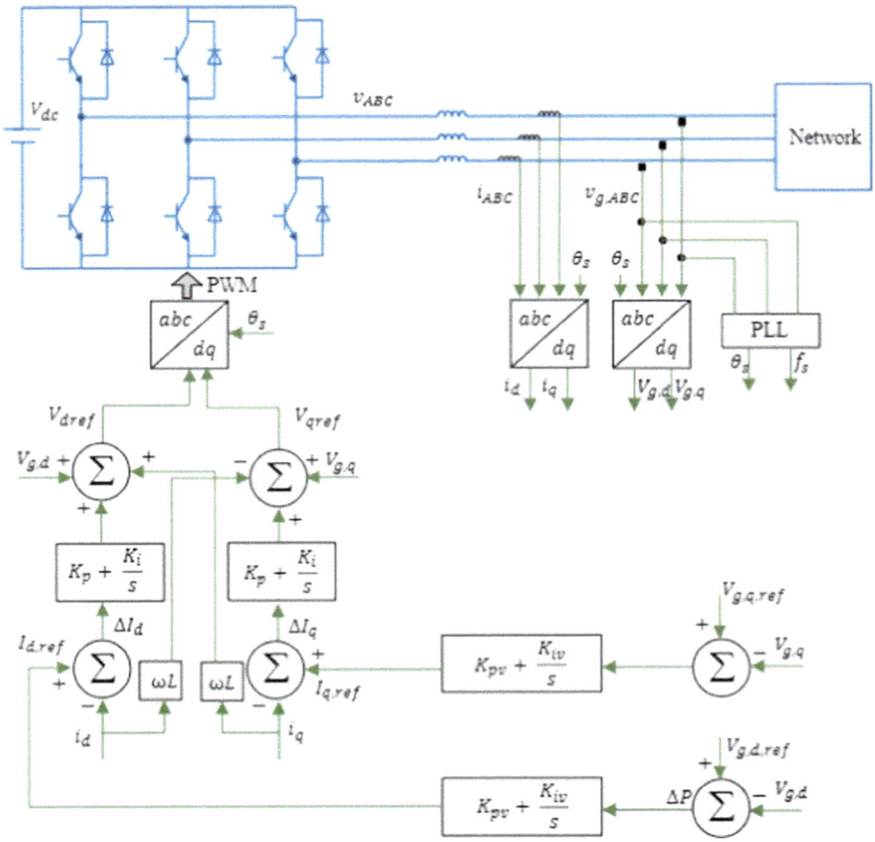

Fig. 2 Power electronics interface and its typical double-loop controller

2.1 Mathematical Model of the Inverter in the d-q Frame

Inverters are commonly connected to the grid via inductive filters to smooth their output waveforms. The dynamic behavior of a grid-connected inverter with an *L*-filter can be described using Kirchhoff's Voltage Law (KVL) in the *d-q* synchronous reference frame:

$$V_d = L\frac{\mathrm{d}i_d}{\mathrm{d}t} - \omega L i_q + V_{g,d} \tag{1}$$

where

- V_d, V_q are the inverter's direct and quadrature control voltages.
- i_d, i_q are the inverter's current components in the *d-q* frame.
- L is the inductance of the output filter.
- ω is the grid angular frequency.
- $V_{g,d}$, $V_{g,q}$ are the *d-q* components of the grid voltage.

The terms $\omega L i_q$ and $\omega L i_d$ represent cross-coupling effects due to the rotating reference frame, which must be compensated for in the control design.

2.2 Double-Loop Control Strategy

Although these power-electronic converters provide a high degree of flexibility, they also present significant challenges, including reduced inertia, nonlinear dynamics, and complex control requirements. Effectively addressing these challenges requires advanced control strategies to maintain system stability and operational efficiency. Given the absence of physical inertia in power-electronics-based DERs, robust control mechanisms are crucial for ensuring stable operation. One of the most widely adopted approaches for inverter control in microgrid applications is the double-loop control scheme, which enhances system response and ensures reliable performance under varying grid conditions.

The double-loop controller is a fundamental strategy for regulating inverter output in both grid-connected and islanded microgrid operations. This control scheme consists of two cascaded loops:

1. Outer Voltage Control Loop—Ensures that the inverter maintains the desired voltage profile by generating appropriate reference signals for the inner loop.
2. Inner Current Control Loop—Responsible for regulating the output current of the inverter to ensure proper power injection while maintaining stability under disturbances.

This hierarchical structure allows for precise regulation of both active and reactive power while maintaining grid stability.

2.2.1 Outer Control Loop

The design of outer loop controller depends on the specific control objectives and the system architecture.

For instance, if the inverter operates in grid-forming mode, it controls the grid voltage. In this case, the outer loop will regulate the grid-side voltage to generate reference current values $I_{d,\mathrm{ref}}$ and $I_{q,\mathrm{ref}}$ based on the error between the measured and reference voltage:

$$I_{d,\mathrm{ref}} = K_{pv}\left(V_{g,d,\mathrm{ref}} - V_{g,d}\right) + K_{iv}\int\left(V_{g,d,\mathrm{ref}} - V_{g,d}\right)\mathrm{d}t, \tag{2}$$

$$I_{q,\mathrm{ref}} = K_{pv}\left(V_{g,q,\mathrm{ref}} - V_{g,q}\right) + K_{iv}\int\left(V_{g,q,\mathrm{ref}} - V_{g,q}\right)\mathrm{d}t. \tag{3}$$

where K_{pv} and K_{iv} are proportional and integral gains of the voltage controller.

2.2.2 Inner Control Loop

The inner current control loop is designed to regulate the output current of the inverter, ensuring it tracks the desired reference values while responding rapidly to disturbances. The design of this loop is closely tied to the output circuit of the inverter, which typically consists of an L-filter (inductor filter) used to smooth the voltage and current waveforms before feeding power into the grid or microgrid. The inner current loop ensures the inverter current follows the reference values generated by the outer voltage loop. The current tracking error is defined as

$$e_d = I_{d,\mathrm{ref}} - i_d, \quad e_q = I_{q,\mathrm{ref}} - i_q. \tag{4}$$

A Proportional-Integral (PI) controller is used to minimize this error:

$$u_d = K_p e_d + K_i\int e_d\mathrm{d}t, \quad u_q = K_p e_q + K_i\int e_q\mathrm{d}t. \tag{5}$$

where K_p and K_i are proportional and integral gains.

To decouple the d and q components and cancel cross-coupling effects, we introduce feedforward compensation:

$$V_{d\mathrm{ref}} = u_d + \omega L i_q + V_{g,d}, \tag{6}$$

$$V_{q\mathrm{ref}} = u_q - \omega L i_d + V_{g,q}. \tag{7}$$

Thus, the final control laws for the inverter voltage are

$$V_{d\text{ref}} = K_p\left(I_{d,\text{ref}} - i_d\right) + K_i \int \left(I_{d,\text{ref}} - i_d\right)\mathrm{d}t + \omega L i_q + V_{g,d}, \tag{8}$$

$$V_{q\text{ref}} = K_p\left(I_{q,\text{ref}} - i_q\right) + K_i \int \left(I_{q,\text{ref}} - i_q\right)\mathrm{d}t - \omega L i_d + V_{g,q}. \tag{9}$$

The inner current control loop plays a critical role in inverter operation by:

- Regulating current flow to maintain power quality and stability.
- Ensuring compliance with grid codes by controlling active and reactive power injections.
- Compensating for disturbances such as load changes or renewable intermittency.
- Providing fast dynamic response, essential for both grid-following and grid-forming inverter operations.

3 Modeling and Control of Microgrids

The increasing penetration of DERs in modern power systems has transformed the operational paradigm of microgrids. Unlike traditional bulk power systems dominated by synchronous generators, microgrids rely heavily on power-electronic-interfaced DERs that exhibit distinct control requirements. To maintain stable and efficient operation, microgrid control strategies must be carefully designed, incorporating both dynamic and algebraic constraints due to the complex interactions among DERs, loads, and the grid.

3.1 Microgrid Modeling with Differential–Algebraic Equations (DAEs)

Microgrids are characterized by a combination of dynamic components, such as inverter-interfaced DERs, and algebraic constraints, such as power flow equations and network constraints. This hybrid nature makes Differential–Algebraic Equations (DAEs) a suitable mathematical framework for modeling microgrid dynamics. A general representation of microgrid behavior can be given as [17]:

$$\dot{x} = f(x, y, u) \text{ (Dynamic equations)} \tag{10a}$$

$$0 = g(x, y, u) \text{ (Algebraic constraints)} \tag{10b}$$

where

- $x \in R^n$ represents the dynamic state variables, such as inverter voltage and current.

- $y \in R^m$ represents the algebraic variables, such as bus voltages and power flow constraints.
- $u \in R^p$ represents disturbances, such as load variations or renewable energy fluctuations.

The dynamic equations govern the evolution of inverter-based resources and other active control elements, while the algebraic equations enforce network constraints like KVL and Kirchhoff's Current Law (KCL). This DAE framework serves as the foundation for analyzing the dynamic microgrid systems.

3.2 Droop Control for Inverter-Based Microgrids

A major challenge in microgrid operation arises from the fundamental differences between traditional synchronous generators and power-electronics-based inverters. In large-scale power systems, synchronous generators inherently regulate system frequency and voltage through their rotational inertia and droop characteristics. However, microgrids composed primarily of inverters lack this natural frequency and voltage regulation mechanism, which presents several operational difficulties:

- Lack of Inertia: In contrast to rotating machines, inverters do not store kinetic energy, making microgrids more vulnerable to rapid frequency deviations and transients.
- Decentralized Control Requirement: Unlike conventional grids, where central authorities coordinate frequency regulation, microgrids often require decentralized control to accommodate multiple DERs operating autonomously.
- Voltage and Frequency Stability: In the absence of proper control, voltage and frequency fluctuations can arise due to load changes, renewable energy intermittency, or disturbances.

To address these issues, droop control has been widely adopted for inverter-based microgrids [18]. By mimicking the frequency and voltage response of synchronous generators, droop control enables decentralized operation while ensuring load-sharing and system stability.

3.2.1 Active Power–Frequency (*P-f*) Droop Control

To regulate system frequency, inverters mimic the behavior of synchronous machines using *P-f* droop control, which relates the active power injection to frequency deviations:

$$f = f_0 - K_p\left(P - P^*\right), \tag{11}$$

where

- f is the local inverter frequency.
- f_0 is the nominal system frequency (e.g., 50 Hz or 60 Hz).
- K_p is the droop coefficient.
- P is the measured active power.
- P^* is the reference active power.

This control mechanism ensures load-sharing among multiple inverters in islanded microgrids.

3.2.2 Reactive Power-Voltage (Q–V) Droop Control

Similarly, Q–V droop control governs voltage regulation based on reactive power injection:

$$V = V_0 - K_q\left(Q - Q^*\right), \tag{12}$$

where

- V is the controlled voltage magnitude.
- V_0 is the nominal voltage setpoint.
- K_q is the reactive power droop coefficient.
- Q is the measured reactive power.
- Q^* is the reference reactive power.

This strategy enables decentralized voltage control, preventing overvoltage issues in microgrid operation.

4 Stability Analysis of Microgrids

Microgrids, particularly those dominated by power-electronic-interfaced DERs, exhibit unique stability characteristics that differ from traditional bulk power systems. Ensuring the stable operation of microgrids under varying operating conditions, including disturbances, load changes, and grid-connected or islanded transitions, is essential for reliable performance. Stability in microgrids can be broadly classified into:

- Transient Stability: The ability of the microgrid to maintain synchronism and recover from large disturbances such as faults or sudden changes in load and generation.
- Small-Signal Stability: The ability of the microgrid to remain stable under small perturbations and prevent oscillatory instability.
- Voltage Stability: The ability of the microgrid to maintain acceptable voltage levels across different nodes.

This section focuses on small-signal stability, which is particularly crucial for microgrids due to the prevalence of inverter-based DERs and the reduced system inertia compared to conventional power grids.

4.1 Small-Signal Stability Analysis of Microgrids

Small-signal stability analysis investigates the system's response to small perturbations around an equilibrium operating point. It provides valuable insights into the dynamic behavior of microgrids and their susceptibility to oscillatory instability.

4.1.1 Small-Signal Stability

Considering the DAE modeling in (10a, b), to analyze small-signal stability, the system is linearized around an equilibrium operating point, denoted as (x_0, y_0, u_0). The first-order Taylor expansion leads to the following linearized state-space representation:

$$\begin{bmatrix} \Delta \dot{x} \\ 0 \end{bmatrix} = \begin{bmatrix} A & B \\ C & D \end{bmatrix} \begin{bmatrix} \Delta x \\ \Delta y \end{bmatrix} + \begin{bmatrix} E \\ F \end{bmatrix} \Delta u \tag{13}$$

where

$$A = \frac{\partial f}{\partial x}|_{(x_0, y_0, u_0)}, \quad B = \frac{\partial f}{\partial y}|_{(x_0, y_0, u_0)}, \tag{14}$$

$$C = \frac{\partial g}{\partial x}|_{(x_0, y_0, u_0)}, \quad D = \frac{\partial g}{\partial y}|_{(x_0, y_0, u_0)}. \tag{15}$$

Solving for the algebraic variables $\Delta \boldsymbol{y}$, we obtain the reduced-order dynamic system:

$$\Delta \dot{x} = A_{\text{red}} \Delta x + E_{\text{red}} \Delta u, \tag{16}$$

where the reduced system matrix is given by

$$A_{\text{red}} = A - BD^{-1}C. \tag{17}$$

The eigenvalues of A_{red} determine the stability characteristics of the system:

- If all eigenvalues have negative real parts, the microgrid is stable.
- If any eigenvalue has a positive real part, the system exhibits instability.
- If eigenvalues have zero real parts, the system is on the stability boundary and may exhibit oscillatory behavior.

4.1.2 Factors Affecting Small-Signal Stability

The small-signal stability of microgrids is influenced by multiple factors, ranging from the physical characteristics of the system components to the interactions between different control mechanisms. Unlike conventional power grids, which are dominated by synchronous generators with well-established stability characteristics, microgrids consist of power-electronics-based DERs that fundamentally alter stability dynamics. Several key aspects contribute to small-signal instability in microgrids:

1. Reduced Inertia in Inverter-Based Microgrids

Traditional power systems rely on the inertia of large synchronous generators to dampen frequency deviations caused by fluctuations in generation and load. However, inverter-based DERs do not inherently provide this mechanical inertia, making microgrids more susceptible to rapid frequency changes. Without sufficient damping mechanisms, small disturbances can lead to significant frequency oscillations or even instability, particularly in islanded microgrid operation.

2. Control-Grid Interactions

Power-electronic inverters are equipped with controllers that regulate voltage, frequency, and power sharing. However, these controllers introduce new dynamic interactions that can either stabilize or destabilize the system. Poorly tuned controllers may cause unintended resonance effects or low-frequency oscillations, leading to loss of stability. The interactions between inner current control loops, outer voltage control loops, and network dynamics must be carefully analyzed to prevent instability.

3. Impact of Network Topology and Loading Conditions

Microgrid stability is strongly influenced by its topology and the real-time power flow conditions. Radial microgrid configurations often exhibit weaker voltage support compared to meshed networks, increasing the risk of voltage instability. Additionally, high penetration of renewable energy sources results in varying power injections, altering system eigenvalues and shifting stability margins. Overloaded microgrids, particularly those operating close to their power capacity limits, are more prone to instability under even small disturbances.

4. Mode Coupling Between Power-Electronic Devices

Since multiple inverters operate concurrently in a microgrid, their control loops may exhibit undesired mode coupling. The interactions between different inverter controllers can lead to stability issues, particularly if they have different control bandwidths. This phenomenon, often referred to as sub-synchronous interactions, can amplify certain oscillatory modes in the system and lead to sustained instability.

5. Communication Delays and Cyber-Physical Interactions

Microgrids often employ communication-based control strategies for distributed coordination of DERs, such as droop control, secondary control, and centralized microgrid energy management. However, communication networks introduce delays and packet losses, which can degrade system stability. Cybersecurity threats, such as data spoofing or denial-of-service (DoS) attacks, can further disrupt stability by altering control signals and triggering cascading failures.

4.2 Mitigating Small-Signal Instability in Microgrids

Given the numerous factors influencing small-signal stability, it is essential to adopt robust control and operational strategies to enhance microgrid resilience. Several techniques have been proposed to mitigate small-signal instability in power-electronics-dominated microgrids:

1. Virtual Inertia Control

One of the most effective approaches to compensate for the lack of rotational inertia in inverter-based microgrids is to implement virtual inertia control. This technique mimics the dynamic response of synchronous generators by introducing an additional control loop that emulates inertia using power-electronic controllers. Common implementations include droop-based virtual inertia, swing equation-based control, and grid-forming inverter strategies. By appropriately tuning virtual inertia, the system can improve its frequency response and stability characteristics.

2. Adaptive Droop Control

Traditional droop control strategies are widely used for decentralized power sharing among multiple DERs. However, static droop gains may not be sufficient to handle dynamic operating conditions. Adaptive droop control introduces real-time adjustments based on system conditions, allowing better frequency and voltage regulation. By dynamically adjusting droop coefficients based on real-time conditions, adaptive droop control enhances microgrid stability.

3. Advanced Model Predictive Control (MPC)

Model Predictive Control (MPC) is a powerful optimization-based control technique that predicts system behavior over a future horizon and computes optimal control actions in real time. MPC is particularly useful in microgrids for stability enhancement because it can:

- Account for multiple constraints, such as voltage limits and power-sharing constraints.
- Predict future disturbances and take preemptive actions.
- Optimize the response of inverters and DERs to enhance small-signal stability.

4. Network Reconfiguration and Optimal Dispatch

In microgrids, dynamic network reconfiguration can be an effective way to mitigate instability. By strategically opening or closing switches, the system can alter power flows to reduce oscillations and improve stability margins. Similarly, optimal DER dispatch ensures that active and reactive power generation is dynamically adjusted to enhance voltage stability and frequency regulation.

5. Cybersecurity Measures to Prevent Control Manipulation

Since microgrids are increasingly reliant on communication networks, ensuring robust cybersecurity measures is critical. Techniques such as anomaly detection, secure communication protocols, and intrusion detection systems can help protect against malicious attacks that could otherwise destabilize the system.

5 Problems

1. Define and explain the concept of a microgrid. Discuss its key components, operational modes, and the advantages it provides over traditional power grids. What are the major challenges associated with microgrid stability and control?
2. Explain the significance of droop control in microgrid operation. How does it help in managing power-sharing among DERs? Why is it essential for islanded microgrid operation?
3. A controllable DER is connected to an AC system via an inverter. The inverter operates under a frequency droop control strategy with the following droop coefficients:

    - Active power droop: 0.04 Hz/kW
    - Reactive power droop: 0.012 kV/kVAR

    The system operates at a nominal frequency of 60 Hz and supplies an initial active power of 40 kW and reactive power of 25 kVAR. The initial bus voltage at the DER connection point is 0.42 kV.

    - If the total power demand decreases to 30 kW and 18 kVAR, what will be the new system frequency?
    - Discuss the impact of increasing the droop coefficients on system frequency and voltage regulation.

4. A microgrid operates under droop control, with an inverter-based DER supplying power. Initially, at 60 Hz, the active power load is 35 kW and the reactive power load is 22 kVAR. The bus voltage is 0.39 kV.

    After a load disturbance, the new system conditions are:

    - Active power load: 50 kW
    - Reactive power load: 18 kVAR
    - System frequency: 59.4 Hz
    - Bus voltage: 0.41 kV
    - Determine the active power droop coefficient.
    - Determine the reactive power droop coefficient.

6 Problem Solutions

Solution 1

Microgrids are localized power networks that integrate distributed energy resources (DERs) such as solar photovoltaic (PV) panels, wind turbines, and battery energy storage systems (BESS) to supply electricity to a specific area. Unlike traditional centralized power grids, microgrids are capable of operating in both grid-connected and islanded modes, making them highly resilient to disturbances and improving power supply reliability.

Microgrids are decentralized power systems that integrate distributed energy resources (DERs), such as solar photovoltaic (PV) panels, wind turbines, and battery storage, to supply localized power. Unlike traditional centralized power systems, microgrids can operate in both grid-connected and islanded modes, enhancing energy resilience and reliability.

Key Components:

- Energy Sources: Solar, wind, hydro, and other renewable sources.
- Energy Storage: Batteries to store and supply power during demand fluctuations.
- Power Electronics: Inverters to regulate voltage and frequency.
- Control System: Manages power flow and ensures stable operation.

Advantages:

- Resilience: Microgrids can continue operation during grid failures.
- Sustainability: Integration of renewable energy reduces carbon footprint.
- Cost Efficiency: Reduces reliance on central utilities and minimizes transmission losses.

Challenges:

- Stability issues due to low inertia of power-electronic-based DERs.
- Complex control and optimization strategies required for efficient operation.
- Higher initial investment compared to traditional grid supply.

Solution 2

Droop control is a decentralized control strategy widely used in microgrids to regulate frequency and voltage while ensuring stable operation. It mimics the natural frequency response of traditional synchronous generators and enables DERs to autonomously share power without the need for a centralized controller.

Microgrids consist of multiple DERs, each with different capacities and dynamic characteristics. Droop control ensures proportional power-sharing among DERs by adjusting their output based on system conditions.

In a microgrid, DERs must proportionally share the active power demand to avoid overloading any single unit. The frequency droop control ensures that DERs with lower droop coefficients contribute a higher share of active power. Proper coordination of active power droop coefficient values among DERs allows load balancing without communication among units.

Similarly, voltage droop control regulates reactive power sharing by adjusting the terminal voltage. However, due to network impedance variations, reactive power-sharing is more challenging than active power-sharing. Proper tuning of reactive power droop coefficient values helps maintain voltage stability across the network.

When a microgrid operates in islanded mode, it is disconnected from the main grid and must maintain its own frequency and voltage stability. In this mode, droop control is essential for:

- Maintaining Frequency Stability: In the absence of a central frequency reference, frequency droop control ensures a balance between generation and demand.
- Voltage Regulation: Reactive power droop helps maintain proper voltage levels and prevents excessive fluctuations.
- Ensuring Seamless Transition: During transitions from grid-connected to islanded mode, droop control enables smooth dynamic response, minimizing transients.
- Preventing Overloading: Without droop control, some DERs may supply excessive power while others contribute minimally, leading to system imbalance.

Solution 3

Given Data:

- Active power droop coefficient: $K_p = 0.04$ Hz/kW
- Reactive power droop coefficient: $K_q = 0.012$ kV/kVAR
- Nominal system frequency: $f_0 = 60$ Hz
- Initial active power: $P_0 = 40$ kW
- Initial reactive power: $Q_0 = 25$ kVAR
- Initial voltage: $V_0 = 0.42$ kV

Step 1: Compute New System Frequency

$$\Delta P = P_{\text{new}} - P_0 = 30 - 40 = -10 \text{ kW}$$

$$\Delta f = k_p \cdot \Delta P = 0.04 \times (-10) = -0.4 \text{ Hz}$$

$$f_{\text{new}} = f_0 + \Delta f = 60 - 0.4 = 59.6 \text{ Hz}$$

Step 2: Compute New System Voltage

$$\Delta Q = Q_{\text{new}} - Q_0 = 18 - 20 = -2 \text{ kVAR}$$

$$\Delta V = K_q \cdot \Delta Q = 0.012 \times (-2) = -0.024 \text{ kV}$$

$$V_{\text{new}} = V_0 + \Delta V = 0.42 - 0.024 = 0.396 \text{ kV}$$

Final Answer:

- New frequency: 59.6 Hz
- New voltage: 0.396 kV

Increasing the active power droop coefficient makes the system more responsive to frequency changes but leads to greater steady-state frequency deviations. Increasing the reactive power droop coefficient leads to faster voltage regulation but may cause larger voltage fluctuations and potential instability. The selection of droop coefficients must balance stability, responsiveness, and grid robustness, particularly in microgrids with high penetration of renewable energy sources.

Solution 4

Given Data:

- Initial frequency: $f_0 = 60$ Hz
- New frequency: $f = 59.4$ Hz
- Initial active power: $P_0 = 35$ kW
- New active power: $P = 50$ kW
- Initial reactive power: $Q_0 = 22$ kVAR
- New reactive power: $Q = 18$ kVAR
- Initial voltage: $V_0 = 0.39$ kV
- New voltage: $V = 0.41$ kV

Step 1: Compute Active Power Droop Coefficient

$$\Delta P = P - P_0 = 50 - 35 = 15 \text{ kW}$$

$$\Delta f = f - f_0 = 59.4 - 60 = -0.6 \text{ Hz}$$

$$K_p = \frac{\Delta f}{\Delta P} = \frac{-0.6}{15} = -0.04 \text{ Hz/kW}$$

Step 2: Compute Reactive Power Droop Coefficient

$$\Delta Q = Q - Q_0 = 18 - 22 = -4 \text{ kVAR}$$

$$\Delta V = V - V_0 = 0.41 - 0.39 = 0.02 \text{ kV}$$

$$K_q = \frac{\Delta V}{\Delta Q} = \frac{0.02}{-4} = -0.005 \text{ kV/kVAR}$$

Final Answer:

- Active power droop coefficient: 0.04 Hz/kW
- Reactive power droop coefficient: 0.005 kV/kVAR

Acknowledgements The authors acknowledge support from the National Science Foundation under the award OAC-2417773 and the use of ChatGPT for language refinement in this chapter.

References

1. Flores N, McBrien H, Do V, Kiang M, Schlegelmilch J, Casey J (2022) The 2021 Texas power crisis: distribution, duration, and disparities. J Expo Sci Environ Epidemiol 1–11
2. Sullivan B, Malik N (2021) 5 million americans have lost power from Texas to North Dakota after devastating winter storm. TIME. https://time.Com/5939633/texas-poweroutageblackouts
3. Aldhous P, Lee S, Hirji Z (2021) The Texas winter storm and power outages killed hundreds more people than the state says. BuzzFeed News
4. Paulson M (2019) Power failure cost Broadway $3.5 million. The New York Times, pp A17-L
5. Tabatabaei N, Kabalci E, Bizon N (2019) Microgrid architectures, control and protection methods. Springer
6. Bossart S (2012) Doe perspective on microgrids. In: Advanced microgrid concepts and technologies workshop
7. Ganjian-Aboukheili M, Shahabi M, Shafiee Q, Guerrero J (2019) Seamless transition of microgrids operation from grid-connected to islanded mode. IEEE Trans Smart Grid
8. Abu-Rub O, Fard A, Umar M, Hosseinzadehtaher M, Shadmands M (2021) Towards intelligent power electronics-dominated grid via machine learning techniques. IEEE Power Electron Mag 8:28–38
9. Roos M, Nguyen P, Morren J, Slootweg J (2019) Modeling and experimental validation of power electronic loads and DERs for microgrid islanding simulations. IEEE Trans Power Syst 35:2279–2288
10. Li Y (2022) Cyber-physical microgrids. Springer
11. Hossain J, Mahmud A (2014) Renewable energy integration: challenges and solutions. Springer Science & Business Media
12. Alhasnawi B, Jasim B, Sedhom B, Hossain E, Guerrero J (2021) A new decentralized control strategy of microgrids in the internet of energy paradigm. Energies 14:2183
13. Abdi H, Mohammadi-ivatloo B, Javadi S, Khodaei A, Dehnavi E (2017) Energy storage systems. Distrib Gener Syst 7:333–368
14. Ganjian-Aboukheili M, Shahabi M, Shafiee Q, Guerrero J (2019) Seamless transition of microgrids operation from grid-connected to islanded mode. IEEE Trans Smart Grid 11:2106–2114
15. Wang X, Guerrero J, Blaabjerg F, Chen Z (2012) A review of power electronics based microgrids. Int J Power Electron 12:181–192
16. Espın-Sarzosa D, Palma-Behnke R, Canizares C, Annakkage U, Elizondo M, Espina E, Du W, Kabalan M, Meegahapola L, Mendoza-Araya P et al (2023) Microgrid modeling for stability analysis. IEEE Trans Smart Grid 15:2459–2479
17. Singhal A, Vu T, Du W (2022) Consensus control for coordinating grid-forming and grid-following inverters in microgrids. IEEE Trans Smart Grid 13:4123–4133
18. Jiang X, Li Y, Huang D (2024) Modularized bilinear Koopman operator for modeling and predicting transients of microgrids. IEEE Trans Smart Grid

19. Wang C, Li Y, Peng K, Hong B, Wu Z, Sun C (2013) Coordinated optimal design of inverter controllers in a micro-grid with multiple distributed generation units. IEEE Trans Power Syst 28:2679–2687

Energy Integration and Sustainable Building Infrastructure

3D Spectral Selective Solar Light Harvesting and Photothermal Energy Generation via Nano Hybrids for Energy Neutral Infrastructure

Donglu Shi

Abstract The rapid urbanization and growth of megacities have intensified the demand for energy solutions that won't harm the planet. Since nearly two out of every five units of energy used worldwide goes to powering buildings, improving how these structures consume energy is critical. Taking action here helps lower emissions and supports international commitments, such as the 2050 net-zero target. Large solar installations have become important in renewable energy, but issues like land demands, expensive setups, heavy machinery, and upkeep challenges slow their spread. Similarly, conventional solar heating systems often require significant upfront investment, ample space, and depend on weather, which limits their use in urban areas. To overcome these hurdles, advanced photothermal technologies offer a fresh approach to harvesting solar power, designed to efficiently support both heating and cooling functions. These transparent multilayer photothermal films are built to capture light from the sun and indoor sources in all directions. They turn that light into heat efficiently without needing big equipment, making them perfect for tight city spaces. During hot seasons, clear solar panels placed on building exteriors and inside—like on walls, floors, and ceilings—help gather energy from sunlight by day and indoor lights at night or on cloudy days. This technology offers subtle but powerful heating and cooling solutions, especially for tall buildings in urban areas. Building on this idea, the Photothermal Solar Tunnel Radiator (PSTR) uses plasmonic nanomaterials to turn both direct sunlight and scattered light into heat efficiently. It fits smoothly with modern building designs, helping meet sustainability goals. Additionally, new methods that capture indoor lighting help create energy-neutral buildings by going beyond traditional solar panels. By using different types of light, these buildings generate electricity more consistently and also provide energy for heating and cooling throughout the year. Together, these approaches offer scalable, affordable, and eco-friendly solutions for sustainable urban energy systems. This chapter is among the latest studies investigating how plasmonic nanoparticles can be

D. Shi (✉)
The Materials Science and Engineering Program, Department of Mechanical and Materials Engineering, Department of Biomedical Engineering, College of Engineering and Applied Science, University of Cincinnati, Cincinnati, OH, USA
e-mail: donglu.shi@uc.edu
URL: https://researchdirectory.uc.edu/p/shid

D. Shi (ed.), *Energy Materials Science and Engineering: From New Devices to AI Power Systems*, Green Energy and Technology, https://doi.org/10.1007/978-3-031-88550-1_10

used for heating and cooling in buildings—a shift from earlier work that focused on medical applications. By taking advantage of transparent thin films, strong absorption in ultraviolet and infrared ranges, and combining multiple panels to boost solar energy capture, this new method offers a fresh route for green energy production. Plasmonic nanoparticles enable efficient harvesting of energy across a wide spectrum while maintaining transparency, which fits well with building designs. This blend of cutting-edge materials and sustainable energy approaches showcases how nanotechnology can play a key role in solving today's energy challenges.

Keywords Solar harvesting · Localized Surface Plasmon Resonance (LSPR) · Photothermal effect · Plasmonic nanoparticles · Building utility · Iron oxide · Superparamagnetic · Optical absorption · Average visible transmittance (AVT) · Power conversion efficiency (PCE) · Energy generation

1 Introduction

The rapid growth of urbanization and the emergence of megacities, marked by extensive high-rise developments and dense populations, pose substantial challenges to energy management and sustainability. Buildings are among the most significant energy consumers, accounting for a considerable portion of global energy demand. In the United States alone, the residential and commercial sectors consumed approximately 40 quadrillion British Thermal Units (BTUs) in 2018, constituting nearly 40% of the nation's total energy consumption [1]. The substantial energy requirements of buildings, especially in densely populated areas, underscore the need for enhanced energy efficiency as a cornerstone of sustainable urban development. Addressing these challenges is crucial for reducing greenhouse gas emissions and achieving international climate goals, such as the net-zero emissions target by 2050 set forth at COP26 [2]. Consequently, innovative approaches and advanced engineering solutions are essential to transform the way buildings harness and utilize energy.

The term "utility-scale solar" refers to solar power facilities designed to produce electricity at a capacity sufficient to supply many customers rather than a single home or business (Fig. 1). These terms generally describe installations that operate on a larger scale than residential or commercial solar systems, contributing substantial amounts of renewable energy to the grid. Utility-scale solar facilities are built specifically to supply power directly to the grid, providing electricity to various customer bases through agreements with utilities or electricity providers. By contrast, large-scale solar can sometimes refer to facilities slightly smaller than utility-scale but still expansive enough to support multiple structures or large communities. The difference between the terms is often situational, with utility-scale solar having formalized connections to utility companies, while large-scale solar may include private installations feeding power to multiple locations.

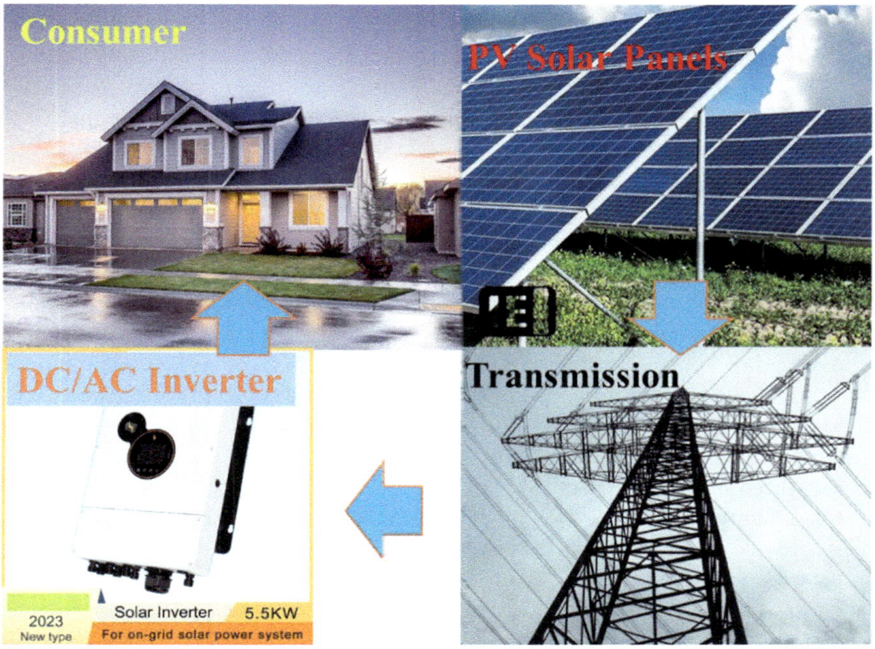

Fig. 1 Schematic diagram of utility-scale solar system

While utility-scale solar systems provide significant benefits in terms of renewable energy production and emissions reduction, they come with several limitations related to facility size, cost, and maintenance complexity. Utility-scale solar installations demand vast areas of land to support the thousands of solar panels needed to produce energy on a large scale. These projects often occupy hundreds or even thousands of acres, leading to potential challenges with land availability and environmental impact. Acquiring land suitable for solar development can be costly and may face local opposition, especially in regions with high land value or competing land uses, such as agriculture. Additionally, the remote locations chosen to avoid populated areas may require extensive infrastructure for electricity transmission.

The construction of utility-scale solar facilities entails a high upfront cost, including expenses related to land acquisition, equipment, labor, and permitting. Solar panels, inverters, and energy storage systems (if incorporated) add to the capital expenditure. Financing these projects often requires complex arrangements with investors or government incentives to make them financially viable. These costs can make utility-scale solar less accessible for some regions or developers, especially where financial support is limited or energy prices are low, which affects return on investment.

The sheer scale of utility-scale solar systems adds complexity to their maintenance and operation. Large facilities require regular cleaning of panels to ensure efficiency, especially in dusty or dry areas. Given their often-remote locations, accessing these sites can be challenging and costly, and unexpected repairs or breakdowns in equipment can disrupt energy production. Additionally, the electrical and mechanical systems (like inverters, wiring, and support structures) require continuous monitoring to maintain system reliability and safety. As these systems age, maintenance requirements increase, which can drive up operational costs over time.

While utility-scale solar systems play a crucial role in renewable energy generation, they face limitations that include substantial land requirements, high capital costs, and complex maintenance demands. Addressing these limitations requires careful planning, financial support, and technological advances to improve efficiency and durability, making utility-scale solar more viable and accessible in the long term. These solar projects are generally located on open land in areas with high sun exposure. They are carefully positioned near existing substations or major transmission infrastructure to minimize power losses during transmission and to reduce the cost and logistical complexity of integrating them into the electrical grid. This setup enables the efficient transfer of electricity from the solar facility to the broader power network, supplying energy to a large, distributed group of consumers.

Traditionally, solar heating systems generally incorporate various components for capturing and converting solar energy, such as solar collectors, heat exchangers, and storage tanks, to ensure a consistent energy supply during low sunlight periods. These systems often involve fluid media (e.g., water or synthetic transfer fluids) to convey heat, insulation to minimize loss, and backup heating systems, adding to the complexity and cost of implementation [3]. While effective, traditional solar heating solutions face challenges, including high initial costs, large space requirements, and seasonal variations in efficiency. Integration with existing heating infrastructures can also be complex and costly. Aesthetic considerations further impede adoption in urban areas, as collectors can appear visually intrusive.

For example, a conventional solar hot water heating system, also known as a solar thermal system, is designed to capture and utilize solar energy to heat water for domestic or commercial use (Fig. 2). This type of system consists of several key components that work together to absorb sunlight, transfer the heat to water, and store the heated water until it is needed. A conventional solar hot water heating system, or solar thermal system, captures solar energy to heat water for domestic or commercial use through several key components working together to absorb, transfer, and store heat. The system's core component, the solar collector, is designed to capture sunlight and transfer heat to water. Two common types of collectors include flat-plate collectors, which consist of a large, insulated box with a transparent cover and a dark-colored absorber plate, and evacuated tube collectors, which are composed of parallel glass tubes with vacuum insulation that reduces heat loss and enhance efficiency, especially in colder climates. Within the collector, a heat transfer fluid, often water or a glycol-based antifreeze solution, absorbs the thermal energy and circulates between the collector and storage tank. In active systems, this fluid circulation is managed by a pump, while passive systems rely on natural convection, a simpler but less efficient method. The heat transfer fluid then flows through a heat exchanger, which transfers the captured heat to water in the storage tank without

Fig. 2 Photograph of a conventional solar hot water heating system

allowing the fluids to mix, which is especially important when using antifreeze. The heated water is stored in an insulated tank to minimize heat loss, with some systems using a separate tank and others integrating the heat exchanger into the storage unit itself. A control system manages the entire process by using temperature sensors in the collector and storage tank to activate the pump when the collector's temperature exceeds that of the stored water, ensuring efficient heat transfer and protecting the system from overheating or freezing.

Despite their effectiveness, conventional solar hot water systems have notable limitations, particularly concerning efficiency and complexity. These systems often experience significant heat losses due to environmental exposure of collectors and heat dissipation from storage tanks, making it difficult to retain heat over long periods, even with insulation. The system's efficiency also heavily relies on sunlight, which is affected by seasonal and weather changes, necessitating a backup heating source during cloudy or winter conditions. Additionally, stagnation losses can occur when the system is not in use, causing the collector temperature to remain high and potentially overheating the fluid, leading to stagnation and damage to system components.

Conventional solar hot water systems are complex due to their reliance on multiple components like pumps, heat exchangers, and controllers, each of which adds to maintenance needs and introduces potential points of failure; pump malfunctions or controller inaccuracies can impact system reliability. Installation can be challenging and costly, as it requires significant roof space positioned at an optimal angle for sunlight exposure, which may not be feasible in all locations. Hard water can cause scaling in collectors and heat exchangers, reducing heat transfer efficiency and necessitating regular descaling. Additionally, these systems must account for freezing and overheating risks: in colder climates, antifreeze is often needed in the heat transfer fluid to prevent freezing, while summer overheating can require cooling measures to avoid system damage. In summary, while conventional solar hot water heating systems can effectively reduce energy consumption and greenhouse gas emissions, they come with inherent drawbacks in terms of system complexity, susceptibility to heat loss, and dependency on weather conditions. These limitations have led to the development of advanced solar hot water systems that aim to improve efficiency and simplify operations.

A promising approach to overcoming these challenges is direct solar-to-thermal energy conversion through photothermal (PT) heating. Unlike conventional PV systems, which convert sunlight into electricity, or thermal collectors requiring extensive infrastructure, PT systems enable more direct and efficient energy conversion [4]. A recent innovation involves transparent multilayer PT thin films designed for 3D solar light harvesting. These films permit solar light to penetrate multiple layers, each converting solar energy into heat. This design significantly enhances energy density and solar harvesting efficiency, offering a valuable path for developing solar technologies that can seamlessly integrate with urban building designs [4]. By utilizing multiple panels within a confined volume, PT films can collect photons across all layers, making them highly suited to building heating applications.

This novel PT-based approach not only increases solar energy efficiency but also overcomes spatial limitations common to traditional systems. Transparent multilayer PT films maximize energy absorption without extensive ground or roof space, making them ideal for high-rise buildings in densely populated areas. By integrating PT films into windows, facades, and structural components, this approach offers a visually unobtrusive solution that aligns with modern architectural aesthetics. Moreover, the reduced complexity of PT systems—requiring no electrical conversion or storage components—simplifies installation and maintenance, while economic advantages make this method more feasible for widespread adoption. The 3D system leverages advanced nanotechnologies to enhance energy density and offers an innovative alternative to traditional PV heating systems. Specifically, for building heating, directly converting solar light to thermal energy using photothermal (PT) heating proves advantageous. Understanding that a larger surface area is essential for high energy density, a transparent multilayer system has been devised to collect solar light in three-dimensional spaces [4, 5].

This novel design, depicted in Fig. 3, uses transparent PT thin films that allow sunlight to pass through multiple layers, activating photothermal heating on each layer. This configuration increases energy density and enhances the efficiency of solar energy capture. Within a compact volume, these PT films can absorb a higher number of photons, paving the way for innovative solar heating technologies in building applications. This approach aims to resolve the limitations of conventional

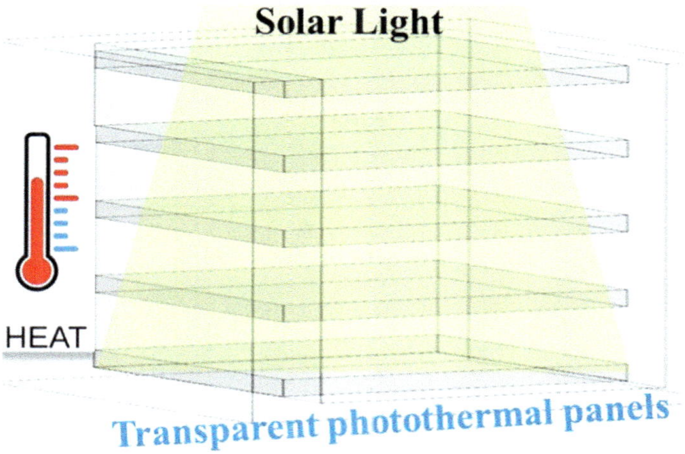

Fig. 3 Schematic diagram illustrating 3D solar harvesting and energy production using multiple transparent photothermal panels

PV heating systems by providing a cost-effective and efficient alternative through transparent PT thin films.

Building on the concept of transparent PT panels shown in Fig. 3, we introduce a new model called the Photothermal Solar Tunnel Radiator (PSTR), a novel approach to solar harvesting. The PSTR concept involves multiple transparent panels coated with photothermal nanomaterials, as shown in Fig. 4. This setup envisions sustainable energy capture by directing sunlight through a series of transparent glass layers, each coated with PT films containing plasmonic Fe_3O_4 nanoparticles (Fig. 4). Figure 4 depicts the key elements of this innovative system, where a solar tunnel functions as a radiator activated by sunlight. The tunnel's dome captures and directs sunlight through transparent glass layers with PT coatings. These coatings convert solar radiation into heat upon exposure, warming the surfaces of the glass layers. Acting as a thermal radiator, the heated glass releases the stored solar energy without external power inputs. This direct energy conversion represents a substantial step forward in achieving energy-neutral systems, moving away from conventional energy-intensive processes. Additionally, a diffuser at the end of the PSTR channels light into interior spaces that lack direct sunlight.

The direct conversion of solar light into thermal energy through PT heating offers a number of advantages over traditional PV heating. PT heating provides a simplified and cost-effective approach, often eliminating the need for complex components such as heat transfer fluids, exchangers, or storage tanks in certain applications. This streamlined process results in reduced installation and maintenance costs, making PT heating a viable and attractive option for building heating with much higher efficiency. By efficiently bypassing the multiple stages typically found in PV-based heating, PT heating minimizes energy loss and enhances energy utilization. Innovations in 3D

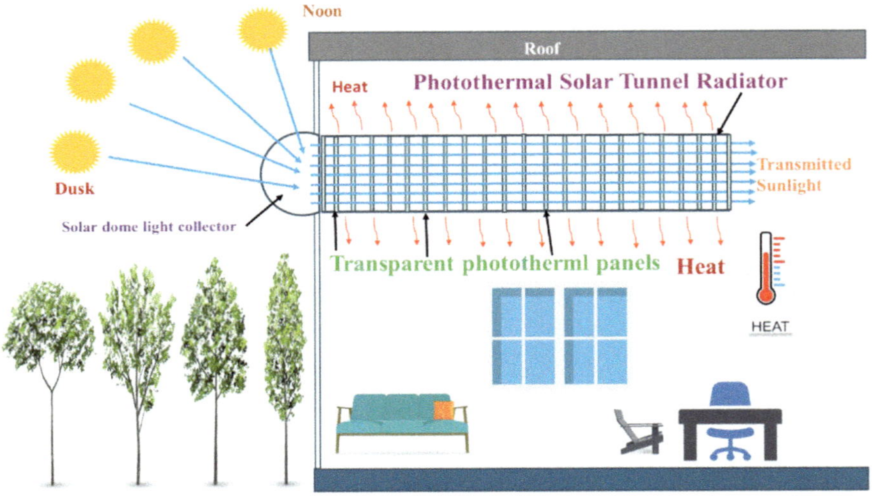

Fig. 4 Conceptual illustration of the Photothermal Solar Tunnel Radiator (PSTR)

solar harvesting, demonstrated by transparent multi-panel systems, tackle the limitations of 2D PV systems by expanding sunlight exposure and significantly boosting energy density for heating applications. The scalability and environmental compatibility of PT heating, especially within 3D solar harvesting frameworks, contribute to its appeal as an eco-friendly and scalable solution well-aligned with sustainability goals.

An additional innovative approach to building energy generation involves harnessing indoor light available at night. The transition toward energy-neutral buildings offers numerous advantages, including utility cost reduction, enhanced grid reliability, improved air quality, and carbon-free environments. While PV solar cells have advanced sustainable energy generation, their dependency on direct sunlight limits energy production during low-light or nighttime hours, posing challenges for uninterrupted renewable energy supply. Traditional solar research has mainly focused on direct sunlight; however, alternative photon sources, such as diffused indoor light, also offer significant potential [5].

Urban areas, especially densely populated cityscapes, require continuous indoor lighting at night for safety and security. Plasmonic nanoparticles, such as $Fe_3O_4@Cu_{2-x}S$, can convert this light, such as light-emitting diode (LED), into heat using the localized surface plasmon resonance (LSPR) effect, which occurs when nanoparticles absorb light and generate heat [6–9]. This effect, largely unexplored in indoor settings, presents a unique opportunity to harness scattered and reflected light for heating applications. By embedding plasmonic nanoparticles within transparent films or coatings, diffused light can be effectively harvested, reducing dependence on conventional heating systems and enhancing building energy efficiency [5].

The LSPR-driven photothermal effect in $Fe_3O_4@Cu_{2-x}S$ nanoparticles enables selective absorption across a broad wavelength range, including ultraviolet (UV) and infrared (IR) spectra, making it ideal for transparent applications in urban environments [10]. Studies show that multiple layers of transparent films coated with these nanoparticles can be stacked to create a powerful photothermal radiator. Each layer absorbs photons from diffused light, converting it into heat, making this setup an efficient solution for energy-efficient building technologies.

2 Fundamentals of the Photothermal Effect

2.1 Photothermal Effect Due to Localized Surface Plasmon Resonance (LSPR)

The photothermal effect, which converts light energy into heat, is a critical phenomenon in energy materials. Among the mechanisms driving this effect, Localized Surface Plasmon Resonance (LSPR) is particularly efficient and occurs when metallic nanoparticles are exposed to specific wavelengths of light [6–9]. LSPR involves the collective oscillation of conduction electrons on the nanoparticle surface when excited by the electromagnetic field of incident light, generating a strong resonance typically in the visible or near-infrared regions (Fig. 5). This process enables efficient absorption of light energy, which is dissipated as heat, resulting in localized heating effects.

When light interacts with metallic nanoparticles, the incident electromagnetic field excites free electrons on their surface, causing them to oscillate collectively at the same frequency as the light. These oscillations amplify light absorption and scattering, with the efficiency depending on factors such as nanoparticle size, shape, material, and surrounding environment. Theoretical analysis of LSPR, governed

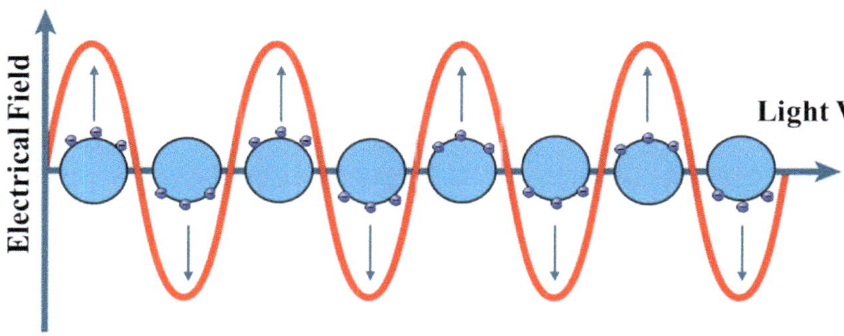

Fig. 5 Schematic diagram showing the Localized Surface Plasmon Resonance (LSPR)

by the dipole approximation of Mie theory [11], provides a framework for understanding how electromagnetic waves interact with sub-wavelength particles, leading to intense localized electromagnetic fields and heat generation. This phenomenon has significant applications in localized energy conversion and heat generation.

LSPR occurs when metallic nanoparticles interact with light of a wavelength larger than the particle size, inducing resonant oscillations of free conduction electrons confined to the nanoparticle's surface [6–9]. These oscillations generate highly localized and intense electromagnetic fields near the particle surface, enhancing light absorption and scattering. Heat is generated as a direct result of energy dissipation from these oscillations. The theoretical framework provided by Mie theory [11] describes this interaction, highlighting the importance of LSPR in nanoscale energy applications.

The significance of LSPR-driven photothermal effects lies in its remarkable ability to convert light into heat with high efficiency, making it a promising mechanism for a range of applications, particularly in solar energy harvesting, drug delivery systems, and thermal management in electronic devices. In the context of solar energy, LSPR can be utilized to improve the performance of photothermal devices, such as solar collectors or photothermal solar cells, by enhancing the absorption of sunlight and converting it into thermal energy. One of the primary advantages of LSPR is its ability to tune the resonance frequency of nanoparticles by varying their size, shape, and material. For example, nanoparticles made from gold, silver, or copper exhibit strong plasmonic behavior and can be engineered to target specific wavelengths of light, optimizing their heat generation for specific applications. This tunability offers the potential for creating tailored materials that can efficiently capture and convert solar radiation across a broad spectrum, further improving the overall efficiency of solar thermal energy systems.

In the dipole approximation, the nanoparticle is treated as a simple dipole that oscillates in response to the electric field [6–9, 11]. The interaction of light with a nanoparticle of radius r, and dielectric function $\varepsilon(\omega)$, can be described by the following general equation:

$$\epsilon_{\text{eff}}(\omega) = [\epsilon_m(\omega) + V(\omega)/\epsilon_m(\omega)][\epsilon(\omega) - \epsilon_m(\omega)] \quad (1)$$

where $\epsilon_{\text{eff}}(\omega)$ is the effective dielectric function of the nanoparticle at a frequency ω, $\epsilon_m(\omega)$ is the dielectric function of the surrounding medium, $\epsilon(\omega)$ is the dielectric function of the metal (typically, this is a complect function that depends on the wavelength of the incident light), and $V(\omega)$ is the nanoparticle volume. The dielectric function of the metal $\varepsilon(\omega)$ determines the frequency at which LSPR occurs. When the frequency of the incident light matches the natural oscillation frequency of the conduction electrons in the nanoparticle, a strong plasmon resonance occurs, leading to maximum absorption of light.

The extinction cross-section $\sigma_{\text{ext}}(\lambda)$ is a key quantity used to describe the light interaction with a nanoparticle. It represents the total energy loss in the direction of light propagation due to the metal and the nanoparticle's geometry. For a spherical nanoparticle, the extinct cross-section can be approximated by the following expression [11]:

$$\sigma_{\text{ext}}(\lambda) = (4\pi r^3/\lambda^4)\,[2\varepsilon(\omega)/\epsilon_m(\omega) + 2\varepsilon(\omega)]^2 \tag{2}$$

where r is the radius of the nanoparticle, l is the wavelength of the incident light, $\varepsilon(\omega)$ and $\epsilon_m\,(\omega)$ are the dielectric functions of the nanoparticle and surrounding medium.

This formulation highlights the dependence of the extinction cross-section on the nanoparticle size and the wavelength of the incident light, with LSPR being most prominent when the resonance frequency of the nanoparticle matches the frequency of the incident light. The enhancement in absorption and scattering leads to increased heat generation, which is a central feature of photothermal applications. In practical applications, the LSPR phenomenon can be tuned by altering the size, shape, and material of the nanoparticles, as well as the surrounding medium, thereby optimizing the absorption and scattering properties for specific light wavelengths [6]. This tunability is what makes LSPR-based systems so effective for photothermal applications, such as solar energy harvesting and photothermal therapy. The theory of LSPR explains how the interaction between light and nanoparticles results in the resonant oscillation of surface electrons, leading to localized electromagnetic field enhancements and efficient energy conversion. By using the dipole approximation of Mie theory and calculating the extinction cross-section, we can quantify and optimize the efficiency of light-to-heat conversion in LSPR-based systems.

2.2 Resonant Oscillation of Surface Electrons and Electromagnetic Field Enhancements

At the heart of LSPR is the resonant oscillation of surface electrons, which is especially pronounced at the particle's surface where the electric field is strongest [6–9, 11]. When the light frequency matches the natural frequency of electron oscillation, a resonance condition is achieved, resulting in enhanced oscillations. These oscillations of electrons collectively behave like an oscillating dipole, creating a localized electric field around the particle's surface. This resonance effect results in a significant amplification of the electromagnetic field near the nanoparticle's surface, often referred to as "near-field" enhancement. This field concentration is the basis for many applications of LSPR, from sensing and imaging to photothermal energy conversion, because it allows for an exceptionally strong interaction between the particle and the incident light.

Using Mie theory [11], we can quantify the interaction of light with spherical particles by calculating parameters such as the extinction cross-section, which combines absorption and scattering effects. For small particles, where the particle size is much smaller than the wavelength of light (typically less than 100 nm), the dipole approximation can simplify the complex Mie calculations. In this approximation, we consider the particle as a simple dipole, and the extinction cross-section σ_{ext} for a particle of polarizability α in an electric field E can be expressed as [6–9, 11]:

$$\sigma_{\text{ext}} = 4\pi k I_m(\alpha)/\varepsilon(\omega)|E|^2 \tag{3}$$

where k is the wavenumber, and $I_m(\alpha)$ denotes the imaginary part of the polarizability, which represents the energy absorbed by the particle. The polarizability α of a particle, which governs how it responds to the electric field, depends on factors such as particle size, shape, and material. For spherical metallic nanoparticles, the polarizability α can be given by:

$$\alpha = 4\pi r^3 \epsilon(\omega) - \epsilon_m(\omega)/\epsilon(\omega) + 2\epsilon_m(\omega) \tag{4}$$

where r is the particle radius, $\epsilon(\omega)$ is the dielectric function of the particle material, and ϵ_m is the dielectric constant of the surrounding medium. When the real part of $\epsilon(\omega)$ approaches $-2\epsilon(\omega)$, the denominator in the polarizability expression minimizes, maximizing α and creating the LSPR condition. This resonant interaction increases the extinction cross-section, resulting in strong absorption and scattering of light, which is key for efficient light-to-heat conversion.

2.3 Light-to-Heat Conversion Efficiency

In LSPR-based systems, the absorbed energy is rapidly converted into heat as the oscillating electrons experience damping, typically within a few femtoseconds. This damping process transforms the electromagnetic energy into kinetic energy, which is then dissipated as thermal energy, leading to a rapid temperature rise in the vicinity of the nanoparticle. By calculating the extinction cross-section, we can optimize factors like particle size, composition, and shape to maximize the absorption of light and thereby the efficiency of light-to-heat conversion [6]. For example, nanoparticles with a larger absorption cross-section relative to their scattering cross-section are generally better suited for photothermal applications, as they absorb more light and convert it to heat rather than scattering it away. Additionally, tuning the particle size and material can shift the LSPR peak to specific wavelengths, allowing for targeted absorption in regions of the spectrum where the available light source (e.g., sunlight or LED) is most intense.

2.4 Optimizing LSPR Efficiency for Applications

By fine-tuning parameters such as the particle's material, size, and shape, we can engineer nanoparticles to achieve maximal light-to-heat conversion. For instance, increasing particle size or employing non-spherical shapes, like nanorods or nano stars, allows us to broaden the LSPR peak or tune it to specific wavelengths, further enhancing photothermal efficiency [6]. Materials such as $Fe_3O_4@Cu_{2-x}S$ exhibit unique optical properties, making them especially effective for photothermal applications, as they display strong LSPR effects across a broad range of wavelengths, including the infrared region. Mie theory and the dipole approximation provide a theoretical framework for understanding and optimizing LSPR effects in nanoparticles, enabling precise control over their light-absorbing properties [11]. By calculating the extinction cross-section and tailoring nanoparticle parameters, we can maximize light absorption and subsequent heat generation, advancing applications in sustainable energy and photothermal technologies.

3 Photothermal Materials: Plasmonic Nanoparticles

3.1 Photothermal Effects on Metallic Nanoparticles

The photothermal effect in metallic nanoparticles arises from the interaction between light and the surface electrons of the metal, where the energy of incident photons is absorbed and then dissipated as heat [9, 11–19]. This process is primarily driven by Localized Surface Plasmon Resonance (LSPR), a phenomenon that occurs when the frequency of incident light matches the natural oscillation frequency of free electrons on the surface of the nanoparticle. These oscillating electrons are confined to the metal's surface and respond to the electromagnetic field of the incoming light, leading to a collective resonance effect. When light strikes the surface of a metallic nanoparticle, its energy is absorbed by the free conduction electrons, which begin to oscillate relative to the fixed lattice of positive ions within the material. This oscillation occurs at the nanoparticle's characteristic resonance frequency, which is determined by the size, shape, and material properties of the nanoparticle. The energy absorption efficiency is maximized when the wavelength of the incident light corresponds to this natural frequency. As the electrons oscillate, they transfer their energy to the lattice, causing atomic vibrations, or phonons, which are ultimately responsible for the dissipation of the energy in the form of heat.

This photothermal effect in metallic nanoparticles is notably distinct from typical thermal absorption mechanisms. In conventional heat absorption, materials absorb electromagnetic radiation over a broad range of wavelengths, and the energy is transferred into the bulk material uniformly. However, in the case of metallic nanoparticles, the localized nature of the plasmonic resonance means that the energy is absorbed and concentrated in very small regions at the surface, leading to highly localized heating. Among the metals that exhibit strong plasmonic resonance and are commonly used in photothermal applications, gold, silver, and copper are particularly prominent [9, 11–19]. These materials are favored due to their favorable optical properties and resonance frequencies that align with the visible and near-infrared regions of the light spectrum, making them ideal for harnessing sunlight. Additionally, their plasmonic behavior can be tuned by varying the size, shape, and surface characteristics of the nanoparticles, allowing for customization of their absorption spectra. For instance, gold nanoparticles can be engineered to strongly absorb light at wavelengths that are optimal for both solar energy harvesting and medical applications, such as photothermal therapy.

The unique properties of LSPR in these metallic nanoparticles enable them to achieve high light-to-heat conversion efficiency, which is critical in applications where photothermal energy conversion is essential. In solar energy systems, such as photothermal solar collectors, metallic nanoparticles can enhance the efficiency of light absorption and conversion to heat, improving the performance of solar thermal devices. In nanomedicine, the localized heating effects can be used for precise thermal treatment of tumors or other targeted tissues without affecting surrounding healthy cells. The photothermal effect in metallic nanoparticles is driven by the resonance of surface electrons with incident light, converting photon energy into localized heat. The ability to control and optimize this effect through material selection and nanoparticle design has led to a wide range of applications, particularly in energy harvesting and medical treatments, where efficient light-to-heat conversion is paramount.

However, the inherent optical properties of metallic nanoparticles limit their use in applications requiring transparency, such as building skins and energy-efficient window designs. Due to their high optical density, metallic nanoparticles exhibit minimal visible light transmittance, often appearing opaque even at relatively low concentrations. While they can generate substantial heat under irradiation, their lack of transparency poses a challenge in building-integrated photothermal systems, where both light transmittance and efficient thermal management are desirable. In building applications, materials that allow natural light transmission while maintaining photothermal functionality are essential to maximize daylighting and reduce the need for artificial lighting. Non-transparent metallic nanoparticles, despite their photothermal efficiency, block visible light, which can reduce interior natural illumination and compromise the aesthetic and functional aspects of building design. Consequently, alternative materials—such as transparent photothermal thin films or nanocomposites with non-metallic photothermal agents like Fe_3O_4—are more suitable for energy-efficient building skins. These materials offer a balance between photothermal heating and visible light transmittance, enabling the construction of building envelopes that improve thermal insulation without sacrificing transparency.

3.2 Electronic Structures of the Superparamagnetic Iron Oxide Nanoparticles

Superparamagnetic Fe_3O_4 nanoparticles, also known as magnetite nanoparticles, have a cubic inverse spinel structure, where Fe ions are distributed between octahedral and tetrahedral sites within an oxygen lattice. This crystalline structure contributes to their magnetic properties and stability, making them suitable for various technological and biomedical applications. At the nanoscale, Fe_3O_4 particles exhibit unique electronic configurations due to their mixed Fe^{2+} and Fe^{3+} oxidation states, which enhance their magnetic behavior and stability. Bulk magnetite (Fe_3O_4) exhibits an inverse spinel structure characterized by a face-centered cubic lattice of O^{2-} ions, where Fe^{3+} ions occupy the tetrahedral sites and both Fe^{3+} and Fe^{2+} ions are distributed in approximately equal amounts within the octahedral sites, represented by the chemical formula $(Fe^{3+})[Fe^{3+}Fe^{2+}](O^{2-})_4$ [19]. Studies indicate that the valence band of the O(2p) orbitals up to the empty Fe(4s) orbitals in Fe_3O_4 is separated by an energy gap of 4–6 eV [19]. Between these primary bands lie the crystal field bands corresponding to the octahedral and tetrahedral sites, composed of 3d metal atomic orbitals (Fig. 6). As illustrated in this figure, the approximate band structure of Fe_3O_4 nanoparticles was estimated through our photoluminescence measurements. Both experimental and theoretical investigations report a crystal-field splitting energy gap of ~2.2 eV at the octahedral sites and ~0.9 eV at the tetrahedral sites. Additionally, the valence band of O(2p) is separated by ~0.9 eV from the crystal field t_{2g} and e bands of both the octahedral and tetrahedral sites, respectively [19–21].

Fe_3O_4 nanoparticles are notable for their superparamagnetic behavior, which only appears at the nanoscale. In this state, the particles demonstrate strong magnetic responses when subjected to an external magnetic field but lose this magnetism in the absence of the field, minimizing agglomeration due to magnetic interactions. This property is crucial in applications where stability and reversible magnetic control are

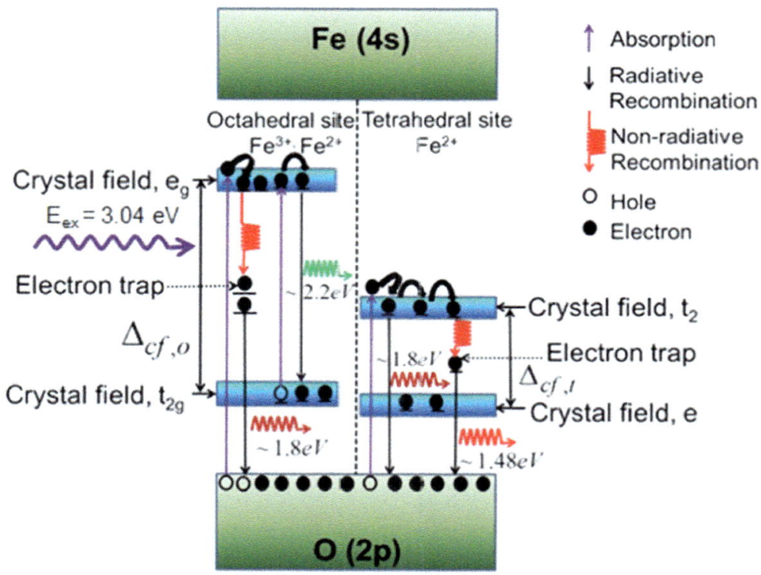

Fig. 6 Schematic of the energy bands of the Fe_3O_4 nanoparticle system [22]

desired, such as in targeted drug delivery and MRI contrast enhancement. Additionally, the particles have high thermal conductivity, allowing efficient heat dissipation and stability under thermal fluctuations. Their small size, high surface-to-volume ratio, and surface chemistry adaptability make them well-suited for functionalization, improving compatibility with biological systems and broadening their application potential.

Fe_3O_4 nanoparticles demonstrate a strong photothermal response by absorbing near-infrared (NIR) light and transforming it into thermal energy [22–32]. This property arises from their efficient light absorption and heat conversion capabilities, which are influenced by factors such as particle size, surface coating, and laser wavelength. Under NIR irradiation, Fe_3O_4 nanoparticles can rapidly increase in temperature, making them effective for applications such as hyperthermia-based cancer therapy, where localized heating can destroy cancer cells with minimal impact on surrounding healthy tissues. The photothermal response of Fe_3O_4 nanoparticles can be tuned by modifying particle size, morphology, and surface properties, allowing optimization for specific applications.

The superparamagnetic and photothermal properties of Fe_3O_4 nanoparticles have driven their use across several fields, particularly in medicine and environmental science. Some of the key applications include biomedical imaging and diagnostics: Fe_3O_4 nanoparticles are widely used as contrast agents in magnetic resonance imaging (MRI) due to their superparamagnetic properties, which enhance image contrast [33]. Their ability to accumulate in specific tissues or organs when functionalized with targeting ligands allows for high-resolution imaging of tumors and other abnormalities. Drug Delivery: Superparamagnetic Fe_3O_4 nanoparticles serve as carriers in targeted drug delivery systems. By attaching therapeutic agents to the nanoparticles and guiding them with external magnetic fields, precise drug delivery to target sites can be achieved, reducing side effects and improving therapeutic efficacy. Photothermal Therapy (PTT): The photothermal effect in Fe_3O_4 nanoparticles enables their use in cancer treatment through photothermal therapy [22–32]. When exposed to NIR light, the nanoparticles generate localized heat that can effectively destroy cancer cells. Functionalized Fe_3O_4 nanoparticles that specifically target tumor cells enhance this treatment's precision and efficacy. In summary, superparamagnetic Fe_3O_4 nanoparticles possess unique structural, magnetic, and photothermal properties that underpin their broad application range. Their versatility in both medical and environmental fields continues to make them a focal point for ongoing research and development.

$Fe_3O_4@Cu_{2-x}S$ nanoparticles represent an innovative class of multifunctional materials that combine the magnetic properties of iron oxide (Fe_3O_4) with the photothermal and optoelectronic properties of copper sulfide ($Cu_{2-x}S$) [12, 28, 34]. These core–shell structures have garnered significant attention in nanotechnology due to their synergistic properties, which make them highly versatile for a wide range of applications in energy, medicine, and environmental science. The Fe_3O_4 core, well-known for its superparamagnetic behavior and biocompatibility, allows for precise control and manipulation of the nanoparticles using external magnetic fields. This property is particularly valuable in applications such as magnetic resonance imaging

(MRI), targeted drug delivery, and magnetic separation [22–32]. On the other hand, the $Cu_{2-x}S$ shell exhibits excellent near-infrared (NIR) absorption, making it ideal for photothermal therapy (PTT), solar energy harvesting, and optoelectronic devices.

One of the key advantages of $Fe_3O_4@Cu_{2-x}S$ nanoparticles lies in their ability to integrate magnetic and photothermal functionalities into a single platform. This enables applications such as magnetically targeted photothermal cancer therapy, where the particles can be directed to tumor sites using a magnetic field and then activated with NIR light to generate localized heat, effectively destroying cancer cells. Moreover, the unique optical properties of $Cu_{2-x}S$, arising from its plasmonic resonance and tunable bandgap, make these nanoparticles highly efficient in converting light to heat. This property has been exploited in energy systems, such as transparent photothermal coatings for solar energy modulation and hybrid energy systems combining photovoltaic, thermoelectric, and photothermal processes [35]. The development of $Fe_3O_4@Cu_2-_xS$ nanoparticles also highlights major progress in synthesis techniques. Methods such as solvothermal synthesis and chemical co-precipitation ensure precise control over the size, morphology, and composition of these nanoparticles, which are critical for optimizing their performance in various applications. In summary, $Fe_3O_4@Cu_{2-x}S$ nanoparticles represent a remarkable convergence of magnetic, optical, and thermal functionalities. Their diverse range of applications highlights their potential to address critical challenges in healthcare, energy, and environmental sustainability, establishing them as a cornerstone in the next generation of multifunctional nanomaterials.

3.3 Superparamagnetic Properties of Fe_3O_4 Nanoparticles and Magnetic Hysteresis Behavior

Superparamagnetic Fe_3O_4 nanoparticles, also referred to as magnetite nanoparticles, exhibit unique magnetic properties due to their nanoscale dimensions and crystalline structure [19–22, 24]. These nanoparticles possess a cubic inverse spinel structure, where Fe ions occupy both octahedral and tetrahedral sites in an oxygen lattice. This well-defined arrangement contributes significantly to their magnetic properties and structural stability, which make them versatile materials for technological and biomedical applications, such as magnetic drug delivery, hyperthermia therapy, and magnetic resonance imaging (MRI) contrast agents. At room temperature, Fe_3O_4 nanoparticles smaller than a critical size (typically less than 20 nm) exhibit superparamagnetic behavior [24]. This phenomenon arises because thermal energy at room temperature is sufficient to overcome the magnetic anisotropy energy barrier that aligns the magnetic moments of the nanoparticles. Consequently, their magnetic moments fluctuate randomly when no external magnetic field is applied. However, under an external magnetic field, the magnetic moments align in the field's direction, resulting in strong magnetization. When the field is removed, the thermal agitation

Fig. 7 Magnetization hysteresis curves of Fe_3O_4 nanoparticle with the sizes indicated

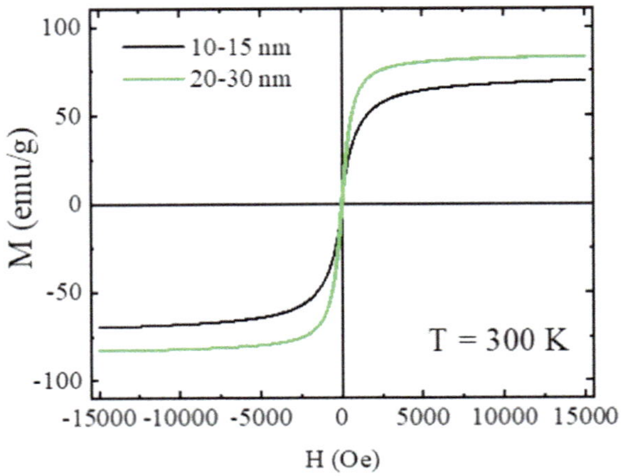

restores the random orientation of the magnetic moments, leading to zero remanence (no residual magnetization).

The magnetic hysteresis behavior of superparamagnetic Fe_3O_4 nanoparticles is characterized by a highly reversible loop (Fig. 7). Unlike ferromagnetic materials, where the hysteresis loop shows remanence and coercivity, superparamagnetic nanoparticles exhibit negligible coercivity and remanence. This means that the magnetization curve for these particles is symmetrical and retraces itself during the magnetization and demagnetization processes [22, 24]. This reversibility indicates minimal energy loss due to magnetic hysteresis, making these nanoparticles ideal for applications requiring efficient and repeatable magnetic performance, such as alternating magnetic fields used for hyperthermia treatments or magnetic fluid applications. In summary, the superparamagnetic properties of Fe_3O_4 nanoparticles stem from their nanoscale dimensions and unique inverse spinel structure. The highly reversible magnetic hysteresis behavior further enhances their suitability for various advanced applications, enabling efficient and energy-saving performance in biomedical and technological fields.

3.3.1 Synthesis and Optical Properties of Fe_3O_4 and $Fe_3O_4@Cu_{2-x}S$ Nanoparticles

Research on the synthesis and characterization of iron oxide nanoparticles have been carried out extensively [22–28]. To synthesize Fe_3O_4 nanoparticles [34], a chemical coprecipitation method was used, yielding particles with an average diameter of approximately 10 nm. The synthesized Fe_3O_4 nanoparticles were dispersed directly in octane, forming a stable ferrofluid. This ferrofluid was then combined with an aqueous solution containing the surfactant sodium dodecyl sulfate (SDS) to create a miniemulsion, effectively encapsulating Fe_3O_4 nanoparticles within dispersed octane droplets. Separately, a second miniemulsion was prepared by dispersing styrene monomer droplets in water, stabilized by SDS to prevent coalescence. The two miniemulsions—one containing the Fe_3O_4 ferrofluid and the other containing styrene monomer droplets—were then mixed thoroughly to ensure a uniform distribution of Fe_3O_4 nanoparticles within the styrene droplets. The combined emulsion was subsequently polymerized by heating the solution to 80 °C and maintaining it for 20 h. This process resulted in the formation of polystyrene nanospheres, within which Fe_3O_4 nanoparticles were embedded, forming a stable composite structure. To enhance the functionality of these nanospheres, a silica coating was applied to the surface by following a previously reported silica-coating method. This final step yielded Si/Polystyrene/Fe_3O_4 composite nanospheres with a well-defined core–shell structure, combining magnetic and surface-functional properties that are advantageous for various biomedical and industrial applications.

To synthesize Fe_3O_4@$Cu_{2-x}S$ nanoparticles, 60 ml of oleylamine was heated to 300 °C in a three-necked flask under a nitrogen atmosphere and stirred for 30 min [34]. Then, a solution of 12 ml oleylamine, 8 ml N-methyl-2-pyrrolidone, and 700 mg of $Fe(acac)_3$ was injected into the flask. The mixture was maintained at 300 °C for 10 min before being cooled to 70 °C and held there for another 10 min. Next, 128.28 mg of sulfur was dissolved in 12 ml of oleylamine and mixed with 10 ml of cyclohexane. This sulfur solution was then injected into the main reaction mixture, followed by stirring at 70 °C for 10 min. Afterward, 523.52 mg of $Cu(acac)_2$ dissolved in a solution of 4 ml oleylamine and 16 ml chloroform was added. The final mixture was kept at 70 °C with continuous stirring for an additional 30 min. The resulting Fe_3O_4@$Cu_{2-x}S$ nanoparticles were collected using a strong magnet, washed three times with methanol to remove impurities, and then freeze-dried. After drying, the nanoparticles were dispersed in toluene for future applications.

Figure 8 shows the transmission electron microscopy (TEM) images of Fe_3O_4 (Fig. 8a) and Fe_3O_4@$Cu_{2-x}S$ (Fig. 8b) nanoparticles. As shown in this figure, the average size of Fe_3O_4 is around 10 nm, while that of $Cu_{2-X}S$ coated nanoparticles is about 15 nm. This considerable size difference suggests the $Cu_{2-X}S$ coating on the nanoparticles of Fe_3O_4, which is consistent with the report by Tian et al. [28].

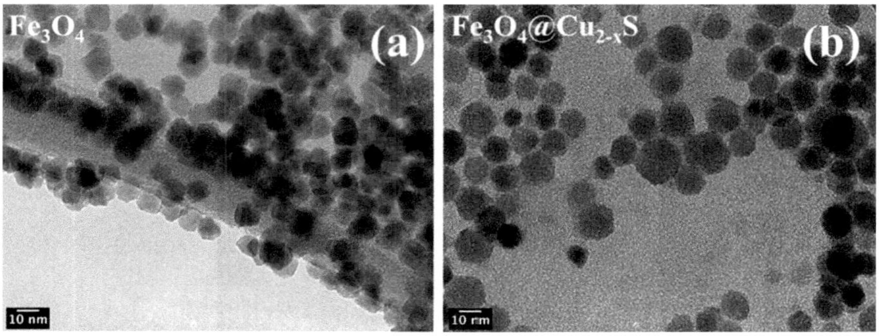

Fig. 8 Transmission electron microscopy images of **a** the Fe_3O_4 and **b** Fe_3O_4@$Cu_{2-x}S$ nanoparticles in solution (0.1 mg/ml) [34]

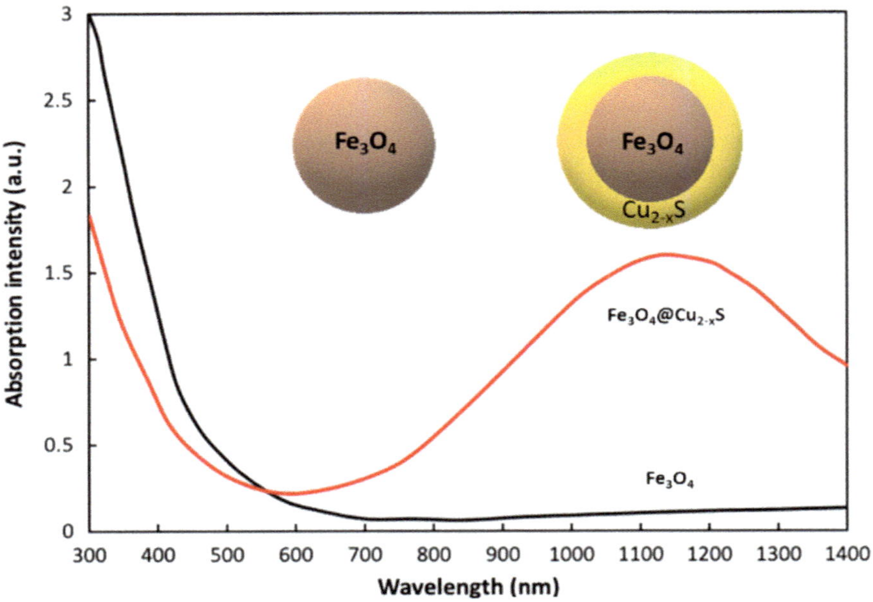

Fig. 9 **a** Absorption spectra of the Fe_3O_4 and **b** $Fe_3O_4@Cu_{2-x}S$ solution (0.1 mg/ml)

Figure 9 shows the absorption spectra of Fe_3O_4 and $Fe_3O_4@Cu_{2-x}S$ suspensions at a concentration of 0.1 mg/ml [34]. The $Fe_3O_4@Cu_{2-x}S$ suspension displays a pronounced absorption peak at 1160 nm, which is shorter than the 1349 nm peak observed in its thin-film counterpart, highlighting the influence of particle interactions on absorption properties. Additionally, the minimum of the characteristic "U"-shaped curve shifts from 639 nm in the thin film to 560 nm in the suspension, as both Fe_3O_4 thin films and suspensions exhibit decreased absorption beyond 700 nm (Fig. 9). This red-shifted peak in $Fe_3O_4@Cu_{2-x}S$ is attributed to the plasmonic coupling between nanoparticles, as previously reported [12, 28, 34]. According to Tian et al. [28], the NIR absorption in $Fe_3O_4@Cu_{2-x}S$ primarily originates from the $Cu_{2-x}S$ shell. To improve NIR absorption, a core–shell structure was designed, with Fe_3O_4 as the core and $Cu_{2-x}S$ as the shell [12, 28, 34].

The $Cu_{2-x}S$ shell on Fe_3O_4 introduces substantial NIR absorption, seen in the strong absorption peak at 1160 nm in $Fe_3O_4@Cu_{2-x}S$, absent in the Fe_3O_4-only solution, underscoring the role of particle interactions. $Cu_{2-x}S$ nanocrystals also exhibit tunable localized surface plasmon resonance (LSPR) in the NIR region [12, 28, 34, 35], a phenomenon influenced by nanoparticle size, shape, and surrounding environment, which several studies have explored as a mechanism for energy conversion [36].

3.3.2 Photothermal Properties of Iron Oxide Nanoparticles

Localized Surface Plasmon Resonance (LSPR) in Fe_3O_4 nanoparticles offers unique advantages, particularly for biomedical applications where deep tissue penetration and efficient heat generation are essential. LSPR is a phenomenon typically associated with metallic nanoparticles, such as gold and silver, where incident light induces coherent oscillations of conduction electrons on the nanoparticle surface, leading to enhanced light absorption and scattering. While Fe_3O_4, being a transition metal oxide, does not have the free electron density that characterizes noble metals, it still exhibits plasmonic-like behavior under certain conditions due to the oscillation of surface-bound electrons. This enables Fe_3O_4 nanoparticles to resonate at specific wavelengths, especially when engineered for near-infrared (NIR) excitation [18, 34, 35].

Figure 10a shows the change in temperature increase as a function of time for the Fe_3O_4 thin films of various concentrations irradiated by white light (i.e., simulated solar light). As shown in this figure, one can see that the maximum temperature increase reaches 6.53 °C under simulated solar. Figure 10b shows the change in temperature increase, ΔT, as a function of time for the $Fe_3O_4@Cu_{2-x}S$ thin films of various concentrations irradiated by white light (i.e. simulated solar light). As shown in this figure, ΔT_{max} reaches 7.10 °C, considerably higher than that of the Fe_3O_4 thin film. The ability of Fe_3O_4 nanoparticles to exhibit LSPR in the NIR range is particularly valuable in medical applications, where NIR wavelengths (typically in the range of 700–1000 nm) can penetrate biological tissues more effectively than visible or ultraviolet light. This range is known as the "biological transparency window," where absorption and scattering by water, hemoglobin, and other tissue components are minimized, allowing deeper penetration of light. When Fe_3O_4 nanoparticles are optimized for absorption within this range, they can efficiently reach and affect deeper tissue layers, a characteristic highly advantageous for treatments such as photothermal therapy.

Through careful synthesis and tuning, Fe_3O_4 nanoparticles can be engineered to enhance their optical absorption in the NIR region. This involves controlling parameters like particle size, shape, and surface modification to manipulate how the nanoparticles interact with light [36]. When optimized for NIR absorption, Fe_3O_4 nanoparticles can convert absorbed light into heat with high efficiency. This photothermal effect is a result of non-radiative relaxation processes, where the absorbed photon energy is dissipated as thermal energy, increasing the local temperature around the nanoparticles. In biomedical contexts, this localized heating is particularly useful for selectively targeting cancer cells.

The LSPR and optical absorption properties of Fe_3O_4 nanoparticles continue to be a subject of research for further enhancement and optimization. Current efforts are focused on synthesizing Fe_3O_4 nanoparticles with precise control over morphology and surface functionalization to maximize their LSPR intensity and photothermal conversion efficiency. Advanced surface modifications, such as coating with biocompatible polymers or conjugating targeting ligands, can also enhance their stability and selectivity for cancer cells, allowing for more effective and safe applications in

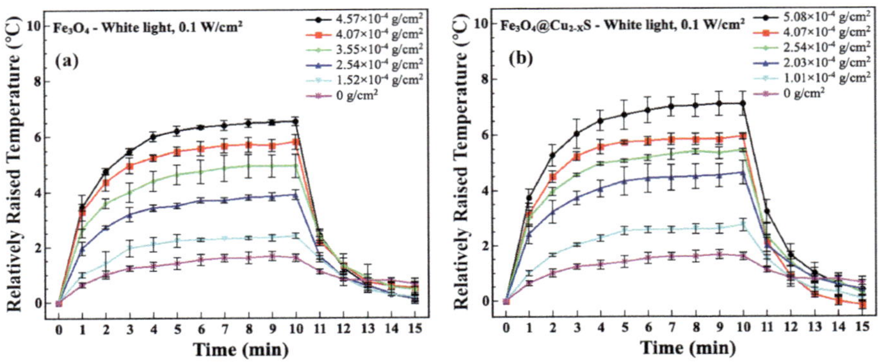

Fig. 10 **a** The change in temperature increase, ΔT, versus time curves under white light irradiation for the Fe_3O_4 thin films of different concentrations. **b** The change in temperature increase, ΔT, versus time curves under white light irradiation for the $Fe_3O_4@Cu_{2-x}S$ thin films of different concentrations. The environmental temperature is 25 °C [34]

clinical settings. Fe_3O_4 nanoparticles, with their NIR-tuned LSPR, biocompatibility, and magnetic properties, represent a promising, multifunctional tool for applications in photothermal cancer therapy and beyond. Their ability to absorb NIR light and convert it to localized heat enables non-invasive treatments, offering a safer, targeted alternative to conventional cancer therapies.

4 Applications of Plasmonic Nanoparticles

4.1 Application of the Photothermal Effect in Medical Therapeutics

Compared to metallic nanoparticles, Fe_3O_4 offers unique advantages in biocompatibility and non-toxicity, making it a preferred material for applications that require long-term stability and minimal adverse effects. In addition to being safer for biological use, Fe_3O_4 nanoparticles are relatively inexpensive and possess magnetic properties, which allow them to be guided or localized within the body using an external magnetic field. This magnetic control, combined with their NIR-tuned LSPR, makes Fe_3O_4 nanoparticles particularly promising for multifunctional therapeutic applications where both imaging and therapy can be achieved. In photothermal cancer therapy, Fe_3O_4 nanoparticles are introduced into the target tissue, where they absorb NIR light and convert it to heat, selectively killing cancer cells [16–18, 23–25]. This approach leverages the fact that cancer cells are more sensitive to temperature increases compared to normal cells. When the local temperature reaches the therapeutic range of 42–45 °C, cancer cells undergo apoptosis or necrosis, effectively reducing tumor mass with minimal impact on surrounding healthy tissue. The deep tissue penetration afforded by NIR-activated Fe_3O_4 nanoparticles allows for non-invasive treatments, even for tumors located beneath several layers of tissue, which would otherwise be challenging to access with visible or ultraviolet light.

The photothermal effect, particularly through Localized Surface Plasmon Resonance (LSPR), has shown immense promise in medical therapeutics, with a primary focus on cancer therapy [6, 16–18, 22–33, 35–42]. In photothermal therapy (PTT), metallic nanoparticles, such as gold or silver, are designed to accumulate specifically in cancerous tissues [20, 32, 33]. Upon exposure to a targeted light source, typically in the near-infrared (NIR) range, these nanoparticles undergo LSPR, resulting in localized heating that can effectively destroy cancer cells without damaging surrounding healthy tissue. This ability to generate precise thermal effects opens up new avenues in non-invasive cancer treatments, offering a targeted approach to eradicate tumor cells with minimal side effects and reduced recovery times compared to conventional treatments.

The efficacy of photothermal therapy lies in the high photothermal conversion efficiency of plasmonic nanoparticles, which can absorb light and convert it into heat at the cellular level. The NIR region is particularly beneficial for cancer therapy, as it allows for deeper tissue penetration due to lower absorption by biological tissues in this wavelength range. By finely tuning the size, shape, and composition of nanoparticles, researchers can control the resonance frequency and optimize the heat generation, allowing the particles to selectively absorb light at specific wavelengths. This selectivity not only enhances precision but also improves the efficacy and safety of the procedure, reducing the likelihood of harm to healthy cells. These "theranostic" nanoparticles can carry therapeutic agents to the tumor site and release them upon heating, resulting in a synergistic effect that enhances treatment outcomes [22, 28–32]. The controlled release of drugs under the influence of localized heating allows for a dual-action approach—direct cell destruction through heating and targeted drug release for sustained therapeutic effects. Such advancements in photothermal applications are revolutionizing the field of nanomedicine, offering new solutions for treating complex diseases like cancer while minimizing side effects and improving patient outcomes.

The unique properties of the photothermal effect via LSPR are thus paving the way for transformative innovations in medical therapeutics and oncology, with potential applications extending to other diseases where targeted cell destruction is beneficial. As research continues, the integration of photothermal therapy with other treatment modalities promises to yield comprehensive, effective therapeutic options that leverage the high energy conversion efficiency of LSPR-based materials.

4.2 Novel Photothermal Materials for Energy Applications

Recent advancements in materials science have led to the development of innovative photothermal materials that offer enhanced efficiency and broaden the range of applications in energy harvesting and thermal management [4, 5, 10]. These novel materials aim to improve upon traditional photothermal systems by offering advantages such as better spectral selectivity, greater transparency, and tunable optical properties. Among the most promising new materials for photothermal applications are magnetic nanocomposites, graphene, and hybrid plasmonic structures, each of which presents unique opportunities for improving the performance of energy systems while maintaining or enhancing other functional properties, such as transparency and mechanical stability.

One example of such a material is the $Fe_3O_4@Cu_{2-x}S$ nanoparticles, which combine the properties of iron oxide (Fe_3O_4) with copper sulfide ($Cu_{2-x}S$) to form hybrid nanostructures [22, 25–29, 34, 35]. These nanoparticles have been found to exhibit strong Localized Surface Plasmon Resonance (LSPR) in the near-infrared (NIR) region, which is highly advantageous for energy applications, as this wavelength range is both less absorbed by the atmosphere and more efficiently transmitted through various materials, making it ideal for applications like photothermal energy harvesting and thermal management in building materials. The hybrid structure of $Fe_3O_4@Cu_{2-x}S$ nanoparticles also ensures that the material retains its photothermal efficiency while potentially improving the transparency in the visible range of the spectrum, which is crucial for applications such as window coatings that need to maintain aesthetic appearance and visual clarity while still harnessing solar energy.

These advanced hybrid materials are especially suitable for applications that require high performance in the NIR spectrum, such as smart windows and building-integrated photovoltaics. In these systems, the goal is often to absorb and convert sunlight into heat while allowing visible light to pass through unimpeded. The $Fe_3O_4@Cu_{2-x}S$ nanoparticles, for instance, could be integrated into transparent coatings that enable a building to absorb solar radiation during the day, effectively keeping the interior warm without the need for additional heating systems. Furthermore, such coatings could also utilize indoor LED lighting to maintain thermal comfort during evening hours, offering an energy-efficient solution that minimizes the need for traditional heating methods [5]. Other novel photothermal materials, such as graphene and hybrid plasmonic structures, provide additional avenues for enhancing energy applications [16]. Graphene, with its high thermal conductivity and optical transparency, can serve as an excellent material for enhancing the photothermal effects of nanoparticles. It can be combined with plasmonic metals like gold or silver to form hybrid structures that offer improved thermal management and spectral tuning. These materials can be designed to selectively absorb light in the optimal wavelengths for energy harvesting, while still allowing for sufficient light transmission in other ranges.

The development of these novel materials for photothermal applications is opening up new opportunities for energy-efficient buildings, solar collectors, and thermal management systems. By combining plasmonic effects, magnetic properties, and advanced nanostructures, researchers are able to design materials that not only optimize the conversion of light to heat but also integrate seamlessly into existing infrastructure. The ability to fine-tune the optical properties of these materials through composition, size, and structure allows for tailored solutions to specific energy challenges, offering the potential for enhanced energy savings and sustainability in diverse applications. In summary, the continued exploration of new photothermal materials, including magnetic nanocomposites like $Fe_3O_4@Cu_{2-x}S$, graphene, and hybrid plasmonic structures, promises significant advances in energy-saving applications. These materials offer improved spectral selectivity, greater transparency, and the ability to efficiently convert light into heat, opening the door to more effective solar energy harvesting, building-integrated energy solutions, and smart thermal management systems.

5 Plasmonic Nanoparticles in Green Energy Applications

Plasmonic nanoparticles are gaining traction in energy applications due to their ability to convert light into heat with high efficiency, offering potential breakthroughs in sustainable building utilities [4, 43–46]. The localized surface plasmon resonance (LSPR) effect in these nanoparticles allows them to absorb light across a range of wavelengths, from visible to near-infrared (NIR), making them particularly useful for harvesting solar energy. When incorporated into building materials, plasmonic nanoparticles can enhance the thermal efficiency of structures by converting absorbed light—whether from the sun or indoor LED sources—into heat, reducing the need for additional heating energy.

For instance, hybrid plasmonic nanoparticles, such as $Fe_3O_4@Cu_{2-x}S$, have shown promise in energy-saving applications due to their strong LSPR response in the NIR region, which corresponds to a significant portion of solar irradiance. When embedded in building windows, coatings, or facades, these nanoparticles can passively heat the interior by converting incident sunlight into thermal energy, which then radiates throughout the space. This capability helps maintain thermal comfort in buildings, reducing dependence on traditional heating systems and lowering energy consumption during colder months.

An additional benefit of plasmonic nanoparticles is their potential to enhance building efficiency by harvesting ambient indoor light. The plasmonic materials can effectively absorb light emitted from indoor LED sources, a key feature for energy-neutral buildings where sustainable energy practices extend to both day and night cycles. By using plasmonic coatings on windowpanes or walls, indoor light that would typically be lost can be absorbed and converted into heat, contributing to the overall thermal balance of the building. This ability to harness both outdoor and indoor lighting minimizes energy losses and maximizes resource utilization, offering a novel solution for sustainable building management.

Furthermore, plasmonic nanoparticles can be engineered for specific applications by fine-tuning their size, shape, and composition to target desired light wavelengths. This adaptability enables designers to customize nanoparticles to work in tandem with a building's lighting and heating requirements, optimizing the balance between transparency, spectral selectivity, and heat generation. In smart building systems, plasmonic nanoparticle-enhanced coatings could be paired with thermochromic materials, enabling dynamic responses to fluctuating environmental conditions and further improving energy efficiency.

Overall, plasmonic nanoparticles present a versatile approach to building utility heating by simultaneously harnessing solar and indoor light, offering a promising pathway for sustainable energy applications in modern architecture. This technology aligns with global goals to reduce carbon emissions and increase energy efficiency, paving the way for innovative building solutions that prioritize thermal comfort, resource conservation, and environmental responsibility.

5.1 Building Utility Heating and Energy Saving Through Solar Light Harvesting

Given that residential and commercial buildings account for a significant portion of total energy use [1], transitioning to energy-neutral structures offers substantial benefits, including reduced utility costs, enhanced grid stability, improved air quality, better public health, and a cleaner environment solution for addressing the energy demands of buildings is the adoption of solar energy, known for its technical, environmental, and ecological advantages. While solar cells have become widely popular, they still present certain limitations. Despite the vast potential, conventional PV systems face economic and environmental challenges. Typical 2D PV systems require flat surfaces for optimal solar harvesting, which can limit efficiency and energy density, especially in densely populated urban areas where space is limited. This reliance on expansive, flat areas limits the potential for maximizing solar absorption and complicates large-scale energy production in urban environments, where available land is increasingly scarce. In traditional solar building utility heating, various components work together to capture and utilize solar energy for heating applications. Central to these systems are solar collectors—either PV panels that convert sunlight to electricity or solar thermal collectors that absorb sunlight to generate heat. Heat transfer fluids collect heat to the building's heating system, while heat exchangers, storage tanks, and control mechanisms optimize and distribute the heat where it is needed. Although sustainable, these systems have notable drawbacks, including high installation costs, substantial space requirements, and seasonal efficiency variations, along with maintenance and aesthetic challenges. Efforts to enhance these systems are ongoing, with advancements aiming to improve cost-effectiveness, efficiency, and integration with existing infrastructure.

In response to the limitations of traditional PV and solar heating systems, there is a shift toward three-dimensional (3D) solar harvesting approaches, which can increase the energy density by expanding the surface area exposed to sunlight. For building heating, a direct solar-to-thermal conversion through photothermal (PT) materials is emerging as an efficient alternative. A recent innovation in this field involves transparent multilayer PT films, arranged in 3D to capture solar energy across multiple layers, thereby increasing solar harvesting efficiency. This approach, illustrated in Figs. 2 and 3, employs transparent PT thin films that allow solar light to pass through each layer, activating the photothermal effect on each surface and enhancing energy density. Multi-layer PT systems present a promising solution for building heating, offering improved efficiency over traditional PV systems and reducing land requirements.

5.1.1 Photothermal Solar Tunnel Radiator (PSTR)

Building on the concept of multi-layer transparent PT panels (Figs. 3 and 4), this research introduces the Photothermal Solar Tunnel Radiator (PSTR), a new paradigm for solar harvesting (Fig. 11) [4, 5]. This innovative system utilizes multiple transparent panels with nano-photothermal coatings to efficiently harvest solar energy [34, 35]. At the core of the PSTR is a solar tunnel, which captures sunlight through the solar dome on the rooftop and channels it through transparent glass substrates coated with a PT film containing plasmonic Fe_3O_4 nanoparticles [4].

As shown in Fig. 11, the solar dome captures sunlight at various angles throughout the day and channels it into the solar tunnel, reaching all the way to the base. When the sunlight contacts the photothermal coatings on each panel, heat is generated, elevating the temperature within the entire tunnel structure. Operating as a thermal radiator, the Photothermal Solar Tunnel Radiator (PSTR) efficiently radiates this collected heat without the need for external power sources or PV panels. Since the sunlight is directly converted to heat, the efficiency is much higher reaching above 65% [34, 35]. At the tunnel's end, a diffuser distributes light within indoor spaces, providing illumination to areas that lack direct sunlight exposure.

Direct photothermal heating offers several advantages over traditional PV-based solar heating methods. Unlike PV systems that require multiple components (such as heat transfer fluids, exchangers, and storage tanks), photothermal systems can convert solar energy directly into heat, potentially simplifying installation and reducing maintenance costs. By streamlining the energy conversion process, photothermal systems minimize energy losses and improve efficiency, making them an economically attractive solution for building heating. Furthermore, 3D solar harvesting through multi-panel PT films enhances energy density and scalability, offering a larger active surface area exposed to sunlight compared to 2D PV systems. This approach aligns with environmental goals, providing a sustainable, scalable, and aesthetically adaptable solution for building heating.

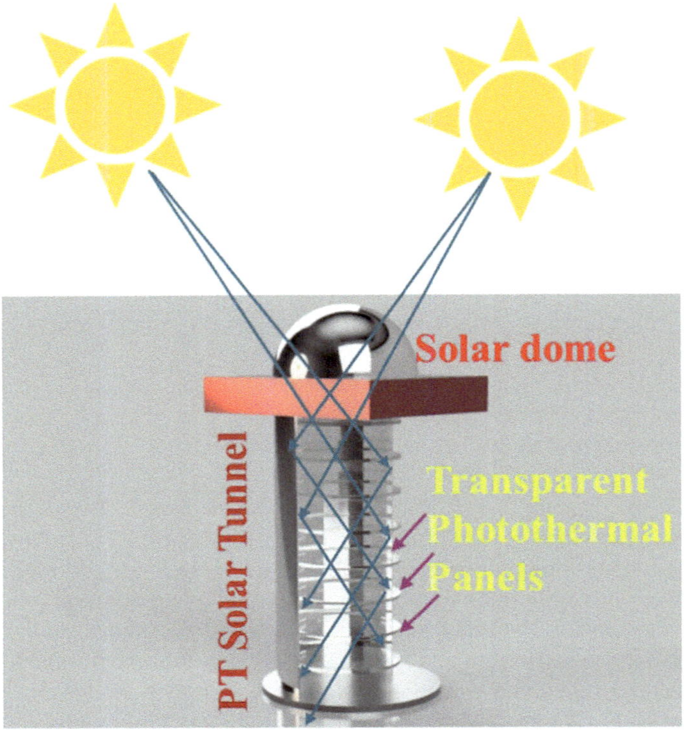

Fig. 11 Diagram showing the Photothermal Solar Tunnel Radiator (PSTR) concept [4, 5]

The primary aim of this research is to develop a groundbreaking solution for building heating that aligns with energy-neutral principles. This goal is exemplified by the so-called "Photothermal Solar House (PSH)," designed to rely entirely on multi-layer PT thin films for heating under natural sunlight. PSH is fundamentally different from the conventional "Solar House" that relies upon PV panels with more complicated system components. By utilizing multilayer PT films, PSH will overcome the limitations of conventional heating methods, which often depend on non-renewable energy sources and contribute to environmental degradation. The PSH serves as a tangible demonstration of the potential for PT thin films to provide a sustainable, energy-efficient solution for building heating, reducing reliance on traditional heating systems and their associated environmental impacts.

This research highlights a critical step toward energy-neutral building heating solutions through the direct conversion of solar light into thermal energy. The lab-scale "photothermal solar house" serves as a model for scalable applications of multi-layer PT thin films in building heating, offering a more sustainable and eco-friendly alternative to conventional methods. By advancing this novel approach, we aim to contribute meaningfully to the transition to carbon–neutral buildings and a more sustainable future.

5.1.2 Architectural Integration of Photothermal Solar Tunnel Radiators (PSTR) in Energy-Neutral Building Design

Integrating Photothermal Solar Tunnel Radiators (PSTR) into building design presents innovative opportunities for architects and engineers focused on creating energy-neutral, sustainable structures [4, 5]. The PSTR system, with its capacity to capture and convert sunlight directly into thermal energy, provides architects with flexible design options that enhance both aesthetic appeal and functional efficiency in buildings. By embedding these solar harvesting structures within the architectural layout, buildings can achieve significant energy savings, lower operational costs, and contribute to net-zero emissions targets.

Rooftops offer an optimal location for integrating Photothermal Solar Tunnel Radiator (PSTR) systems, as a solar dome positioned here can capture abundant sunlight from multiple angles throughout the day (Fig. 12). This rooftop installation allows the PSTR to maximize solar absorption and direct sunlight down into the building through a specially designed tunnel. By incorporating a skylight adaptation of the PSTR, the system fulfills a dual purpose: not only does it deliver natural lighting throughout the structure, but it also captures solar heat for use in building utility applications, including space heating and water preheating.

This dual functionality serves as both a heating system and a dynamic architectural feature, enhancing the aesthetic and thermal comfort of common areas such as atriums, lobbies, or open-plan office spaces. The system is designed to channel sunlight through a series of photothermal coatings within the tunnel, which absorb solar energy and convert it into heat. This heat radiates from the PSTR panels to warm the surrounding environment, reducing the need for external heating sources and improving energy efficiency. As illustrated in Fig. 12, the design also allows sunlight to travel through the tunnel and reach a diffuser at the tunnel's endpoint. This diffuser disperses the light, illuminating interior spaces that might otherwise require artificial lighting. Thus, the PSTR not only provides targeted heating but also enhances daylight access, creating a visually appealing interplay of light and warmth in interior spaces. This integrated system adds a unique visual component to the building design, offering occupants a closer connection to natural light and thermal comfort while advancing energy-neutral building goals.

Facade Integrated PSTR: Incorporating PSTR systems within building facades introduces a vertical, space-efficient solar harvesting solution, transforming external walls into energy-generating surfaces (Fig. 13) [4, 5]. By utilizing multiple transparent or semi-transparent glass panels coated with photothermal materials, the building facade can become a significant source of renewable heat energy, blending function with modern architectural aesthetics. These photothermal panels are strategically positioned to capture solar light across a larger surface area than traditional rooftop installations, enabling a more extensive energy collection and distribution

Fig. 12 Schematic diagram showing the dual purpose of the Photothermal Solar Tunnel Radiators (PSTR)

mechanism throughout the building's skin. One key advantage of this design lies in the panels' ability to significantly raise the facade's surface temperature, thereby narrowing the temperature gradient between the building's interior and exterior. This reduction in temperature difference in the room interior and building skin minimizes heat loss, as less thermal energy escapes from the warmer indoor environment to the cooler exterior, leading to a more energy-efficient building. This effect is particularly beneficial in colder climates or during winter months, as the heated facade acts as an insulation barrier, retaining warmth and reducing the building's reliance on internal heating systems.

For high-rise buildings, where rooftop space is often limited, the facade-integrated PSTR system leverages vertical space to provide ample sunlight exposure throughout the day. This approach maximizes solar harvesting potential, especially for structures with extensive exterior wall surfaces. The integration of photothermal panels in facades not only meets heating needs but also enhances the building's visual appeal, providing a sleek, transparent, or semi-transparent design that aligns with contemporary architectural trends. Additionally, this facade-based approach allows buildings to capitalize on solar energy without sacrificing rooftop or ground space,

Fig. 13 Schematic diagram showing the integration of PSTR into the Façade. Incorporating PSTR systems within building facades introduces a vertical, space-efficient solar harvesting solution, transforming external walls into much larger energy-generating surfaces

making it a practical solution for urban environments where land is limited. The multi-panel PSTR system, with its transparent photothermal coatings, also promotes a more uniform temperature distribution on the building envelope. By reducing the thermal flux across the facade, it enhances occupant comfort near windows and exterior walls, preventing cold drafts and creating a more consistent indoor environment. Overall, this innovative facade design not only contributes to the energy-neutral objectives of the building but also advances the integration of renewable energy solutions into urban architecture, setting a new standard for sustainable, aesthetically pleasing building designs.

Atrium PSTR Installations: In buildings with expansive central atriums, the Photothermal Solar Tunnel Radiator (PSTR) system can be strategically installed to maximize solar energy harvesting and thermal efficiency. Due to the large spatial volume and high ceilings typical of atriums, these areas often require substantial heating to maintain comfortable indoor temperatures, especially in colder seasons. Traditional heating methods in such vast spaces can be energy-intensive, relying heavily on electrical or gas heating systems to meet the demand. However, integrating a PSTR within an atrium presents a more sustainable solution that reduces dependence on conventional heating, capitalizing on the abundant natural light flooding the space throughout the day. By suspending the PSTR system or embedding it within the

Fig. 14 Diagram showing the installation of PSTR in an atrium

open central area, the design allows sunlight to filter through multiple layers of transparent photothermal panels (Fig. 14). These panels can capture sunlight from various angles as it enters the atrium, ensuring efficient absorption of solar energy across a larger area and optimizing heating potential. The multi-panel configuration raises the air temperature in the atrium by radiating heat evenly across the space, creating a more comfortable environment with minimized need for additional electrical heating.

The large vertical dimension of an atrium enables the PSTR system to function effectively as a thermal radiator, diffusing the captured heat downward into lower levels and promoting a natural convection current that distributes warmth throughout the entire area. This approach not only addresses heating demands but also enhances the energy efficiency of the building by reducing thermal losses and lowering heating costs. Moreover, by introducing warmth from a renewable source, the PSTR system contributes to the building's overall energy neutrality, aligning with sustainable design objectives.

An added benefit of installing a PSTR in atriums is the enhanced aesthetic and experiential value it brings to the space. As a visible architectural element, the PSTR provides occupants with a direct connection to natural light and warmth, fostering a sense of openness and connection to the environment. This quality is especially appealing in commercial or institutional buildings, where creating inviting, comfortable common areas is essential for enhancing user experience. The natural light streaming through the photothermal panels adds ambient warmth and a visually dynamic element, turning the atrium into a focal point that reinforces the building's commitment to sustainability and architectural innovation. In summary, installing PSTR systems in atriums not only optimizes energy use for large indoor spaces but also enriches the spatial quality and sustainability of the building. The combination of efficient solar heating and natural lighting helps reduce the environmental impact, while the design's aesthetic appeal offers a welcoming, nature-inspired atmosphere for occupants.

5.1.3 Integration of Thermoelectric Generators (TEGs) with Photothermal Solar Tunnel Radiator (PSTR) for Enhanced Resiliency and Energy Independence

The growing prevalence and intensity of natural disasters—such as hurricanes, floods, wildfires, and droughts—fueled by climate change underscore the need for resilient infrastructure capable of ensuring essential services during times of crisis. These events often disrupt building systems, impacting living conditions and safety. As a response, the novel design integrates a Photothermal Solar Tunnel Radiator (PSTR) with thermoelectric generators (TEGs) to provide energy for heating and lighting, even during prolonged power outages caused by natural or human-made disasters [35, 43, 44]. This system not only offers continuous functionality but also serves as a sustainable solution that aligns climate adaptation and mitigation strategies.

In this design, the PSTR serves as the primary heating and lighting source. It collects sunlight through a horizontal solar tunnel coated with highly reflective and photothermal materials, effectively transporting light and heat into building interiors. By attaching multiple TEG modules to the PSTR surface, the design harnesses the thermal energy produced by the PSTR for electricity generation through thermoelectric conversion (Fig. 15). This system provides a dual advantage: while the PSTR delivers heat, the TEGs generate electricity to support essential indoor lighting via energy-efficient LED systems. The self-sustaining nature of this setup enhances energy resilience, particularly valuable during times of grid failure. The solar tunnel of the PSTR is strategically placed within a building's plenum space, capturing sunlight through transparent photothermal nanomaterial or parallel photothermal films. These advanced materials selectively absorb ultraviolet (UV) and near-infrared (NIR) radiation while reflecting visible light, ensuring that transmitted light remains visually comfortable and beneficial for occupants. As sunlight travels through the tunnel, the absorbed energy is converted into heat, which is subsequently transferred to the TEGs attached to the PSTR surface. The resulting temperature differential between the hot side of the TEGs (heated by the PSTR) and the cooler interior generates electricity continuously throughout the day.

Heating and Comfort: During colder months or in low-temperature climates, the heat generated by the PSTR can be utilized to warm interior spaces by distributing thermal energy from the solar tunnel to floors or interior walls. This reduces reliance

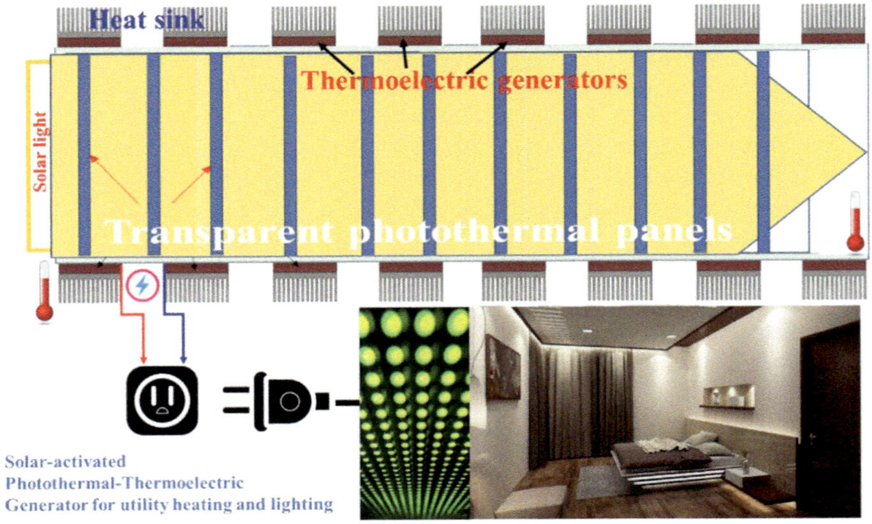

Fig. 15 Schematic diagram showing solar-activated photothermal-thermoelectric Generator for utility heating and lighting [35, 43, 44]

on traditional heating systems, conserving energy and lowering greenhouse gas emissions associated with conventional heating methods. When heating demand is low, the thermal energy absorbed by the PSTR is converted by TEGs into electrical energy, which can be stored in battery systems. This electricity can then power LED lighting or other essential systems, especially valuable when external power sources are unavailable due to disaster-related power outages. By providing energy for basic lighting, the PSTR-TEG system ensures safe and functional indoor environments during emergencies.

Excess electricity produced by TEGs can be stored in dedicated battery systems, creating a backup power supply for critical needs. This stored energy supports indoor lighting or small plug-in devices, ensuring continued building operability during prolonged power disruptions. The system's energy independence offers a safeguard against the vulnerabilities introduced by climate-induced disasters. The integrated PSTR-TEG system provides an autonomous energy source for heating and lighting, essential during power outages caused by extreme weather or natural disasters, thereby supporting continuous building operations and occupant safety. By reducing dependence on grid electricity and fossil-fuel-based heating, the PSTR-TEG system lowers greenhouse gas emissions and contributes to climate-positive building designs. This system supports climate adaptation goals by ensuring that essential building functions are maintained, even as external energy supplies fluctuate. The dual heating and electricity generation functions of the PSTR-TEG system facilitate energy-neutral or net-zero energy designs, aligning with green building standards and supporting long-term sustainability objectives.

Designed to integrate with architectural elements, the PSTR-TEG system can be installed within facades, atriums, or rooftop installations without disrupting building aesthetics. Its multifunctional design enhances both utility and visual appeal, making it a versatile solution for modern, energy-conscious buildings. This design not only serves as a core system for day-to-day operations but also offers critical energy independence during emergency scenarios, enhancing the resilience of buildings in the face of climate change. The application of PSTR systems in architectural designs offers numerous benefits that align with the principles of energy-neutral and sustainable construction: Direct Thermal Harvesting and Efficiency: Photothermal heating systems are inherently efficient, as they bypass several stages typical of photovoltaic systems by converting solar energy directly into heat. This direct conversion process minimizes energy loss and optimizes heating output, enabling buildings to meet their thermal requirements with minimal reliance on external power sources. Reduction in Carbon Footprint and Operational Costs: By reducing dependence on grid energy or fossil fuels for heating, PSTR systems contribute to significant reductions in a building's carbon footprint. The low operational costs and long lifespan of photothermal heating installations add economic benefits, making them cost-effective for building owners and developers aiming to achieve net-zero energy targets.

The integration of diffusers within PSTR systems provides controlled indoor lighting, offering natural illumination in spaces that lack direct sunlight exposure. This soft distribution of light not only creates a welcoming environment but also reduces the need for artificial lighting, further conserving energy. Aesthetic Versatility and Architectural Harmony: PSTR systems offer flexible design options that can be harmoniously integrated into a building's architectural vision. Whether as part of the facade, skylights, or canopies, these installations blend form and function, allowing architects to enhance visual appeal while meeting energy performance objectives. Transparent or semi-transparent photothermal panels can maintain a building's openness, promoting both daylight access and connectivity with outdoor spaces.Adaptability to Urban Constraints: Given the limited space available in urban settings, PSTR's vertical and multi-layer designs are particularly advantageous. By extending solar capture beyond traditional rooftop PV systems, PSTR-equipped facades, and skylights allow urban buildings to harness solar energy without occupying valuable ground or roof space, thereby addressing the spatial challenges of dense city environments.

The integration of Photothermal Solar Tunnel Radiators (PSTR) in architectural designs represents a forward-thinking approach to achieving energy neutrality in buildings. Through innovative placements in rooftops, facades, atriums, and canopies, PSTR systems provide sustainable heating, energy savings, and aesthetic appeal. These designs maximize the benefits of photothermal heating, helping buildings reach ambitious environmental goals while providing comfortable, light-filled spaces for occupants. As cities and developers increasingly prioritize sustainable building practices, PSTR technology offers architects and engineers a powerful tool to drive the future of energy-neutral construction.

5.1.4 Photothermal Utility Heating with Diffused Indoor Light via PSTR

Transitioning to energy-neutral buildings offers far-reaching advantages, such as lower utility costs, improved grid stability, enhanced air quality, and numerous public health and environmental benefits [5]. This transition is critical in addressing the challenges posed by climate change and increasing global energy demand. Photothermal energy presents a promising approach within this field due to its substantial technical, environmental, and ecological benefits. While photovoltaic (PV) solar cells have made notable progress in sustainable energy, their dependency on direct sunlight limits their effectiveness during adverse weather conditions or nighttime, which interrupts continuous energy generation. This reliance on sunlight for energy capture has sparked exploration into alternative, more resilient energy sources, including diffused light within buildings—an often-overlooked resource.

Public buildings and high-rises, particularly in urban environments, demand constant illumination for security and emergency purposes, even after daylight fades. Utilizing this diffused light for energy generation could transform these buildings' efficiency. Research into photothermal heating with plasmonic nanoparticles has demonstrated effective energy conversion under direct sunlight, yet their potential to capture diffused light remains largely unexplored. Unlike direct solar exposure, indoor or shaded environments provide a more consistent, dispersed light, which may benefit photothermal processes. By incorporating plasmonic nanoparticles like $Fe_3O_4@Cu_{2-x}S$ into transparent films or coatings, we can harness diffused light within buildings, opening new pathways for energy-efficient indoor heating solutions.

The unique ability of $Fe_3O_4@Cu_{2-x}S$ nanoparticles to absorb across broad wavelengths, especially in the UV and infrared (IR) regions, makes them ideal for photothermal applications in environments where sunlight is scarce. This material exhibits a broad IR absorption band alongside a minimal visible spectrum absorption, facilitating efficient visible light transmission, which can be advantageous in transparent film applications. Coatings made from these nanoparticles can capture energy from available diffused light, effectively triggering photothermal conversion. By implementing multi-layer transparent films embedded with $Fe_3O_4@Cu_{2-x}S$ nanoparticles, entire panel arrays can serve as a photothermal radiator, capturing and converting light into heat, thereby reducing the need for traditional heating systems and supporting building energy efficiency.

By integrating photothermal spectral modulators (PSTRs) under indoor lighting, substantial heat generation becomes possible, offering a transformative approach to energy-efficient indoor heating [5]. Just as PSTRs convert direct sunlight into heat, these panels can also efficiently harvest energy from artificial indoor lighting sources such as LEDs. This capability makes PSTRs highly versatile and applicable across various indoor environments, enabling their deployment in numerous strategic locations within buildings to generate heat precisely where it is needed.

The core principle of PSTRs is based on their ability to absorb diffused or ambient light and convert it into thermal energy through the photothermal effect. When placed under indoor lighting, PSTRs utilize light from commonly used fixtures, turning it into heat that can supplement traditional utility heating systems. The strategic placement of PSTRs around high-traffic areas, along walls, or near heating-demand locations in public buildings can provide localized heating, thus reducing the load on centralized heating systems and lowering overall energy consumption. This approach is particularly beneficial in colder climates or during winter months, where consistent indoor heating is necessary.

Figure 16 illustrates this concept, depicting how multiple transparent photothermal panels can be used under diffused indoor lights to produce heat. Each panel operates by harnessing light from surrounding LED fixtures or other indoor sources. The panels are coated with plasmonic nanoparticles that can absorb light across a range of wavelengths, including visible and infrared. When exposed to indoor lighting, these nanoparticles become phononically active, resonating in response to the light's electric field. This activity drives localized heat production across the panel's surface, providing a reliable and consistent heating effect. The transparent nature of the PSTR panels allows for their installation in various areas without obstructing visibility, making them ideal for spaces like hallways, lobbies, and work areas where lighting is abundant, and there is a steady need for warmth. By stacking multiple PSTR panels, the heat generated can be amplified, effectively creating a photothermal radiator system capable of generating substantial warmth from standard indoor lighting.

This innovative heating strategy not only leverages existing lighting infrastructure but also supports energy conservation goals by repurposing ambient light for thermal applications. As demonstrated in Fig. 16, the use of photothermal panels with indoor lighting setups represents a new frontier in sustainable energy utilization, turning everyday lighting systems into dual-purpose tools for both illumination and heating.

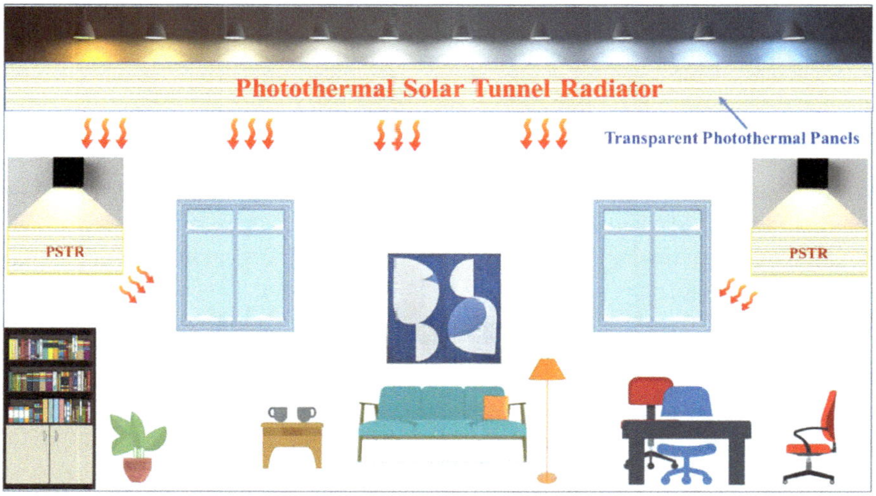

Fig. 16 Schematic diagram showing the concept of PSTR utility heating via indoor lights

6 Experimental Investigation on Solar Harvesting and Photothermal Energy Generation via Multiple Transparent Plasmonic Thin Films for Building Utility Heating

6.1 Introduction

Conventional solar building heating systems often rely on photovoltaic (PV) panels, which, while effective in converting sunlight into electricity, face several limitations. These include high installation costs, seasonal inefficiencies, spatial constraints, and integration challenges in urban environments. Traditional PV systems typically operate in two-dimensional configurations, which limit solar energy absorption and are impractical for densely populated regions. Moreover, the complex infrastructure required for heat storage, transfer, and distribution adds to operational inefficiencies and costs. In response to these challenges, three-dimensional (3D) solar harvesting technologies offer a promising alternative, focusing on the direct conversion of solar light to thermal energy via photothermal heating. Photothermal systems are inherently simpler, bypassing the need for intermediary processes such as electrical conversion, heat transfer fluids, and extensive storage systems. By employing transparent photothermal (PT) thin films, these systems enable enhanced photon capture and energy density through multi-panel configurations, effectively addressing the shortcomings of traditional PV-based heating.

This study introduces the Photothermal Solar Tunnel Radiator (PSTR) as a novel paradigm in solar energy harvesting and building utility heating [4]. Central to this innovation is the utilization of multiple transparent photothermal glass panels (TPGP) coated with plasmonic $Fe_3O_4@Cu_{2-x}S$ nanoparticles. The PSTR harnesses solar energy through a tunnel-like structure where sunlight is channeled, converted into thermal energy by photothermal coatings, and radiated to heat-enclosed spaces. This direct and efficient approach not only optimizes energy utilization but also offers a scalable, eco-friendly alternative to conventional systems. A key milestone in this research is the development of a lab-scale "Photothermal Solar Box" (PSB), which demonstrates the potential of the PSTR under diverse environmental conditions. With minimal external insulation and ambient temperatures as low as 8 °C, the system achieves internal temperatures up to 40 °C, delivering an impressive thermal output of 9.1×10^5 J using eight photothermal panels. These results highlight the system's resilience, efficiency, and suitability for energy-neutral building heating, even in cold and partially cloudy weather. This research marks a significant step towards revolutionizing solar energy applications for sustainable building utilities. By leveraging the unique advantages of 3D photothermal harvesting, the PSTR system aims to reduce dependency on traditional energy sources, minimize environmental impact, and pave the way for scalable adoption of energy-neutral heating technologies.

6.2 Experimental Procedures

Nanoparticle synthesis and thin film deposition: Following established protocols for the synthesis of $Fe_3O_4@Cu_{2-x}S$ nanoparticles [34], both the Fe_3O_4 and $Fe_3O_4@Cu_{2-x}S$ nanoparticles were produced and characterized as shown in Figs. 7 and 8. For thin film deposition, 10 ml of $Fe_3O_4@Cu_{2-x}S$ nanoparticle solution was blended with 80 ml of epoxy resin. The polymer solution was stirred for five minutes to ensure uniformity and the removal of air bubbles. Ten milliliters of this solution were evenly distributed on each of the eight glass panels, achieving a nanoparticle concentration of 0.006 mg/cm^2 with an average visible transmittance (AVT) of 75%. The panels were cured overnight to ensure complete polymer solidification.

Design of a Photothermal Solar Box (PSB): A lab-scale Photothermal Solar Box (PSB) was developed to evaluate the heating performance of $Fe_3O_4@Cu_{2-x}S$ thin films under simulated building conditions [4]. The PSB consists of a transparent PMMA enclosure measuring 50 × 37 × 50 cm, featuring a vertical solar tunnel lined with mirrors to enhance solar energy utilization. Eight removable transparent photothermal glass panels (TPGPs), each measuring 18 × 13 × 3 cm, were vertically stacked inside the tunnel. Uncovered PSB: TPGPs were exposed to direct sunlight at varying angles (Fig. 17a). Covered PSB: The box was insulated with white Styrofoam, leaving only the top exposed to mimic a window-like setup and to reflect sunlight for reduced surface heating (Fig. 17b). Photographs of these configurations are shown in Fig. 17c, d. This design allows the investigation of solar energy harvesting under diverse environmental conditions, accounting for the time of day and seasonal variations. The primary objective is to evaluate whether the TPGPs can generate sufficient thermal energy to maintain comfortable indoor temperatures in cold climates (~0 °C).

Temperature Monitoring and Light Power Density: Temperatures of the photothermal panels and the PSB were monitored using K-type thermocouples and thermal images captured by a FLIR camera, providing detailed insights into thermal distribution. The AVT of the TPGPs was maintained at 75%, while the PMMA box allowed 93% transmittance. Experiments were conducted at 2 PM on the University of Cincinnati campus, with light power density (LPD) ranging from 0.048 W/cm^2 on sunny days to 0.0001 W/cm^2 in overcast conditions, measured using a Newport optical power meter (Model 1600-R). To analyze thermal performance, the PSB was insulated with Styrofoam, leaving a top opening for light entry. This setup facilitated comparative analysis between the covered and uncovered configurations, investigating the influence of light intensity and angle on photothermal heating. Insulation was also employed in winter and rainy conditions to systematically examine heat loss rates.

Time-Dependent Solar Angle Effects on Photothermal Performance: Experiments were conducted between 2 and 6 PM in October and November 2023, as depicted in Fig. 18a. During this period, the solar angle ranged from 11° to 73° (Fig. 18b). Corresponding variations in LPD were recorded, with a linear decrease from 48 to 30 mW/cm^2 between 2 and 5 PM, followed by a more rapid decline between 5 and

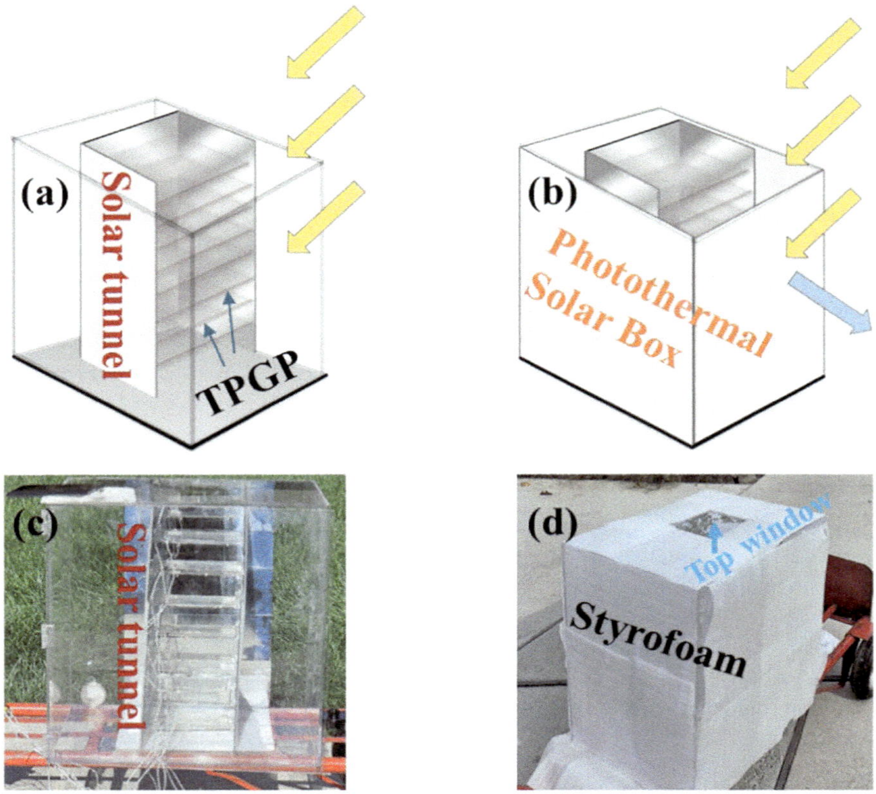

Fig. 17 **a** Schematic diagram of uncovered Photothermal Solar Box (PSB) with exposed Transparent Photothermal Glass Panel (TPGP); **b** schematic of PSB covered with Styrofoam; **c** photograph of the uncovered PSB; and **d** photograph of the covered PSB with Styrofoam [4]

6 PM (Fig. 18c). The PSB was oriented with its top "window" facing upward, and the vertical walls aligned in cardinal directions, enabling comprehensive assessment of the impact of solar angle and shadows on photothermal performance. Figure 19a presents an optical photograph of a single-layer transparent photothermal glass panel (TPGP), which demonstrates significant transparency with an average visible transmittance (AVT) of 75%. The UV–Vis spectrum of the TPGP, shown in Fig. 19b, displays the characteristic "U"-shaped curve that accounts for its high AVT. As illustrated in Fig. 19c, eight TPGPs are arranged in parallel. On October 2, 2023, light power density (LPD) measurements were conducted on the University of Cincinnati's main campus under partially sunny and cloudy weather conditions. As shown in Fig. 19d, AVT of the panels decreases from 75% to below 10% as the number of stacked TPGPs increases from 1 to 8.

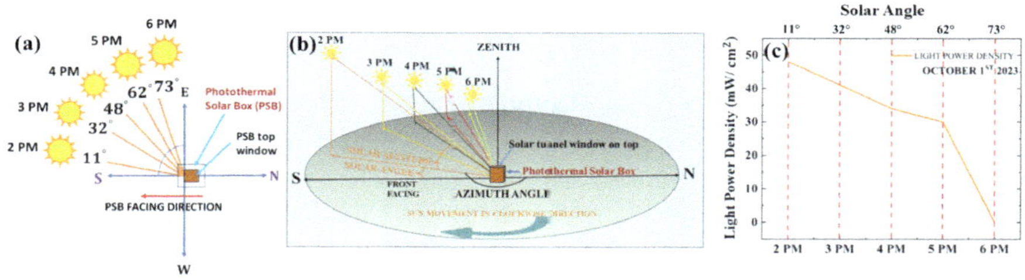

Fig. 18 **a** Variation in solar angle over time in relation to the orientation of the Photothermal Solar Box (PSB); **b** three-dimensional depiction showing the spatial alignment of solar angles throughout time, correlated with the orientation of the PSB; and **c** time and solar angle dependence of light power density [4]

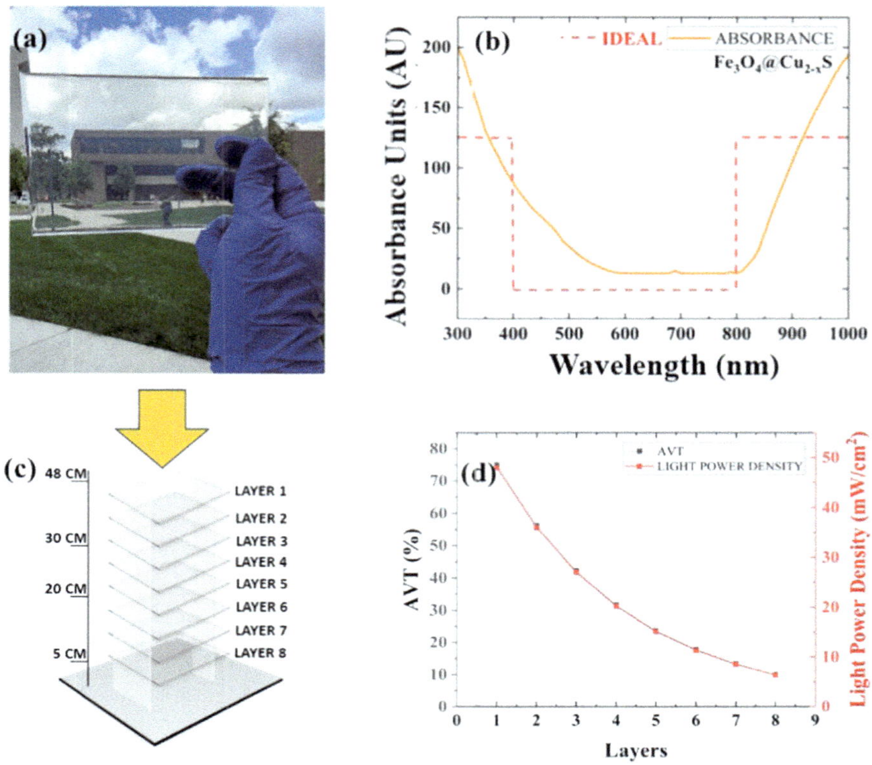

Fig. 19 **a** Photograph of a transparent photothermal glass panel (TPGP); **b** UV–Vis spectrum of TPGP; **c** multiple TPGP arranged in parallel in the solar tunnel; and **d** light power density as a function of number of TPGP [4]

6.3 Experimental Results

Photothermal experiments with the Photothermal Solar Box (PSB) were conducted on October 2, 2023, at the University of Cincinnati's main campus under the conditions detailed in Fig. 5. The solar intensity on the day ranged between 0.048 and 0.030 W/cm^2. Figure 20a illustrates a schematic of the uncovered PSB, allowing sunlight to directly strike the transparent photothermal panels, generating thermal energy. The corresponding heating curves are presented in Fig. 20b, showing that all eight TPGPs experience a rapid initial temperature increase before leveling off into plateaus until sunlight is blocked at 150 min.

As indicated by the arrows in Fig. 20a, sunlight enters at an angle (illustrated in Fig. 17), starting at 11° at 2 PM when the experiment began. This variation in solar angle results in different photon intensities received by each TPGP, leading to distinct temperatures. Notably, the middle panel (layer 4, green) achieves the highest temperature, approximately 65 °C, while the top panel (layer 8, brown) reaches the

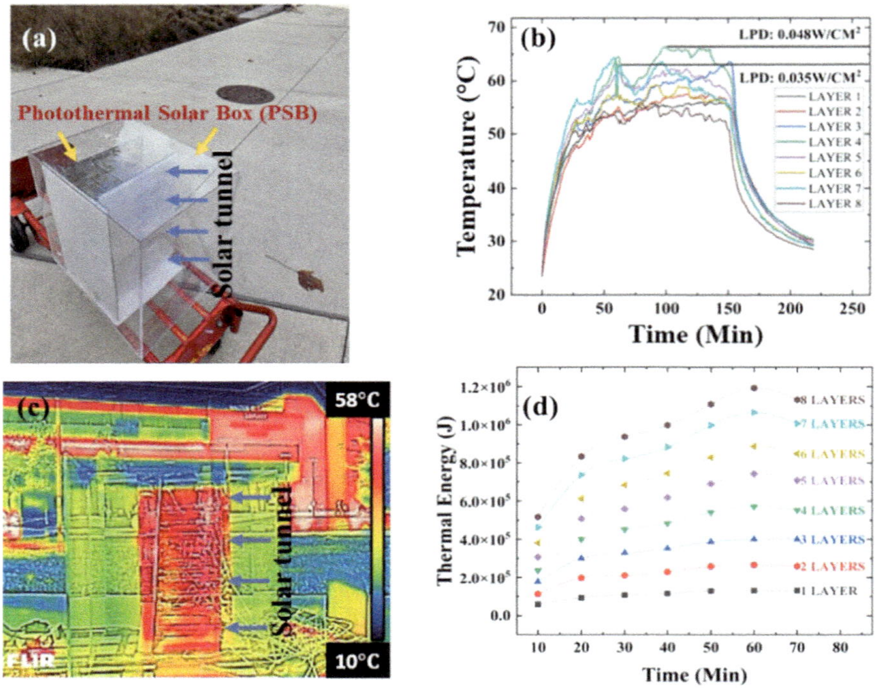

Fig. 20 **a** Schematic diagram of the uncovered Photothermal Solar Box (PSB); **b** heating curves of the Transparent Photothermal Glass Panels (TPGPs); **c** infrared thermal image of the PSB with eight $Fe_3O_4@Cu_{2-x}S$ panels (75% AVT) under natural sunlight; **d** thermal energy versus number of panels. The experiment was conducted on October 2, 2023, at 2 PM in sunny weather with light power densities of 0.048 and 0.030 W/cm^2. Ambient temperature: 25 °C [4]

lowest at 54 °C. The changing solar angle and light power density, as shown in Fig. 18, reveal that the middle layers receive greater solar exposure when the PSB is uncovered. Additionally, the reflective mirrors inside the solar tunnel enhance the photothermal heating of the middle layers compared to the top and bottom ones. Figure 20b also marks two specific light power density values: 0.048 mW/cm^2 at 3 PM and 0.035 mW/cm^2 at 4 PM.

Infrared thermal images in Fig. 19c depict the temperature distribution within the PSB under natural sunlight, showing uniform heating due to the photothermal properties of the films. Thermal energy Q is defined as $Q = \Delta T_{\text{max}} \times m \times Cp$, where ΔT represents the maximum temperature difference caused by a given number of thin films, mmm is the mass of the sample (film + substrate), and Cp is the specific heat of the glass panel (neglecting the negligible contribution from the ultra-thin photothermal films). Thermal energy calculations, based on the heating curves of the $Fe_3O_4@Cu_{2-x}S$ films in the PSB, are plotted against the number of layers in Fig. 20d. This figure demonstrates that heat energy generation increases as additional TPGPs are added (e.g., 1, 1 + 2, 1 + 2 + 3, etc.). The calculation focuses solely on the heating phase to highlight maximum heat energy generation for various panel configurations. A single TPGP in the PSB generates approximately 1.3×10^5 J thermal energy. As more panels are added, the thermal energy generated is consistently increasing up to 1.13×10^6 J for a total of 8 panels.

An experiment was conducted on January 20, 2024, during the winter in Cincinnati, Ohio, to evaluate the photothermal heating performance under low ambient temperatures (−10 °C). Figure 21a shows a snapshot of the outdoor photothermal setup featuring a partially insulated Photothermal Solar Box (PSB). The insulation includes Styrofoam covering three side walls, leaving one side wall, the top, and the bottom exposed for increased sunlight exposure. Figure 21b illustrates the placement of thermometers inside and outside the solar tunnel, while Fig. 21c displays the heating curves of the Transparent Photothermal Glass Panels (TPGPs) with thermometers positioned between the layers. Impressively, the TPGPs achieved a temperature of 40 °C at layers 3–4 despite the outdoor temperature being −10 °C. The lower plateaus observed in the heating curves correspond to moving clouds, which reduced the Light Power Density (LPD). The interior temperature of the PSB reached 35 °C at a depth of 30 cm, demonstrating effective photothermal heating even under extreme cold conditions. Figure 21c, d reveals a sharp temperature drop within 20 min after sunlight was obstructed at the 80-min mark, with an even steeper cooling rate observed at 100 min. This highlights the significant temperature differential between the interior of the PSB and the external environment.

The Photon Conversion Efficiency (PCE), η, is calculated using the following equation [23]:

$$\eta = \frac{\left(C_g m_g + C_{\text{PTmaterial}} m_{\text{PTmaterial}} + C_{\text{polymer}} m_{\text{polymer}}\right) \Delta T_{\text{max}}}{IA\Delta t} \approx \frac{C_g m_g \Delta T_{\text{max}}}{IA\Delta t} \times 100\% \tag{5}$$

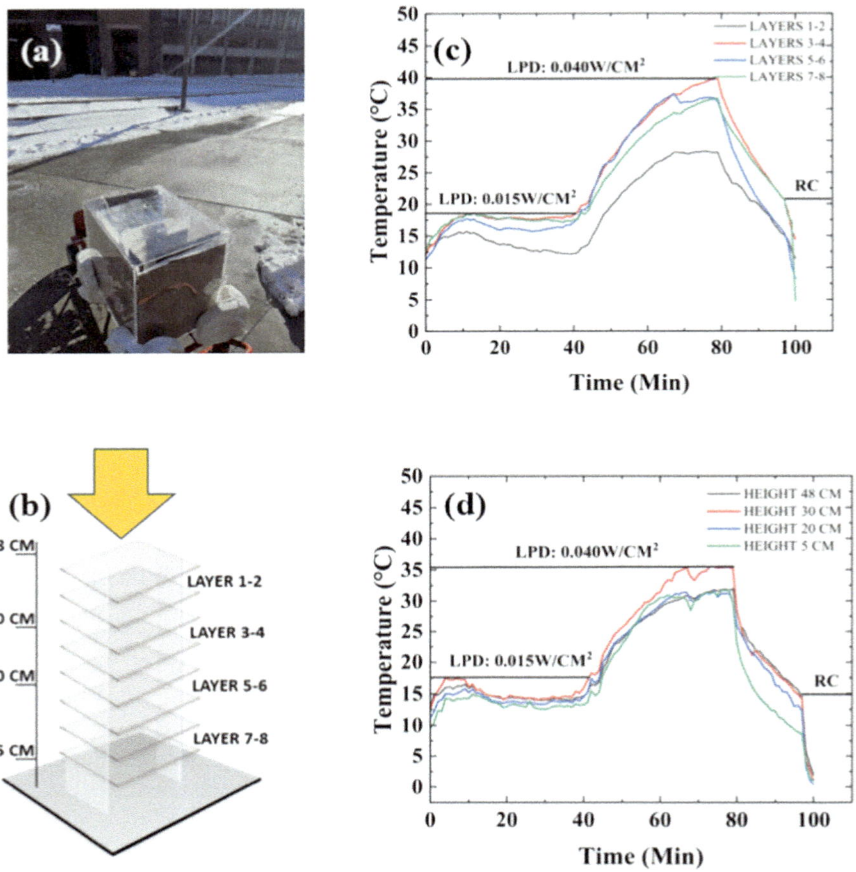

Fig. 21 **a** Photothermal Solar Box (PSB), partially insulated, in open air at −10 °C ambient temperature; **b** the positions of thermometers within and outside the solar tunnel; **c** heating curves with thermometers between different TPGP as shown in (**b**); **d** heating curves at different heights outside the solar tunnel as shown in (**b**). The experiment was carried out on January 20th, 2024, at 2 PM. Sunny weather. Light power density: 0.040 and 0.002 W/cm^2. Ambient temperature: −10 °C [4]

where g is glass, c is the specific heat capacity (J/g °C), m is mass (g), $\Delta T_{\max}$ is the maximum change in temperature increase in the sample (°C), I is the incident light power density (W/cm^2), A is the surface area of the sample, and Δt is the time required for a sample to reach the maximum temperature (s) [45].

When uncovered, the solar tunnel captures photons from both the top and front sides (Fig. 16a), providing favorable Light Power Densities (LPD) across all Transparent Photothermal Glass Panels (TPGP). The top panel achieves a PCE of 34% and combining all eight layers results in an impressive 321% net PCE increase compared to a single layer, primarily due to consistent light attenuation through the layers. When the PSB is covered, the solar tunnel collects photons only from the top side

(Fig. 16b), reducing the single-layer PCE to 28.5%. However, the mirrored surfaces inside the solar tunnel enhance photon redirection to the lower layers, yielding a cumulative PCE improvement of 198% across all eight layers compared to a single layer. This demonstrates the critical role of the mirrored solar tunnel in significantly boosting the PSB's overall PCE.

6.4 Summary

The experiments and analysis underscore the innovative use of transparent multilayer photothermal thin films for solar energy harvesting, particularly in building utility heating. The Photothermal Solar Tunnel Radiator (PSTR) presents a groundbreaking method for sustainable energy generation, leveraging an array of transparent glass substrates coated with photothermal materials. The $Fe_3O_4@Cu_{2-x}S$-coated glass exhibits high transparency and a U-shaped optical absorption profile, enabling efficient photothermal energy conversion within the Photothermal Solar Building (PSB). Figure 19d illustrates the thermal energy output relative to the number of Transparent Photothermal Glass Panels (TPGP), with the system achieving a maximum of 9.1×10^5 J of thermal energy even when the PSB is covered. This performance is comparable to the uncovered Photothermal Solar Glass (PSG). Further, Fig. 20c depicts the heating behavior of TPGPs in colder conditions, despite accelerated heat loss in colder weather, the panels reached maximum temperatures within 70 min without exhibiting sustained plateaus.

The findings emphasize the balance between transparency and photothermal efficiency, critical for optimizing system performance. Experimental results demonstrate the feasibility of achieving comfortable room temperatures using solar energy alone, with the potential for energy storage and utilization during periods of reduced sunlight. This innovative transparent photothermal technology offers a scalable, cost-effective alternative to traditional photovoltaic (PV) systems for building heating, addressing challenges such as high costs, space constraints, and system complexity. The transparent multilayer system and PSTR provide a promising pathway toward energy-neutral heating utilities for buildings, contributing to sustainable and eco-friendly urban infrastructure. These advancements align with environmental conservation goals and net-zero energy consumption, representing a significant step forward in practical solar energy applications for urban settings.

7 Experimental Investigation on Photothermal Utility Heating with Indoor LEDs via Multiple Transparent $Fe_3O_4@Cu_{2-x}S$ Thin Films

7.1 Introduction

Achieving sustainable indoor heating under conditions of limited direct sunlight represents a critical challenge in modern solar energy technologies. This study presents a groundbreaking solution: a photothermal utility heating system that leverages transparent $Fe_3O_4@Cu_{2-x}S$ thin films to harness diffused indoor light efficiently. As depicted in Fig. 15, by transforming UV and IR light into thermal energy through the plasmonic properties of $Fe_3O_4@Cu_{2-x}S$ nanoparticles, this approach offers a sustainable pathway for maintaining comfortable indoor temperatures without dependence on conventional solar resources. The core innovation lies in the unique capabilities of $Fe_3O_4@Cu_{2-x}S$ nanoparticles, which exhibit strong UV and IR absorptions via Localized Surface Plasmon Resonance (LSPR). These nanoparticles enable the photothermal activation of transparent thin films even in low-light scenarios, such as cloudy days or nighttime conditions. By integrating multiple transparent thin films, this system maximizes energy capture and conversion, making it ideally suited for energy-neutral building applications.

To evaluate its effectiveness, the photothermal radiator was tested in a custom-designed Diffused Light Photothermal Box (DLPB) that simulates indoor environments. The experiments demonstrated that the system could elevate temperatures above 50 °C, showcasing its potential to generate substantial thermal energy from diffused light sources. This capability underscores its versatility and effectiveness in enhancing energy efficiency for indoor heating applications. This innovative photothermal radiator represents a scalable, cost-effective solution for sustainable heat generation, significantly reducing reliance on external energy grids. By utilizing ubiquitous diffused light, it aligns seamlessly with the principles of energy-neutral and eco-friendly building design. This research offers a transformative contribution to renewable energy technologies, paving the way for sustainable indoor heating solutions that meet the growing demands of modern infrastructure.

7.2 Experimental Design

To evaluate heating efficiency under diffused light conditions, a custom-built Diffused Light Photothermal Box (DLPB) was designed to replicate indoor environments (Fig. 21) [5]. The experimental setup utilized Transparent Photothermal Panels (TPTPs) coated with $Fe_3O_4@Cu_{2-x}S$ thin films. The DLPB featured a transparent PMMA enclosure with dimensions of 50 cm × 37 cm × 50 cm, housing a vertical photothermal tunnel. Eight removable TPTPs were arranged horizontally in parallel within the tunnel to maximize photon collection (Fig. 21a). To enhance photon capture, the tunnel's interior was lined with mirrors, reflecting incident light toward the TPTPs. Each TPTP measured 18 cm × 13 cm × 3 mm, ensuring uniform surface area across all panels. Figure 3b provides a photograph of the constructed DLPB, including its photothermal tunnel utilized for experimental analysis. The TPTPs were designed with an Average Visible Transmittance (AVT) of 75%, while the enclosing PMMA box maintained an AVT of 93%. Figure 21c depicts the lighting setup, employing IR and UV lamps to simulate diffused light conditions. The IR lamps emitted light in the 660–850 nm range, while the UV lamps operated in the 380–420 nm range. Light power density (LPD) was maintained at 15 mW/cm^2, measured using a Newport Optical Power Meter Model 1600-R. The experiments were conducted under three lighting conditions:

IR-only: Three IR lamps (56 W each) are positioned 30 cm from the TPTPs.
UV-only: Three UV lamps (100 W each) are positioned 30 cm from the TPTPs.
Combined IR + UV: Three IR and three UV lamps positioned 20 cm from the front and 5 cm from the top of the TPTPs.

Figure 22 provides a schematic of the DLPB, showing the parallel arrangement of TPTPs. K-type thermocouples were strategically placed to monitor temperatures at various heights within the DLPB (e.g., 5, 20, 30, and 40 cm) and between the TPTPs. Upon activation of the IR and UV lamps, temperature readings were recorded and analyzed using temperature versus time heating curves. This experimental configuration allowed for a comprehensive evaluation of the photothermal performance of the TPTPs under simulated indoor lighting conditions.

7.3 Experimental Results

The photothermal performance of the Transparent Photothermal Panels (TPTPs) was assessed under various conditions using the custom-designed Diffused Light Photothermal Box (DLPB). Figure 23a provides an image of the DLPB setup, showcasing three IR lamps positioned at the front to simulate diffused light exposure. The corresponding heating profiles are illustrated in Fig. 23b and c, which depict the temperature variations between the TPTPs and within the DLPB at different heights, respectively. Thermocouple placement for these measurements is shown schematically in Fig. 22d. The data reveals that all eight TPTPs exhibit a rapid temperature rise during the initial phase of light exposure, stabilizing into plateau regions until the IR lamps are switched off at 400 min. Notably, the temperature between the first and second TPTP layers reaches 37.3 °C within 100 min, demonstrating the efficient photothermal conversion facilitated by the IR lamps. Similarly, the internal temperature within the DLPB shows consistent heating trends across different heights, with the maximum temperature recorded at 29.2 °C at a height of 20 cm. These findings underscore the capability of the DLPB and TPTPs to effectively harness diffused IR

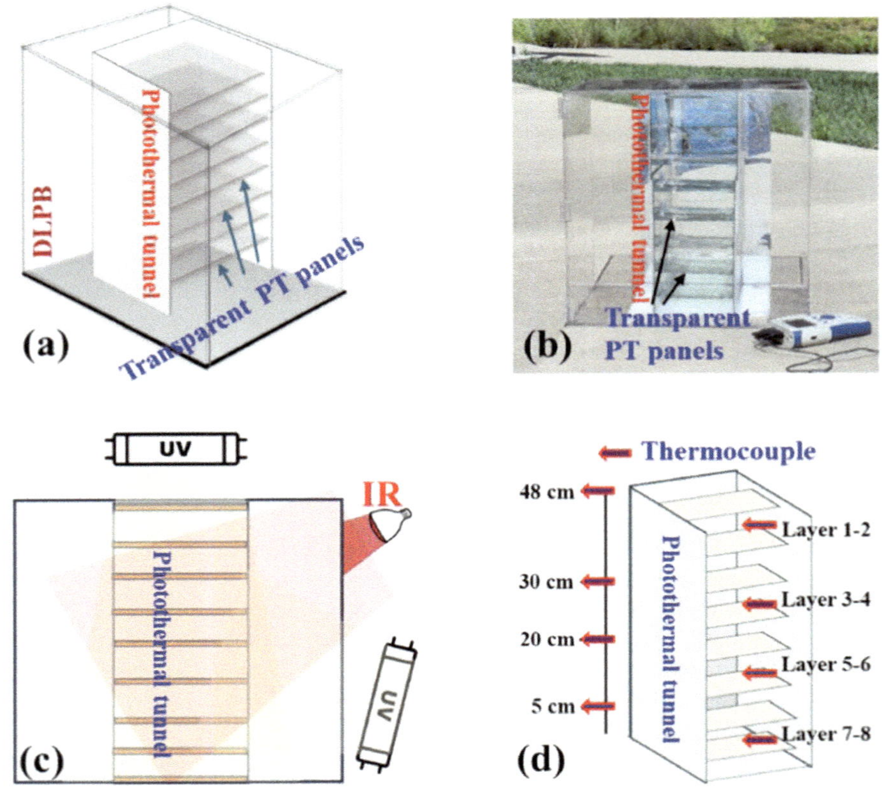

Fig. 22 **a** Schematic diagram showing Diffused Light Photothermal Box (DLPB), **b** photograph of DLPB, **c** schematic diagram showing the IR and UV lamps outside the DLPB, **d** schematic diagram showing the photothermal tunnel with the locations of the thermocouples. Note that the thermocouples are placed within and outside the photothermal tunnel inside the DLPB at different heights as indicated [5]

light for photothermal heating, offering promising implications for energy-efficient indoor temperature regulation.

The calculation of thermal energy follows the formula $Q = \Delta T_{\max} \times m \times Cp$, where ΔT represents the maximum temperature difference produced by a given number of thin films, m stands for the mass of the sample (including the thin film and substrate), and Cp denotes the specific heat of the glass panel (notably, the PT films are exceedingly thin). This calculation, derived from the heating curves of $Fe_3O_4@Cu_{2-x}S$ films within the DLPB, is illustrated in Fig. 23d, which plots thermal energy output against the number of TPTP layers. The results demonstrate a clear trend: thermal energy output increases significantly as additional TPTP layers are incorporated. Importantly, only the heating phase of the temperature curves is considered in this analysis, as the focus is on evaluating maximum thermal energy

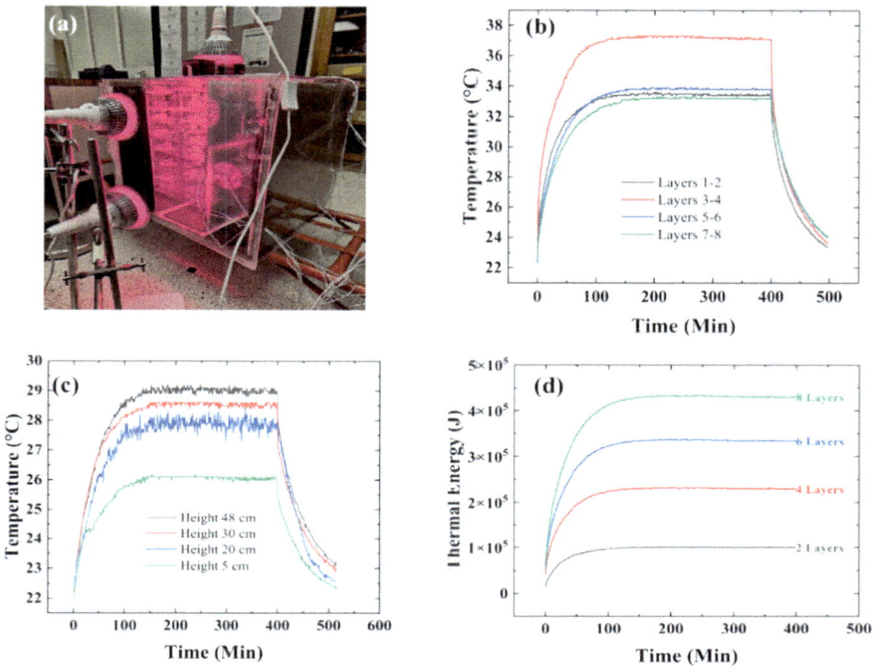

Fig. 23 **a** Schematic diagram showing the Diffused Light Photothermal Box (DLPB) under 3 IR lights only; **b** heating curves taken between different Transparent Photothermal Panels (TPTP); **c** heating curves taken at different heights within the Diffused Light Photothermal Box (DLPB), and **d** thermal energy versus number of panels. Ambient temperature: 21 °C. Light power density is 15 mW/cm^2 [5]

generation. As delineated in Fig. 23d, the thermal energy produced aligns consistently with the heating curves presented in Fig. 23b. For two panels within the DLPB, the system yields approximately 1.0×10^5 J of thermal energy. With the inclusion of more panels, the generated thermal energy continues to rise steadily, reaching 4.3×10^5 J for a total of 8 panels.

The structural configuration of the Diffused Light Photothermal Box (DLPB), equipped with three UV lights—two positioned at the front and one at the top—is shown in Fig. 24a. The heating performance across various locations of the Transparent Photothermal Panels (TPTP) is illustrated in Fig. 24b, with temperature measurements corresponding to positions identified in Fig. 22d. The data indicate a rapid rise in temperatures for all eight TPTPs during the first 100 min of UV light exposure, reaching their respective peaks before the light source is deactivated. Compared to results obtained under IR light alone, the temperatures achieved under UV illumination are markedly higher. Specifically, the temperature between the first and second layers of TPTP reaches a maximum of 48.1 °C, showcasing the effectiveness of UV light in driving photothermal heating. Similar trends are observed

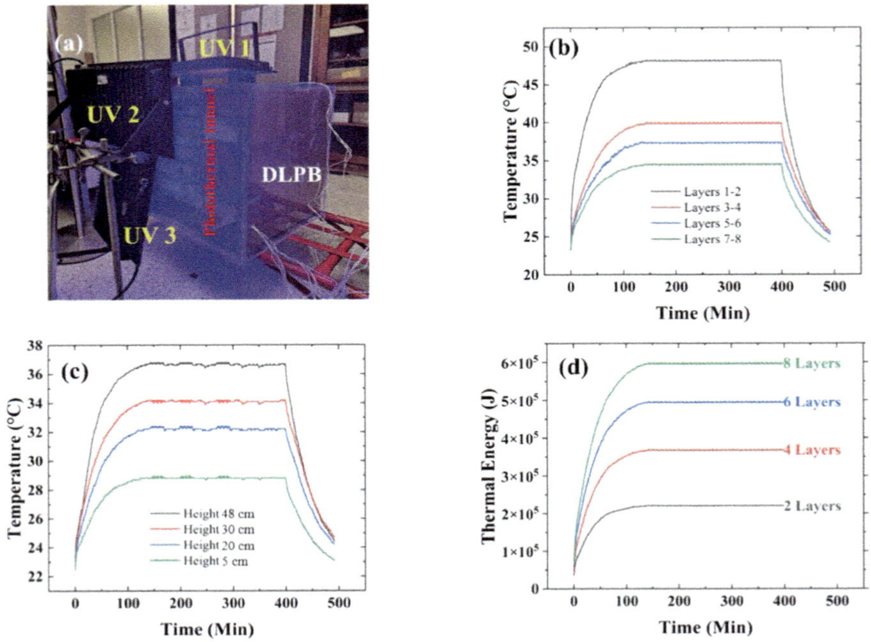

Fig. 24 **a** Photograph of the Diffused Light Photothermal Box (DLPB) with the UV light only; **b** heating curves taken between different Transparent Photothermal Panels (TPTP); **c** heating curves taken at different heights within the Diffused Light Photothermal Box (DLPB), and **d** thermal energy versus number of panels. Ambient temperature: 21 °C. Light power density is 15 mW/cm^2 [5]

in the temperature profiles recorded at varying heights within the DLPB system, reinforcing the uniformity and efficiency of the UV-driven photothermal process. Thermal energy generation as a function of the number of TPTPs is depicted in Fig. 24d. The results reveal that with eight panels, the system produces a maximum thermal energy output of approximately 6.0 × 105 J, a significant improvement over the thermal energy values recorded under IR light conditions (Fig. 24d).

These findings emphasize the enhanced photothermal efficiency of the DLPB system when utilizing UV light. The notable increase in both peak temperatures and thermal energy output highlights UV illumination as a more effective approach compared to IR light, offering substantial potential for optimizing energy generation in photothermal applications.

To maximize the absorption spectrum of Fe_3O_4@$Cu_{2-x}S$ nanoparticles, as depicted in Fig. 8, a combination of ultraviolet (UV) and infrared (IR) light sources was employed to enhance photothermal energy output. As shown in Fig. 25a, the experimental setup included three IR lights positioned in front of the Diffused Light Photothermal Box (DLPB) and three UV lights, with two placed at the front and one mounted on top of the DLPB. Despite the increased number of light sources, the overall light power density was carefully maintained at 15 mW/cm^2 to ensure

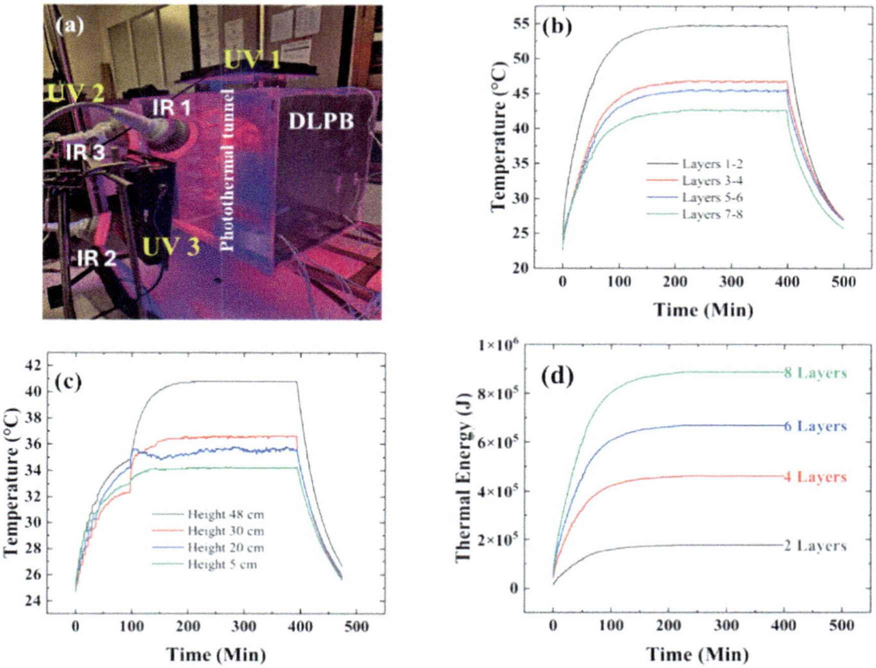

Fig. 25 **a** Heating curves taken between different Transparent Photothermal Panels (TPTP) under both UV and IR lights; **b** heating curves taken at different heights within the Diffused Light Photothermal Box (DLPB) under both UV and IR lights; **c** thermal energy versus number of panels, and **d** the heat energy generated and absorbed by each panel over time (data draw from Fig. 7a). Ambient temperature: 25 °C. Light power density is 15 mW/cm^2 [5]

consistent experimental conditions. The heating behavior of the system, observed between the Transparent Photothermal Panels (TPTPs), is presented in Fig. 25b. The combined UV and IR lighting induced a rapid rise in temperature across all eight TPTPs, reaching a maximum of 54.8 °C. Similarly, the heating profiles recorded at various heights within the DLPB, shown in Fig. 25c, display consistent temperature increases, confirming the efficacy of the diffused light sources spanning a broad wavelength range from UV to IR. The thermal energy output, calculated for varying numbers of TPTPs, is summarized in Fig. 25d. The combined lighting setup enabled the system to generate a maximum thermal energy of approximately 8.9×10^5 J with 8 photothermal panels. This energy output significantly surpasses the thermal energy generated under IR-only (Fig. 23d) conditions, underscoring the synergistic advantage of utilizing both UV and IR light sources for photothermal energy generation. These findings highlight the potential of such hybrid lighting configurations for optimizing indoor heating applications in energy-efficient systems.

In these experimental setups, both infrared (IR) and ultraviolet (UV) lamps were employed to emit light within specific wavelength ranges. The IR lamps operated in the near-infrared (NIR) spectrum of 660–850 nm, while the UV lamps emitted light in the 380–420 nm range. Although the UV light closely matched the absorption spectrum of $Fe_3O_4@Cu_{2-x}S$, the IR lamps were limited to wavelengths below the material's absorption peak at 1150 nm. This mismatch in wavelength alignment reduced the photothermal efficiency of the IR lamps, resulting in lower performance compared to UV light, as demonstrated in Figs. 23 and 24. Aligning the light source's emission spectrum with the peak absorption characteristics of $Fe_3O_4@Cu_{2-x}S$ could substantially enhance photothermal energy conversion, suggesting a clear path for optimizing system performance. The use of Light Emitting Diodes (LEDs) for photothermal heating introduces a novel approach to sustainable energy generation in indoor settings. While LEDs predominantly emit light within the visible spectrum, where $Fe_3O_4@Cu_{2-x}S$ exhibits lower absorption rates compared to UV and IR wavelengths, they still demonstrate notable heating potential. This resilience highlights the feasibility of using LEDs for photothermal applications, even with their reduced efficiency during the initial phases of light irradiation. These observations underscore the critical role of spectral selectivity in plasmonic particle-mediated photothermal heating, emphasizing the need to customize absorption properties for improved energy conversion.

The incorporation of Light Emitting Diodes (LEDs) into the infrastructure of buildings and high-rises has become a cornerstone in the pursuit of energy efficiency, cost-effectiveness, and environmental sustainability. LEDs stand out for their exceptional energy efficiency, consuming significantly less power compared to traditional lighting systems while delivering equivalent or superior levels of illumination. This efficiency translates into considerable reductions in electricity consumption, leading to lower operational costs for building owners and managers. Additionally, the extended lifespan of LEDs reduces the frequency and expense of replacements, offering long-term economic advantages. Their low maintenance requirements further enhance their appeal as a cost-saving solution. LEDs are also highly versatile, enabling tailored lighting designs that meet the functional and aesthetic requirements of diverse architectural spaces. From an environmental perspective, LEDs align with sustainability objectives by producing minimal heat and containing no harmful substances, such as mercury. This makes them safer for both occupants and the environment, while also helping buildings comply with green building standards and regulations.

The widespread adoption of LEDs in contemporary construction projects not only delivers financial and operational benefits but also advances energy conservation and eco-friendly practices. Their combination of efficiency, durability, and design flexibility makes LEDs an indispensable component of sustainable building strategies.

7.4 Summary

This study highlights the innovative use of diffused light to generate heat through the photothermal effect of plasmonic $Fe_3O_4@Cu_{2-x}S$ nanoparticles. Transparent thin films incorporating these nanoparticles demonstrated exceptional performance, achieving temperature increases of up to 54.8 °C within the Diffused Light Photothermal Box (DLPB). By leveraging the nanoparticles' UV and IR absorption properties, photothermal heating is driven through Localized Surface Plasmon Resonance (LSPR). The research aims to advance energy efficiency in buildings by developing a photothermal radiator capable of utilizing diffused light for heat generation. This approach is particularly advantageous for energy-neutral buildings, offering enhanced performance during nighttime or adverse weather conditions such as rain or overcast skies. The integration of $Fe_3O_4@Cu_{2-x}S$ nanoparticles into transparent thin films provides a sustainable means of maintaining comfortable indoor temperatures, reducing reliance on conventional energy sources.

The system's design optimizes energy capture by activating the transparent photothermal thin films with UV and IR light sources, ensuring functionality even in low-light environments. Within the DLPB, temperatures exceeding 50 °C are achieved using diffused UV and IR light, showcasing the system's potential for sustainable indoor heat generation. By exploiting the robust UV and IR absorption capabilities of $Fe_3O_4@Cu_{2-x}S$ through LSPR, this photothermal radiator sets a new benchmark for energy-efficient building heating and climate control. The incorporation of Light Emitting Diodes (LEDs) into the photothermal energy generation system adds significant advantages, including energy efficiency, prolonged operation using diffused indoor light, and cost-effectiveness. LEDs consume far less energy than traditional lighting sources while maintaining high illumination levels, contributing to sustainable building operations.

One of the key benefits of LEDs lies in their ability to emit light within specific wavelengths. This precision enables the optimization of spectral characteristics for photothermal heating by aligning with the absorption spectra of plasmonic nanoparticles such as $Fe_3O_4@Cu_{2-x}S$, thereby maximizing energy conversion efficiency. Furthermore, the long lifespan and low power consumption of LEDs result in substantial cost savings over time, making them an economically viable solution for integrating sustainable heating technologies in buildings. Together, the robust performance of $Fe_3O_4@Cu_{2-x}S$ nanoparticles and the strategic use of LEDs underscore a forward-looking approach to energy-efficient, environmentally friendly indoor heating. This novel system paves the way for sustainable, cost-effective solutions to meet the heating demands of modern building environments.

Future research could focus on refining LED configurations and tailoring their spectral outputs to align more closely with the absorption properties of photothermal materials. Such advancements would further enhance the viability of LED-based photothermal systems for building applications, contributing to improved energy efficiency and sustainability. This study also underscores the swift thermal response, efficient heat dissipation, and synergistic heat generation capabilities of transparent photothermal panels (TPTPs). These properties significantly impact the overall performance of the Diffused Light Photothermal Box (DLPB). The dynamic interplay of heat within the system showcases the potential of TPTP technology as a robust solution for sustainable energy applications, offering a promising pathway for optimizing indoor energy systems.

8 Harnessing Sunlight and Indoor Light for the Energy-Efficient Cooling Mode of Buildings

The previous sections primarily addressed heating modes for cold climates, focusing on direct solar energy harvesting and the utilization of indoor lights. However, as seasons shift and temperatures rise, the demand for cooling in summer and warm climates becomes a critical factor, often leading to significant energy consumption. Cooling requirements for buildings not only strain energy resources but also pose challenges for maintaining sustainability and reducing environmental impact.

To tackle these challenges, innovative strategies that harness both sunlight and indoor light are emerging as promising solutions. Advanced technologies such as Multiple Transparent Photovoltaic (MTPV) systems and Indoor Photovoltaic (IDPV) systems offer practical and energy-efficient approaches to address cooling demands. By integrating transparent PV materials as part of the building envelope—such as in windows, façades, and skylights—MTPV systems capture and convert sunlight into electricity while maintaining the aesthetic and functional transparency of the building design (Fig. 26). Concurrently, IDPV systems leverage indoor light sources, such as LEDs, to generate electricity even in the absence of direct sunlight.

These technologies work synergistically to create a dynamic and versatile energy framework that reduces reliance on traditional grid power while supporting cooling needs. By strategically combining outdoor and indoor light harvesting, buildings can achieve a higher degree of energy efficiency and sustainability, paving the way for innovative solutions in green building design tailored to warm climates.

8.1 Building Skins with Multiple Transparent PVs (MTPV)

Building Skins with Multiple Transparent PVs (MTPV) incorporate photovoltaic materials directly into the building envelope, replacing conventional construction materials such as roofs, façades, or skylights. MTPV systems offer several advantages, including offsetting initial installation costs by reducing expenses for traditional building materials and labor. Beyond economic benefits, MIPV systems can integrate into the architectural design of buildings, maintaining aesthetics while enabling widespread solar adoption, even in projects where traditional solar panels might disrupt the intended look.

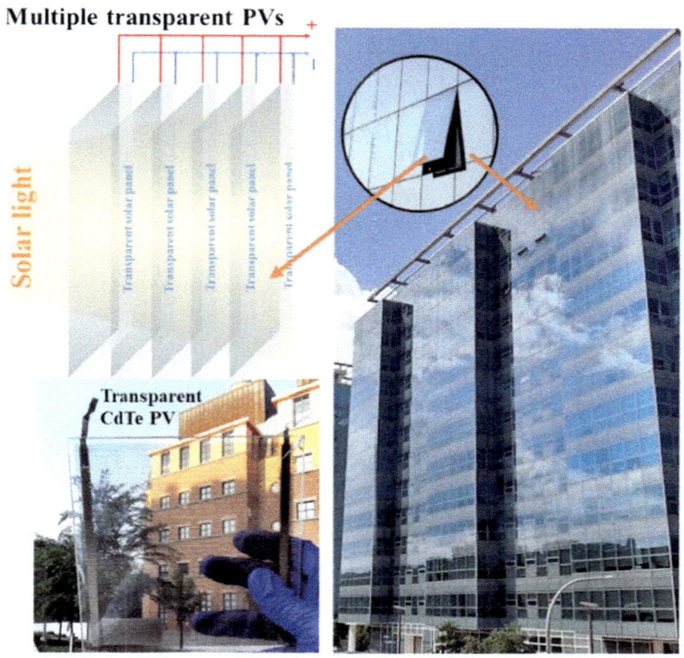

Fig. 26 Schematical diagram showing the multiple transparent photovoltaic building skin

In cooling applications, MTPV systems harvest sunlight and convert it into electricity to power air conditioning systems and other energy-intensive appliances. By strategically incorporating these materials into the building's structure, they provide an effective way to generate renewable energy while minimizing the building's dependence on traditional grid electricity. MTPV systems, whether installed during construction or retrofitted as Building-Integrated Photovoltaics (BIPV), serve as a crucial element in designing energy-efficient buildings that meet cooling demands during the summer months.

Among the various energy conversion methods, photovoltaic (PV) solar cells have emerged as one of the most widely adopted technologies due to their proven efficiency and versatility across numerous applications. Traditional PV panels, however, are typically limited by their two-dimensional (2D) design, which relies on expansive surface areas to capture sunlight effectively. This limitation is particularly problematic in densely populated urban environments, where available land and rooftop space are scarce. To overcome this constraint, innovative three-dimensional (3D) solar harvesting systems have been developed to enhance energy density by extending solar harvesting capabilities beyond the conventional x–y plane into the z-axis.

One promising approach involves integrating multiple transparent CdTe PV panels as building skins, such as windows, façades, and skylights, to convert sunlight into electricity for summer cooling and air-conditioning. These CdTe panels, known for their high transparency and efficient energy conversion, can be arranged in a multi-layer configuration, with panels stacked in parallel along the z-axis. This design allows sunlight to partially penetrate through the layers, significantly increasing the effective solar harvesting surface area.

The advantages of this multilayer CdTe system are threefold [47]:

1. **Enhanced Solar Harvesting Area and Energy Density**: By employing multiple transparent PV panels, the system dramatically increases the total solar light collection area compared to a single-panel configuration. This results in greater energy density and higher power output from the same surface footprint.
2. **Higher Power Conversion Efficiency (PCE)**: When arranged in descending order of transparency—placing the panel with the highest transparency on top and the lowest at the bottom—the multilayer CdTe system has demonstrated a collective output power and PCE increase of 233% compared to a single panel under identical surface area constraints.
3. **Reduced Space Requirements**: This innovative 3D solar harvesting system is particularly advantageous for densely populated urban areas, where the availability of large, flat surfaces is limited. The ability to stack transparent panels vertically maximizes energy production while minimizing spatial requirements.

For building cooling applications, the electricity generated by these multilayer CdTe PV systems can be directly utilized to power air-conditioning units, reducing the reliance on conventional grid electricity. The transparent nature of the panels also allows for aesthetic integration into building designs, enabling architects and engineers to seamlessly incorporate energy-generating components into façades and windows without compromising natural lighting or building aesthetics. This shift from 2 to 3D solar harvesting represents a transformative advancement in sustainable energy technologies. By employing multiple transparent CdTe PV panels as multifunctional building skins, the system addresses both the energy needs and space constraints of urban environments, providing an efficient and practical solution for cooling mode air-conditioning in warm climates.

8.2 Harnessing Indoor Lighting via Photovoltaic Technology: An Innovative Approach to Renewable Energy and Building Climate Control

To address the pressing global energy challenges, solar energy continues to emerge as one of the most promising strategies for clean energy generation, offering a wealth of technical, environmental, and ecological benefits. However, in warm climates and during summer months, a major challenge arises from the heating effect of intense sunlight, which increases indoor temperatures and significantly escalates cooling demands in buildings. Conventional solar energy systems often exacerbate this issue by absorbing and retaining heat, contributing to the thermal load of buildings. To mitigate these challenges, advanced PV technologies must evolve to not only harness sunlight efficiently but also address the heating effects by integrating cooling strategies and thermal management solutions. Beyond traditional outdoor applications, an innovative and underutilized solution lies in leveraging diffuse and low-intensity indoor lighting—especially in urban high-rises where artificial illumination is ubiquitous—to generate electricity while reducing heat gain and improving the overall energy balance of buildings.

While solar energy remains a leading solution for clean and sustainable energy generation, its reliance on direct sunlight presents significant challenges, particularly during nighttime and cloudy or rainy days. The absence or reduction of sunlight during these periods severely limits the effectiveness of traditional photovoltaic (PV) systems, creating gaps in energy generation when demand may still be high. This intermittency issue poses a critical challenge to the reliability and consistency of solar power, especially in regions prone to extended periods of overcast weather or during evening hours when buildings require substantial energy for lighting, cooling, and other operations. In such scenarios, the potential of utilizing indoor lighting as a renewable energy source becomes increasingly compelling. Artificial lighting, particularly in urban environments, remains operational around the clock for various reasons, including security, safety, and aesthetic purposes. Public buildings, offices, residential towers, and commercial spaces are often illuminated at night to ensure safety for occupants and passersby. This constant presence of indoor light, much of which is emitted by energy-efficient LEDs, represents a largely untapped resource for renewable energy generation.

By integrating advanced PV technologies capable of capturing and converting diffuse indoor light into electricity, buildings can achieve a dual benefit: harnessing an alternative energy source while addressing the intermittency challenges of traditional solar systems. Indoor PV systems can operate independently of weather conditions and sunlight availability, providing a reliable and consistent energy supply even during nighttime or overcast days. Moreover, such systems can complement outdoor solar panels by ensuring round-the-clock energy generation, effectively bridging the gap in renewable energy production caused by unfavorable environmental conditions. The application of indoor PV systems is particularly advantageous in high-rise buildings, where artificial lighting is a constant necessity. By strategically positioning PV panels to capture photons emitted by indoor lights, such as LEDs, energy can be harvested efficiently and used to power essential systems like air conditioning, ventilation, and other electrical appliances. This approach not only reduces reliance on traditional grid power but also contributes to a building's overall energy efficiency and sustainability. Additionally, it offers a unique opportunity to transform wasted indoor light into a valuable energy source, further enhancing the role of renewable energy technologies in modern urban environments.

The Potential of Indoor Lighting as an Energy Source Modern urban landscapes are dominated by high-rise buildings that consume colossal amounts of energy for indoor lighting. These energy-intensive systems, including public lighting for safety, architectural illumination for aesthetics, and round-the-clock office and residential lighting, represent a largely untapped energy reservoir. The diffuse light emitted from these systems, particularly LEDs, holds immense potential for photovoltaic electricity generation. By strategically integrating PV systems into indoor environments, we can transform this overlooked energy source into a consistent and reliable supply of renewable energy. Transparent or semi-transparent PV materials, such as cadmium telluride (CdTe) solar panels, can be embedded into building façades, windows, and walls to capture photons emitted by indoor lights. Commercial Silicon PV panels can also be utilized on building floors and sealings for harvesting indoor lights to generate additional electricity for air conditioning (Fig. 27). This approach maximizes energy conversion efficiency by utilizing light with wavelengths close to the bandgap of the PV material, such as LEDs emitting at optimal wavelengths.

As shown in Fig. 27, the integration of PV systems for harvesting indoor light has transformative implications for building energy management, particularly in summer and warm climates where cooling systems are a significant energy burden. Utilizing indoor lights, particularly energy-efficient LEDs [48–51], for photovoltaic (PV) electricity generation offers a novel and sustainable energy solution (Fig. 28). By installing PV panels strategically within the indoor environment—on walls, ceilings, or floors—buildings can effectively capture photons emitted by LED lighting. When the wavelengths of LED light align with the bandgap of the PV material, efficient energy conversion is achieved. This innovative approach not only optimizes

Fig. 27 Experimental setup for harvesting LED light with silicon PV panels

Fig. 28 A schematic diagram showing the concept of the wall-mounted PV panels for harvesting diffused LED lights to power electrical appliances such as air conditioning

space utilization by leveraging existing surfaces but also ensures consistent energy generation independent of outdoor weather conditions. The electricity generated can directly power indoor systems, such as air conditioning, ventilation units, and electronic devices, thereby reducing reliance on grid electricity and lowering the building's environmental footprint. Furthermore, the flexibility in the placement of PV panels within indoor spaces enables seamless integration into building aesthetics and functionality. This system provides a scalable and practical solution for both residential and commercial applications, promoting energy efficiency while enhancing sustainability. By harnessing indoor lighting for renewable energy generation, this approach represents a significant advancement in achieving energy-efficient and self-sustaining buildings.

By coupling PV-generated electricity with energy-efficient cooling technologies, buildings can achieve dual benefits:

1. **Powering Cooling Systems**: Electricity generated from indoor lighting can be used to power air conditioning systems, which are typically the largest energy consumers in warm climates. This approach reduces reliance on grid electricity, especially during peak hours when cooling demands are highest.
2. **Enhancing Thermal Comfort**: Semi-transparent PV panels integrated into windows and façades can serve as solar modulators, reducing heat gain by filtering and converting a portion of incoming sunlight into electricity. This not only lowers indoor temperatures but also minimizes the workload on cooling systems, resulting in significant energy savings.
3. **Energy Storage for Nighttime Cooling**: The excess energy generated during the day from indoor and outdoor light can be stored in integrated battery systems.

This stored energy can then be used to power cooling systems during nighttime hours, maintaining thermal comfort while ensuring a steady, renewable power supply.

4. **Reducing Heat Island Effect**: High-rise buildings in urban areas often contribute to the heat island effect due to their absorption of solar radiation. By utilizing PV systems that reflect and convert sunlight into electricity, these buildings can mitigate heat accumulation, promoting cooler urban microclimates.

Advantages of Indoor PV Systems

In addition to supporting building climate control, the integration of PV technology to harvest indoor lighting offers several broader advantages:

1. **All-Weather and All-Day Energy Generation**: Unlike traditional PV systems that rely on sunlight, indoor PV systems harvest energy continuously, regardless of external weather or time of day. By converting "wasted" light energy from indoor sources, these systems provide uninterrupted power, overcoming the intermittency issues of conventional PV systems.
2. **Efficient Space Utilization**: PV panels can be seamlessly integrated into walls, ceilings, and other interior surfaces without requiring additional land or rooftop space. This is particularly advantageous in densely populated urban areas where space is limited.
3. **Resilience and Reliability**: Indoor PV systems are shielded from extreme weather conditions, ensuring their durability and reliability over time.
4. **Energy Diversification**: Harvesting diffuse indoor light diversifies energy sources for PV generation, enhancing the overall resilience of renewable energy systems.

9 Photothermal-Photovoltaic Dual-Modality Building Skin

One of the key challenges in energy efficiency is addressing thermal insulation, particularly in large-surface area, high-rise buildings located in cold climates. Traditional approaches have focused on optimizing thermal insulation materials, but these are often limited by conflicting properties such as the trade-off between transparency and insulation performance. Conventional window technologies, such as double-pane insulated glass units (IGUs) with low-emissivity coatings, have been the industry standard for reducing heat loss. Thermal performance is commonly measured by the U-factor, which quantifies thermal transmittance due to conduction, convection, and radiation. The U-factor is defined as the ratio of heat flux per unit area to the temperature gradient across the material and is expressed in watts per kelvin per meter. Lower U-factor values indicate improved insulation. However, current methods of improving the U-factor through double-pane IGUs are constrained by the limited insulating properties of transparent gases like argon.

An innovative approach to overcoming these limitations is the development of Optical Thermal Insulation (OTI), which bypasses traditional methods by leveraging optical mechanisms [34, 43–46, 52]. This concept eliminates the reliance on thermal intervention media such as air or argon. Through the application of spectrally selective thin film coatings, single-pane windows can absorb sunlight in the UV and NIR regions while maintaining high visible light transmittance (AVT). The absorbed UV and NIR radiation is converted to heat through photothermal (PT) effects, elevating the surface temperature of the window and reducing the temperature difference between the window surface and interior. This reduction in temperature gradient effectively minimizes heat transfer, achieving significant energy savings without requiring multiple panes. Research has demonstrated that single-pane windows with photothermal coatings achieve U-factors comparable to or better than double-pane windows. For example, chlorophyll-coated single panes have exhibited efficient photothermal heating while maintaining high visible light transmittance, significantly reducing U-factor values. This approach offers considerable potential for energy and materials savings, supporting sustainability goals.

Despite its effectiveness, OTI presents challenges for year-round use in cold climates. While it works well in winter by reducing heat loss, the same photothermal heating can increase cooling demands in summer. To address this, we developed a dual-function building skin combining photothermal (PT) and photovoltaic (PV) modalities [52]. This design transforms the building façade from a passive insulator into an active energy device capable of seasonal adaptability. The smart building skin reduces heat loss during winter through optical insulation and generates electricity year-round via photovoltaic conversion. Additionally, the PT layer absorbs IR radiation, lowering the surface temperature of the PV cells and enhancing their efficiency under prolonged solar exposure.

The PT-PV dual-modality building skin operates as follows (Fig. 29): During winter, the PT film on the outer pane photothermally increases the window surface temperature, reducing heat loss through the window. In this mode, the building skin functions primarily as an optical thermal insulator with significantly reduced U-factors. Conversely, in summer, when exterior temperatures are higher than interior room temperatures, the PV film on the inner pane becomes the primary energy-generating device, converting sunlight into electricity. Simultaneously, the PT film removes the IR portion of the solar light, effectively reducing the surface temperature of the PV film and enhancing its efficiency. This dual functionality ensures that the PT-PV building skin fully utilizes IR light for two purposes: (1) heating the outer pane during winter to minimize heat loss and (2) filtering out IR during summer to maintain optimal PV performance.

The dual-modality structure, illustrated in Fig. 1, is integrated into a double-pane window design. The outer pane features a spectrally selective PT coating, while a porphyrin-based dye-sensitized solar cell (DSSC) is applied to the inner pane. Both coatings are highly transparent in the visible spectrum, making them suitable for window applications. The PT coating absorbs UV and IR light, converting it to thermal energy in winter to lower the U-factor, while the DSSC generates electricity. The IR absorption by the PT layer also mitigates temperature-related efficiency losses in the DSSC, as PV cells are known to suffer from decreased power conversion efficiency (PCE) with rising temperatures.

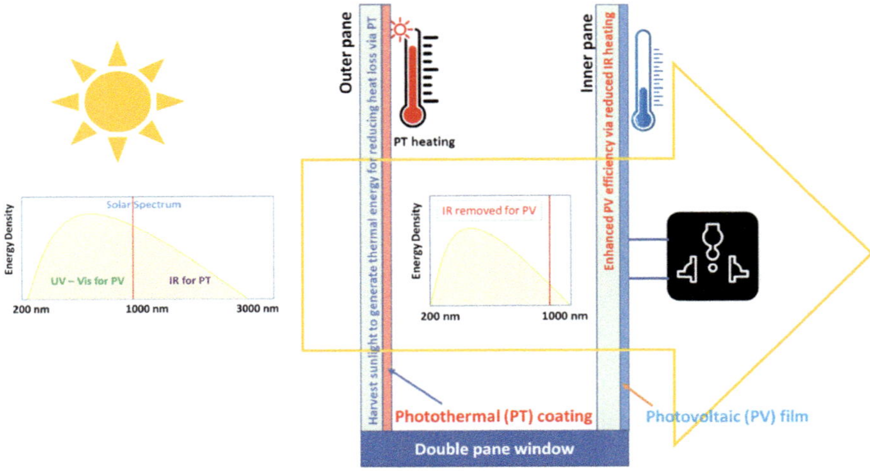

Fig. 29 The PT-PV dual modality operates in the following fashion: in the wintertime, the PT film on the outer pane photothermally increases the window surface temperature to reduce heat loss through the window. Under this condition, the PT-PV modality building skin functions mainly in the optical thermal insulation mode with lowered U-factors. In the summertime, since the exterior temperature is higher than the interior room temperature, the PV film on the inner pane is the major device to generate electricity. Meanwhile, the PT film can significantly remove the IR portion of the solar light to reduce the surface temperature of PV for enhanced efficiency. In this way, the PT-PV dual modality building skin fully utilizes IR from solar light for two purposes: (1) heating the outer pane in the winter to reduce heat loss, and (2) removing IR in the summer to lower the PV temperature on the inner pane [52]

The dual-modality design addresses two key objectives: reducing heat transfer in winter through photothermal heating and improving PV efficiency by minimizing IR-induced temperature increases. This chapter explores the fundamental mechanisms of PT and PV conversions using highly transparent porphyrin-based thin films, focusing on their spectral selectivity and energy conversion efficiencies. Materials such as $Fe_3O_4@Cu_{2-x}S$, chlorophyllin, phthalocyanine, and their hybrids are examined for their photon-to-heat and photon-to-electricity conversion efficiencies, paving the way for advanced, sustainable building technologies.

10 Future Research Objectives for Practical Applications of Indoor PV Systems

The potential of indoor photovoltaic (PV) systems as a sustainable energy solution is gaining increasing attention, particularly for large buildings where artificial lighting operates continuously. Future research objectives aim to address the challenges, optimize efficiency, and enhance the economic and practical viability of these systems.

Indoor PV systems offer unique advantages by leveraging existing infrastructure, such as LED lighting, as a reliable photon source. Below are key future research directions for advancing indoor PV technologies and integrating them into practical applications.

1. Addressing Challenges and Optimizing Efficiency

One of the primary challenges for indoor PV systems lies in the lower light intensity and spectral differences compared to natural sunlight. Unlike outdoor PV systems, indoor setups rely on diffuse and artificial light, which requires advanced materials and designs to maximize light absorption and conversion efficiency. Future research will focus on:

- **Developing High-Efficiency Materials**: Advancing semitransparent PV materials, such as CdTe and organic PVs, optimized for artificial light sources like LEDs.
- **Enhancing Multi-layer Panel Systems**: Investigating stacked panel configurations to increase power conversion efficiency (PCE) by capturing photons across a broader spectrum.
- **Customizing Spectral Responses**: Tailoring PV materials to efficiently convert the specific wavelengths emitted by LEDs, which are often concentrated in the visible spectrum.

2. Harnessing Indoor Lighting During Night and Adverse Weather

Indoor PV systems offer a practical solution to the limitations of outdoor PVs, particularly during nighttime or adverse weather conditions such as cloudy or rainy days. Since indoor lights, especially in high-rise buildings and public spaces, remain operational around the clock for security and safety reasons, they present a consistent energy source. Research objectives include:

- **Photon Recycling**: Exploring the feasibility of harvesting light that would otherwise be wasted during nighttime in high-occupancy buildings.
- **Continuous Energy Generation**: Designing PV systems that complement outdoor solar panels by ensuring round-the-clock electricity production.
- **Energy Modeling and Optimization**: Developing models to estimate energy savings and electricity generation potential in various building types and operating conditions.

3. Economic and Practical Viability

While indoor and outdoor PV panels share similar manufacturing costs, indoor systems benefit from reduced installation complexities and longer lifespans due to the controlled environment. Future research will address:

- **Cost-Effectiveness Analysis**: Comparing lifecycle costs of indoor PV systems to outdoor installations, including maintenance, structural support, and operational expenses.

- **Integration with Building Design**: Investigating how PV panels can be seamlessly incorporated into walls, ceilings, and fixtures to minimize installation costs and maximize energy generation.
- **Scalability for Urban High-Rises**: Focusing on large-scale deployment in densely populated cities, where high-rise buildings and public facilities provide ideal conditions for indoor PV applications.

4. Energy Savings and Building Applications

A case study for Willis Tower demonstrates the significant energy-saving potential of indoor PV systems:

- **Estimated Energy Savings**: By installing PV panels on 30% of the building's floor area (125,400 m^2) and leveraging 24-h LED lighting, estimate the annual electricity generation of approximately and potential savings per year.
- **Energy-Neutral Design**: Indoor PV systems align with energy-neutral building concepts by harvesting both natural sunlight during the day and artificial light at night, ensuring continuous energy generation.

5. Advancing Building-Integrated Photovoltaics (BIPV)

Indoor PV systems complement ongoing advancements in Building-Integrated Photovoltaics, particularly for transparent PV applications. Transparent CdTe PV panels can serve dual purposes such as windows and energy harvesters, capturing sunlight during the day and artificial light at night. This dual functionality expands the effective surface area for energy generation, making high-rises and modern urban buildings more energy-efficient.

6. Pioneering Sustainable Energy Solutions

Future research will explore innovative architectural designs and PV system configurations to maximize energy harvesting in indoor environments. Emphasis will be placed on:

- **Adaptive PV Technologies**: Developing systems that dynamically adjust to varying light intensities and sources.
- **Energy Storage Integration**: Pairing indoor PV systems with advanced energy storage technologies to store surplus electricity for later use.
- **Smart Building Integration**: Incorporating PV systems into smart building designs to optimize energy management and reduce overall consumption.

By addressing these objectives, indoor PV systems have the potential to revolutionize sustainable energy solutions in urban environments, providing reliable and consistent electricity generation while reducing dependency on outdoor solar installations. As PV technology advances and costs decline, the integration of indoor PV systems into modern architecture will play a pivotal role in achieving energy-neutral buildings and enhancing global energy sustainability.

11 Summary

The integration of solar and indoor light harvesting photovoltaic (PV) systems for cooling mode applications represents a transformative approach to energy efficiency and sustainable building design. By leveraging the dual capabilities of outdoor sunlight and indoor LED lighting, PV systems overcome the limitations of traditional sunlight-dependent systems, enabling continuous energy generation. This innovative strategy not only diversifies renewable energy sources but also supports the growing energy demands for cooling in warm climates, where electricity consumption for air conditioning and ventilation dominates.

The ability of semi-transparent PV materials to modulate heat while generating electricity further enhances building thermal comfort and energy efficiency. By coupling PV electricity generation with cooling technologies, such as air conditioning and ventilation systems, indoor PV systems provide a sustainable solution to address both energy and climate control needs. Indoor PV systems also capitalize on the consistent and predictable photon source provided by LED lighting, ensuring stable energy generation independent of weather conditions. This integration reduces reliance on grid power, minimizes the environmental footprint, and promotes space efficiency by utilizing existing infrastructure. Advances in PV technology, combined with strategic design innovations, unlock untapped energy potential within modern buildings, transforming them into energy-efficient and climate-resilient infrastructures.

As urban areas continue to grow, indoor PV systems will play a critical role in achieving energy-neutral buildings, where renewable energy generation is seamlessly integrated with climate control strategies. This paradigm shift highlights the future of sustainable energy systems, positioning solar and indoor light harvesting PV technologies as foundational pillars of energy-efficient urban environments.

Acknowledgements We acknowledge the financial support from the National Science Foundation CMMI-1953009 and the Michelman Green, Clean and Sustainable Technology Research Innovation Program (F103484).

Competing Interests The author has no conflicts of interest to declare that are relevant to the content of this chapter.

Ethical Approval Approval was granted by the Ethics Committee of the University of Cincinnati.

References

1. https://www.eia.gov/consumption/commercial/
2. Dwivedi YK, Hughes L, Kar AK, Baabdullah AM, Grover P, Abbas R, Andreini D, Abumoghli I, Barlette Y, Bunker D, Kruse LC, Constantiou I, Davison RM, De' R, Dubey R, Fenby-Taylor H, Gupta B, He W, Kodama M, Mäntymäki M, Metri B, Michael K, Olaisen J, Panteli N, Pekkola S, Nishant R, Raman R, Rana NP, Rowe F, Sarker S, Scholtz B, Sein M, Shah JD, Teo TSH, Tiwari MK, Vendelø MT, Wade M (2022) Climate change and COP26: are digital

technologies and information management part of the problem or the solution? An editorial reflection and call to action. Int J Inf Manag 63:102456. ISSN 0268-4012. https://doi.org/10.1016/j.ijinfomgt.2021.102456
3. Khan MI, Asfand F, Al-Ghamdi SG (2022) Progress in research and technological advancements of thermal energy storage systems for concentrated solar power. J Energy Storage 55(Part D):105860. ISSN 2352-152X. https://doi.org/10.1016/j.est.2022.105860
4. Katepalli A, Wang Y, Lin J, Harfmann A, Bonmarin M, Krupczak J, Shi D (2024) A photothermal solar tunnel via multiple transparent $Fe_3O_4@Cu_{2-x}S$ thin films for heating utility application. Solar Energy 271:112444. ISSN 0038-092X. https://doi.org/10.1016/j.solener.2024.112444
5. Katepalli A, Tene NK, Wang Y, Harfmann A, Bonmarin M, Krupczak J, Shi D (2024) Photothermal utility heating with diffused indoor light via multiple transparent $Fe_3O_4@Cu_{2-x}S$ thin films. Energy Technol 12(8):2400703. https://doi.org/10.1002/ente.202400703
6. Shabaninezhad M, Ramakrishna G (2019) Theoretical investigation of size, shape, and aspect ratio effect on the LSPR sensitivity of hollow-gold nanoshells. J Chem Phys 150(14):144116
7. Agrawal A, Cho SH, Zandi O, Ghosh S, Johns RW, Milliron DJ (2018) Localized surface plasmon resonance in semiconductor nanocrystals. Chem Rev 118(6):3121–3207. https://doi.org/10.1021/acs.chemrev.7b00613
8. Chen Y, Ming H (2012) Review of surface plasmon resonance and localized surface plasmon resonance sensor. Photonic Sens 2:37–49. https://doi.org/10.1007/s13320-011-0051-2
9. Zalyubovskiy SJ, Bogdanova M, Deinega A, Lozovik Y, Pris AD, An KH, Hall WP, Potyrailo RA (2012) Theoretical limit of localized surface plasmon resonance sensitivity to local refractive index change and its comparison to conventional surface plasmon resonance sensor. J Opt Soc Am A 29:994–1002
10. Julian Wang J, Shi D (2018) Spectral selective and photothermal nano structured thin films for energy efficient windows. Appl Energy. https://doi.org/10.1016/j.apenergy.2017.10.066
11. Wriedt T (2012) Mie theory: a review. In: Hergert W, Wriedt T (eds) The mie theory. Springer series in optical sciences, vol 169. Springer, Berlin, Heidelberg. https://doi.org/10.1007/978-3-642-28738-1_2
12. Wei T, Liu Y, Dong W, Zhang Y, Huang C, Sun Y, Chen X, Dai N (2013) Surface-dependent localized surface plasmon resonances in CuS nanodisks. ACS Appl Mater Interfaces 5(21):10473–10477
13. Zaitoun MA, Mason WR, Lin C-T (2001) Magnetic circular dichroism spectra for colloidal gold nanoparticles in xerogels at 5.5 K. J Phys Chem B 105(29):6780–6784
14. Melinger JS, Kleiman VD, McMorrow D, Gröhn F, Bauer BJ, Amis E (2013) Ultrafast dynamics of gold-based nanocomposite materials. J Phys Chem A 107(18):3424–3431
15. Petryayeva E, Krull UJ (2011) Localized surface plasmon resonance: nanostructures, bioassays and biosensing—review. Analytica Chimica Acta 706(1):8–24. ISSN 0003-2670. https://doi.org/10.1016/j.aca.2011.08.020
16. Yang K, Zhang S, Zhang G, Sun X, Lee S-T, Liu Z (2010) Graphene in mice: ultrahigh in vivo tumor uptake and efficient photothermal therapy. Nano Lett 10(9):3318–3323
17. Link S, El-Sayed MA (2003) Optical properties and ultrafast dynamics of metallic nanocrystals. Annu Rev Phys Chem 54(1):331–366
18. Chu M, Li H, Wu Q, Wo F, Shi D (2014) Pluronic-encapsulated natural chlorophyll nanocomposites for in vivo cancer imaging and photothermal/photodynamic therapies. Biomaterials 35:8357–8373
19. Fontijn WFJ, Van der Zaag PJ, Feiner LF, Metselaar R, Devillers MAC (1999) A consistent interpretation of the magneto-optical spectra of spinel type ferrites. J Appl Phys 85:5100
20. Boxall C, Kelsall G, Zhang Z, Boxall C, Kelsall G, Zhang Z (1996) J Chem Soc Faraday Trans 92:791
21. Balberg I, Pinch HL (1978) The optical absorption of iron oxides. J Magn Magn Mater 7:12
22. Sadat ME, Baghbador MK, Dunn AW, Wagner HP, Ewing RC, Zhang J, Xu H, Pauletti GM, Mast DB, Shi D (2014) Photoluminescence and photothermal effect of Fe_3O_4 nanoparticles for medical imaging and therapy. Appl Phys Lett 105(9):091903

23. Han X, Deng Z, Yang Z, Wang Y, Zhu H, Chen B, Cui Z, Ewing RC, Shi D (2017) Biomarkerless targeting and photothermal cancer cell killing by surface-electrically-charged superparamagnetic Fe_3O_4 composite nanoparticles. Nanoscale 9:1457–1465
24. Shi D, Sadat ME, Dunn AW, Mast DB (2015) Photo-fluorescent and magnetic properties of iron oxide nanoparticles for biomedical applications. Nanoscale 7(18):8209–8232
25. Dunn A, Ehsan SM, Mast D, Pauletti GM, Xu H, Zhang J, Ewing RC, Shi D (2015) Photothermal effects and toxicity of Fe_3O_4 nanoparticles via near infrared laser irradiation for cancer therapy. Mater Sci Eng C 46:97–102
26. Zhao Y, Sadat ME, Dunn AW, Xu H, Chen C-H, Nakasuga W, Ewing RC, Shi D (2017) Photothermal effect on Fe_3O_4 nanoparticles irradiated by white-light for energy-efficient window applications. Solar Energy Mater Solar Cells 161:247–254
27. Zhao Y, Dunn A, Lin J, Shi D (2019) Photothermal effect of nanomaterials for efficient energy applications. In: Novel nanomaterials for biomedical, environmental and energy applications. Elsevier, pp 415–434
28. Tian Q, Hu J, Zhu Y, Zou R, Chen Z, Yang S, Li R, Su Q, Han Y, Liu X (2013) Sub-10 nm Fe_3O_4@$Cu_{2-x}S$ core-shell nanoparticles for dual-modal imaging and photothermal therapy. J Am Chem Soc 135(23):8571–8577
29. Yang HW, Liu HL, Li ML, Hsi IW, Fan CT, Huang CY, Lu YJ et al (2013) Magnetic gold-nanorod/PNIPAAmMA nanoparticles for dual magnetic resonance and photoacoustic imaging and targeted photothermal therapy. Biomaterials 34(22):5651–5660
30. Peer D, Karp JM, Hong S, Farokhzad OC, Margalit R, Langer R (2007) Nanocarriers as an emerging platform for cancer therapy. Nat Nanotechnol 2(12):751–760
31. Crezee J, Franken NAP, Oei AL (2021) Hyperthermia-based anti-cancer treatments. Cancers (Basel) 13:1240. https://doi.org/10.3390/cancers13061240
32. Dunn AW, Zhang Y, Mast D, Pauletti GM, Xu H, Zhang J, Ewing RC, Shi D (2016) In-vitro depth-dependent hyperthermia of human mammary gland adenocarcinoma. Mater Sci Eng C 69:12–16. https://doi.org/10.1016/j.msec.2016.06.026
33. Wang Z, Qiao R, Tang N, Lu Z, Wang H, Zhang Z, Xue X, Huang Z, Zhang S, Zhang G et al (2017) Active targeting theranostic iron oxide nanoparticles for MRI and magnetic resonance-guided focused ultrasound ablation of lung cancer. Biomaterials 127:25–35. https://doi.org/10.1016/j.biomaterials.2017.02.037
34. Lin J, Zhao Y, Shi D (2020) Optical thermal insulation via the photothermal effects of Fe_3O_4 and Fe_3O_4@$Cu_{2-x}S$ thin films for energy-efficient single-pane windows. MRS Commun 1 of 9. https://doi.org/10.1557/mrc.2020.4
35. Lyu M, Lin J, Wang Y, Krupczak J, Shi D (2023) Synergistic photothermal-thermoelectric-photovoltaic energy generation via a transparent spectral modulating solar module. Solar Energy 258:220–231. ISSN 0038-092X. https://doi.org/10.1016/j.solener.2023.05.001.R
36. Zhao, Sun X, Sun J, Wang L, Han J (2017) Polypyrrole-modified CuS nanoprisms for efficient near-infrared photothermal therapy. RSC Adv 7(17):10143–10149
37. Liu T-M, Conde J, Lipiński T, Bednarkiewicz A, Huang C-C (2016) Revisiting the classification of NIR-absorbing/emitting nanomaterials for in vivo bioapplications. NPG Asia Mater 8:295
38. Feng L, Wu L, Qu X (2013) New horizons for diagnostics and therapeutic applications of graphene and graphene oxide. Adv Mater 25(2):168–186
39. Xu Y, Qin Y, Palchoudhury S, Bao Y (2011) Water-soluble iron oxide nanoparticles with high stability and selective surface functionality. Langmuir 27:8990–8997. https://doi.org/10.1021/la201652h
40. Sun S, Wang Y, Zhou R, Deng Z, Han Y, Han X, Tao W, Yang Z, Shi C, Hong D et al (2017) Targeting and regulating of an oncogene via nanovector delivery of MicroRNA using patient-derived tumor xenografts. Theranostics 7:677–693. https://doi.org/10.7150/thno.16357
41. Wang Y, Sun S, Zhang Z, Shi D (2018) Nanomaterials for cancer precision medicine. Adv Mater 30:1705660. https://doi.org/10.1002/adma.201705660
42. Wang Z, Chang Z, Lu M, Shao D, Yue J, Yang D, Zheng X, Li M, He K, Zhang M et al (2018) Shape-controlled magnetic mesoporous silica nanoparticles for magnetically-mediated suicide gene therapy of hepatocellular carcinoma. Biomaterials 154:147–157. https://doi.org/10.1016/j.biomaterials.2017.10.047

43. Lyu M, Lin J, Krupczak J, Shi D (2021) Solar harvesting through multilayer spectral selective iron oxide and porphyrin transparent thin films for photothermal energy generation. Adv Sustain Syst 5(6):2100006. https://doi.org/10.1002/adsu.202100006
44. Lin J, Lyu M, Wang Y, Webster B, Shi D (2022) Advanced materials for energy efficient building skins. In: Wang J, Shi D, Song Y (eds) Advanced materials in smart building skins for sustainability: nano to macroscale. Springer Nature. http://www.lavoisier.eu/books/civil-engineering/advanced-materials-in-smart-building-skins-for-sustainability/description_4731023
45. Lin J, Shi D (2021) Photothermal and photovoltaic properties of transparent thin films of porphyrin compounds for energy applications. Appl Phys Rev 8:011302. https://doi.org/10.1063/5.0036961
46. Lin J, Wang Y, Lyu M, Shi D (2022) Spectral selective solar harvesting and energy generation via porphyrin-based hybrid films, Boston. In: Materials research society fall meeting, Nov 27–Dec 2, 2022, Boston
47. Katepalli A, Wang Y, Shi D (2023) Solar harvesting through multiple semi-transparent cadmium telluride solar panels for collective energy generation. Solar Energy 264:112047. ISSN 0038-092X. https://doi.org/10.1016/j.solener.2023.112047
48. Kıyak İ, Oral B, Topuz V (2017) Smart indoor LED lighting design powered by hybrid renewable energy systems. Energy Build 148:342–347. https://doi.org/10.1016/j.enbuild.2017.05.016
49. Byun J, Hong I, Lee B, Park S (2013) Intelligent household LED lighting system considering energy efficiency and user satisfaction. IEEE Trans Consum Electron 59(1):70–76. https://doi.org/10.1109/TCE.2013.6490243
50. Yun GY, Kim H, Kim JT (2012) Effects of occupancy and lighting use patterns on lighting energy consumption. Energy Build 46:152–158. https://doi.org/10.1016/j.enbuild.2011.10.034
51. Yun GY, Jung H, Kim JT (2013) Energy-saving potential of LED lighting systems. Indoor Built Environ 22(1):235–241. https://doi.org/10.1177/1420326X12470298
52. Lin J, Krupczak J, Shi D (2022) Solar harvesting and energy generating building skins with photothermal–photovoltaic dual-modality based on porphyrin thin films. MRS Commun 12:1225–1234. https://doi.org/10.1557/s43579-022-00299-x

Numerical Methods for Integrating Nanoscale Material Features into Building Scale Energy Performance

Julian Wang

Abstract The pursuit of energy-efficient buildings has driven significant advancements in material science, particularly at the nanoscale, where materials exhibit unique optical, thermal, and mechanical properties. However, the challenge remains in effectively integrating these nanoscale features into whole-building energy performance. This chapter explores the pivotal role of numerical methods in bridging the gap between nanoscale material innovations and macroscale architectural applications. A range of computational techniques, including molecular dynamics (MD), finite element analysis (FEA), computational fluid dynamics (CFD), and machine learning-based predictive modeling, are discussed to illustrate how material properties can be accurately represented across scales. By leveraging multi-scale simulations, this research enables the translation of nanoscale behaviors—such as localized surface plasmon resonance (LSPR)-induced photothermal effects—into real-world energy performance assessments. A central case study investigates the role of plasmonic nanomaterials in spectrally selective glazing systems. The study examines how nanoparticle-induced photothermal effects enhance solar heat gain while maintaining high optical transparency, thereby improving building energy efficiency. A physics-based numerical framework is developed to model the nanoscale-to-building-scale thermal transfer, with results validated through parametric energy simulations in EnergyPlus. The findings demonstrate that LSPR-driven glazing systems can achieve heating energy savings comparable to double-pane windows while offering additional daylighting benefits. By integrating numerical modeling with real-world applications, this chapter highlights how computational methods are indispensable for optimizing advanced materials in sustainable architecture. The research underscores the necessity of multi-scale numerical simulations to unlock the full potential of nano-engineered building components, providing a pathway for bridging fundamental material science with large-scale implementation.

J. Wang (✉)
College of Engineering, Pennsylvania State University, University Park, PA, USA
e-mail: julian.wang@psu.edu
URL: https://archilambda.com

D. Shi (ed.), *Energy Materials Science and Engineering: From New Devices to AI Power Systems*, Green Energy and Technology, https://doi.org/10.1007/978-3-031-88550-1_11

Keywords Numerical modeling · Multi-scale simulation · Nanoscale materials · Building energy performance · Computational methods · Solar-responsive glazing · Photothermal effects · Parametric energy simulation

1 Introduction

The demand for energy-efficient buildings has driven extensive research into advanced building materials that can enhance energy performance. One of the most promising frontiers in this field is the integration of nanoscale materials into building applications. These materials, with their unique thermal, optical, and mechanical properties, offer transformative potential for improving insulation, thermal regulation, and overall energy efficiency. However, bridging the gap between nanoscale material discoveries and their macroscopic effects at the whole-building scale remains a significant challenge. Understanding and quantifying the impact of these nanoscale materials on building energy performance requires sophisticated numerical and simulation methods that operate across multiple scales.

1.1 Building Materials for Energy Performance Across Scales

Building materials play a crucial role in determining the energy efficiency of structures. From traditional insulation materials to innovative phase change materials (PCMs) and nanocoatings, advancements in material science have paved the way for enhanced thermal performance, reduced energy consumption, and improved occupant comfort. At the nanoscale, materials exhibit unique properties such as enhanced thermal conductivity, radiative cooling capabilities, and phase transition behaviors that can be leveraged to optimize building performance. However, the direct implementation of these nanoscale enhancements at the whole-building scale necessitates computational modeling to capture the complex interactions between materials, environmental conditions, and building operations.

Multi-scale modeling approaches have emerged as essential tools to evaluate the performance of nano-enhanced materials. These methods integrate nanoscale characteristics into macroscale simulations, allowing researchers to predict their impact on overall building energy consumption. Examples include nano-enhanced PCMs for thermal storage, nanocoatings for dynamic or spectral selective window technologies, and aerogels for superior insulation properties. Each of these applications requires rigorous computational analysis to assess their real-world effectiveness in improving energy efficiency. One of the major reasons for the necessity of these numerical and computational methods is the high cost and complexity of conducting large-scale experiments, which makes it impractical to test every material variation physically. Additionally, uncertainties in real-world environmental conditions and material behaviors require extensive simulations to ensure accurate predictions and robust performance assessments before large-scale implementation.

1.2 The Role of Numerical and Simulation Methods

To accurately model and predict the influence of nanoscale materials on building-scale energy performance, researchers rely on various numerical and simulation techniques. These methods facilitate the translation of material properties from the atomic level to full-scale buildings, providing insights into thermal behavior, energy savings potential, and long-term durability. Molecular dynamics (MD) simulations are widely used to model the interactions of atoms and molecules within nanoscale materials, helping researchers understand thermal conductivity, phase transitions, and mechanical properties that impact building-scale applications [1, 2]. Similarly, finite element analysis (FEA) is employed to simulate heat transfer and structural integrity, allowing for the assessment of how nanoscale modifications influence building components such as walls, roofs, and windows [3]. Furthermore, computational fluid dynamics (CFD) plays a crucial role in evaluating the effects of nanocoatings and insulation materials on airflow and heat distribution within buildings. CFD simulations help optimize passive and active climate control strategies, as demonstrated in research assessing the energy performance of building windows with nanoscale photothermal effects [4]. Additionally, the co-simulated material-component-system-district framework proposed in prior work enables synthetical and comprehensive analysis of building sustainability [5].

Last, it is also worth mentioning that machine learning and physics-informed neural networks have recently emerged as powerful tools in this domain. By leveraging large datasets and predictive modeling techniques, machine learning enhances the accuracy of simulations, reducing computational costs while improving the reliability of energy performance assessments. A pertinent example is the study by Liu et al., where a hierarchical multi-scale model utilizing Physics-Informed Neural Networks (PINNs) was proposed to predict the thermal conductivity of Polyurethane incorporated with Phase Change Materials. This approach allowed for accurate predictions at both meso-scale and macro-scale levels, demonstrating the effectiveness of PINNs in modeling complex thermal behaviors in building materials [6].

1.3 Solar Responsive Nanomaterials and Glazed Façade Applications

Heating and cooling energy loss through windows represents 3.95 quads of primary energy use in the U.S. [7]. A window's thermal insulation is quantified by its U-factor; lower U-factors correspond to better insulation. The current methods for developing windows with low U-factors come from two primary directions: either structural or spectral design. From the structural design perspective, adding layers of windowpane separated by insulating barriers minimizes conductive heat transfer. However, one cannot simply add an unlimited number of thermal barriers and windowpane layers, as this would result in unacceptable transparency, haze, thickness, and other practical issues. Consequently, research has focused on addressing spectral characteristics in terms of thermal radiation and solar energy utilization. For temperatures relevant to building energy efficiency, thermal radiation flux occurs in four basic bands: ultraviolet irradiance, visible light, infrared irradiance from the sun, and long-wave irradiance from interior or outdoor temperatures. About 51% of the power of solar irradiance is infrared [8], so regulating a window's spectral properties to control the passage of thermal radiation in these ranges can significantly improve building energy efficiency.

Emerging nanoscale materials and technologies play a critical role in optimizing the spectral and thermal properties of glazed façades. Innovations such as nanocoatings, spectrally selective glazing, and photothermal-responsive materials enable dynamic control over heat and light transmission. In particular, in recent years, many studies on the optical properties of nanoparticles (NPs) have found that due to the phenomenon of localized surface plasmon resonance (LSPR) [9]—the collective oscillation of conduction electrons on the surface of NPs in the presence of incident light—a few metallic NPs can show strong extinction peaks in the NIR regions of the electromagnetic spectrum. (In simple terms, they are capable of scattering or absorbing large amounts of light when illuminated.) [10–14]. It is evident that the wavelengths at which plasmonic NPs interact with light, as well as their ratio of absorption and scattering abilities, can be tuned by varying the composition, shape, and size of the particles [15–17]. These NPs are capable of manipulating light on a nanoscale with the possibility of being incorporated into spectrally selective glazing systems.

The LSPR-mediated spectrally selective behaviors have been extensively studied in recent years using both theoretical and experimental methods. Their absorption and scattering abilities and LSPR frequencies can be tuned by controlling particle size, shape, orientation, distribution, concentration, and composition [11, 15, 18]. Regarding the potential of incorporating LSPR effects into glazing structures, the most encouraging work was conducted by Milliron, which integrated the LSPR effect and electrochromism into a dynamic glazing film design, achieving 39% NIR transmission modulations with 82–85% T_{vis} [19]. Although many studies on the nanoscale have claimed that the LSPR effect can be used in building glazing systems and several conducted simple simulations or offered modest predictions [20, 21], much less attention has been paid to analyzing the thermal and optical behaviors of the LSPR employed in complex fenestrations (e.g., with low-e coatings, air cavity, and a variety of structures). Therefore, some of the results of these attempts have **led to conflicting conclusions**, even though they addressed the same phenomenon. Studies by Li et al. [22], Xu et al. [23], Pardiñas-Blanco et al. [24], and Besteiro et al. [25] were focused on the plasmonic NPs' NIR absorption, and suggested their use in blocking solar heat for cooling purposes. Zhao et al. [26], Khashan et al. [27], and Wang and Shi [28] also focused on the NIR absorption induced by the LSPR, concluding it offered the potential for heat-related energy savings in winter. The controversy is due to the complexity of building windows' thermal and optical transfers.

In particular, highly absorptive coating schemes may block solar heat transmission, but conduct, convect, and/or radiate the absorbed energy inwards and/or outwards at relative proportions, depending on the glazing structure, low-e coating's spectral features, and positions relative to the plasmonic films, as well as the corresponding external and internal environments. Therefore, compared to the relatively established theories, models, and experimental validations in the nanomaterial field of LSPR-induced absorption and scattering, **little has changed in the physics-based numerical method** at the architectural scale that incorporates new nanoscale physical relations and models employing LSPR in the thermal and optical performances of glazing.

Although these nanoparticles have demonstrated the ability to manipulate light at the nanoscale and show promise for integration into spectrally selective glazing systems, a critical challenge remains in transitioning from fundamental material studies to practical architectural applications. The interaction of LSPR effects with multi-layer glazing system properties—such as spectral emissivity, thermal insulation, and structural placement—must be thoroughly understood through comprehensive numerical modeling and analysis. A precise evaluation of how these interactions affect the overall thermal and optical performance of glazing systems is necessary to unlock their full potential in energy-efficient building applications.

In this chapter, using the above research challenge about numerical modeling and analysis in nanoscale glazing systems as an example, the focus will be on how emerging nanoscale materials and their newly discovered physical relationships can be incorporated into building-scale performance analysis. This chapter will illustrate how advanced numerical methods can bridge the gap between fundamental nanoscale physics and practical architectural applications, demonstrating key approaches for evaluating energy transmission, spectral selectivity, and thermal efficiency in complex fenestration structures.

2 Fundamentals of the Numerical Modeling and Analysis

2.1 Thermal Transfer Modeling of Glazing Systems

Figure 1 shows the analytical glazing structures consisting of n layers; each of the layers has four key variables: $T_{f,i}$, $T_{b,i}$, $J_{f,i}$, and J_{bi}, which refer to the temperatures of the outdoor and indoor-facing surfaces, and radiant heat fluxes leaving the front and back-facing surfaces, respectively.

The heat flux across the ith gap (q_i) is

$$q_i = h_{c,j}\left(T_{f,j} - T_{b,j-1}\right) + J_{f,j} - J_{b,j-1} \tag{1}$$

Similarly, the heat flux across the $(i + 1)$th gap is

$$q_{i+1} = h_{c,j+1}(T_{f,j+1} - T_{b,j}) + J_{f,j+1} - J_{b,j} \tag{2}$$

The solution (i.e., the temperature at each glazing surface and corresponding radiant fluxes) is generated by applying the following four equations at each layer:

$$q_i = S_i + q_{i+1} \tag{3}$$

$$J_{f,j} = \varepsilon_{f,j}\,\sigma T_{f,j}^4 + \tau_i J_{f,j+1} + \rho_{f,j} J_{b,j-1} \tag{4}$$

$$J_{b,j} = \varepsilon_{b,j}\,\sigma T_{b,j}^4 + \tau_i J_{b,j-1} + \rho_{b,j} J_{f,j+1} \tag{5}$$

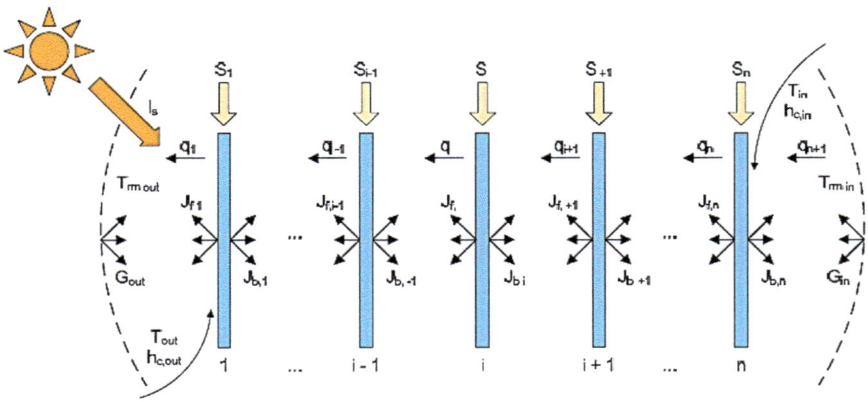

Fig. 1 Numbering system, boundary conditions and energy balance for N-layer glazing system

$$T_{b,j} - T_{f,j} = \frac{t_{g,j}}{2k_{g,j}}[2q_{i+1} + S_i] \tag{6}$$

Equation 3 describes the energy balance imposed on the surface of the ith glazing layer. Equations 4 and 5 define the radiosity at the ith glazing, where $\rho_{f,j} = 1-\varepsilon_{f,j}-\tau_i$ and $\rho_{b,j} = 1 - \varepsilon_{b,j} - \tau_i$, while the temperature difference across the ith glazing layer is given by Eq. 6. In all, four n-equations can be written for a glazing system; these can either be solved using an iterative solution algorithm with convergence checking on the sets of glazing-layer temperatures [29, 30], or by setting equations in matrix form and then decomposing the matrix to find sets of glazing-layer temperatures [31, 32]. In response to such basic methods, some studies have focused on improving the analytical models, including correcting the convective heat coefficient [33, 34], and internal shading effect models [35], conjugating the heat transfer analysis after considering both the conduction and convection processes [36], and comparing empirical correlations to long-wave reflectivity [37].

Within a macroscopic framework, a temperature field is assumed not only to be a continuum, but also to hold a thermal equilibrium at every location [38, 39]. Conventional window heat transfer studies are built upon such assumptions [40]. However, when the thickness scale is comparable to or smaller than the mean free path of the material, the macroscopic model is of question [41 –44]. Since there are no sufficient energy carriers in the interested direction, the temperature field is discontinuous. Once the concept of a temperature gradient fails, the classic Fourier's law is also questionable, as is the associated diffusion equation. A similar situation exists in the response time for temperature [45, 46]. The macroscopic heat equation assumes that the temperature gradient follows simultaneously with the heat flux vector [38]. Due to the quick thermal effects activated by the LSPR, the response time of primary concern is of the same order of magnitude as the mean free time; the lagging behavior caused by phonon–electron interaction in films or phonon scattering must be taken into account [47, 48]. On the nanoscale, some models analyzing thermal behaviors that are strongly enhanced in the presence of plasmonic NPs have been developed and experimentally verified [46–49], but little research has studied the micro-to-macroscale heat transfer mechanism of multi-layer glazing structures with air layer and low-e coatings.

2.2 *Spectral Solar Irradiance Modeling*

Accurate knowledge of locally available solar radiation is essential for the proper sizing, design, and dynamic simulation of solar energy systems [50]. Initial approaches used current weather conditions as input for mathematical models with dynamic atmosphere parameters and solar physics to predict global solar irradiance [51–53]. The typical meteorological year (TMY) data files, developed from extensive datasets, serve as the most widely used weather files in building envelope simulations. Each TMY file consists of 12 months selected as most representative of long-term climate patterns. These files include three primary broadband solar components: global horizontal irradiance (GHI), direct normal irradiance (DNI), and diffuse horizontal irradiance (DHI). These components encompass ultraviolet (UV), visible light (VIS), and near-infrared radiation (NIR), which are critical for solar energy and building performance simulations.

Recent research has highlighted the necessity of decomposing broadband solar radiation into narrowband components or specific spectral bands for more accurate analyses. Solar VIS radiation can enhance indoor lighting efficiency and circadian health [54] but also causes glare issues. NIR radiation can reduce heating loads in cold climates but is undesirable in hot regions. Additionally, studies have shown that VIS and NIR transmission through glazing affect occupants' thermal comfort differently near windows [55]. With the advent of spectral-selective materials, independent modulation of solar radiation across different spectral bands has become feasible. For instance, metallic nanoparticle-based nanocomposites leverage plasmonic resonance effects to selectively modulate VIS and NIR transmission. Jahid et al. demonstrated reversible photothermal windows utilizing nanoscale solar-induced plasmonic photothermal effects to regulate solar heat independent of visible light conditions [56, 57]. Shen et al. explored silver nanorods (AgNRs) that allow for over 50% luminous transmittance while blocking up to 80% of solar radiation via tunable plasmonic resonance [58]. Forrest et al. reviewed semitransparent organic photovoltaics, which selectively absorbs UV and NIR while maintaining visible transparency, making them ideal for power-generating windows [59]. Other researchers have also investigated light and heat-splitting materials by designing spectral transmittance and absorptance properties of glazing materials across different solar spectra [25, 60–62].

To assess photovoltaic systems and solar-selective glazing that incorporate nanoscale spectral-selective features, several solar decomposition models—such as the BRL, Perez, Boland, Maxwell, and Lauret models—have been developed to predict spectral irradiance from broadband global radiation using statistical techniques. However, accurately modeling visible and NIR components for the glazing surface remains essential for optimizing NIR-modulated window systems. Due to the high cost of spectroradiometers, most regions lack direct measurements of spectral components, relying instead on pyranometer-derived broadband data. As a result, many conventional weather files lack spectral information needed for precise simulation of spectral-selective materials.

To address this gap, a reconstruction algorithm is needed to decompose solar VIS and NIR irradiance from broadband radiation data for building simulations. Given that solar simulation relies on GHI, DHI, and DNI components, all three must be decomposed into narrowband data. We have conducted two specific studies to develop models for decomposing broadband solar irradiance data into its visible (VIS) and near-infrared (NIR) components. The first study focused on developing a predictive model using the Classification and Regression Tree (CART) algorithm for estimating VIS and NIR components in global horizontal irradiance (GHI) based on typical meteorological year (TMY) weather data. The second study extended the above-mentioned work by building spectral solar radiation models for direct normal irradiance (DNI) and diffuse horizontal irradiance (DHI), using extracted and derived features from existing weather files without requiring additional measurements or sensor installations.

(1) Decomposing VIS and NIR from GHI

The model development utilized two years of solar spectral and TMY-compliant hourly weather data from the Solar Radiation Research Laboratory (SRRL) Baseline Measurement System (BMS) database at NREL. The dataset, covering wavelengths from 300 to 1700 nm, underwent rigorous data cleaning, including the removal of missing, outliers, and erroneous values based on physics-based calculations. After processing, the finalized dataset consisted of 7,583 observations. To enhance model generalizability across different locations, we incorporated ten local meteorological parameters, such as humidity and temperature, along with five newly derived predictor variables informed by solar radiation and building physics principles. Among these, the cloud transmittance parameter, derived from cloudiness values in standard weather files, was particularly effective in improving model accuracy for the NIR component. In total, 15 predictor variables were used to develop the estimation models for hourly VIS/GHI and NIR/GHI. Model validation yielded strong results, with mean absolute error (MAE) between 2.13 and 2.41% and root mean square error (RMSE) between 3.88 and 4.25%, confirming the reliability of the classification tree models [63].

(2) Decomposing VIS and NIR from DNI and DHI

For DNI and DHI components, we employed an extreme boosting regression tree method, leveraging readily available TMY weather data. Multiple databases from the NREL Solar Radiation Research Laboratory were curated and processed [64]. After a thorough data cleaning procedure, which involved removing missing values, outliers, and physically erroneous data, the finalized dataset comprised 11,862 observations. To ensure the model's generalizability across different locations, easily obtainable meteorological parameters were incorporated into the modeling process. The full estimation model utilized 20 predictive variables, while a simplified version, designed for scenarios with limited meteorological data, relied on 15 predictive variables.

Four widely used supervised, non-parametric machine learning algorithms—Decision Tree, M5' Tree, Random Forest, and XGBoost—were employed to train and compare the models. The comparative analysis revealed that XGBoost outperformed the other techniques in accuracy and reliability for DNI decomposition across both full and simplified models. It achieved the lowest RMSE (18.080 for V1, 18.985 for V2, 15.868 for N1, and 17.606 for N2) and MAE (9.192 for V1, 10.181 for V2, 8.383 for N1, and 9.506 for N2), along with the highest R^2 values (0.983 for V1, 0.980 for V2, 0.993 for N1, and 0.990 for N2). Among the predictive variables, DNI and Kb were identified as the most influential for both VIS and NIR decomposition. Additionally, certain atmospheric parameters, including aerosol optical depth (AOD), precipitable water vapor (PWV), and total and opaque sky cover, contributed significantly to the predictions. Through feature importance and interaction analyses, previously hidden correlations between predictors, as well as between predictors and responses, were uncovered.

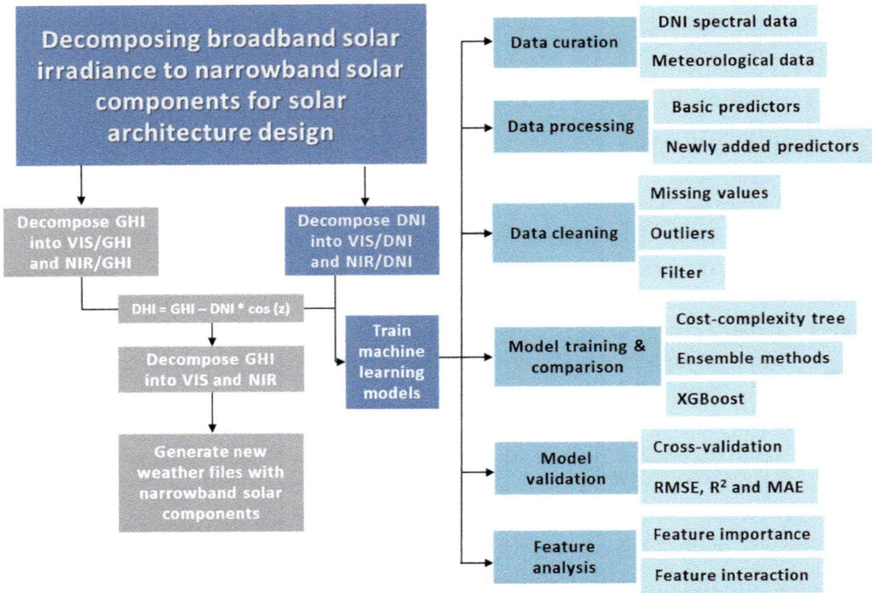

Fig. 2 The workflow of generating new weather files with narrowband solar components

By leveraging the above two models, it becomes feasible to effectively break down conventional, broad-spectrum weather files into distinct, narrowband weather files. Figure 2 presents the workflow of generating new weather files with narrowband solar components. Subsequently, this can be fed into typical building energy simulation programs for further analysis of integrating nanoscale spectral selective materials into building scale.

2.3 Parametric Energy Simulation and Performance Analysis

Parametric energy simulation is different from the traditional energy simulation procedure and incorporates more complicated parametric relationships between the design variables, construction states, and environmental boundaries. The original development of the overall workflow of parametric energy simulation was for dynamic building envelope's performance simulation. Throughout the history of architecture, dynamic building envelopes have attracted tremendous attention and effort due to their aesthetic characteristics, pleasant visual impact, design uniqueness, and interaction with users [65–67]. The dynamic envelope concept has long been established and integrated into the current architecture curriculum via a variety of generative design tools and fabrication techniques [68–70]. Nevertheless, until recently, advances in and the availability of new materials and structures such as vacuum insulation [71, 72] and gas-filled panels [73, 74]; phase-change materials [75, 76]; and thermotropic, photochromic, and electrochromic materials have focused the attention of architects and designers on such envelopes' performance [77–79]. One of the fundamental steps in establishing performance-driven dynamic envelopes is an analytical approach that determines the range of dynamic properties necessary to satisfy the adaptability requirements, as well as their amplitude of variability. Since the development of this kind of analysis requires a "parametric energy" approach (i.e., environmental conditions and envelope behaviors must be parametrically connected), the simulation instruments currently available are often inadequate; thus, researchers must develop ad hoc numerical models [80] (such as the CABS approach) for wall-specific heat and thermal conductivity [81, 82], external absorption coefficients of the surface, glazing typologies, the sensitivity analysis framework for the envelope's thermal properties [83–85], and the BKE method for the envelope's thermal conductivity and SHGC [80, 86–88]. To integrate the new dynamic structures into glazing systems (e.g., LSPR-mediated dynamic attachments), a computationally efficient method is needed to quantitatively ideate and optimize the characteristics of a system.

Here we introduce the parametric energy simulation method and procedure utilizing the Energy Management System (EMS) in EnergyPlus. EnergyPlus is a widely used, open-source building energy simulation engine developed by the U.S. Department of Energy (DOE). It enables the modeling of energy consumption for heating, cooling, ventilation, lighting, and other building-related processes. EnergyPlus EMS module is particularly advantageous for integrating advanced simulation techniques, such as parametric energy modeling, to account for complex, dynamic physical relationships not inherently embedded in conventional building simulations. The EMS-based parametric energy simulation method allows users to define custom control algorithms and dynamic parameter adjustments, making it ideal for advanced energy modeling scenarios. Similar EMS-based approaches have been applied in previous studies to incorporate dynamic parameters into energy simulations [87, 89–96].

An EMS program in EnergyPlus comprises several essential components:

- **Sensor**—Extracts information from the model (e.g., temperature, humidity, solar radiation).
- **Actuator**—Modifies specific model components or elements based on sensor input.
- **Program Calling Manager**—Determines when the EMS program is executed.
- **Program**—Defines the logic for modifying model parameters using conditional statements.
- **Construction Index Variable**—Links EMS-based changes to material properties within the simulation.

In particular, sensors in EnergyPlus can be used to track key environmental parameters relevant to window technology simulations, such as indoor air temperature, relative humidity, and incident solar radiation. The actuator then modifies window properties based on these sensor readings, allowing for dynamic adjustments in response to external conditions. For example, if window properties vary based on outdoor temperature, the EMS sensor should monitor "Site Outdoor Air Drybulb Temperature," while the actuator should control "Construction State" to modify the window properties dynamically. The Program Calling Manager should be set to "Begin Timestep Before Prediction," ensuring that adjustments occur before EnergyPlus predicts energy performance for the next timestep. The program logic is implemented using "IF" statements and "While" loops, allowing windows to respond dynamically to changing outdoor air temperature conditions.

A crucial advantage of EMS-based parametric simulations is the ability to bridge the gap between nanoscale material properties and large-scale building energy performance. Parametric relationships can be derived from: experimental data on advanced glazing systems incorporating nanomaterials; regression models based on numerical simulations of nanoscale materials. Using the "Construction Index Variable" section in EnergyPlus, optical and thermal properties of an advanced glazing system are linked to the EMS program. These properties are defined under the "Window Material: Glazing" section and are dynamically adjusted based on the assigned outdoor weather conditions in the EMS Program section.

The following diagram outlines the workflow for implementing an EMS-based parametric energy simulation, demonstrating how advanced parametric relationships are incorporated into building-scale energy analysis:

- **Define sensors** to extract real-time weather and indoor environmental data.
- **Select actuators** to modify dynamic material or system properties.
- **Develop EMS logic** using conditional programming to enable real-time responsiveness.
- **Integrate nanoscale material properties** by linking EMS with advanced material simulations.
- **Run parametric simulations** in EnergyPlus to evaluate energy performance.

This EMS-based approach enhances the accuracy of building energy simulations by integrating real-time, complicated physical relationships, particularly for new glazing systems and other advanced building envelope technologies that incorporate newly discovered nanoscale materials and features (Fig. 3).

3 Case Study: Numerical Analysis of the Energy Performance of Solar NIR-Driven Nano Plasmonic Photothermal Effects

3.1 Theoretical Analysis and Physical Relationship Derivation from Experiments

The overall heat transfer process under solar radiation is illustrated in Fig. 4. For building windows, solar radiation is transmitted, absorbed, and reflected in varying proportions based on the window's optical properties and environmental conditions.

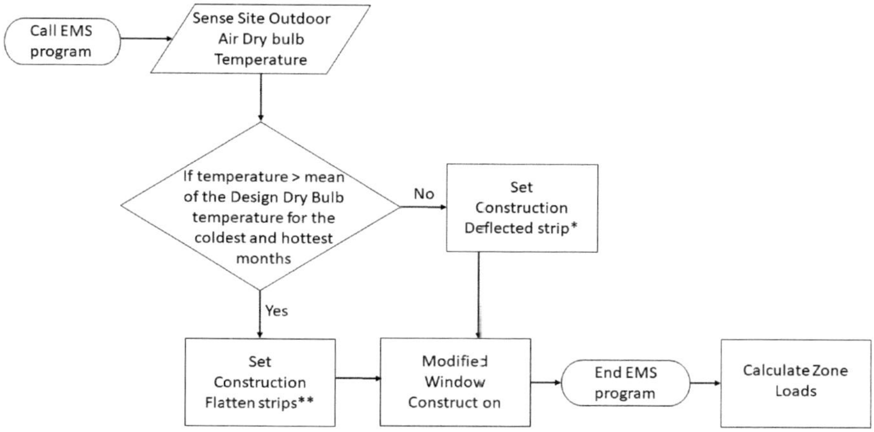

Fig. 3 Exemplary EMS workflow at each timestep for yearly energy use simulation [97]

Transmitted solar radiation directly reduces space heating demands, while absorbed radiation partially flows inward from the panes, increasing indoor temperatures and enhancing heating energy savings in winter. The distribution of these effects is quantified by the solar heat gain coefficient (SHGC) in fenestration analysis. The absorbed solar irradiance (q_{abs}) was calculated by Eq. 7. Similarly, the solar irradiance (q_{trans}) transmitted by the glazing system was obtained by Eq. 8. Solar irradiance can increase the glazing system's temperature, especially in photothermal windows.

$$q_{\text{abs}} = \int G_{\lambda}\alpha_{\lambda}d_{\lambda} \quad (7)$$

$$q_{\text{trans}} = \int G_{\lambda}\tau_{\lambda}d_{\lambda} \quad (8)$$

where G_{λ} is the incident spectral irradiance from solar light or simulated solar light and α_{λ} is the spectral absorptance of the photothermal-coated windows.

The solar energy absorbed by a glazing system contributes to a spontaneous temperature rise in the glazing. In a conventional steady-state heat transfer model, this temperature change is governed by

$$\frac{\mathrm{d}T}{\mathrm{d}t} = \frac{1}{c * m} * \left(\sum Q_{\text{in}} - \sum Q_{\text{out}}\right) \quad (9)$$

$$Q_{\text{in}} = q_{\text{abs}} * A_s \quad (10)$$

$$Q_{\text{out}} = Q_{\text{conv}} + Q_{\text{cond}} + Q_{\text{rad}} \quad (11)$$

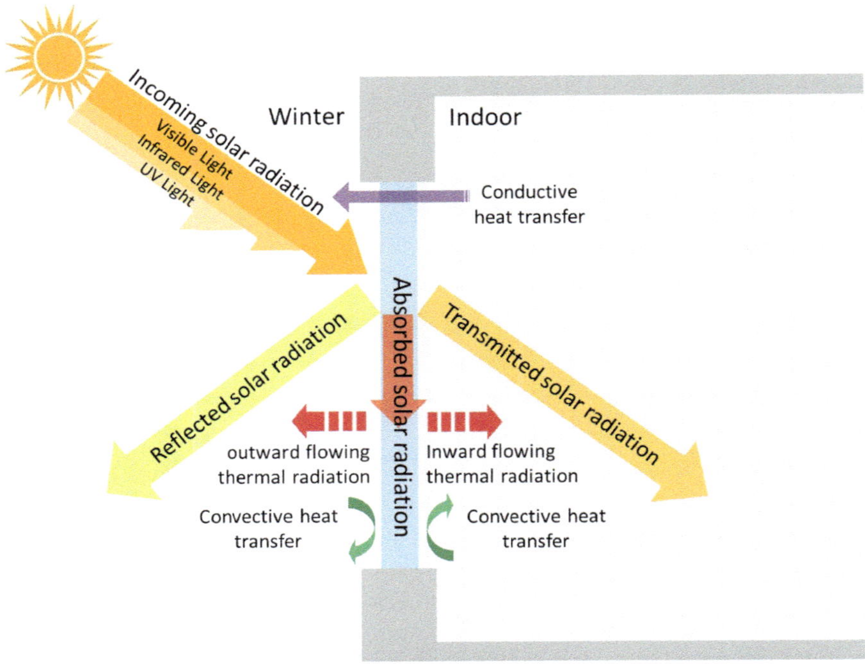

Fig. 4 Solar radiation and window thermal transfer

Here, c is the heat capacity (J/g K), and m is the material mass (g). Under predefined internal and external boundary conditions, Q_{in} corresponds to the absorbed solar radiation on the window surface (Eq. 10), while Q_{out} accounts for conductive, convective, and radiative heat transfer from the glazing's interior and exterior surfaces (Eq. 11). This process is complex due to convection flow and surface emissivity variations. With stable ambient air conditions and continuous solar irradiance, an energy balance model can be developed to analyze thermal behavior and temperature distribution. Traditional calculations assume the glazing system as a homogeneous bulk, leading to an equal temperature rise on both surfaces.

However, as highlighted in the introduction, when plasmonic NPs coat the glazing surface, LSPR-induced photothermal effects can significantly elevate the coating surface temperature while reducing heat transfer to the substrate. This phenomenon is attributed to a quasi-ballistic effect, where the NP size is much smaller than the phonon mean free path in the substrate material [98]. Inspired by previous studies on LSPR-induced solar vapor generation [99], an interfacial insulation layer (R_h) was hypothesized between the NP-coated layer and the glazing substrate [57]. This layer was incorporated into the conventional heat transfer model to account for heat dissipation variations.

Under stable boundary conditions and identical ambient temperatures, the absorbed solar energy Q_{in} is dissipated through the NP-coated surface and via conductive transfer through the interfacial layer into the glass. The heat dissipation Q_{out} is expressed as

$$Q_{out} = h_{con\text{-}NP}(T_{NP,s} - T_{\infty}) + A_s + \varepsilon_{NP}\sigma(T_{NP,s}^4 - T_{\infty}^4)A_s + \left(\frac{1}{R_h}\right)(T_{NP,s} - T_{glass,s})A_s \tag{12}$$

where

- $T_{NP,s}$ is the coating surface temperature.
- $T_{glass,s}$ is the soda-lime glass surface temperature.
- $(T_{NP,s} - T_{glass,s})$ represents the temperature differential across the interfacial insulation.
- T_{∞} is the ambient temperature.
- R_h is determined by fitting the analytical surface temperature model to measured data.
- $h_{con\text{-}NP}$, ε_{NP}, σ are heat transfer parameters specific to the photothermal coating.

Notably, unlike conventional macroscopic heat transfer models that are used in building scale heat transfer and energy balance calculations, this function assumes the interfacial insulation layer acts as a thermal barrier, moderating heat transfer from the NP layer to the glass while enhancing heat dissipation at the coating surface. By integrating Eqs. (9–12), the two-dimensional temperature evolution from initial irradiation to steady-state conditions can be analytically solved.

The key hypothesized interfacial insulation R_h needs to be obtained through nanoscale material experiments. In this case study, $Fe_3O_4@Cu_{2-x}S$ was particularly selected, which has been found to have robust spectrally selective absorption of NIR radiation with only a minor reduction in the visible region. The material demonstrates highly efficient photothermal heating conversion under continuous illumination at room temperature. The synthesis and fundamental characterization of $Fe_3O_4@Cu_{2-x}S$ photothermal nanoparticles (NPs) have been previously reported [100]. They revealed that the absorption peak of $Fe_3O_4@Cu_{2-x}S$ NPs shifts from 960 nm to 1,150 nm as particle size increases from sub-10 to 15 nm. Based on this, $Fe_3O_4@Cu_{2-x}S$ NPs with an average size of 15 nm were synthesized and utilized in this study.

The NP solutions were spin-coated onto glass panes ($\sim$2.5 × 2.5 cm^2) for 10 s at 1,000 rpm, with varying concentrations. The resulting thin-film NP mass per unit coating area was 2.54×10^{-4} g/cm^2. The spectral properties, including transmissivity and reflectivity, were measured using a LAMBDA 900 UV/Vis/NIR spectrophotometer (Fig. 5b). The coated samples were exposed to an irradiance of 1,000 W/m^2 using a Newport 150W solar simulator, while the surface temperature was continuously monitored with a FLIR E6 infrared camera. Nanoscale materials experimental results indicate that the surface temperature of the $Fe_3O_4@Cu_{2-x}S$ NP-coated glass

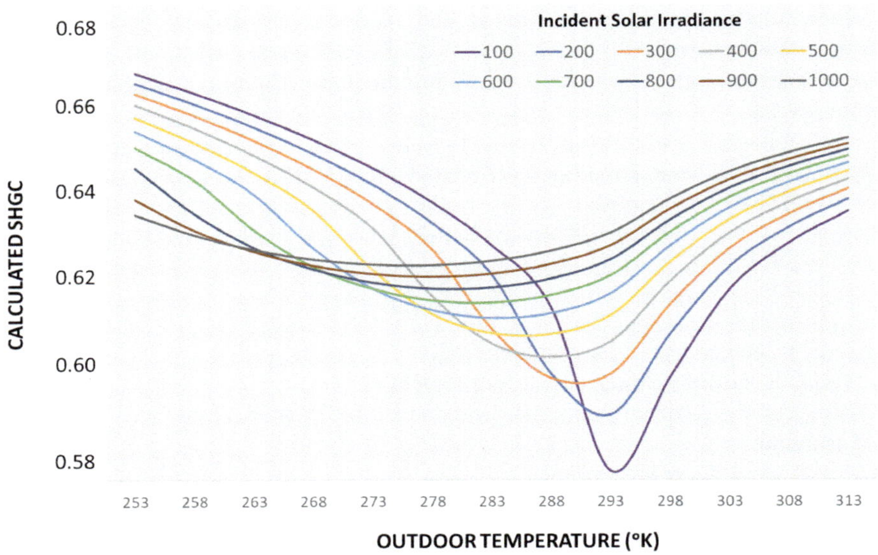

Fig. 5 SHGC variations driven by nanoscale PPE with different outdoor boundaries

increased rapidly under simulated solar radiation, reaching a thermal equilibrium within 10 min. Upon turning off the solar simulator at the 10-min mark, temperature decay was monitored throughout the cooling period.

As mentioned above, the presence of the interfacial insulation was assumed to match the enhanced temperature rise by the nanoscale plasmonic photothermal effect (PPE). By fitting the analytically calculated surface temperature to the measured data above, $R_h = 0.03$ m^{2}K/W was obtained. In other words, a thermal barrier appears when it comes to the surface plasmonic resonance of the Fe_3O_4@$Cu_{2-x}S$ NPs and reduces the effective thermal conductance to another side of the glazing. Notably, such interfacial insulating ability is only activated when solar radiation is incident on the materials.

3.2 Influence of Nanoscale PPE on SHGC

In conventional photothermal processes, the temperature rise of building glazing systems increases with higher absorbed solar energy. However, the fraction of thermal energy transferred indoors may be reduced by the low emissivity of the inner surface. According to the NFRC SHGC calculation standard, the Solar Heat Gain Coefficient (SHGC) is defined as:

$$\text{SHGC} = \tau_s + N_i \alpha_s \tag{13}$$

where τ_s is the solar transmittance of window systems, α_s is the solar absorptance, and N_i is the inward-flowing fraction of absorbed radiation.

The value of N_i is influenced by the heat transfer coefficients on the glazing's inner and outer surfaces. When solar absorptance is low, the inward-flowing heat fraction remains minimal, particularly when the inner surface has low emissivity. Under such conditions, boundary conditions have little to no impact on the SHGC. However, for high solar absorptance scenarios, boundary conditions can significantly alter N_i. Moreover, the temperature increase in this study is not solely attributed to increased photon absorption but also to the strong LSPR-induced plasmonic heating effect. The LSPR-induced PPE is governed by the electric field intensity generated by the collective oscillation of energetic electrons upon resonant excitation. At higher irradiation intensities, a greater number of energetic electrons are excited within the plasmonic NPs, leading to an enhanced PPE.

The mathematical model developed in Sect. 3.1 was used to examine the relationship between solar irradiance levels, PPE effects, and the inward-flowing fraction of absorbed solar radiation under varying outdoor air temperatures. Solar irradiance levels ranging from 100 to 1,000 W/m^2 were selected to represent the incident radiation on a vertical window surface. The outdoor temperature varied between 258 and 313 K in 5 K increments, while the indoor temperature remained constant at 297 K. This setup resulted in 130 different boundary condition combinations.

Using the analytical model, temperature distributions across window surfaces were determined, followed by the calculation of inward-flowing thermal radiation, inward-flowing fraction, and SHGC. These calculations adhered to the NFRC 201 solar calorimetric method, as described by the following equations:

$$Q_{\text{solar gains}} = h_i A_s (T_{\text{in},s} - T_i) + \varepsilon_i \sigma A_s (T_{\text{in},s}^4 - T_i^4) \quad (14)$$

$$\text{SHGC} = (Q_{\text{solar gains}} - Q_{\text{U-factor}}) / \int A_s G_\lambda \quad (15)$$

where $Q_{\text{solar gains}}$ is the overall heat flux flowing through the window under solar irradiation and given temperature boundaries, $T_{\text{in},s}$ is the inner surface temperature of the window, T_i represents the indoor air temperature, and T_∞ is the ambient air temperature. h_i, ε_i, k_{in}, σ are parameters or constants involved in the convective and radiative heat transfer. $Q_{\text{U-factor}}$ is the heat flux due to indoor and outdoor air temperature differences and calculated upon the overall U-factor of the window system. G_λ is the spectral solar irradiance incident on the window. To simplify the analysis, the effects of incident solar angles, window frames, glazing edges, and condensation were not considered.

The SHGC variations of the window with nanoscale PPE are complicated. This feature can be clearly seen in Fig. 5. Since solar transmittance remained constant across all conditions, variations in SHGC were primarily influenced by changes in the inward-flowing heat fraction. Theoretically, when the temperature difference between the inner window surface and indoor air is minimal, the inward-flowing heat fraction is negligible, resulting in the lowest SHGC value. For example, at an incident solar irradiance of 100 W/m^2, the inner window surface temperature reached 297.6 K when the outdoor temperature was approximately 293 K, yielding a minimum SHGC of 0.58. Conversely, when the temperature increase caused by 100 W/m^2 solar irradiation brought the outer window surface temperature closer to the outdoor conditions, the SHGC reached its maximum value of 0.67. Furthermore, since SHGC variations are driven by changes in inward-flowing heat rather than solar transmittance, the visible transmittance of the photothermal windows remains unaffected.

3.3 Parametric Energy Simulation with Nanoscale PPE-Driven SHGC

Although such SHGC variations with the nanoscale PPE observed in Sect. 3.2 are complicated, it can be integrated into the building simulation program via the EMS module of EnergyPlus. To integrate this dynamic SHGC into an energy simulation platform for analyzing photothermal windows, a simplified mathematical model is necessary. Given the data distribution and physical significance of SHGC, a parabolic function of the form $y = a(x - b)^2 + c$ as used to model SHGC for the photothermal window.

In this model, parameter b represents the outdoor temperature at which y equals c, corresponding to the minimum SHGC. The final regression model is expressed in Eq. 16. To distinguish this from the standard SHGC concept, we introduce the term $\mathrm{SHGC}_{\mathrm{PPE}}$, which specifically accounts for the nanoscale PPE-induced solar heat gain characteristics.

$$\mathrm{SHGC}_{\mathrm{PPE}} = a(T_o - b)^2 + c \tag{16}$$

$$a = 4.5 \times 10^{-5}$$

$$b = -0.02 \times \int G_\lambda \mathrm{d}\lambda + 297.17$$

$$c = 4.13 \times 10^{-5} \times \int G_\lambda \mathrm{d}\lambda + 0.58$$

where $\int G_\lambda \mathrm{d}\lambda$ is the total incident solar irradiance and T_o is the outdoor temperature. This model can be simply interpreted from the physical perspective and also estimate the photothermal window SHGC with high accuracy (mean absolute percentage error 1.7%, root mean square error 0.01).

Furthermore, to evaluate whole-building energy performance while accounting for the dynamic characteristics of $\mathrm{SHGC}_{\mathrm{PPE}}$, the parametric energy simulation approach described in Sect. 2.3 was required to model the evolving properties of building

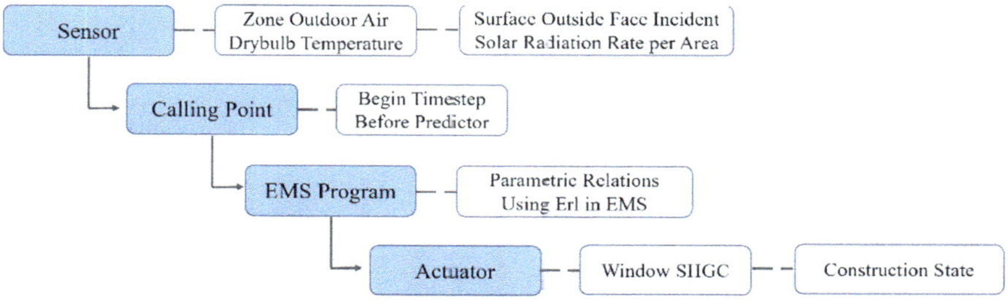

Fig. 6 EMS parametric energy simulation workflow for windows with nanoscale PPE

windows. In this simulation, the following parameters were selected as sensors: "Zone Outdoor Air Dry Bulb Temperature" (to track external temperature conditions) and "Surface Outside Face Incident Solar Radiation Rate per Area" (to capture solar irradiance on window surfaces). The actuator used was "Construction State," which determines the window type and its associated properties. Regarding the simulation workflow, the EMS calling point "Begin Timestep Before Predictor" was used, ensuring calculations occurred at the start of each timestep, before zone loads were determined. The $SHGC_{PPE}$ model (Eq. 16) was implemented to dynamically compute SHGC values for different window configurations. The parametric energy simulation workflow incorporating $SHGC_{PPE}$ is illustrated in Fig. 6.

The DOE prototypical small office building model (single-story, ~510 m^2 floor area) was selected as the case study, following the most recent ASHRAE 90.1 standards. The window-to-wall ratio was approximately 24.4% on the south-facing façade and 19.8% on the other three orientations. In this model, windows were the only control variable, allowing for a focused analysis of photothermal window performance. To evaluate energy performance, the windows with nanoscale PPE were compared against three conventional window types: Single-pane clear windows, Single-pane low-e coated windows, and Double-pane windows. The thermal and optical properties of each window type used in this analysis are detailed in Table 1. The construction configurations for single-pane clear, single-pane low-e, single-pane photothermal, and double-pane windows were implemented in EnergyPlus, replacing the original windows in the prototypical model. To ensure a fair performance comparison, all other building components—including roofs, walls, lighting, and HVAC system configurations—were kept identical across all simulations.

The prototypical model was designed with a core zone and four perimeter zones, one for each cardinal direction. A series of annual whole-building energy simulations were conducted across four climate zones (Zones 3–6). Representative cities selected for each climate zone included: Atlanta, GA (Zone 3), Seattle, WA (Zone 4), Chicago, IL (Zone 5), and Great Falls, MT (Zone 6).

Additionally, given the potential risk of overheating associated with photothermal windows, an additional design modification was introduced. Fixed overhangs were added to the photothermal window models to mitigate excess solar heat gain during summer. The depth of these overhangs was determined based on solar angles during

Table 1 Window properties used in whole building energy analysis

Window	U-factor (W/ Km^2)	Solar heat gain coefficient	Visible transmittance	Emissivity
Single-pane clear	5.88	0.87	0.90	0.84 (surface 2)
Single-pane low-e	3.28	0.55	0.76	0.05 (surface 2)
Single-pane photothermal	3.30	0.58–0.67	0.75	0.06 (surface 2)
Double-pane low-e	2.42	0.53	0.68	0.05 (surface 3)

the summer solstice for each representative city, ensuring an optimized balance between winter solar gain and summer shading. This setup allowed for a comprehensive comparison of photothermal window performance across different climatic conditions while addressing potential thermal comfort concerns.

3.4 Whole Building's Energy Analysis

Since the case study in this chapter was designed and developed for heating energy saving purposes, this section presents the heating loads energy at the whole building scale particularly. The energy use of a single-pane clear window was established as the baseline for comparison. Annual energy savings percentages were calculated for single-pane low-e, single-pane photothermal, and double-pane windows relative to this baseline. The annual heating loads and corresponding energy savings across the four studied cities are presented in Fig. 7. The addition of a low-e coating led to 3.3–11.9% annual heating energy savings compared to single-pane clear windows. This improvement is primarily attributed to enhanced thermal insulation, though it also results in some reduction in solar heat gain due to the low-e coating's reflective properties. In contrast, the designed windows with nanoscale PPE exhibited significantly greater heating energy savings in winter, achieving 16.2–20.8% reductions relative to the baseline model. Compared to single-pane low-e windows, the photothermal systems provided an additional 7.6–13.2% heating energy savings due to their enhanced nanoscale PPE-induced heat gain.

The previously mentioned heating energy savings were calculated based on whole-building energy use, incorporating the combined effects of all building envelope elements and systems. However, this holistic approach may dilute the true impact of window performance improvements. A more precise evaluation can be obtained by isolating the energy performance of the window components themselves. Figure 8 presents solar heat gains and total heat losses through window components for each window type across different climate conditions. The results show that photothermal windows significantly reduce heat loss by 34.9–40.8% compared to baseline clear

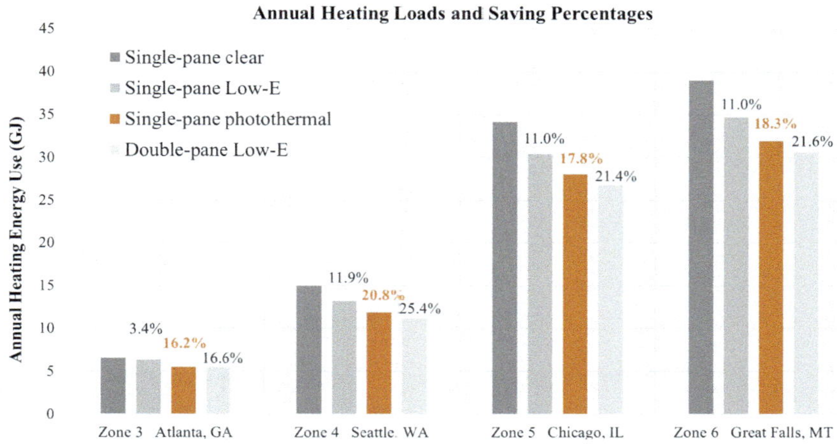

Fig. 7 Annual heating loads and associated savings percentages based on clear single-pane

single-pane windows. Additionally, when compared to low-e windows, photothermal windows still achieved 12.2–20.1% lower heat loss. The solar heat gain performance of photothermal windows also aligns with this observation. While the transmitted solar radiation levels of low-e and photothermal windows remain nearly identical, the PPE-enhanced inward-flowing heat effect in photothermal windows resulted in 14.2–16.7% greater solar heat gain than low-e windows. The combination of reduced heat loss and enhanced solar heat gain is responsible for the superior energy savings of nanoscale PPE-induced windows, making them a promising alternative for improving building energy efficiency.

3.5 *Discussion of Numerical Analysis and Findings*

This case study explored the energy-saving potential of surface plasmon-induced NP photothermal effects applied to single-pane windows—a solution aimed at addressing energy inefficiencies in building retrofits. While previous research has investigated the feasibility of photothermal coatings for windows, a comprehensive energy analysis has not been conducted until now. To bridge this gap, a simplified analytical model was developed to account for localized surface heating on one side of the glass pane. This model incorporated interfacial insulation effects and was validated through photothermal experiments. The validated model enabled the computation of temperature distributions under various design and environmental conditions.

Using this framework, the study analyzed the thermal performance improvements of photothermal windows compared to Low-E single-pane windows. The research focused on the dynamic Solar Heat Gain Coefficient (SHGC), specifically introducing

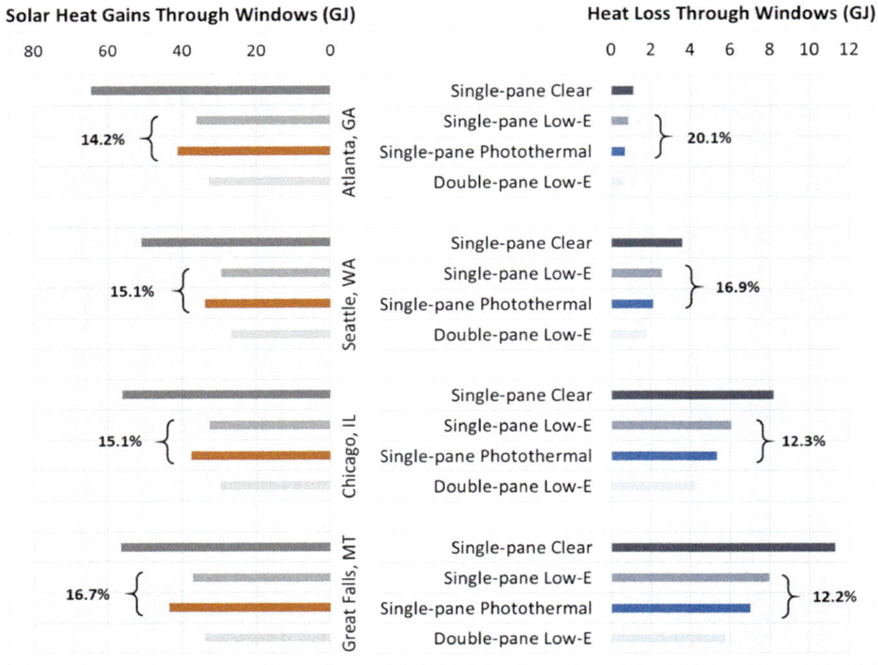

Fig. 8 Solar heat gains and building heat loss through windows

$SHGC_{PPE}$ to quantify the plasmonic photothermal effect. A physics-based mathematical model was developed through numerical analysis across multiple boundary conditions (solar irradiance levels and outdoor temperatures).

To assess the whole-building energy performance, a parametric energy simulation approach was employed, integrating the temperature- and solar-dependent SHGC. The results demonstrated that photothermal coatings offer a highly efficient energy retrofit solution for single-pane windows, achieving:

- 16.2–20.8% overall energy savings relative to single-pane clear windows.
- 12.2–21.9% reduction in heat loss compared to low-e single-pane windows.
- 14.2–16.7% greater solar heat gain due to enhanced plasmonic photothermal effects.
- 7.6–13.2% whole-building heating energy savings, making performance comparable to double-pane windows.

Unlike traditional energy-saving window technologies that rely on added insulation layers (e.g., air gaps in double-pane windows), the energy efficiency of photothermal windows is driven by nanoparticle-induced LSPR and solar NIR energy absorption. This method improves thermal performance without compromising visible transmittance (VT), making it a viable alternative to double-pane systems. Additionally, this new nanomaterial-embedded window system could offer comfort and condensation control benefits over low-e coated single panes, addressing local thermal discomfort issues often associated with low-emissivity coatings. Considering lighting energy use, photothermal windows further narrow the performance gap with double-pane windows, offering a promising retrofit solution for enhancing solar energy utilization in buildings.

4 Summary

In the ongoing effort to enhance building energy efficiency, integrating nanoscale materials into architectural applications presents a groundbreaking opportunity. Nanomaterials, with their distinctive optical, thermal, and phase-changing properties, offer promising solutions for improving insulation, solar control, and energy retention. However, the transition from nanoscale properties to full-building performance remains complex, requiring sophisticated numerical methods to model interactions across multiple scales accurately.

This chapter systematically examines the computational frameworks necessary to bridge this gap. Molecular dynamics (MD) simulations allow for detailed analysis of thermal conductivity, phase transitions, and radiative properties at the atomic level. Finite element analysis (FEA) extends this understanding to the material component scale, helping predict how nano-enhanced materials affect the performance of building envelopes. Computational fluid dynamics (CFD) further refines the analysis by modeling heat transfer and airflow interactions within built environments. Recently, machine learning techniques, such as Physics-Informed Neural Networks (PINNs), have emerged as powerful tools for improving numerical accuracy while reducing computational costs.

A key focus of this chapter is the application of numerical simulations to solar-responsive nanomaterials in glazing systems. Windows account for a significant portion of a building's energy loss, and spectrally selective nanocoatings have the potential to modulate solar heat gain while maintaining transparency. The chapter presents a case study on localized surface plasmon resonance (LSPR)-induced photothermal effects in nanoparticle-coated glazing systems. By analyzing nanoparticle absorption, heat transfer mechanisms, and optical modulation, the study develops a physics-based numerical model that is then integrated into whole-building parametric energy simulations using EnergyPlus.

Results from the case study reveal that LSPR-mediated nanocoatings can enhance passive solar heat gain in colder climates while maintaining transparency, effectively reducing heating loads by up to 20% compared to traditional single-pane windows. These nano-enhanced windows demonstrate comparable performance to double-pane glazing, providing an alternative energy retrofit solution that avoids the added weight, cost, and installation complexity of conventional insulating glass units.

Beyond the immediate case study, this chapter underscores a broader message: numerical modeling is essential for translating nanoscale material innovations into practical building applications. The methodologies discussed—ranging from molecular simulations to full-building energy analysis—demonstrate how multi-scale numerical approaches enable precise performance assessments before large-scale implementation. The findings contribute to the growing body of knowledge on nano-integrated architectural solutions, helping pave the way for sustainable, high-performance building materials through computational optimization. By leveraging numerical simulations, researchers and engineers can predict, optimize, and implement nano-enhanced materials in energy-efficient buildings, ensuring that these advancements move beyond laboratory studies to real-world impact in architectural engineering and urban sustainability.

Acknowledgements We acknowledge the financial support from National Science Foundation #1953004 and #2001207.

Competing Interests The author has no conflicts of interest to declare that are relevant to the content of this chapter.

References

1. Bao H, Chen J, Gu X, Cao B (2018) A review of simulation methods in micro/nanoscale heat conduction. ES Energy Environ 1:16–55
2. Arasteh H, Maref W, Saber HH (2024) 3D numerical modeling to assess the energy performance of solid–solid phase change materials in glazing systems. Energies 17(15)
3. Mastino C, Concu G, Frattolillo A (2024) Energetic performance of natural building materials: numerical simulation and experimental evaluation. Energies 17(768)
4. Duan Q, Wang J (2019) A parametric study of the combined effects of window property and air vent placement. Indoor Built Environ 28(3)
5. Zhou Y, Zheng S (2024) A co-simulated material-component-system-district framework for climate-adaption and sustainability transition. Renew Sustain Energy Rev 192
6. Liu B, Wang Y, Rabczuk T, Olofsson T, Lu W (2024) Multi-scale modeling in thermal conductivity of polyurethane incorporated with phase change materials using physics-informed neural networks. Renew Energy 220
7. Mi R, Li T, Dalgo D, Chen C, Kuang Y, He S, Zhao X, Xie W, Gan W, Zhu J, Srebric J, Yang R, Hu L (2020) A clear, strong, and thermally insulated transparent wood for energy efficient windows. Adv Funct Mater 30(1)
8. Song Y, Duan Q, Feng Y, Zhang E, Wang J, Niu S (2020) Solar infrared radiation towards building energy efficiency: measurement, data, and modeling. Environ Rev 28(4):457–465
9. Huang X, Neretina S, El-Sayed MA (2009) Gold nanorods: from synthesis and properties to biological and biomedical applications. Adv Mater 21(48):4880–4910
10. Abendroth T, Schumm B, Alajlan SA, Almogbel AM, Mäder G, Härtel P, Althues H, Kaskel S (2017) Optical and thermal properties of transparent infrared blocking antimony doped tin oxide thin films. Thin Solid Films 624:152–159
11. Liu H, Zeng X, Kong X, Bian S, Chen J (2012) A simple two-step method to fabricate highly transparent ITO/polymer nanocomposite films. Appl Surf Sci 258(22):8564–8569
12. Xu X, Stevens M, Cortie MB (2004) In situ precipitation of gold nanoparticles onto glass for potential architectural applications. Chem Mater 16(11):2259–2266
13. Xiao L, Su Y, Ran J, Liu Y, Qiu W, Wu J, Lu F, Shao F, Tang D, Peng P (2016) First-principles prediction of solar radiation shielding performance for transparent windows of GdB. J Appl Phys 119(16)

14. Schelm S, Smith GB (2003) Dilute LaB6 nanoparticles in polymer as optimized clear solar control glazing. Appl Phys Lett 82(24)
15. Hou H, Wang P, Zhang J, Li C, Jin Y (2015) Graphene oxide-supported Ag nanoplates as LSPR tunable and reproducible substrates for SERS applications with optimized sensitivity. ACS Appl Mater Interfaces 7(32):18038–18045
16. Pallavicini P, Chirico G, Collini M, Dacarro G, Donà A, D'Alfonso L, Falqui A, Diaz-Fernandez Y, Freddi S, Garofalo B, Genovese A, Sironi L, Taglietti A (2011) Synthesis of branched Au nanoparticles with tunable near-infrared LSPR using a zwitterionic surfactant. Chem Commun 47(4):1315–1317
17. Nehl CL, Hafner JH (2008) Shape-dependent plasmon resonances of gold nanoparticles. J Mater Chem 18:2415–2419
18. Jana NR, Gearheart L, Murphy CJ (2001) Seed-mediated growth approach for shape-controlled synthesis of spheroidal and rod-like gold nanoparticles using a surfactant template. Adv Mater 13(18):1389–1393
19. Garcia G, Buonsanti R, Llordes A, Runnerstrom EL, Bergerud A, Milliron DJ (2013) Near-infrared spectrally selective plasmonic electrochromic thin films. Adv Opt Mater 1(3):215–220
20. Mendelsberg RJ, Garcia G, Milliron DJ (2012) Extracting reliable electronic properties from transmission spectra of indium tin oxide thin films and nanocrystal films by careful application of the Drude theory. J Appl Phys 111
21. Khandelwal H, Loonen RCGM, Hensen JLM, Debije MG, Schenning APHJ (2015) Electrically switchable polymer stabilised broadband infrared reflectors and their potential as smart windows for energy saving in buildings. Sci Rep 3
22. Li Y, Liu J, Liang J, Yu X, Li D (2015) Tunable solar-heat shielding property of transparent films based on mesoporous Sb-doped SnO_2 microspheres. ACS Appl Mater Interfaces 7(12):6574–6583
23. Xu X, Cortie MB, Stevens M (2005) Effect of glass pre-treatment on the nucleation of semi-transparent gold coatings. Mater Chem Phys 94(2–3):266–274
24. Pardiñas-Blanco I, Hoppe CE, Piñeiro-Redondo Y, López-Quintela MA, Rivas J (2008) Formation of gold branched plates in diluted solutions of poly(vinylpyrrolidone) and their use for the fabrication of near-infrared-absorbing films and coatings. Langmuir ACS J Surf Colloids 24(3):983–990
25. Besteiro LV, Kong X-T, Wang Z, Rosei F, Govorov AO (2018) Plasmonic glasses and films based on alternative inexpensive materials for blocking infrared radiation. Nano Lett 18(5):3147–3156
26. Zhao Y, Sadat M, Dunn A, Xu H, Chen C-H, Nakasuga W, Ewing RC, Shi D (2017) Photothermal effect on Fe_3O_4 nanoparticles irradiated by white-light for energy-efficient window applications. Sol Energy Mater Sol Cells 161:247–254
27. Khashan S, Dagher S, Omari SA, Tit N, Elnajjar E, Mathew B, Hilal-Alnaqbi A (2017) Photothermal characteristics of water-based $Fe_3O_4@SiO_2$ nanofluid for solar-thermal applications. Mater Res Express 4
28. Wang J, Shi D (2017) Spectral selective and photothermal nano structured thin films for energy efficient windows. Appl Energy 208:83–96
29. Korpela SA, Lee Y, Drummond JE (1982) Heat transfer through a double pane window. ASME J Heat Mass Transf 104(3):539–544
30. Rubin M (1982) Calculating heat transfer through windows. Int J Energy Res
31. Siegel R, Howell JR (2002) Thermal radiation heat transfer
32. Carli I (2006) Technical report: Tarcog: mathematical models for calculation of thermal performance of glazing systems with or without shades. Technology
33. Lü X, Lu T, Viljanen M (2006) A new analytical method to simulate heat transfer process in building. Appl Therm Eng 26(16):1901–1909
34. Cuevas C, Fissore A, Fonseca N (2010) Natural convection at an indoor glazing surface with different window blinds. Energy Build 42(10):1685–1691
35. Abel E (1994) Low-energy buildings. Energy Build 21(3):169–174

36. Aydın O (2006) Conjugate heat transfer analysis of double pane windows. Build Environ 41(2):109–116
37. Littlefair P (1982) Effective glass transmission factors under a CIE sky. Light Res Technol 14(4):232–235
38. Kreith F, Bohn M, Kirkpatrick A (1997) Principles of heat transfer. J Solar Energy Eng 119(2)
39. Modest MF (2013) Radiative heat transfer
40. Arıcı M, Karabay H, Kan M (2015) Flow and heat transfer in double, triple and quadruple pane windows. Energy Build 86:394–402
41. Qiu T, Tien C (1994) Femtosecond laser heating of multi-layer metals—I. Analysis. Int J Heat Mass Transf 37(17):2789–2797
42. Qiu T, Juhasz T, Suarez C, Bron W, Tien C (1994) Femtosecond laser heating of multi-layer metals—II. Experiments. Int J Heat Mass Transf 37(17):2799–2808
43. Tzou DY (1990) Three dimensional structures of the thermal shock waves around a rapidly moving heat source. Int J Eng Sci 28(10):1003–1017
44. Majumdar A (1993) Microscale heat conduction in dielectric thin films. J Heat Transfer 115(1):7–16
45. Keith Roper D, Ahn W, Hoepfner M (2007) Microscale heat transfer transduced by surface Plasmon resonant gold nanoparticles. J Phys Chem C 111(9):3636–3641
46. Govorov AO, Richardson HH (2007) Generating heat with metal nanoparticles. NanoToday 2(1):30–38
47. Tzou D (1995) Experimental support for the lagging behavior in heat propagation. J Thermophys Heat Transf 686–693
48. Tang D, Araki N (1999) Wavy, wavelike, diffusive thermal responses of finite rigid slabs to high-speed heating of laser-pulses. Int J Heat Mass Transf 42(5):855–860
49. Ho J-R, Kuo C-P, Jiaung W-S (2003) Study of heat transfer in multilayered structure within the framework of dual-phase-lag heat conduction model using lattice Boltzmann method. Int J Heat Mass Transf 46(1):55–69
50. Huld T, Paietta E, Zangheri P, Pinedo Pascua I (2018) Assembling typical meteorological year data sets for building energy performance using reanalysis and satellite-based data. Atmosphere 9(53)
51. Lorenz E, Hurka J, Karampela G, Heinemann D, Beyer H, Schneider M (2008) Qualified forecast of ensemble power production by spatially dispersed grid-connected PV systems. In: 23rd European photovoltaic solar energy conference
52. Rabl A (1976) Comparison of solar concentrators. Sol Energy 18(2):93–111
53. Bacher P, Madsen H, Aalborg Nielsen H (2009) Online short-term solar power forecasting. Solar Energy 83(10):1772–1783
54. Roberts JE (1995) Visible light induced changes in the immune response through an eye-brain mechanism (photoneuroimmunology). J Photochem Photobiol B 29(1):3–15
55. Wang N, Wang J (2021) A spectrally-resolved method for evaluating the solar effect on user thermal comfort in the near-window zone. Build Environ 202
56. Jahid MA, Wang J, Zhang E, Duan Q, Feng Y (2022) Energy savings potential of reversible photothermal windows with near infrared-selective plasmonic nanofilms. Energy Convers Manag 263
57. Zhang E, Duan Q, Wang J, Zhao Y, Feng Y (2021) Experimental and numerical analysis of the energy performance of building windows with solar NIR-driven plasmonic photothermal effects. Energy Convers Manag 245
58. Pu J, Shen C, Wang J, Zhang Y, Zhang C, Kalogirou SA (2023) Near-infrared absorbing glazing for energy-efficient windows: a critical review and performance assessments from the building requirements. Nano Energy 110
59. Li Y, Huang X, Sheriff HKM, Forrest SR (2023) Semitransparent organic photovoltaics for building-integrated photovoltaic applications. Nat Rev Mater 8(3)
60. Du W-C, Xie J, Xia L, Liu Y-J, Yang H-W, Zhang Y (2021) Study of new solar film based on near-infrared shielding. J Photochem Photobiol A Chem 418

61. Gao Q, Wu X, Huang T (2021) Novel energy efficient window coatings based on In doped CuS nanocrystals with enhanced NIR shielding performance. Sol Energy 220:1–7
62. Xu Q, Xiao L, Ran J, Tursun R, Zhou G, Deng L, Tang D, Shu Q, Qin J, Lu G, Peng P (2018) Cs0.33WO3 as a high-performance transparent solar radiation shielding material for windows. J Appl Phys 124
63. Duan Q, Feng Y, Wang J (2021) Clustering of visible and infrared solar irradiance for solar architecture design and analysis. Renew Energy 668–677
64. Chen C, Duan Q, Feng Y, Wang J, Ghaeili Ardabili N, Wang N, Hosseini SM, Shen C (2023) Reconstruction of narrowband solar radiation for enhanced spectral selectivity in building-integrated solar energy simulations. Renew Energy 219
65. Perino M, Serra V (2015) Switching from static to adaptable and dynamic building envelopes: a paradigm shift for the energy efficiency in buildings. Eng Environ Sci
66. Fox M (ed) (2016) Interactive architecture: adaptive world. Chronicle Books
67. Moloney J (2011) Designing kinetics for architectural facades: state change. Routledge, London
68. Kolarevic B (2003) Architecture in the digital age-design and manufacturing. Spon Press, London, New York
69. Oxman R (2008) Digital architecture as a challenge for design pedagogy: theory, knowledge, models and medium. Des Stud 29(2):99–120
70. McCormack J, Dorin A, Innocent T (2004) Generative design: a paradigm for design research. In: Futureground—DRS international conference, Melbourne
71. Baetens R, Jelle BP, Thue JV, Tenpierik MJ, Grynning S, Uvsløkk S, Gustavsen A (2010) Vacuum insulation panels for building applications: a review and beyond. Energy Build 42(2):147–172
72. Alam M, Singh H, Limbachiya M (2011) Vacuum Insulation Panels (VIPs) for building construction industry—a review of the contemporary developments and future directions. Appl Energy 88(11):3592–3602
73. Baetens R, Jelle BP, Gustavsen A, Grynning S (2010) Gas-filled panels for building applications: a state-of-the-art review. Energy Build 42(11):1969–1975
74. Griffith BT, Arashteh D, Türler D (1995) Gas-filled panels: an update on applications in the building thermal envelope superinsulations and the building envelope. In: BETEC Fall symposium
75. Kuznik F, David D, Johannes K, Roux J-J (2011) A review on phase change materials integrated in building walls. Renew Sustain Energy Rev 15(1):379–391
76. Loonen R, Trčka M, Cóstola D, Hensen J (2013) Climate adaptive building shells: state-of-the-art and future challenges. Renew Sustain Energy Rev 25:483–493
77. Baetens R, Jelle BP, Gustavsen A (2010) Properties, requirements and possibilities of smart windows for dynamic daylight and solar energy control in buildings: a state-of-the-art review. Sol Energy Mater Sol Cells 94(2):87–105
78. Addington M, Schodek D (2012) Smart materials and technologies in architecture: for the architecture and design professions. Routledge
79. Wang Y, Runnerstrom EL, Milliron DJ (2016) Switchable materials for smart windows. Annu Rev Chem Biomol Eng 7:283–304
80. . Wang J, Beltran L, Kim J (2014) A parametric simulation study on kinetic envelopes. In: 43rd ASES national solar conference 2014, SOLAR 2014, including the 39th national passive solar conference and the 2nd meeting of young and emerging professionals in renewable energy
81. Kasinalis C, Loonen R, Cóstola D, Hensen J (2014) Framework for assessing the performance potential of seasonally adaptable facades using multi-objective optimization. Energy Build 79:106–113
82. Loonen R (2014) Bio-inspired adaptive building skins. In: Biotechnologies and biomimetics for civil engineering. Springer, pp 115–134
83. Goia F, Cascone Y (2014) The impact of an ideal dynamic building envelope on the energy performance of low energy office buildings. Energy Procedia 58:185–192

84. Goia F, Haase M, Perino M (2013) Optimizing the configuration of a façade module for office buildings by means of integrated thermal and lighting simulations in a total energy perspective. Appl Energy 108:515–527
85. Favoino F, Goia F, Perino M, Serra V (2014) Experimental assessment of the energy performance of an advanced responsive multifunctional façade module. Energy Build 68:647–659
86. Wang J, Beltran L, Kim J (2014) From static to kinetic: a review of acclimated kinetic building envelopes. In: SOLAR 2014 conference, American Solar Energy Society, San Francisco, CA
87. Wang J, Beltran L (2016) A method of energy simulation for dynamic building envelopes. In: ASHRAE and IBPSA-USA SimBuild 2016: building performance modeling conference, Salt Lake City, UT
88. Miao Z, Li J, Wang J (2011) Kinetic building envelopes for energy efficiency: modeling and products. In: 2011 international conference on green building, materials and civil engineering, GBM, Shangri-La
89. Duan Q, Zhang E, Hinkle L, Wang J (2021) Parametric energy simulation methods for solar-NIR selective glazing systems. In: 8th international building physics conference (IBPC 2021), Copenhagen
90. Duan Q, Hinkle L, Wang J, Zhang E, Memari A (2021) Condensation effects on energy performance of building window systems. Energy Rep 7:7345–7357
91. Hinkle L, Ghaeili Ardabili N, Wang J (2022) Parametric simulation of dynamic glazing system with changeable NIR response. In: Conference on sustainable development of energy, water, and environment system
92. Hinkle LE, Wang J, Brown NC (2022) Quantifying potential dynamic façade energy savings in early design using constrained optimization. Build Environ 221
93. Wang N, Ghaeili N, Wang J, Feng Y, Zhang E, Chen C (2023) Using architectural glazing systems to harness solar thermal potential for energy savings and indoor comfort. Renew Energy 219
94. Duan Q, Hinkle L, Wang J (2021) Parametric energy simulation considering the effects of condensation on glazing systems. In: 2021 ASHRAE virtual annual conference, ASHRAE 2021
95. Wang J, Beltran L (2021) Energy performances of future dynamic building envelopes. In: 3rd IBPSA-England conference BSO 2016, Newcastle
96. Ghaeili Ardabili N, Feng Y, Wang J (2023) Design and optimization of thermally responsive autonomous dynamic glazed attachment systems for building solar heat gain control. Build Simul 16:1971–1986
97. Ardabili NG, Feng Y, Wang J (2023) Design and optimization of thermally responsive autonomous dynamic glazed attachment systems for building solar heat gain control. In: Building Simulation vol 16, no. 10. Tsinghua University Press, Beijing, pp 1971–1986
98. Chen G (1996) Nonlocal and nonequilibrium heat conduction in the vicinity of nanoparticles. J Heat Transf 118(3):539–545
99. Neumann O, Urban AS, Day J, Lal S, Nordlander P, Halas NJ (2013) Solar vapor generation enabled by nanoparticles. ACS Nano 7(1):42–49
100. Tian Q, Hu J, Zhu Y, Zou R, Chen Z, Yang S, Li R, Su Q, Han Y, Liu X (2013) Sub-10 nm $Fe_3O_4@Cu_{2-x}S$ core-shell nanoparticles for dual-modal imaging and photothermal therapy. J Am Chem Soc 135(23):8571–8577

Index

D. Shi (ed.), *Energy Materials Science and Engineering: From New Devices to AI Power Systems*, Green Energy and Technology, https://doi.org/10.1007/978-3-031-88550-1